A. Brandt · P. Franken · B. Lisek

Stationary Stochastic Models

Mathematische Lehrbücher und Monographien
Herausgegeben von der Akademie der Wissenschaften der DDR
Karl-Weierstraß-Institut für Mathematik

II. Abteilung
Mathematische Monographien
Band 78

Stationary Stochastic Models

von A. Brandt, P. Franken und B. Lisek

Stationary Stochastic Models

by Andreas Brandt, Peter Franken and Bernd Lisek

with 4 figures

Akademie-Verlag Berlin
1990

Verfasser:

Dr. rer. nat. Andreas Brandt
Prof. Dr. rer. nat. habil. Peter Franken †
Dr. sc. nat. Bernd Lisek

Humboldt-Universität zu Berlin
Sektion Mathematik

ISBN 3-05-500605-4
ISSN 0076-5430

Erschienen im Akademie-Verlag Berlin, Leipziger Str. 3—4, Berlin, DDR-1086

Lizenznummer: 202 · 100/408/89
Printed in the German Democratic Republic
Gesamtherstellung: VEB Druckhaus „Maxim Gorki", Altenburg, DDR-7400
Lektor: Dr. Reinhard Höppner
LSV 1075
Bestellnummer: 7639135 (9160)
06800

Contents

Abbreviations

a.s. almost surely
d.f. distribution function
iff if and only if
i.i.d. independent identically distributed
LST Laplace-Stieltjes transform (of a distribution function $\mathsf{F}(t)\colon \mathscr{F}(s) = \int_0^\infty e^{-sx}\, d\mathsf{F}(x)$).
MPP marked point process
PEMP process with an embedded marked point process
PGF probability generating function
r.v. random variable
SGS solution generating system

Preface

One of the basic problems arising again and again in the stochastic modelling of systems is the existence and uniqueness of stationary (limiting) distributions of system characteristics. Moreover, one is interested in:

- relationships between the stationary distributions with respect to continuous time and to certain embedded epochs (e.g. the arrival instants of customers in a queue),
- model continuity, i.e. continuity of system characteristics with respect to the input,
- insensitivity of the stationary distributions with respect to the form of the probability distributions of certain input characteristics.

Traditionally, these problems are treated by means of Markov processes, renewal, and Markov renewal processes. However, these methods require some independence and distributional assumptions which are often not adequate for the problems stated above. During the last two decades, there have been some successful attempts in weakening such assumptions, see Borovkov (1972a, 1980), Franken, König, Arndt, and Schmidt (1981), Baccelli and Brémaud (1987), and the references therein. The last two books used concepts and results from the theory of random point processes. At first glance one could come to the conclusion that point process theory is the up-to-date tool for treating these questions. However, comparing the arguments in the monographs mentioned above, one observes that the key point is merely the fact that stationarity of a random sequence or point process involves some interesting consequences (without any additional independence or distributional assumption), cf. e.g. the Appendix of this book, Section A 1. Moreover, irrespective of the use of random sequences or point processes, one realizes that the common principle of all approaches is the description of the system dynamics "step by step", i.e., given the present state and the external influence, the future state can be calculated. The simplest form of such a working manner is the recursive stochastic equation

$$X_{n+1} = f(X_n, U_n), \qquad n = \ldots, -1, 0, 1, 2, \ldots, \tag{0.1}$$

cf. Borovkov (1978), Lisek (1982). Here (U_n) is a given stationary sequence of random variables. We may interprete (0.1) as the mathematical model of the time evolution of a stochastic system at some suitably chosen embedded epochs. (U_n), which is called the input, represents the influence of the environment, whereas the X_n are the consecutive system states at these epochs. Consider, for example, a water reservoir receiving U_n units of water during the n-th day. Daily, the constant amount u of water is taken away as long as there is any water in the reservoir. When the capacity $x_{\max}$ is exceeded, the excess water is drained by means of an overflow. If X_n denotes the water level on the morning of the n-th day, then the equation

$$X_{n+1} = \max\left(0, \min\left(x_{\max}, X_n + U_n - u\right)\right)$$

describes the evolution of the system.

Systems working in continuous time can be described by an analogue of equation (0.1). The methods of investigation of such continuous time equations are quite similar to those of equation (0.1), cf. Section 4.5. However, there is a more elegant way to construct and investigate continuous time stationary state processes, namely, the use of relations between stationary sequences and stationary point processes. These relations make it possible to understand the differences between and relations between the stationarity with respect to embedded epochs and that with respect to continuous time, cf. Chapters 3, 4, and 6.

In recent years some fundamental results concerning equations of the type (0.1) have been obtained, and new applications of these results inside and outside queueing theory have been provided. (Cf. the survey papers Brandt, Franken and Lisek (1984, 1986), Fichtner (1979a, 1979b), Nawrotzki (1981a). It is the aim of this book to present some basic methods for treating equation (0.1) as well as some results obtained by means of these methods for stochastic models. The applications given here are centred around the investigation of qualitative properties of queueing models. However, compared with Franken, König, Arndt, and Schmidt (1981) which can be considered as a direct predecessor of this book, some general methods for treating stationary stochastic models are explained more broadly (Chapters 1–4).

Of course, the restriction to stationary models is a great disadvantage, which has often been disputed. However, there are good reasons for investigating this case so extensively. First of all we believe that, sooner or later each process in real life tends to some equilibrium (ergodic hypothesis), which is also called the steady state. It is often more important to consider quantities of economic relevance (such as costs, gain, and loss) in the average over a certain time interval, rather than at a fixed time instant. If the ergodic hypothesis is true and the time interval is sufficiently long, then the averages of system quantities can be calculated by using the stationary system. Other reasons for the stationarity assumption lie with the available mathematical tools. The analysis of non-stationary models is sometimes very hard and unsatisfactory. A further point is the statistical analysis of data from stochastic models varying in time. Almost invariably we have only a single sample path of the process, rather than the many replications of measurements assumed in classical statistics. To get some idea of the stochastic mechanism, we are forced to assume stationarity of the underlying process.

We point out that we shall only assume the stationarity of the input. First results for systems with non-stationary inputs that tend in some sense to a stationary one can be found in BOROVKOV (1972a), ROLSKI (1981a), SZCZOTKA (1986a, b).

Concerning the queueing theoretical considerations, some topics, which are already discussed in a sufficiently deep manner elsewhere, are omitted, in particular the intensity conservation principle, systems of balance equations, priority queues, the insensitivity of stationary state probabilities, cf. e.g. FRANKEN, KÖNIG, ARNDT, and SCHMIDT (1981), and the martingale approach, cf. e.g. BRÉMAUD (1981). The queueing theoretical investigations are directed at two targets. On the one hand, they show how the general methods work, on the other hand, the results are of interest in themselves. For this reason we handle the standard queueing systems $G/G/1/\infty$, $G/G/m/\infty$, $G/G/m/0$, and in addition we make accessible some non-standard applications of our methods.

Throughout this book batch arrivals are allowed. This requires a modification of the usual notion of a random marked point process. We use "ordered point processes", i.e. the particles (e.g. customers) arising at the same time are a-priori ordered. Concerning batch arrival queues some results (in particular those connected with the point process approach) are proved in full rigour here for the first time.

The book is divided into 9 chapters. Chapter 1 provides some useful methods for proving the existence, uniqueness, and continuity of stationary solutions of equation (0.1). We introduce the notions of strong and weak solutions of equation (0.1). First we discuss two methods for constructing stationary strong solutions. These methods are due to LOYNES (1962), and FRANKEN (1969, 1970) and BOROVKOV (1972a, b, 1978, 1980), respectively. The idea of both methods consists in constructing the states X_n as a function of the past $(U_k, k < n)$ of the input. For each sample path of the input, exactly one sequence of states is given in this way. We then construct stationary weak solutions having finitely many sequences of states for every fixed sample path of the input. As an application of this approach, we obtain a generalization of Wald's identity for dependent summands. Under certain assumptions, we give a constructive description of the decomposition of a stationary weak solution into ergodic components. Finally we discuss continuity properties of the model (0.1), i.e. we investigate the conditions under which small perturbations of the input lead to small deviations of a particular stationary (weak or strong) solution of (0.1). The ideas of Chapter 1 will be the basis of most results of Chapters 5, 8, and 9.

In Chapter 2 equation (0.1) is discussed under the additional assumption that the input variables U_n are i.i.d. Of course, there are close links with classical results for Markov chains. In particular, we provide a new proof of Rosenblatt's well-known representation theorem.

Chapter 3 introduces (ordered) marked point processes on the real line. The emphasis is on relations between stationary point processes and stationary sequences. The very useful inversion formula (3.4.1) plays a key role in the book. The detailed proofs given here are suitable modifications of those of the corresponding theorems for simple marked point processes given e.g. in KÖNIG, MATTHES and NAWROTZKI (1971).

In Chapter 4 we deal with continuous time models. As a technical tool, we introduce the notions of stationary and synchronous stochastic processes with an embedded marked point process. For many systems working in continuous time, equation (0.1) describes the dynamics at certain embedded points, e.g. the arrival instants of customers. The state of the system at an arbitrary instant t is often determined by the state at the last embedded point T_n prior to t and the distance $t - T_n$. In this situation the stationary state process in continuous time can be constructed using the results of Chapter 1 and the relations between synchronous and stationary stochastic processes with an embedded marked point process. A somewhat different approach for constructing stationary state processes in continuous time consists in investigating a continuous time analogue of equation (0.1).

Chapters 5–8 contain some results for standard and non-standard queues. We have tried to make these chapters intelligible to the reader without detailed knowledge of Chapters 1–4. For first reading it should be enough to look into the introductions 1.1, 2.1, 3.1, and 4.1, in which all essential notions and results of the chapters are summarized.

In Chapter 5 we prove the existence and (if possible) uniqueness of a stationary state process for some standard queues as well as for single server queues with warming-up, for loss systems with repeated calls, and for queueing networks consisting of loss systems.

In Chapter 6 we consider arrival-, time-, and departure-stationary queueing quantities. Using the ideas from Chapter 4, we prove several relationships among the distributions of these quantities and their moments. These relationships are also known as conservation laws. They are of great practical importance in the simulation, optimization, and statistical evaluation of queueing systems. Here we give only selected results, since this topic was extensively discussed in earlier publications, cf. e.g. Franken, König, Arndt, and Schmidt (1981) and the references therein.

In Chapter 7 some special problems concerning batch arrival queues are treated. We introduce the batch-arrival-stationary quantities (i.e. steady state viewed by the customer in front of an arbitrary batch). The techniques given in the book are applied to infinite server and single server queues with batch arrivals, as well as feedback queues. In some particular cases formulae for the binomial moments of the stationary number of customers in the systems are obtained. Relations between batch delays and customer delays are discussed.

The general ideas of Chapter 1 for proving model continuity are illustrated by means of standard queueing systems in Chapter 8. For the systems G/G/∞, G/G/m/0, G/G/1/∞, and G/GI/m/∞, we prove continuity theorems for the arrival- and time-stationary state processes; for G/G/∞ estimates of the rate of convergence are obtained.

Chapter 9 contains some applications of the general methods from Chapter 1 to two models outside queueing theory. For the linear filter equation $X_{n+1} = A_n X_n + B_n$ the continuous dependence of the solution on the coefficients $([A_n, B_n], n = \ldots, -1, 0, 1, \ldots)$ is proved. Finally, we discuss the stability of certain robust filters for cleaning time series from outliers.

The appendix summarizes some mathematical prerequisites for the readers convenience. We recommand the reader to read thoroughly Section A 1.3 on conditional distributions of stationary sequences; it will help him to understand the main results of Chapters 1, 3, 4, 6, and 7.

The book is intended to be for specialists in applied probability and statistics, computer science and operations research, especially those working in queueing, reliability and inventory theory, as well as for students of higher courses. Some parts of the book can be used as a textbook; other are at the monograph level. The level of mathematical difficulty varies considerably. The basics of probability, queueing theory, and topology are assumed. An acquaintance with the basics of ergodic theory also would be helpful.

Some technical hints: The symbols $\mathbb{Z}$, $\mathbb{Z}_+$, $\mathbb{R}$, $\mathbb{R}_+$ denote the sets of integers, nonnegative integers, real numbers, and nonnegative real numbers, respectively. Most spaces occuring in this book are complete, separable metric spaces (Polish spaces is another word for this). Usually they will be denoted by special roman letters $\mathbb{U}, \mathbb{X}, \mathbb{Y}, \ldots$, and their Borel σ-fields by the corresponding script letters $\mathcal{U}, \mathcal{X}, \mathcal{Y}, \ldots$ The set $\mathbb{U}^{\mathbb{Z}}(\mathbb{X}^{\mathbb{Z}}, \mathbb{Y}^{\mathbb{Z}}, \ldots)$ of all sequences of elements of $\mathbb{U}(\mathbb{X}, \mathbb{Y}, \ldots)$ will be endowed with the σ-field $\mathcal{U}_{\mathbb{Z}}(\mathcal{X}_{\mathbb{Z}}, \mathcal{Y}_{\mathbb{Z}}, \ldots)$ generated by the measurable cylinders. We do not speak much about measurability; the measurability of most functions is easy to see, since they are simple algebraic compositions of measurable functions or limits of sequences of measurable functions.

By $F(\mathbb{Y})$ $\big(F_+(\mathbb{Y})\big)$ we denote the set of all (non-negative) real-valued measurable functions on $\mathbb{Y}$.

All results presented in this book can be formulated in a pure measure-theoretical manner, i.e. they concern probability distributions on some suitably chosen spaces. However, as is usual in applied probability, we think of a probability measure as the distribution of a random variable (r. v.) taking values in the corresponding space. The reader who is used to working with a basic probability space may imagine that all random variables appearing in a particular problem are defined on a common basic probability space (which has to be rich enough). In general, the random variables will be denoted by (latin or greek) capitals, and the sample paths will be denoted by the corresponding small letters. (However, there are exceptions: For instance, stopping times and other random indices of sequences will be denoted by τ, ν, ν_0, $\tilde{\nu}$, $\nu_j, \ldots$) The symbol (Y, P) means a random variable Y and its (induced) probability distribution P. The notations $\mathsf{P}(Y \in A)$ and $\mathsf{P}(A)$ will be used synchronously. In general, to each random variable there will be attached a particular symbol for its distribution. So we have (Φ, P), (Φ', P'), $(\tilde{\Phi}, \tilde{\mathsf{P}})$, $(\bar{\Phi}, \bar{\mathsf{P}})$, $(\bar{\chi}, \bar{\mathsf{P}}), \ldots$ In order to avoid fussy notation, in some parts of the book we shall use P as a universal symbol for probability. If Y and Y' are identically distributed (i.e. for (Y, P), (Y', P') we have $\mathsf{P} = \mathsf{P}'$), then we shall write $Y \overset{\mathcal{D}}{=} Y'$. The symbol $\xrightarrow{\mathcal{D}}$ stands for convergence in distribution. The indicator function $\mathbf{1}_A(x)$ of a set A will often be written in the form $\mathbf{1}\,\{x \in A\}$, which is more convenient when the event $\{x \in A\}$ has a complicated structure. For a sequence $(x_n, n \in \mathbb{Z})$ we shall merely write (x_n). (Analogously, $\big(x(t)\big)$ is an abbreviation of $\big(x(t), t \in \mathbb{R}\big)$). A sequence (Φ, P), $\Phi = (X_n)$, is called stationary if $\theta\Phi \overset{\mathcal{D}}{=} \Phi$, where θ is the usual shift operator: $\theta(X_n) = (X_{n+1})$. By P^+ we denote the distribution of the half sequence $(X_n, n \geqq 0)$; i.e. $\mathsf{P}^+(A) = \mathsf{P}\big((X_n, n \geqq 0) \in A\big)$. To shorten the notation, sometimes we shall write $[a_n, b_n, c_n]$ instead of $\big[[a_n, b_n], c_n\big]$ or $\big[a_n, [b_n, c_n]\big]$. The numbering of formulae and statements (definitions, lemmas, theorems, corollaries, remarks) is by section.

For the most frequently used references we introduce the following abbreviations:

[KMN] – König, Matthes, and Nawrotzki (1971)
[KMM] – Kerstan, Matthes, and Mecke (1974)
[FKAS] – Franken, König, Arndt, and Schmidt (1981)

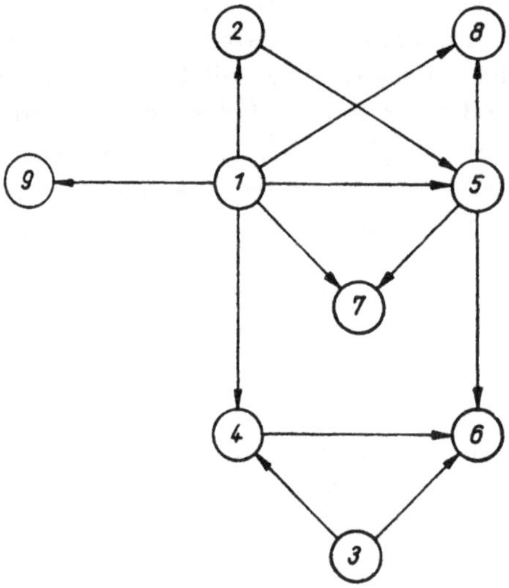

Fig. 1. Interdependence graph

At this point we should like to thank all those people whose interest, encouragement and criticism have stimulated our work. First of all, our thanks go to S. Asmussen, F. Baccelli, Yu. Belyaev, G. Last, A. Lesanovsky, O. Nerman, Yu. Sukhov and A. Tikhomirov for useful discussions and numerous hints. We are also indebted to A. D. Barbour and T. Lindvall who read parts of the manuscript and helped with some comments.

As concerns the technical completion of the book we are most grateful to J. Kerger, Ch. Reimann, S. Schmidt and, especially, to H. Kahlert. Our thanks go also to Dr. R. Höppner of Akademie-Verlag for his co-operation.

1. Recursive stochastic equations $X_{n+1} = f(X_n, U_n)$

1.1. Introduction

In this chapter we deal with recursive stochastic equations of the type

$$X_{n+1} = f(X_n, U_n), \qquad n \in \mathbb{Z}, \tag{1.1.1}$$

where (U_n) is a given sequence of r.v.'s. We may interpret (1.1.1) as the mathematical model of the temporal behaviour of a given stochastic system working in discrete time. Here the sequence (U_n) represents the influence of the environment, whereas X_n, $n \in \mathbb{Z}$, are the consecutive system states. In what follows we assume that U_n and X_n take values in Polish spaces $(\mathbb{U}, \mathcal{U})$ and $(\mathbb{X}, \mathcal{X})$, respectively. As we are interested in the steady state behaviour of the system, we always assume that (U_n) is stationary, i.e. $(U_n) \overset{\mathcal{D}}{=} \theta(U_n)$, where θ is the usual shift operator and $\overset{\mathcal{D}}{=}$ denotes equality in distribution. The form of the (measurable) function f depends on the system considered and on the choice of the state space $\mathbb{X}$, too. For brevity we use the notation (Φ, P) for the sequence $\Phi = (U_n)$ with the probability distribution P and call (Φ, P) the *input* of the system.

A *strong solution* of (1.1.1) for the input (Φ, P), $\Phi = (U_n)$, is a sequence $([U_n, X_n])$, where $X_n = x_n(\Phi)$ are $\mathbb{X}$-valued r.v.'s defined on the probability space $(\mathbb{U}^{\mathbb{Z}}, \mathcal{U}_{\mathbb{Z}}, \mathsf{P})$, satisfying the relation

$$x_{n+1}(\varphi) = f(x_n(\varphi), u_n), \qquad n \in \mathbb{Z},$$

for P-a.e. sample path $\varphi = (u_k) \in \mathbb{U}^{\mathbb{Z}}$ (Definition 1.2.1). At first sight the concept of strong solution seems to be reasonable. However, it is too restrictive in some cases, cf. Example 1.2.2. For this reason we define: A sequence $\Phi' = ([U_n', X_n'])$ with the distribution P' is called a *weak solution* of (1.1.1) for the input (Φ, P), $\Phi = (U_n)$, iff

$$(U_n) \overset{\mathcal{D}}{=} (U_n') \quad \text{and} \quad \mathsf{P}'\big(X'_{n+1} = f(X_n', U_n'), \quad n \in \mathbb{Z}\big) = 1$$

(Definition 1.2.3). Obviously, each strong solution is a weak solution, too. In the following we do not distinguish weak solutions which are stochastically equivalent. In this sense we understand uniqueness of weak solutions of (1.1.1) as uniqueness up to stochastic equivalence.

In this book we are interested in the steady state behaviour of systems described by (1.1.1). Obviously, for each *stationary weak solution* (Φ', P'), $\Phi' = ([U_n', X_n'])$, we can interpret (X_n') as a stationary state process of the system considered, whereas $([U_n', X_n'])$ describes a steady state of the pair "input – system". Hence we search for stationary weak (or strong) solutions of equation (1.1.1).

First, we discuss two methods for constructing stationary strong solutions, cf. Section 1.3. For given $x \in \mathbf{X}$ and $\varphi = (u_n) \in \mathbf{U}^{\mathbb{Z}}$ let

$$x_0(x) = x, \qquad x_1(x, u_0) = f(x, u_0),$$

$$x_2(x, u_0, u_1) = f\big(f(x, u_0), u_1\big), \ \ldots$$

$$x_n(x, u_0, \ldots, u_{n-1}) = f\big(x_{n-1}(x, u_0, \ldots, u_{n-2}), u_{n-1}\big).$$

Obviously, $x_n(x, u_0, \ldots, u_{n-1})$ is the state at time n of the system starting in the state x at time 0 if the sample path of the input is φ. The idea in each method consists of constructing states X_n as a function of the past $(U_k, k < n)$ of the input at time n. More precisely,

$$X_n = \lim_{r\to\infty} x_r(x, U_{n-r}, \ldots, U_{n-1}) \qquad \mathsf{P}\text{-a.s.} \tag{1.1.2}$$

for a suitably chosen fixed state x.

In the first method it is assumed that $\mathbf{X}$ is endowed with a semi-ordering relation and possesses a minimal element O. If $f(\cdot, u)$ is non-decreasing for each u, then $x_r(O, U_{n-r}, \ldots, U_{n-1})$ is non-decreasing in r. Assuming the existence and finiteness of the limits $x_n^{\min}(\Phi) = \lim_{r\to\infty} x_r(O, U_{n-r}, \ldots, U_{n-1})$, $n \in \mathbb{Z}$, and the left continuity of $f(\cdot, u)$, we obtain a stationary strong solution $\big([U_n, x_n^{\min}(\Phi)]\big)$ of (1.1.1). Furthermore, it is minimal (Theorem 1.3.1). If, in addition, $f(\cdot, u)$ is contractive and some further assumptions are satisfied, a maximal stationary strong solution can be constructed, cf. Theorem 1.3.2. Thus we have a unique stationary weak solution if the minimal and maximal stationary solutions coincide. Under the same assumptions, the uniqueness of a stationary solution is equivalent to the following limiting property, cf. Theorem 1.3.3: For an arbitrary pair $[X, U_n']$ where X is an $\mathbf{X}$-valued r.v. and $(U_n') \overset{\mathcal{D}}{=} (U_n)$:

$$\big(x_{n+k}(X, U_0', \ldots, U'_{n+k-1}), n \geqq 0\big) \xrightarrow[k\to\infty]{\mathcal{D}} \big(\lim_{r\to\infty} x_r(O, U'_{n-r}, \ldots, U'_{n-1}), n \geqq 0\big).$$

The method described above is based on the monotone convergence of $x_r(O, U_{n-r}, \ldots, U_{n-1})$ as $r \to \infty$. The second method consists of finding conditions ensuring the convergence of $x_r(x, U_{n-r}, \ldots, U_{n-1})$ in a finite (random) number of steps. These conditions (the existence of so-called renewing epochs) guarantee that $\big([U_n, \hat{x}_n(x, \Phi)]\big) = \big(\big[U_n, \lim_{r\to\infty} x_r(x, U_{n-r}, \ldots, U_{n-1})\big]\big)$ is a stationary strong solution, cf. Theorem 1.3.9. If, in addition $\hat{x}_n(\Phi) = \hat{x}_n(x, \Phi)$ does not depend on x, then

$$\mathsf{P}\left(\bigcup_{m=1}^{\infty} \{x_k(x, U_0, \ldots, U_{k-1}) = \hat{x}_k(\Phi), k \geqq m\}\right) = 1$$

for every $x \in \mathbf{X}$ (Theorem 1.3.11).

The point of both methods is the construction of exactly one sequence of system states associated with the given sample path of the input. In Section 1.4

we construct stationary weak solutions having finitely many state sequences for almost every fixed sample path of the input Φ. The starting point of this method is the notion of *solution generating system* (SGS). An SGS is a collection of finite nonempty sets $A(n, \varphi) \subseteqq \mathbb{X}$ defined for almost all φ and $n \in \mathbb{Z}$ satisfying the conditions

$$f(A(n, \Phi), U_n) \subseteqq A(n+1, \Phi), \qquad n \in \mathbb{Z}, \qquad \mathsf{P}\text{-a.s.}, \tag{1.1.3}$$

$$A(n, \Phi) = A(0, \theta^n\Phi), \qquad n \in \mathbb{Z}, \qquad \mathsf{P}\text{-a.s.},$$

and some regularity conditions, cf. Definition 1.4.2. The elements of $A(n, \varphi)$ can be interpreted as *potential states* of the system considered at time n, which are associated with the given sample path φ of the input Φ. The sets $A(n, \varphi)$ are related according to equation (1.1.1), cf. (1.1.3). If there is an SGS, then there exists a stationary weak solution (Ψ, Q), $\Psi = ([\hat{U}_n, \hat{X}_n])$, of (1.1.1) for (Φ, P) (Theorem 1.4.3). In case of an ergodic input (Φ, P) we have the representation

$$\mathsf{Q}\left(([\hat{U}_n, \hat{X}_n],\ n \geqq 0) \in (\cdot)\right)$$

$$= \frac{1}{q} \sum_{j=1}^{q} \mathsf{P}\left(\left([U_n, x_n(x(j, \Phi), U_0, \ldots, U_{n-1})],\ n \geqq 0\right) \in (\cdot)\right),$$

where $q \geqq 1$ is a finite number and $x(1, \varphi), \ldots, x(q, \varphi)$ are measurable functions (Corollary 1.4.6).

The existence of a stationary weak solution does not imply the existence of an SGS in general, cf. Example 1.4.10. However, if there is a stationary weak solution (Φ', P'), $\Phi' = ([U_n', X_n'])$ possessing the recurrence properties $\{n: X_n' = x\} = \infty$ P'-a.s. and $\mathsf{E} \inf \{n > 0: x_n(x, U_0', \ldots, U_{n-1}') = x\} < \infty$ for some x, then there exists an SGS, cf. Theorem 1.4.11. Under the assumption of an ergodic input (Φ, P) and a fixed SGS (consisting of sets $A(n, \varphi)$), in Section 1.5 we give a constructive description of all stationary ergodic weak solutions that are concentrated on

$$\left\{([u_n, x_n]): x_n \in A(n, (u_k)),\ n \in \mathbb{Z}\right\}. \tag{1.1.4}$$

It turns out that there is a finite number of such ergodic weak solutions; each stationary weak solution that is concentrated on (1.1.4) can be represented as a mixture of these ergodic solutions (Theorem 1.5.4). In case of a denumerable state space and ergodic input, each stationary weak solution is a mixture of at most countably many ergodic weak solutions (Theorem 1.5.8).

In Section 1.6 we deal with a generalization of Wald's identity. The well-known result by Wald is:

$$\mathsf{E}_\mathsf{P} \sum_{n=1}^{\tau} U_n = \mathsf{E}_\mathsf{P}\tau \cdot \mathsf{E}_\mathsf{P} U_1, \tag{1.1.5}$$

where the U_n are i.i.d. real valued r.v.'s and τ is a stopping time with respect to $(U_n, n \geqq 1)$. For an arbitrary stationary sequence (U_n) we cannot expect the validity of (1.1.5) in general, cf. Example 1.6.1. However, we can prove the following result, cf. Theorem 1.6.3: Let (Φ, P), $\Phi = (U_n)$, be a stationary sequence of real valued r.v.'s, such that the expectation $\mathsf{E}_{\mathsf{P}} U_0$ exists. Furthermore, let $\tau(\Phi) \geqq 1$ be a stopping time with respect to $(U_n, n \geqq 1)$, and $\mathsf{E}_{\mathsf{P}}\tau(\Phi) < \infty$. Then there exists a stationary sequence (χ', R'), $\chi' = ([U_n', Y_n'])$, $Y_n' \in \{0, 1\}$, such that $(U_n') \overset{\mathfrak{D}}{=} (U_n)$, and

$$\mathsf{E}_{\mathsf{R}'}\left(\sum_{n=1}^{\tau((U_k'))} U_n' \mid Y_0' = 1\right) = \mathsf{E}_{\mathsf{R}'}\left(\tau((U_k')) \mid Y_0' = 1\right) \mathsf{E}_{\mathsf{P}} U_1. \qquad (1.1.6)$$

The existence of such a sequence (χ', R') can be ensured by solving a certain recursive stochastic equation via the construction of an SGS. Formula (1.1.6) reduces to (1.1.5) in the case of i.i.d. r.v.'s (Example 2.2.6). Thus, (1.1.6) appears as a natural generalization of Wald's identity.

In Section 1.7 we discuss continuity properties of the model (1.1.1) in the steady state, i.e. we investigate the conditions under which small perturbations of the input lead to small deviations of a particular stationary (weak or strong) solution of (1.1.1).

1.2. Weak and strong solutions

In the following let $\mathbb{U}$ and $\mathbb{X}$ be Polish spaces, $\mathcal{U}$ and $\mathcal{X}$ their respective Borel σ-fields, and

$$f: \mathbb{X} \times \mathbb{U} \to \mathbb{X}$$

a measurable function. We consider the recursive stochastic equation

$$X_{n+1} = f(X_n, U_n), \qquad n \in \mathbb{Z}. \qquad (1.2.1)$$

Here $\Phi = (U_n)$ is a given stationary sequence of r.v.'s taking values in $\mathbb{U}$. We use the notation (Φ, P), where P is the distribution of Φ (P is a probability measure on $\mathcal{U}_{\mathbb{Z}}$), and call (Φ, P) the input of the system described by (1.2.1). A reasonable notion of a solution of equation (1.2.1) is given by the following definition.

1.2.1. Definition. Let $x_n: \mathbb{U}^{\mathbb{Z}} \to \mathbb{X}$, $n \in \mathbb{Z}$, be measurable functions satisfying

$$x_{n+1}(\Phi) = f\big(x_n(\Phi), U_n\big), \qquad n \in \mathbb{Z}, \quad \mathsf{P}\text{-a.s.} \qquad (1.2.2)$$

Then the sequence $([U_n, X_n])$, $X_n = x_n(\Phi)$, of $\mathbb{U} \times \mathbb{X}$-valued r.v.'s on $(\mathbb{U}^{\mathbb{Z}}, \mathcal{U}_{\mathbb{Z}}, \mathsf{P})$ is called a *strong solution of* (1.2.1).

The sequence $X_n = x_n(\Phi)$ is a state process of the system, whereas the strong solution $([U_n, X_n])$ describes the temporal behaviour of the pair "input — system". At first sight it seems to be more natural to define a sequence (X_n) satisfying (1.2.2) as a strong solution of (1.2.1). However, there are good reasons to use $([U_n, X_n])$, as we shall see below.

The following example illustrates the fact that the concept of strong solution may be too restrictive in some cases.

1.2.2. Example. Let $U_n = [A_n, S_n]$ and $\mathbb{U} = \mathbb{R}_+ \times \mathbb{R}_+$, $\mathbb{X} = \mathbb{R}_+$. Consider the equation

$$X_{n+1} = \max (X_n + S_n 1\{X_n = 0\} - A_n, 0), \qquad n \in \mathbb{Z}, \tag{1.2.3}$$

describing the dynamics of a single server loss system with inter-arrival times A_n and service times S_n if we regard X_n as the residual service time at the arrival epoch of the n-th customer, cf. Section 5.3. In case $A_n \equiv 1$ and $S_n \equiv 3/2$, there are exactly two strong solutions $([U_n, X_n^1])$ and $([U_n, X_n^2])$, namely

$$X_n^1 = \begin{cases} 0 & \text{if } n \text{ is even}, \\ 1/2 & \text{if } n \text{ is odd} \end{cases} \quad \text{and } X_n^2 = \begin{cases} 0 & \text{if } n \text{ is odd}, \\ 1/2 & \text{if } n \text{ is even}. \end{cases}$$

Both $([U_n, X_n^1])$ and $([U_n, X_n^2])$ are non-stationary. In order to obtain stationary state processes we use another description of the system. Obviously the residual service time X_n, $n \in \mathbb{Z}$, takes only two values: 1/2 if the previous customer is being served, and 0 otherwise. The sequence (X_n) is a Markov chain with the transition probability matrix

$$\begin{pmatrix} 0 & 1 \\ 1 & 0 \end{pmatrix}. \tag{1.2.4}$$

For every $\alpha \in [0, 1]$ there is a Markov chain $\big((X_n^\alpha), \mathsf{P}_\alpha\big)$ with transition probability matrix (1.2.4) and

$$\mathsf{P}_\alpha(X_0^\alpha = 0) = 1 - \mathsf{P}_\alpha(X_0^\alpha = 1/2) = \alpha .$$

For each α the sequence (X_n^α) describes a sequence of consecutive states of the system considered for the given input. Notice that $(X_n^{1/2})$ is stationary.

Definition 1.2.1 of a strong solution of (1.2.1) does not allow to interpret any sequence (X_n^α), $0 < \alpha < 1$, of the previous example as a state process, since (X_n^α) are not defined as measurable functions of the input. The following notion makes such an interpretation possible.

1.2.3. Definition. Consider an input (Φ, P), $\Phi = (U_n)$, and a sequence $\Phi' = ([U_n', X_n'])$ of $\mathbb{U} \times \mathbb{X}$-valued r.v.'s. Denote the distribution of Φ' by P'. We call (Φ', P') a *weak solution of* (1.2.1) *for the input* (Φ, P) iff

$$(U_n) \overset{\mathcal{D}}{=} (U_n') \tag{1.2.5}$$

and

$$\mathsf{P}'\big(X_{n+1}' = f(X_n', U_n'), n \in \mathbb{Z}\big) = 1. \tag{1.2.6}$$

The marginal sequence (X_n') of a weak solution $([U_n', X_n'])$ will be called a *state process* of the system described by (1.2.1).

1.2.4. Example. We continue Example 1.2.2. For each $\alpha \in [0, 1]$ the sequence $\Phi_\alpha' = \big([[1, 3/2], X_n^\alpha]\big)$ with the distribution P_α' induced by P_α is a weak solution of (1.2.3) for $\Phi = ([1, 3/2])$.

In the following we do not distinguish between two weak solutions (Φ', P') and (Φ'', P'') of (1.2.1) for (Φ, P) if $\mathsf{P}' = \mathsf{P}''$.

Consequently, we say that equation (1.2.1) has a uniquely determined weak solution for (Φ, P) if all weak solutions for (Φ, P) are stochastically equivalent. Similarly, uniqueness of the state process means that all state processes have the same distribution.

Notice that each strong solution is a weak solution of (1.2.1) for the input (Φ, P).

In the following we are interested in the steady state of the system considered. Each stationary weak solution $\big(([U_n', X_n']), \mathsf{P}'\big)$ describes a steady state of the pair "input — system". Note that the stationarity of $([U_n', X_n'])$ is stronger than the stationarity of the marginal sequences (U_n') and (X_n'). The same is true for stationary strong solutions, too. This will be illustrated by the following example.

1.2.5. Example. Consider the input (Φ, P), $\Phi = (U_n)$, defined by $\mathsf{P}(\Phi = \theta^i\eta) = 1/4$, $i = 0, 1, 2, 3$, $\eta = (n \bmod 4,\ n \in \mathbb{Z})$, where $\theta(u_n) = (u_{n+1})$. Let $\mathbb{X} = \{0, 1\}$ and $f(0, 0) = f(0, 1) = f(0, 2) = f(0, 3) = 1$, $f(1, 0) = f(1, 1) = f(1, 2) = f(1, 3) = 0$. Then the sequence $([U_n, X_n])$ determined by

$$X_n(\eta) = X_n(\theta\eta) = n \bmod 2\,,$$

$$X_n(\theta^2\eta) = X_n(\theta^3\eta) = (n + 1) \bmod 2$$

is a strong solution. Obviously, $\Phi = (U_n)$ and (X_n) are stationary but $([U_n, X_n])$ is not stationary.

1.2.6. Remark. Let $\big([U_n, x_n(\Phi)]\big)$ be a strong solution. Then $\big([U_n, x_n(\Phi)]\big)$ is stationary if and only if

$$x_{n+1}(\varphi) = x_n(\theta\varphi) \quad \text{for} \quad \mathsf{P}\text{-a.e. } \varphi\,. \tag{1.2.7}$$

Namely, let $\big([U_n, x_n(\Phi)]\big)$ be stationary. Then

$$\mathsf{P}\Big(\big([U_{n+1}, x_{n+1}(\Phi)]\big) \in (\cdot)\Big) = \mathsf{P}\Big(\big([U_n, x_n(\Phi)]\big) \in (\cdot)\Big)$$
$$= \mathsf{P}\Big(\big([U_{n+1}, x_n(\theta\Phi)]\big) \in (\cdot)\Big),$$

i.e. $\big([U_{n+1}, x_n(\theta\Phi)]\big)$ is concentrated on

$$\Big\{([u_{n+1}, y_{n+1}]) : y_{n+1} = x_{n+1}\big((u_k)\big),\ n \in \mathbb{Z}\Big\}.$$

Thus, $x_{n+1}(\Phi) = x_n(\theta\Phi)$ a.s.

1.3. Construction of stationary strong solutions

In this section we present two methods for constructing stationary strong solutions of the equation (1.2.1). The first one is based on monotonicity and contractivity assumptions of f.

Assume that $\mathbb{X}$ may be endowed with a metric ϱ and a partial ordering $\leqq$ satisfying the following regularity conditions.

(A 1.3.1) The metric ϱ makes $\mathbb{X}$ a Polish space in which every bounded sequence contains a convergent subsequence.

(A 1.3.2) There exists a smallest element $O \in \mathbb{X}$, i.e. $O \leqq x$ for all $x \in \mathbb{X}$.

(A 1.3.3) For all $x_1, x_2, x_3 \in \mathbb{X}$ with $x_1 \leqq x_2 \leqq x_3$:

$$\varrho(x_1, x_2) \leqq \varrho(x_1, x_3), \quad \varrho(x_2, x_3) \leqq \varrho(x_1, x_3).$$

(A 1.3.4) Consider $x, x' \in \mathbb{X}$, $(x_n) \in \mathbb{X}^{\mathbb{Z}}$ with the properties $x_n \uparrow x$ and $x_n \leqq x'$ for all n. Then $x_n \leqq x \leqq x'$ holds for all n.

(A 1.3.5) $f(x, u)$ is non-decreasing and left continuous in x.

The conditions (A 1.3.1)–(A 1.3.4) are fulfilled e.g. for $\mathbb{X} = \mathbb{R}_+{}^k$ endowed with the maximum metric $\varrho(u, v) = \max_i |u_i - v_i|$ and the partial ordering $u \leqq v$ iff $u_i \leqq v_i$, $i = 1, \ldots, k$. In Section 5.2 we give an example of an infinite dimensional state space $\mathbb{X}$ where (A 1.3.1) is fulfilled. However, (A 1.3.1) is not valid in general.

Assumption (A 1.3.3) means compatibility of the semiordering relation $\leqq$ with the metric ϱ. Obviously (A 1.3.4) allows us to take limits of non-decreasing sequences in inequalities.

For an $x \in \mathbb{X}$, $\varphi = (u_n) \in \mathbb{U}^{\mathbb{Z}}$, denote

$$x_0(x) = x, \qquad x_1(x, u_0) = f(x, u_0),$$

$$x_2(x, u_0, u_1) = f\big(f(x, u_0), u_1\big), \ldots,$$

$$x_n(x, u_0, \ldots, u_{n-1}) = f\big(x_{n-1}(x, u_0, \ldots, u_{n-2}), u_{n-1}\big).$$

Obviously $x_n(x, u_0, \ldots, u_{n-1})$ are the consecutive states of the system starting with the state x at time 0, if the sample path of the input is φ. In view of (A 1.3.2) and (A 1.3.5) the sequence $x_r(O, u_{n-r}, \ldots, u_{n-1})$, $r \geqq 1$, is non-decreasing for every fixed n and every sample path $\varphi = (u_j)$ of the input. The assumptions (A 1.3.1) and (A 1.3.3) imply that either the limit

$$\lim_{r\to\infty} x_r(O, u_{n-r}, \ldots, u_{n-1})$$

exists or

$$\varrho\big(O, x_r(O, u_{n-r}, \ldots, u_{n-1})\big) \xrightarrow[r\to\infty]{} \infty.$$

In the first case we use the notation

$$x_n^{\min}(\varphi) = \lim_{r\to\infty} x_r(O, u_{n-r}, \ldots, u_{n-1}).$$

1.3.1. Theorem. *Let the assumptions* (A 1.3.1)–(A 1.3.5) *be fulfilled. Further we assume that for every* n

$$X_n^{\min} = x_n^{\min}(\Phi) = \lim_{r\to\infty} x_r(O, U_{n-r}, \ldots, U_{n-1}) \tag{1.3.1}$$

is a proper r.v. Then $([U_n, X_n^{\min}])$ *is a stationary strong solution of* (1.2.1). *For an arbitrary (not necessarily stationary) weak solution* $\big(([U_n', X_n']), \mathsf{P}'\big)$ *we have*

$$X_n' \geqq x_n^{\min}\big((U_j')\big) \qquad \mathsf{P}'\text{-}a.s.$$

Proof: Equation $x_{n+1}^{\min}(\Phi) = f\big(x_n^{\min}(\Phi), U_n\big)$ follows immediately from (A 1.3.5). Obviously, $([U_n, X_n^{\min}])$ is stationary and thus it is a stationary strong solution of (1.2.1).

Let $([u_n, x_n])$ fulfil $x_{n+1} = f(x_n, u_n)$, $n \in \mathbb{Z}$. From (A 1.3.2) we get $O \leqq x_{n-r}$ and thus, from (A 1.3.5),

$$x_r(O, u_{n-r}, \ldots, u_{n-1}) \leqq x_r(x_{n-r}, u_{n-r}, \ldots, u_{n-1}) = x_n \quad \text{for all } r \geqq 1.$$

Then

$$x_n^{\min}\big((u_k)\big) \leqq x_n, \qquad n \in \mathbb{Z},$$

follows immediately from (A 1.3.4). ∎

The crucial point in application of Theorem 1.3.1 is the verification of whether or not $X_n^{\min}$ is proper. Sometimes this can be very difficult, as we shall see in Section 5.5. In general, one needs additional assumptions on the input, which depend on the system considered.

The sequence $([U_n, X_n^{\min}])$ – if it exists – defines the minimal weak solution of (1.2.1). We next search for a maximal solution. To do so, we need some additional assumptions:

(A 1.3.6) Consider $x, x' \in \mathbf{X}$, $(x_n) \in \mathbf{X}^{\mathbb{Z}}$, with the properties $x_n \downarrow x$ and $x_n \geqq x'$ for all n. Then $x_n \geqq x \geqq x'$ holds for all n.

(A 1.3.7) For every $x \in \mathbf{X}$ and $\varepsilon > 0$ there is an element $x_{\max}(x, \varepsilon)$ of the ε-ball around x satisfying $x' \leqq x_{\max}(x, \varepsilon)$ for all x' with $\varrho(x, x') \leqq \varepsilon$.

(A 1.3.8) $f(x, u)$ is a contraction as a function of x, i.e.

$$\varrho\big(f(x, u), f(x', u)\big) \leqq \varrho(x, x') \quad \text{for } x, x' \in \mathbf{X}, \quad u \in \mathbf{U}.$$

Consider the following construction. We choose a $\delta > 0$ and a $\varphi \in \mathbf{U}^{\mathbb{Z}}$ with $x_n^{\min}(\varphi) \in \mathbf{X}$ for all n. For fixed n

$$x_r\left(x_{\max}\big(x_{n-r}^{\min}(\varphi), \delta\big), u_{n-r}, \ldots, u_{n-1}\right)$$

is non-increasing for $r \to \infty$, because of (A 1.3.5), (A 1.3.7) and (A 1.3.8). Thus, in view of (A 1.3.3) and (A 1.3.1) the limit

$$x_n^{\delta}(\varphi) = \lim_{r \to \infty} x_r\left(x_{\max}\big(x_{n-r}^{\min}(\varphi), \delta\big), u_{n-r}, \ldots, u_{n-1}\right)$$

does exist. It is easy to see that under the conditions of Theorem 1.3.1 and (A 1.3.7), (A 1.3.8), the sequence $\big(x_n^{\delta}(\Phi)\big)$ defines a stationary strong solution for every fixed δ. For every fixed n, $x_n^{\delta}(\varphi)$ is non-decreasing for $\delta \to \infty$, cf. (A 1.3.6).

1.3.2. Theorem. *Let* (A 1.3.1)–(A 1.3.8) *be fulfilled. Further assume that* $X_n^{\min} = x_n^{\min}(\Phi)$ *and*

$$X_n^{\max} = x_n^{\max}(\Phi) = \lim_{\delta \to \infty} x_n^{\delta}(\Phi)$$

are proper r.v.'s. Then $([U_n, X_n^{\max}])$ *is a stationary strong solution of* (1.2.1). *Moreover, for an arbitrary weak solution* (Φ', P'), $\Phi' = ([U_n', X_n'])$, *for which*

(X_n') is stationary, we have

$$X_n' \leqq x_n^{\max}\big((U_n')\big) \qquad \mathsf{P}'\text{-a.s.}$$

for all n. That means, $(X_n^{\max})$ is the maximal stationary state process.

Proof: Since f is continuous, it follows that $\big([U_n, x_n^{\max}(\Phi)]\big)$ is a stationary strong solution. Let $\Phi' = ([U_n', X_n'])$ be an arbitrary weak solution for which (X_n') is stationary. The distance $\varrho\left(X_n', x_n^{\min}\big((U_k')\big)\right)$ is non-increasing in n. Assume

$$\mathsf{P}'\left(\lim_{n\to-\infty} \varrho\left(X_n', x_n^{\min}\big((U_k')\big)\right) = \infty\right) > 0.$$

Then, from $O \leqq x_n^{\min}\big((U_k')\big) \leqq X_n'$, it follows that

$$\mathsf{P}'\left(\lim_{n\to-\infty} \varrho(O, X_n') = \infty\right) > 0,$$

which contradicts the stationarity of (X_n'). Thus,

$$\lim_{n\to-\infty} \varrho\left(X_n', x_n^{\min}\big((U_k')\big)\right) < \infty \qquad \mathsf{P}'\text{-a.s.}$$

Since $\left(\varrho\left(X_n', x_n^{\min}\big((U_k')\big)\right), n \in \mathbb{Z}\right)$ is non-increasing,

$$\mathsf{P}'\left(\bigcup_{r=1}^{\infty} \bigcap_{n=-\infty}^{\infty} \left\{X_n' \leqq x_{\max}\left(x_n^{\min}\big((U_k')\big), r\right)\right\}\right) = 1.$$

Now the maximality of (X_n') follows immediately:

$$1 = \mathsf{P}'\left(\bigcup_{r=1}^{\infty} \bigcap_{n=-\infty}^{\infty} \bigcap_{k=0}^{\infty} \left\{X_n' \leqq x_k\left(x_{\max}\left(x_{n-k}^{\min}\big((U_k')\big), r\right), U'_{n-k}, \ldots, U'_{n-1}\right)\right\}\right)$$

$$\leqq \mathsf{P}'\left(\bigcup_{r=1}^{\infty} \bigcap_{n=-\infty}^{\infty} \left\{X_n' \leqq x_n^r\big((U_k')\big)\right\}\right) \leqq \mathsf{P}'\left(X_n' \leqq x_n^{\max}\big((U_k')\big)\right). \quad \blacksquare$$

Now we show that there is a close relation between the uniqueness of the stationary solution of (1.2.1) and limiting behaviour of the system considered.

1.3.3. Theorem. *Assume* (A 1.3.1)–(A 1.3.8) *and that $\big(x_n^{\min}(\Phi)\big)$ is proper. Then the following statements are equivalent.*

(i) $\big([U_n, x_n^{\min}(\Phi)]\big)$ *is the unique stationary weak solution of* (1.2.1).

(ii) *For an arbitrary pair $[X, (U_n')]$ such that $(U_n) \overset{\mathfrak{D}}{=} (U_n')$ and X is an $\mathbf{X}$-valued r.v.:*

$$\varrho\left(x_n(X, U_0', \ldots, U'_{n-1}), x_n^{\min}\big((U_j')\big)\right) \xrightarrow[n\to\infty]{} 0 \qquad a.s.$$

(iii) *For an arbitrary pair $[X, (U_n')]$ such that $(U_n) \overset{\mathfrak{D}}{=} (U_n')$ and X is an $\mathbf{X}$-valued r.v.:*

$$\left(x_{n+k}(X, U_0', \ldots, U'_{n+k-1}), n \geqq 0\right) \xrightarrow[k\to\infty]{\mathfrak{D}} \left(x_n^{\min}(\Phi), n \geqq 0\right).$$

Proof: For brevity we use the notation $x_n(x, \varphi) = x_n(x, u_0, \ldots, u_{n-1})$.

(i) → (ii). Let $\delta > 0$. The sequence $\left(\varrho\left(x_{n-k}^{\min}(\Phi), x_n(O, U_{-k}, \ldots, U_{n-k-1})\right), n \geqq 0\right)$ is nonincreasing for each k. This yields

$$\begin{aligned}
&\mathsf{P}\left(\lim_{n\to\infty} \varrho\left(x_n^{\min}(\Phi), x_n(O, \Phi)\right) \geqq \delta\right) \\
&= \lim_{k\to\infty} \mathsf{P}\left(\lim_{n\to\infty} \varrho\left(x_n^{\min}(\theta^{-k}\Phi), x_n(O, \theta^{-k}\Phi)\right) \geqq \delta\right) \\
&= \lim_{k\to\infty} \mathsf{P}\left(\lim_{n\to\infty} \varrho\left(x_{n-k}^{\min}(\Phi), x_n(O, U_{-k}, \ldots, U_{n-k-1})\right) \geqq \delta\right) \\
&\leqq \lim_{k\to\infty} \mathsf{P}\left(\varrho\left(x_0^{\min}(\Phi), x_k(O, U_{-k}, \ldots, U_{-1})\right) \geqq \delta\right) = 0
\end{aligned}$$

and thus

$$\mathsf{P}\left(\lim_{n\to\infty} \varrho\left(x_n^{\min}(\Phi), x_n(O, \Phi)\right) \geqq \delta\right) = 0.$$

This implies

$$\lim_{n\to\infty} \varrho\left(x_n^{\min}(\Phi), x_n(O, \Phi)\right) = 0 \qquad \mathsf{P}\text{-a.s.} \tag{1.3.2}$$

In a similar way it can be proved that

$$\lim_{n\to\infty} \varrho\left(x_n^{\delta}(\Phi), x_n\left(x_{\max}\left(x_0^{\min}(\Phi), \delta\right), \Phi\right)\right) = 0, \qquad \delta > 0, \quad \mathsf{P}\text{-a.s.} \tag{1.3.3}$$

Assume (ii) were not true. Then, in view of the fact that f is a contraction there is a pair $[X, (U_n')]$, $(U_n') \overset{\mathfrak{D}}{=} \Phi$, and positive real numbers d and d' such that

$$\mathsf{R}\left(d \leqq \varrho\left(x_n\left(X, (U_j')\right), x_n^{\min}\left((U_j')\right)\right) \leqq d', n \in \mathbb{Z}_+\right) > 0, \tag{1.3.4}$$

where R denotes the distribution of the pair $[X, (U_j')]$. Consider a fixed pair $[x, \varphi]$, $\varphi = (u_n)$ satisfying

$$\left.\begin{aligned}
&\varrho\left(x_n(O, \varphi), x_n^{\min}(\varphi)\right) \xrightarrow[n\to\infty]{} 0, \\
&\varrho\left(x_n\left(x_{\max}\left(x_0^{\min}(\varphi), d'\right), \varphi\right), x_n^{d'}(\varphi)\right) \xrightarrow[n\to\infty]{} 0,
\end{aligned}\right\} \tag{1.3.5}$$

$$d \leqq \varrho\left(x_n(x, \varphi), x_n^{\min}(\varphi)\right) \leqq d', \qquad n \in \mathbb{Z}_+. \tag{1.3.6}$$

From (1.3.6) and (A 1.3.7) it follows that

$$O \leqq x \leqq x_{\max}\left(x_0^{\min}(\varphi), d'\right)$$

and hence from the monotonicity of f

$$x_n(O, \varphi) \leqq x_n(x, \varphi) \leqq x_n\left(x_{\max}\left(x_0^{\min}(\varphi), d'\right), \varphi\right). \tag{1.3.7}$$

By means of (A 1.3.3), and (1.3.7) we obtain the following inequality:

$$\begin{aligned}\varrho\big(x_n^{\min}(\varphi), x_n(x, \varphi)\big) &\leqq \varrho\big(x_n^{\min}(\varphi), x_n(O, \varphi)\big) + \varrho\big(x_n(O, \varphi), x_n(x, \varphi)\big)\\ &\leqq \varrho\big(x_n^{\min}(\varphi), x_n(O, \varphi)\big)\\ &\quad + \varrho\left(x_n(O, \varphi), x_n\left(x_{\max}\big(x_0^{\min}(\varphi), d'\big), \varphi\right)\right)\\ &\leqq 2\varrho\big(x_n^{\min}(\varphi), x_n(O, \varphi)\big) + \varrho\big(x_n^{\min}(\varphi), x_n^{d'}(\varphi)\big)\\ &\quad + \varrho\left(x_n^{d'}(\varphi), x_n\left(x_{\max}\big(x_0^{\min}(\varphi), d'\big), \varphi\right)\right).\end{aligned}$$

From this and (1.3.5), (1.3.6) we deduce

$$\liminf_{n\to\infty} \varrho\big(x_n^{\min}(\varphi), x_n^{d'}(\varphi)\big) \geqq d.$$

Thus, from (1.3.2), (1.3.3), and (1.3.4) we obtain

$$\mathsf{P}\left(\liminf_{n\to\infty} \varrho\big(x_n^{\min}(\Phi), x_n^{d'}(\Phi)\big) \geqq d\right) > 0. \tag{1.3.8}$$

Since $\big(\varrho\big(x_n^{\min}(\Phi), x_n^{d'}(\Phi)\big), n \in \mathbb{Z}\big)$ is non-increasing and stationary, we get (cf. Lemma A 1.1.4)

$$\varrho\big(x_n^{\min}(\Phi), x_n^{d'}(\Phi)\big) = \varrho\big(x_0^{\min}(\Phi), x_0^{d'}(\Phi)\big), \qquad n \in \mathbb{Z}, \quad \text{a.s.}$$

Thus we get

$$\mathsf{P}\left(\varrho\big(x_n^{\min}(\Phi), x_n^{d'}(\Phi)\big) \geqq d, n \in \mathbb{Z}\right) > 0 \tag{1.3.9}$$

from (1.3.8). We shall use the following notations:

$$\Phi^{\min} = ([U_n, X_n^{\min}]), \qquad \Phi^{\delta} = \big([U_n, x_n^{\delta}(\Phi)]\big), \qquad \delta > 0.$$

Since $\Phi^{d'}$ is a strong solution, Theorem 1.3.1 implies $x_n^{d'}(\Phi) \geqq x_n^{\min}(\Phi)$ a.s. From this and (1.3.9),

$$\Phi^{d'} \overset{\mathcal{D}}{\neq} \Phi^{\min},$$

which is a contradiction to (i).

(ii) → (iii). The set $\mathbb{N} = \{v = (v_n, n \geqq 0)\colon v_n \in \mathbf{X}, n \in \mathbb{Z}_+\}$ endowed with the metric

$$\tilde{\varrho}(v^1, v^2) = \sum_{i=0}^{\infty} \frac{\varrho(v_i^1, v_i^2)}{1 + \varrho(v_i^1, v_i^2)}\, 2^{-i},$$

$v^1 = (v_i^1, i \geqq 0)$, $v^2 = (v_i^2, i \geqq 0)$, becomes a Polish space. Denote the distribution of $[X, (U_j')]$ by R. Since $\varrho\left(x_n\big(X, (U_j')\big), x_n^{\min}\big((U_j')\big)\right)$ is non-increasing, (ii) leads to

$$\begin{aligned}&\mathsf{R}\left(\tilde{\varrho}\left(\left(x_{n+k}\big(X, (U_j')\big), n \geqq 0\right), \left(x_{n+k}^{\min}\big((U_j')\big), n \geqq 0\right)\right) \geqq \varepsilon\right)\\ &= \mathsf{R}\left(\sum_{n=0}^{\infty} \frac{\varrho\left(x_{n+k}\big(X, (U_j')\big), x_{n+k}^{\min}\big((U_j')\big)\right)}{1 + \varrho\left(x_{n+k}\big(X, (U_j')\big), x_{n+k}^{\min}\big((U_j')\big)\right)} \cdot 2^{-n} \geqq \varepsilon\right)\end{aligned}$$

$$\leqq \mathsf{R}\left(\sum_{n=0}^{\infty} \frac{\varrho\left(x_k(X, (U_j')), x_k^{\min}((U_j'))\right)}{1 + \varrho\left(x_k(X, (U_j')), x_k^{\min}((U_j'))\right)} \cdot 2^{-n} \geqq \varepsilon\right)$$

$$= \mathsf{R}\left(\varrho\left(x_k(X, (U_j')), x_k^{\min}((U_j'))\right) \geqq \frac{\varepsilon}{2 - \varepsilon}\right) \xrightarrow[k\to\infty]{} 0 \qquad (1.3.10)$$

for every $\varepsilon > 0$. In view of (1.3.10) and $\left(x_{n+k}^{\min}((U_j')), n \geqq 0\right) \stackrel{\mathcal{D}}{=} \left(x_n^{\min}((U_j')), n \geqq 0\right)$ we can apply Theorem 4.1 of BILLINGSLEY (1968) and obtain (iii).

(iii) → (i). Assume $\Phi^{\min}$ is not the only stationary weak solution, i.e. there is a stationary weak solution (Φ', P'), $\Phi' = ([U_n', X_n'])$,

$$\Phi' \stackrel{\mathcal{D}}{\neq} \Phi^{\min}. \qquad (1.3.11)$$

In view of the minimality of $\left([U_n', x_n^{\min}((U_j'))]\right)$ and (1.3.11), there exists a positive real number d such that

$$\mathsf{P}'\left(\varrho\left(X_0', x_0^{\min}((U_j'))\right) \geqq d\right) > 0. \qquad (1.3.12)$$

Applying again the minimality property we obtain

$$x_0^{\min}((U_j')) \stackrel{\mathcal{D}}{\neq} X_0'. \qquad (1.3.13)$$

On the other hand,

$$X_0' \stackrel{\mathcal{D}}{=} X_n' = x_n(X_0', (U_j')) \xrightarrow[n\to\infty]{\mathcal{D}} x_0^{\min}((U_j'))$$

in view of (iii). This provides $X_0' \stackrel{\mathcal{D}}{=} x_0^{\min}((U_j'))$, which is a contradiction to (1.3.13). ∎

The method of constructing strong solutions described above is based on the monotone convergence of $x_r(O, u_{n-r}, \ldots, u_{n-1})$ as $r \to \infty$. The second method consists in finding conditions ensuring the convergence of $x_r(x, u_{n-r}, \ldots, u_{n-1})$ as $r \to \infty$ for some $x \in \mathbf{X}$ in the following strong sense: For given $\varphi = (u_k)$ and $n \in \mathbb{Z}$, $x_r(x, u_{n-r}, \ldots, u_{n-1})$, $r = 1, 2, \ldots$, is constant for sufficiently large r.

1.3.4. Definition. Let $x \in \mathbf{X}$, $\varphi = (u_k) \in \mathbf{U}^{\mathbb{Z}}$ and let $l \geqq 0$. The number n is called an (x, l)-*renewing epoch of* φ if for all $r > 1$

$$x_r(x, u_{n-r}, \ldots, u_{n-1}) = \begin{cases} x \text{ if } l = 0, \\ x_l(x, u_{n-l}, \ldots, u_{n-1}), \text{ otherwise.} \end{cases}$$

This notion can be interpreted as follows: If the system starts with the state x at the epoch $n - r$, $r > l$, its state at the epoch n (and later) does not depend on the "past of φ viewed from $n - l$" $u_{n-r}, \ldots, u_{n-l-1}$ of φ.

1.3.5. Example. Consider equation (1.2.3) of the single server loss system with the input $\left(([A_n, S_n]), \mathsf{P}\right)$ satisfying $\mathsf{P}\left(([A_n, S_n]) = ([1, s_{1,n}])\right) = \mathsf{P}\left(([A_n, S_n]) = ([1, s_{2,n}])\right) = 1/2$,

$$s_{1,n} = \begin{cases} 1/2 \text{ if } n \text{ is even}, \\ 3/2 \text{ if } n \text{ is odd}, \end{cases} \qquad s_{2,n} = \begin{cases} 3/2 \text{ if } n \text{ is even}, \\ 1/2 \text{ if } n \text{ is odd}. \end{cases}$$

Then, for arbitrary $x \geqq 0$, $l \geqq x$, and $r \geqq l$ we obtain

$$x_r(x, [1, s_{1,n-r}], \dots, [1, s_{1,n-1}]) = 0 \text{ if } n \text{ is odd}$$

and

$$x_r(x, [1, s_{2,n-r}], \dots, [1, s_{2,n-1}]) = 0 \text{ if } n \text{ is even}.$$

Thus, all even numbers n are (x, l)-renewing epochs of $([1, s_{2,k}])$, and all odd numbers n are (x, l)-renewing epochs of $([1, s_{1,k}])$ if $0 \leqq x \leqq l$. Furthermore, all $n \in \mathbb{Z}$ are $(0, 1)$-renewing epochs of both $([1, s_{1,k}])$ and $([1, s_{2,k}])$, since for all $r \geqq 1$

$$x_r(0, [1, s_{1,n-r}], \dots, [1, s_{1,n-1}]) = \begin{cases} 1/2 & \text{if } n \text{ is even}, \\ 0 & \text{if } n \text{ is odd} \end{cases}$$

$$= x_1(0, [1, s_{1,n-1}])$$

and

$$x_r(0, [1, s_{2,n-r}], \dots, [1, s_{2,n-1}]) = \begin{cases} 0 & \text{if } n \text{ is even}, \\ 1/2 & \text{if } n \text{ is odd} \end{cases}$$

$$= x_1(0, [1, s_{2,n-1}])$$

holds.

For fixed x and l the events
$B_n = \{\varphi : n \text{ is an } (x, l)\text{-renewing epoch of } \varphi\}$ are stationary renewing events in the terminology of Borovkov (1978, 1980).

1.3.6. Theorem. *If*

$$\mathsf{P}\left(\bigcup_{n \in \mathbb{Z}} B_n\right) = 1, \tag{1.3.14}$$

then there exists a stationary sequence $(\hat{x}_n(x, \Phi), n \in \mathbb{Z})$ *such that* $\Phi'(x) = ([U_n, \hat{x}_n(x, \Phi)])$ *is a stationary strong solution of* (1.2.1) *and*

$$\mathsf{P}\left(\bigcup_{m=1}^{\infty} \{x_k(x, U_0, \dots, U_{k-1}) = \hat{x}_k(x, \Phi), k \geqq m\}\right) = 1. \tag{1.3.15}$$

In case of an ergodic input (Φ, P), *condition* (1.3.14) *is equivalent to* $\mathsf{P}(B_0) > 0$.

Proof: Define $B = \{\varphi : \text{for every } k \text{ there are } i \text{ and } j \text{ with } i < -k, j > k, \text{ and } \varphi \in B_i \cap B_j\}$. Every sample path $\varphi \in B$ possesses an infinite number of (x, l)-renewing epochs left and right of 0. In view of the assumed stationarity of (U_n), condition (1.3.14) yields

$$\mathsf{P}(B) = 1. \tag{1.3.16}$$

For a fixed $\varphi \in B$, $n \in \mathbb{Z}$, and $i \leqq n$ with $\varphi \in B_i$, define

$$\hat{x}_n(x, \varphi) = x_{n-i+l}(x, u_{i-l}, \dots, u_{n-1}). \tag{1.3.17}$$

According to Definition 1.3.4, the right side of (1.3.17) does not depend on the choice of the (x, l)-renewing epoch $i \leqq n$. Condition (1.3.16) ensures that $\hat{x}_n(x, \Phi)$ is a proper r.v.

The stationarity of $(\hat{x}_n(x, \Phi))$ is an immediate consequence of the construction. For all

$$\varphi = (u_n) \in \bigcup_{l \leq j \leq m} B_j$$

we have

$$x_k(x, u_0, \ldots, u_{k-1}) = \hat{x}_k(x, \varphi), \qquad k \geq m.$$

Thus, (1.3.16) implies (1.3.15). ∎

The statement of Theorem 1.3.6 remains valid if one uses a somewhat weakened notion of a renewing epoch.

1.3.7. Definition. Let $x \in \mathbb{X}$, $\varphi = (u_n) \in \mathbb{U}^{\mathbb{Z}}$. The number n is called an *x-renewing epoch of φ* if there is a number $l(n, x, \varphi) \geq 0$ such that for all $r > l(n, x, \varphi)$

$$x_r(x, u_{n-r}, \ldots, u_{n-1}) = \begin{cases} x, & \text{if } l(n, x, \varphi) = 0, \\ x_{l(n,x,\varphi)}(x, u_{n-l(n,x,\varphi)}, \ldots, u_{n-1}), & \text{otherwise}. \end{cases} \tag{1.3.18}$$

In the following let $l(n, x, \varphi)$ be the smallest number satisfying (1.3.18).

1.3.8. Example. (Continuation of 1.3.5) For fixed $x \geq 0$ and $\varphi = ([1, s_{1,n}])$, $l(n, x, \varphi)$ is obviously defined by

$$l(n, x, \varphi) = \begin{cases} \min \{l : x \leq l; l \in \mathbb{Z}\} \text{ if } n \text{ is odd}, \\ \min \{l : x + 1 \leq l; l \in \mathbb{Z}\} \text{ if } n \text{ is even}. \end{cases}$$

Define

$$C_n = \{\varphi : n \text{ is an } x\text{-renewing epoch of } \varphi\}.$$

1.3.9. Theorem. *If*

$$\mathsf{P}\left(\bigcup_{n \in \mathbb{Z}} C_n\right) = 1, \tag{1.3.19}$$

then the statements of Theorem 1.3.6 *are valid.*

In case of an ergodic input (Φ, P) *condition* (1.3.19) *is equivalent to*

$$\mathsf{P}(C_0) > 0.$$

Proof: Let

$$B_n^l = \{\varphi : n \text{ is an } (x, l)\text{-renewing epoch of } \varphi\}, \qquad l \in \mathbb{Z}_+,$$

and

$$L = \left\{l : l \in \mathbb{Z}_+, \mathsf{P}\left(\bigcup_{n \in \mathbb{Z}} B_n^l \setminus \left(\bigcup_{j=0}^{l-1} \bigcup_{n \in \mathbb{Z}} B_n^j\right)\right) > 0\right\}.$$

From (1.3.19),

$$1 = \mathsf{P}\left(\bigcup_{n \in \mathbb{Z}} C_n\right) = \mathsf{P}\left(\bigcup_{l=0}^{\infty} \bigcup_{n \in \mathbb{Z}} B_n^l\right)$$

$$= \mathsf{P}\left(\bigcup_{l \in L}\left(\bigcup_{n \in \mathbb{Z}} B_n^l \setminus \bigcup_{j=0}^{l-1} \bigcup_{n \in \mathbb{Z}} B_n^j\right)\right). \tag{1.3.20}$$

The events

$$\bigcup_{n\in\mathbb{Z}} B_n{}^l \setminus \bigcup_{j=0}^{l-1} \bigcup_{n\in\mathbb{Z}} B_n{}^j, \qquad l \in L, \tag{1.3.21}$$

are invariant and mutually exclusive, and they have positive probability. For $l \in L$ consider the input (Φ^l, P^l), $\Phi^l = (U_n{}^l, n \in \mathbb{Z})$, defined by

$$\mathsf{P}^l\big(\Phi^l \in (\cdot)\big) = \mathsf{P}\left(\Phi \in (\cdot) \mid \Phi \in \bigcup_{n\in\mathbb{Z}} B_n{}^l \setminus \bigcup_{j=0}^{l-1} \bigcup_{n\in\mathbb{Z}} B_n{}^j\right).$$

Obviously, (Φ^l, P^l) is stationary, and

$$\mathsf{P}^l\left(\bigcup_{n\in\mathbb{Z}} B_n{}^l\right) = 1.$$

Theorem 1.3.6 yields the existence of a stationary strong solution $\big([U_n{}^l, \hat{x}_n(x, \Phi^l)]\big)$ and

$$\mathsf{P}^l\left(\bigcup_{m=1}^{\infty} \{x_k(x, U_0{}^l, \ldots, U_{k-1}^l) = \hat{x}_k(x, \Phi^l), k \geqq m\}\right) = 1. \tag{1.3.22}$$

Obviously, P-a.e. φ belongs to one of the sets (1.3.21), cf. (1.3.20). Hence $\hat{x}_n(x, \varphi)$ is defined for P-a.e. φ, and

$$\mathsf{P}\left(\bigcup_{m=1}^{\infty} \{x_k(x, U_0, \ldots, U_{k-1}) = \hat{x}_k(x, \Phi), k \geqq m\}\right) = 1$$

holds in view of (1.3.22). Obviously, $\big([U_n, \hat{x}_n(x, \Phi)]\big)$ is a stationary strong solution. ∎

The definitions of (x, l)- and x-renewing epochs have the disadvantage that the stationary solution $\Phi'(x)$, in Theorem 1.3.6 and 1.3.9 respectively, depends on the fixed state x. In particular, the conditions (1.3.14) and (1.3.19) do not ensure the uniqueness of the stationary strong or weak solution of (1.2.1) as the following trivial example shows: Consider the equation $X_{n+1} = f(X_n, U_n) = X_n$ with the input $\Phi = (U_n)$, $U_n \equiv 1$. Then all $n \in \mathbb{Z}$ are x-renewing epochs with $l(n, x, \Phi) = 0$ for arbitrary x. Obviously, for each $x \in \mathbb{R}$ the sequence $([U_n, X_n])$, $U_n \equiv 1$, $X_n \equiv x$, is a stationary strong solution. More interesting examples occur in the framework of queueing theory, cf. Section 5.3.

Now we introduce a stronger version of x-renewing epochs.

1.3.10. Definition. The number n is called a *renewing epoch of* $\varphi = (u_n)$ if n is an x-renewing epoch for every $x \in \mathbf{X}$, and for arbitrary x' and x''

$$x_r(x', u_{n-r}, \ldots, u_{n-1}) = x_r(x'', u_{n-r}, \ldots, u_{n-1})$$

for $r > \max\big(l(n, x', \varphi), l(n, x'', \varphi)\big)$.

Define $D_n = \{\varphi\colon n \text{ is a renewing epoch of } \varphi\}$. Note, that D_n is not measurable in general.

1.3.11. Theorem. *Assume there is a measurable set D' satisfying $\mathsf{P}(D') = 1$ and $D' \subseteqq \{\varphi\colon \varphi$ possesses an infinite number of renewing epochs left and right of $0\}$*

(1.3.23)

Then there is a stationary strong solution $\Phi' = ([U_n, \hat{x}_n(\Phi)])$ *and for every* x

$$\mathsf{P}\left(\bigcup_{m=1}^{\infty} \{x_k(x, U_0, \dots, U_{k-1}) = \hat{x}_k(\Phi), k \geqq m\}\right) = 1. \tag{1.3.24}$$

Proof: For every fixed x the renewing epochs are also x-renewing epochs. The assumptions of Theorem 1.3.9 are satisfied. Thus, for every x there is a stationary strong solution $\Phi'(x) = ([U_n, \hat{x}_n(x, \Phi)])$ with the property (1.3.15).

For a $\varphi = (u_j) \in D'$, $n \in \mathbb{Z}$, and $i \leqq n$ with $(u_{j+i}, j \in \mathbb{Z}) \in D_0$ we can define $\hat{x}_n(x, \varphi)$ in the following way:

$$\hat{x}_n(x, \varphi) = x_{n-i+l(i,x,\varphi)}(x, u_{i-l(i,x,\varphi)}, \dots, u_{n-1}).$$

Using Definition 1.3.10 we find that the solution $\Phi'(x) = ([U_n, \hat{x}_n(x, \Phi)])$ does not depend on x. Denote this solution by $\Phi' = ([U_n, \hat{x}_n(\Phi)])$. For an arbitrary x formula (1.3.15) becomes

$$\mathsf{P}\left(\bigcup_{m=1}^{\infty} \{x_k(x, U_0, \dots, U_{k-1}) = \hat{x}_k(\Phi), k \geqq m\}\right) = 1. \quad \blacksquare \tag{1.3.25}$$

1.3.12. Remark. Let (Φ, P) be an ergodic input and $D_0' \subseteqq D_0$ a measurable subset with the property $\mathsf{P}(D_0') > 0$. Then the assumptions of Theorem 1.3.11 are fulfilled.

1.4. Construction of stationary weak solutions

The point of both methods discussed in the preceding section is the construction of exactly one sequence of system states associated with the given sample path of the input. However, in Example 1.2.2, 1.2.4 the stationary weak solution $(\Phi'_{1/2}, \mathsf{P}'_{1/2})$ is concentrated on exactly two sample paths, whereas the input has only one sample path. It seems a natural generalization of the previous methods to search for stationary weak solutions having finitely many state sequences for every fixed sample path of the input. The idea of the method presented below consists in constructing finitely many strong solutions, which are not necessarily stationary. Mixing them appropriately provides a stationary weak solution.

First we will give a more detailed analysis of the single server loss system, cf. Example 1.2.2.

1.4.1. Example. Let (Φ, P), $\Phi = ([A_n, S_n])$, be a stationary input, where the service times have finite expectation $\mathsf{E}S_0 > 0$ and

$$\lim_{n\to\infty} \frac{1}{n} \cdot \sum_{i=0}^{n-1} A_i > 0 \qquad \mathsf{P}\text{-a.s.} \tag{1.4.1}$$

(Note, that (1.4.1) is valid if Φ is ergodic and $\mathsf{E}A_0 > 0$.) For a fixed sample path $\varphi = ([a_j, s_j])$ and $n \in \mathbb{Z}$ define

$$I(n, \varphi) = \left\{j : j < n, s_j > \sum_{i=j}^{n-1} a_i\right\}$$

and

$$A(n, \varphi) = \left\{x : x = s_j - \sum_{i=j}^{n-1} a_i,\, j \in I(n, \varphi)\right\} \cup \{0\}.$$

The sets $I(n, \varphi)$ and $A(n, \varphi)$ are finite for P-a.e. φ. Indeed, by the individual ergodic theorem, we deduce for P-a.e. φ:

$$\lim_{j\to-\infty} \frac{s_j}{n-j} = \lim_{j\to-\infty} \frac{1}{n-j} \sum_{i=j}^{n} s_i - \lim_{j\to-\infty} \left(\frac{n-j-1}{n-j}\right) \frac{1}{n-j-1} \sum_{i=j+1}^{n} s_i = 0$$

and from (1.4.1)

$$\lim_{j\to-\infty} \frac{1}{n-j} \sum_{i=j}^{n-1} a_i > 0 \quad \text{for P-a.e. } \varphi.$$

Thus

$$\# A(n, \varphi) = \# I(n, \varphi) + 1 = \# \left\{j : j < n, \frac{s_j}{n-j} > \frac{1}{n-j} \sum_{i=j}^{n-1} a_i\right\} + 1 < \infty.$$

We can interpret the sets $I(n, \varphi)$ and $A(n, \varphi)$ as follows: Each stationary weak solution (Φ', P'), $\Phi' = \left([[A_n', S_n'], X_n']\right)$, of

$$X_{n+1} = \max\left(X_n + S_n 1\{X_n = 0\} - A_n, 0\right), \qquad n \in \mathbb{Z}, \tag{1.4.2}$$

has the following property: For P′-almost-every sample path $\varphi' = \left([[a_n, s_n], x_n]\right)$ there are infinitely many indices $n < 0$ with $x_n = 0$. Indeed, there are infinitely many indices $n > 0$ with $x_n = 0$, since the state x_0 (residual service time at the origin) and all s_n, $n \geqq 0$, are finite. Now the desired property follows from the stationarity of Φ', cf. Lemma A 1.1.1. This property implies that $\sup\{j : j < n, x_j = 0\}$ is finite for P′-a.e. φ'. Obviously, P′-almost surely

$$\sup\{j : j < n, x_j = 0\} \in I(n, \varphi) \quad \text{or} \quad x_n = 0,$$

where $\varphi = ([a_n, s_n])$. This means that for a.e. sample path φ' of a stationary weak solution the last index j before n, where $x_j = 0$ (i.e. the system is empty), belongs to $I(n, \varphi)$, if $x_n > 0$, and thus

$$x_n \in A(n, \varphi). \tag{1.4.3}$$

We call $A(n, \varphi)$ the set of *potential states* of the loss system considered for a given sample path φ. This is justified by property (1.4.3). Summarizing the preceding considerations we obtain

(i) $0 < \# A(n, \varphi) < \infty$ for P-a.e. φ.
(ii) The elements of $A(n, \varphi)$ can be arranged in ascending order and denoted by $x(1, n, \varphi), \ldots, x\left(\# A(n, \varphi), n, \varphi\right)$. The mappings $\varphi \to x(i, n, \varphi)$, $i = 1, \ldots, \# A(n, \varphi)$, $n \in \mathbb{Z}$, defined on $\{\varphi : i \leqq \# A(n, \varphi) < \infty\}$ are measurable.
(iii) $f\left(A(n, \varphi), [a_n, s_n]\right) \subseteqq A(n+1, \varphi)$, $n \in \mathbb{Z}$, for P-a.e. φ.
(iv) $A(n, \varphi) = A(0, \theta^n \varphi)$, $n \in \mathbb{Z}$, for P-a.e. φ.

A family of finite sets possessing properties (i)–(iv) is the basis for the construction of stationary weak solutions of (1.2.1).

1.4.2. Definition. Let (Φ, P) be a stationary input. A family of sets $\mathcal{S} = \{A(n, \varphi): n \in \mathbb{Z}, \varphi \in M\}$ is called a *solution generating system* (abbreviation SGS) if M is an invariant measurable subset of $\mathbb{U}^{\mathbb{Z}}$ satisfying $\mathsf{P}(M) = 1$ and if the following conditions are fulfilled:

(A 1.4.1) $A(n, \varphi) \subseteqq \mathbb{X}$ and $1 \leqq \# A(n, \varphi) < \infty$ for $\varphi \in M$, $n \in \mathbb{Z}$.

(A 1.4.2) The sets $\{\varphi: \varphi \in M, i \leqq \# A(n, \varphi)\}$, $i = 1, 2, \dots$, are measurable. There exist $\mathbb{X}$-valued measurable mappings $\varphi \to x(i, n, \varphi)$ defined on $\{\varphi: \varphi \in M, i \leqq \# A(n, \varphi)\}$ satisfying $A(n, \varphi) = \{x(1, n, \varphi), \dots, x(\# A(n, \varphi), n, \varphi)\}$.

(A 1.4.3) $f(A(n, \varphi), u_n) \subseteqq A(n + 1, \varphi)$ for $\varphi = (u_j) \in M$, $n \in \mathbb{Z}$.

(A 1.4.4) $A(n, \varphi) = A(0, \theta^n \varphi)$ for $\varphi \in M$, $n \in \mathbb{Z}$.

The notion of an SGS will be justified by the following theorem and its proof.

1.4.3. Theorem. *Assume that there is an SGS $\mathcal{S}$ for the given stationary input (Φ, P). Then there exists a stationary weak solution (Ψ, Q), $\Psi = ([\hat{U}_n, \hat{X}_n])$, of* (1.2.1).

Proof. Consider the sets $B(n, \varphi) = \bigcap_{r \leqq n} x_{n-r}(A(r, \varphi), u_r, \dots, u_{n-1})$, $n \in \mathbb{Z}$, $\varphi \in M$. In view of (A 1.4.3) $x_{n-r}(A(r, \varphi), u_r, \dots, u_{n-1})$ is a monotone decreasing sequence of finite sets as $r \to -\infty$. Thus the intersection $B(n, \varphi)$ is not empty. Furthermore,

$$B(n, \varphi) \subseteqq A(n, \varphi). \tag{1.4.4}$$

It is easy to see that $f(B(n, \varphi), u_n) \subseteqq B(n + 1, \varphi)$ and thus

$$\# B(n, \varphi) \geqq \# B(n + 1, \varphi). \tag{1.4.5}$$

In view of (A 1.4.4) the sequence $(\# B(n, \Phi), n \in \mathbb{Z})$ is stationary. A stationary sequence with monotone sample paths has almost surely constant sample paths, cf. Lemma A 1.1.4. Thus for a.e. φ there is a finite number $i(\varphi)$,

$$i(\varphi) = \# B(n, \varphi), \qquad n \in \mathbb{Z}. \tag{1.4.6}$$

The set $B(0, \varphi)$ can be expressed in the form

$$B(0, \varphi) = \{x(k_1, 0, \varphi), \dots, x(k_{i(\varphi)}, 0, \varphi)\}$$
$$\subseteqq \{x(1, 0, \varphi), \dots, x(\# A(0, \varphi), 0, \varphi)\} = A(0, \varphi),$$

where

$$k_l = \inf \{k: x(k, 0, \varphi) \in A(0, \varphi),\ \# \{i: i \leqq k, x(i, 0, \varphi) \in B(0, \varphi)\} = l\}.$$

That means the elements of $B(0, \varphi)$ are numbered according to their arrangement within $A(0, \varphi)$. From this fact one can easily obtain that $i(\varphi)$ is a measurable function of φ and that the functions

$$\varphi \to x(k_l, 0, \varphi), \qquad l = 1, \dots, i(\varphi) \tag{1.4.7}$$

defined for almost all φ are measurable.

Define the distribution Q^+ of a sequence $([\hat{U}_n, \hat{X}_n], n \geqq 0)$ by

$$\mathsf{Q}^+\left(([\hat{U}_n, \hat{X}_n], n \geqq 0) \in (\cdot)\right)$$
$$= \int_M \frac{1}{i(\varphi)} \sum_{j=1}^{i(\varphi)} 1\left\{\left(\left[u_n, x_n\left(x(k_j, 0, \varphi), u_0, \ldots, u_{n-1}\right)\right], n \geqq 0\right) \in (\cdot)\right\} \mathsf{P}(d\varphi). \tag{1.4.8}$$

In view of (A 1.4.4) and (1.4.6)

$$B(1, \varphi) = B(0, \theta\varphi) \quad \text{and} \quad i(\varphi) = i(\theta\varphi). \tag{1.4.9}$$

Thus

$$\sum_{j=1}^{i(\varphi)} 1\left\{\left(\left[u_{n+1}, x_{n+1}\left(x(k_j, 0, \varphi), u_0, \ldots, u_n\right)\right], n \geqq 0\right) \in (\cdot)\right\}$$
$$= \sum_{j=1}^{i(\theta\varphi)} 1\left\{\left(\left[u_{n+1}, x_n\left(x(k_j, 0, \theta\varphi), u_1, \ldots, u_n\right)\right], n \geqq 0\right) \in (\cdot)\right\}.$$

This implies the stationarity of $([\hat{U}_n, \hat{X}_n], n \geqq 0)$. From (1.4.8)

$$\hat{X}_{n+1} = f(\hat{X}_n, \hat{U}_n), \qquad n \geqq 0, \quad \mathsf{Q}^+\text{-a.s.}$$

and

$$(\hat{U}_n, n \geqq 0) \stackrel{\mathcal{D}}{=} (U_n, n \geqq 0).$$

The sequence $\left(([\hat{U}_n, \hat{X}_n], n \geqq 0), \mathsf{Q}^+\right)$ can be extended in the usual way, cf. Lemma A 1.1.1, to a stationary weak solution (Ψ, Q), $\Psi = ([\hat{U}_n, \hat{X}_n], n \in \mathbb{Z})$. ∎

1.4.4. Remark. It should be mentioned that the distribution Q constructed in the proof depends on the given SGS $\mathcal{S}$. On the other hand, for a given SGS Q is independent of the special choice of functions $x(j, n, \varphi)$ arising in the definition of an SGS. Indeed Q^+ (cf. (1.4.8)) can be written in the form

$$\mathsf{Q}^+\left(([\hat{U}_n, \hat{X}_n], n \geqq 0) \in (\cdot)\right)$$
$$= \int_M \frac{1}{i(\varphi)} \cdot \sum_{x \in B(0,\varphi)} 1\left\{([u_n, x_n(x, u_0, \ldots, u_{n-1})], n \geqq 0) \in (\cdot)\right\} \mathsf{P}(d\varphi).$$

1.4.5. Remark. If for a given input (Φ, P) and an SGS $\mathcal{S}$

$$i(\Phi) = 1 \qquad \mathsf{P}\text{-a.s.},$$

then

$$\Psi = ([U_n, X_n]) = \left([U_n, x(k_1, 0, \theta^n\Phi)]\right)$$

is a stationary strong solution.

For an ergodic input (Φ, P) the stationary weak solution (Ψ, Q) constructed in the proof of Theorem 1.4.3 has a simpler structure.

1.4.6. Corollary. *Let (Φ, P) be a stationary ergodic input and $\mathcal{S}$ an SGS. Then there are a finite number $q \geqq 1$ and measurable mappings $x(j, \varphi)$, $j = 1, \ldots, q$*

defined for almost all φ such that

$$\mathsf{Q}^+\left(\left([\hat{U}_n, \hat{X}_n], n \geqq 0\right) \in (\cdot)\right)$$

$$= \frac{1}{q} \sum_{j=1}^{q} \mathsf{P}\left(\left(\left[U_n, x_n\left(x(j, \Phi), U_0, \ldots, U_{n-1}\right)\right], n \geqq 0\right) \in (\cdot)\right). \tag{1.4.10}$$

determines the distribution Q *of a stationary weak solution* $\Psi = ([\hat{U}_n, \hat{X}_n], n \in \mathbb{Z})$ *of* (1.2.1).

Proof. The events $\{\varphi: i(\varphi) = i\}$, $i = 1, \ldots,$ are invariant and thus there is a number q with $\mathsf{P}(i(\Phi) = q) = 1$. Define

$$x(j, \varphi) = x(k_j, 0, \varphi), \qquad j = 1, \ldots, q, \tag{1.4.11}$$

cf. (1.4.7). Then (1.4.8) reduces to (1.4.10).
Notice that every other choice of measurable functions $x(j, \varphi)$ with $\{x(j, \varphi): j = 1, \ldots, q\} = B(0, \varphi)$ leads to (1.4.10), too. ∎

In order to apply Theorem 1.4.3 or Corollary 1.4.6 to special systems one has to construct an SGS, in particular the corresponding set M. (Often it suffices to define the set M as the set of all sample paths of the input satisfying (A 1.4.1)–(A 1.4.4).)

1.4.7. Example. (Continuation of 1.4.1.) Consider the set of sample paths of the input

$$M = \left\{\varphi: \varphi = ([a_l, s_l]): \# \left\{j: j < n, s_j > \sum_{i=j}^{n-1} a_i\right\} < \infty \text{ for all } n \in \mathbb{Z}\right\}$$

$$= \{\varphi: \# I(n, \varphi) < \infty \text{ for all } n \in \mathbb{Z}\}.$$

Obviously M is invariant. In Example 1.4.1 it is shown that under the given conditions $\mathsf{P}(M) = 1$ and (A 1.4.1)–(A 1.4.4) are satisfied, where

$$A(n, \varphi) = \left\{x: x = s_j - \sum_{i=j}^{n-1} a_i, j \in I(n, \varphi)\right\} \cup \{0\}.$$

From Theorem 1.4.3 we obtain the stationary weak solution (Ψ, Q).

1.4.8. Example. For finite $\mathbb{X} = \{x^{(1)}, \ldots, x^{(m)}\}$ there always exists an SGS, namely $M = \mathbb{X}^{\mathbb{Z}}$ and $A(n, \varphi) \equiv \mathbb{X}$, $x(i, n, \varphi) = x^{(i)}$.

1.4.9. Example. Let (Φ', P'), $\Phi' = \left([U_n, x_n(\Phi)]\right)$, be a stationary strong solution of (1.2.1) for (Φ, P), cf. Definition 1.2.1. Define

$$M = \left\{\varphi: \varphi = (u_j) \in \mathbb{U}^{\mathbb{Z}}, x_{n+1}(\varphi) = f\left(x_n(\varphi), u_n\right), \ x_n(\varphi) = x_{n-1}(\theta\varphi) \text{ for all } n \in \mathbb{Z}\right\}$$

and

$$A(n, \varphi) = \{x_n(\varphi)\}.$$

Then (A 1.4.1)–(A 1.4.4) are fulfilled. Obviously M is measurable and invariant. In view of Remark 1.2.6 we have $x_n(\varphi) = x_{n-1}(\theta\varphi)$ for P-almost all φ and thus $\mathsf{P}(M) = 1$. Thus $\{A(n, \varphi), n \in \mathbb{Z}, \varphi \in M\}$ is an SGS with $q = 1$.

Now we discuss the question how restrictive the assumption of the existence of an SGS is. In the following example, we have a stationary weak solution but no SGS.

1.4.10. Example. Consider the equation

$$X_{n+1} = X_n + 1 - \pi \cdot 1\{X_n + 1 > \pi\},$$

where $\mathbf{X} = [0, \pi)$. The input $\Phi = (U_n)$ is not relevant in this case; we put $U_n \equiv 0$. The distribution P' of a stationary weak solution $\Phi' = ([0, X_n'], n \in \mathbf{Z})$ is determined by

$$\mathsf{P}'\big(([0, X_n'], n \geqq 0) \in (\cdot)\big) = \frac{1}{\pi} \int_0^\pi 1\big\{([0, x_n(x, 0, \ldots, 0)], n \geqq 0\big) \in (\cdot)\big\}\, dx,$$

$\varphi = (\ldots, 0, 0, 0, \ldots)$, $\pi = 3,14\ldots$ However, there is no SGS, which becomes obvious by the following: Assume there is an SGS $\{A(n, \varphi), n \in \mathbf{Z}, \varphi \in M\}$. Since there is only one sample path of the input, all sets $A(n, \varphi)$ have to be equal, cf. (A 1.4.4) in Definition 1.4.2. From this and (A 1.4.3),

$$x_n\big(x(j, 0, \varphi), 0, \ldots, 0\big) \in A(n, \varphi) = A(0, \varphi), \qquad n \in \mathbf{Z}_+.$$

On the other hand $\big(x_n\big(x(j, 0, \varphi), 0, \ldots, 0\big), n \geqq 1\big)$ leaves every finite set.

However, if a stationary weak solution possesses some recurrence properties, then there exists an SGS.

1.4.11. Theorem. *Let (Φ, P) be a stationary input and (Φ', P'), $\Phi' = ([U_n', X_n'])$, be a stationary weak solution. Assume there is a state $x \in \mathbf{X}$ such that*

$$\left.\begin{aligned} &\#\{n\colon X_n' = x\} = \infty \qquad \mathsf{P}'\text{-a.s.} \\ &\text{and} \\ &\mathsf{E} \inf\{n\colon n > 0, x_n(x, U_0, \ldots, U_{n-1}) = x\} < \infty. \end{aligned}\right\} \qquad (1.4.12)$$

Then there is an SGS $\mathcal{S} = \{A(n, \varphi)\colon n \in \mathbf{Z}, \varphi \in M\}$ with the properties $\mathsf{P}'\big(X_n' \in A\big(n, (U_k')\big), n \in \mathbf{Z}\big) = 1$ and

$$\text{(A 1.4.5)} \begin{cases} \text{There exist } \mathbf{X}\text{-valued measurable mappings } \varphi \to x(i, n, \varphi),\ i = 1, 2, \\ \ldots, n \in \mathbf{Z}, \text{ defined on } \{\varphi\colon \varphi \in M, i \leqq \#A(n, \varphi)\}, \text{ satisfying } A(n, \varphi) \\ = \big\{x(1, n, \varphi), \ldots, x\big(\#A(n, \varphi), n, \varphi\big)\big\}, \varphi \in M,\ n \in \mathbf{Z}, \text{ and } x(i, 0, \varphi') \\ = x(i, 0, \varphi'') \text{ for all } \varphi' = (u_k') \in M,\ \varphi'' = (u_k'') \in M \text{ with } u_k' = u_k'' \\ \text{for } k < 0. \end{cases}$$

1.4.12. Remark. The property (A 1.4.5) has the obvious physical interpretation: The set of "potential states" $A(n, \varphi)$ and the measurable numeration, cf. (A 1.4.2), depend on $\varphi = (u_k)$ only via its past $(u_k, k < n)$, $n \in \mathbf{Z}$.

Proof of Theorem 1.4.11. For $\varphi = (u_n)$ define

$$\nu(\varphi) = \inf\{n\colon n > 0, x_n(x, u_0, \ldots, u_{n-1}) = x\}.$$

The sequence $(\nu(\theta^n\Phi))$ is stationary. Consider

$$L_n = \sum_{m \leq n} \mathbf{1}\{m + \nu(\theta^m\Phi) > n\}.$$

We obtain

$$\mathsf{E}_\mathsf{P} L_n = \mathsf{E}_\mathsf{P} L_0 = \sum_{m \leq 0} \mathsf{P}\big(m + \nu(\theta^m\Phi) > 0\big)$$

$$= \sum_{m=0}^{\infty} \mathsf{P}\big(\nu(\theta^{-m}\Phi) > m\big) = \sum_{m=0}^{\infty} \mathsf{P}\big(\nu(\Phi) > m\big) = \mathsf{E}\nu(\Phi) < \infty$$

in view of (1.4.12). In particular

$$\mathsf{P}(L_n < \infty) = 1 \quad \text{for all } n. \tag{1.4.13}$$

Now we define

$$I(n, \varphi) = \{m \leqq n : m + \nu(\theta^m\varphi) > n\}.$$

In view of (1.4.13) $I(n, \varphi)$ is a finite set for each $n \in \mathbb{Z}$ and a.e. φ. Denote the elements of $I(n, \varphi)$ by $m(1, n, \varphi)$, $m(2, n, \varphi)$, ..., $m\big(\#I(n, \varphi), n, \varphi\big)$, $m(1, n, \varphi) > m(2, n, \varphi) > \ldots$, and consider $x(i, n, \varphi) = x_{n-m(i,n,\varphi)}(x, u_{m(i,n,\varphi)}, \ldots, u_{n-1})$, $i = 1, \ldots, \#I(n, \varphi)$. Define

$$A(n, \varphi) = \big\{x(1, n, \varphi), \ldots, x\big(\#I(n, \varphi), n, \varphi\big)\big\}.$$

It follows immediately from the construction of $A(n, \varphi)$ that $\mathcal{S} = \{A(n, \varphi) : n \in \mathbb{Z}, \varphi \in \mathbb{U}^{\mathbb{Z}}, L_k(\varphi) < \infty \text{ for all } k \in \mathbb{Z}\}$ is an SGS satisfying (A 1.4.5). From the recurrence property (1.4.12) it follows that $\mathsf{P}'\big(\Phi' \in A(n, \Phi)\big) = 1$. ∎

Property (A 1.4.5) will be discussed in Section 2.2.

The existence of an SGS does not imply the existence of a stationary weak solution with property (1.4.12) in general, as can be seen by the following trivial example.

1.4.13. Example. Consider the equation

$$X_{n+1} = U_n, \qquad n \in \mathbb{Z},$$

where U_n is a sequence of i.i.d. real valued r.v.'s with a continuous distribution function. Then

$$\{A(n, \varphi) : A(n, \varphi) = \{u_{n-1}\}, \varphi = (u_i) \in \mathbb{R}^{\mathbb{Z}}, n \in \mathbb{Z}\}$$

is an SGS, and $([U_n, U_{n-1}])$ is the only stationary weak solution. Property (1.4.12) does not hold.

1.4.14. Remark. There is a close relation between the property $q = 1$ and the existence of (x, l)- and x-renewing epochs. For the ergodic input (Φ, P), let $\mathcal{S} = \{A(n, \varphi) : n \in \mathbb{Z}, \varphi \in M\}$ be an SGS with $q = 1$. Assume that for fixed x

$$\mathsf{P}\left(\bigcup_{m=1}^{\infty} \{x_r(x, U_{-r}, \ldots, U_{-1}) \in A(0, \Phi), r \geqq m\}\right) = 1. \tag{1.4.14}$$

The proof of Theorem 1.4.3 shows that for a.e. $\varphi = (u_n)$ there is a finite number $m(0, \varphi) \leqq 0$ with the property

$$x_{-m(0,\varphi)}\left(A\big(m(0, \varphi), \varphi\big), u_{m(0,\varphi)}, \ldots, u_{-1}\right) = B(0, \varphi). \tag{1.4.15}$$

The assumption (1.4.14) ensures the existence of a finite number $l_x(\varphi)$ satisfying

$$x_r(x, u_{m(0,\varphi)-r}, \ldots, u_{m(0,\varphi)-1}) \in A\big(m(0, \varphi), \varphi\big), \qquad r \geqq l_x(\theta^{m(0,\varphi)}\varphi) \tag{1.4.16}$$

for a.e. φ. For every φ with the properties (1.4.15) and (1.4.16) the value

$$l(0, x, \varphi) = -m(0, \varphi) + l_x(\theta^{m(0,\varphi)}\varphi)$$

satisfies the conditions of Definition 1.3.7. Thus, 0 is an x-renewing epoch for a.e. φ. From the ergodicity of (Φ, P) it follows that there is a number l such that $l(n, x, \Phi) \leqq l$ for infinitely many n P-a.s. Hence Φ possesses infinitely many (x, l)-renewing epochs. In an analogous way one can show that 0 is a renewing epoch for a.e. φ if $q = 1$ and if there is a measurable subset $M' \subseteqq M$ with the properties $\mathsf{P}(M') = 1$ and

$$M' \subseteqq \{\varphi\colon \text{for every } x \text{ there is an } l \text{ with } x_r(x, u_{-r}, \ldots, u_{-1}) \in A(0, \varphi), r \geqq l\}.$$

1.5. Ergodic weak solutions

In the following we always assume that $\Phi = (U_n)$ is stationary and ergodic. Let $\mathscr{S} = \{A(n, \varphi)\colon n \in \mathbb{Z}, \varphi \in M\}$ be a fixed SGS. In this section we want to describe constructively all stationary ergodic weak solutions, (Φ', P'), $\Phi' = \{[U_n', X_n']\}$ that are concentrated on

$$\mathcal{A}(\mathscr{S}) = \left\{([u_n, x_n])\colon x_n \in A\big(n, (u_k)\big), n \in \mathbb{Z}\right\}, \tag{1.5.1}$$

that means

$$\mathsf{P}'\left(X_n' \in A\big(n, (U_k')\big), n \in \mathbb{Z}\right) = 1.$$

Recall that, in view of Corollary 1.4.6 there are measurable functions $x(j, \varphi)$, $j = 1, \ldots, q$, such that

$$\mathsf{Q}^+\big(([\hat{U}_n, \hat{X}_n], n \geqq 0) \in (\cdot)\big)$$

$$= \frac{1}{q}\sum_{j=1}^{q} \mathsf{P}\left(\left([U_n, x_n\big(x(j, \Phi), U_0, \ldots, U_{n-1}\big)], n \geqq 0\right) \in (\cdot)\right) \tag{1.5.2}$$

defines a stationary weak solution (Ψ, Q), $\Psi = ([\hat{U}_n, \hat{X}_n])$.

1.5.1. Lemma. *Let (Φ, P) be stationary and ergodic and $\mathscr{S}$ an SGS. Then Q is ergodic iff (Ψ, Q) is the only stationary weak solution concentrated on $\mathcal{A}(\mathscr{S})$.*

Proof. 1. Let (Ψ, Q), $\Psi = ([\hat{U}_k, \hat{X}_k])$, be the only stationary weak solution satisfying $\hat{X}_n \in A\big(n, (\hat{U}_k)\big)$ a.s. Assume Q is not ergodic. Then Q is a non-trivial mixture of stationary weak solutions which are concentrated on $\mathcal{A}(\mathcal{S})$, cf. Theorem A 1.2.3. This contradicts the assumption. Thus Q is ergodic.

2. Let (Ψ, Q) be ergodic. Consider an arbitrary stationary weak solution (Φ', P'), $\Phi' = ([U_n', X_n'])$, satisfying

$$\mathsf{P}'\left(X_n' \in A\left(n, (U_k')\right)\right) = 1.$$

According to the construction of $x(j, \varphi) = x(k_j, 0, \varphi)$ in the proof of Theorem 1.4.3, for each sequence (x_n) with the properties $x_n \in A(n, \varphi)$, and $x_{n+1} = f(x_n, u_n)$, $n \in \mathbb{Z}$, we have $x_0 \in \{x(1, \varphi), \ldots, x(q, \varphi)\}$. Thus

$$\begin{aligned}
&\mathsf{P}'\big(([U_n', X_n'], n \geqq 0) \in (\cdot)\big)\\
&= \sum_{j=1}^{q} \mathsf{P}'\left(([U_n', X_n'], n \geqq 0) \in (\cdot), X_0' = x\left(j, (U_k')\right)\right)\\
&= \sum_{j=1}^{q} \mathsf{P}'\left(X_0' = x\left(j, (U_k')\right) \,\middle|\, \left(\left[U_n', x_n\left(x\left(j, (U_k')\right), U_0', \ldots, U_{n-1}'\right)\right], n \geqq 0\right) \in (\cdot)\right)\\
&\quad \times \mathsf{P}\left(\left(\left[U_n, x_n\left(x(j, \Phi), U_0, \ldots, U_{n-1}\right)\right], n \geqq 0\right) \in (\cdot)\right).
\end{aligned}$$

Assume $\mathsf{P}' \neq \mathsf{Q}$, and let $q > 1$ (for $q = 1$ Lemma 1.5.1 is trivial). Define

$$\begin{aligned}
b(j, \cdot) = \frac{1}{q-1}\Bigg(1 - \mathsf{P}'\bigg(X_0' = x\left(j, (U_k')\right) \,\bigg|\, \Big(\Big[U_n', x_n\left(x\left(j, (U_k')\right),\right.\\
\left. U_0', \ldots, U_{n-1}'\right)\Big], n \geqq 0\Big) \in (\cdot)\bigg)\Bigg).
\end{aligned}$$

Consider the probability measure

$$\mathsf{P}''(\cdot) = \sum_{j=1}^{q} b(j, \cdot)\, \mathsf{P}\left(\left(\left[U_n, x_n\left(x(j, \Phi), U_0, \ldots, U_{n-1}\right)\right], n \geqq 0\right) \in (\cdot)\right) \quad (1.5.3)$$

(P'' is indeed a probability measure since $\sum_{j=1}^{q} b\left(j, (\mathbb{U} \times \mathbb{X})^{\mathbb{Z}_+}\right) = 1$ and $b(j, \cdot) \geqq 0$). It holds that

$$\begin{aligned}
&\frac{1}{q}\, \mathsf{P}'\big(([U_n', X_n'], n \geqq 0) \in (\cdot)\big) + \frac{q-1}{q}\, \mathsf{P}''(\cdot)\\
&= \mathsf{Q}\big(([\hat{U}_n, \hat{X}_n], n \geqq 0) \in (\cdot)\big). \qquad (1.5.4)
\end{aligned}$$

Thus P'' is stationary. Hence (1.5.4) contradicts the ergodicity of Q. ∎

How can ergodic weak solutions be found if (Ψ, Q) itself is not ergodic? We start with an example.

1.5.2. Example. Consider the equation (1.2.3) of the single server loss system with the input $\left(([A_n, S_n]), \mathsf{P}\right)$ satisfying

$$\mathsf{P}\left(([A_n, S_n]) = \varphi_1\right) = \mathsf{P}\left(([A_n, S_n]) = \varphi_2\right) = 1/2.$$

$$\varphi_1 = ([1, s_{1,n}]), \qquad \varphi_2 = ([1, s_{2,n}]).$$

$$s_{1,n} = \begin{cases} 4 & \text{if } n \text{ is even,} \\ 6 & \text{if } n \text{ is odd,} \end{cases} \qquad s_{2,n} = \begin{cases} 6 & \text{if } n \text{ is even,} \\ 4 & \text{if } n \text{ is odd.} \end{cases}$$

Define

$$x(1, \varphi_1) = 0, \quad x(2, \varphi_1) = 2, \quad x(3, \varphi_1) = 3, \quad x(4, \varphi_1) = 5, \quad x(5, \varphi_1) = 1,$$

$$x(1, \varphi_2) = 3, \quad x(2, \varphi_2) = 1, \quad x(3, \varphi_2) = 4, \quad x(4, \varphi_2) = 0, \quad x(5, \varphi_2) = 2,$$

$$A(0, \varphi) = \{x(1, \varphi), \ldots, x(5, \varphi)\}, \quad \varphi \in \{\varphi_1, \varphi_2\}, \quad A(n, \varphi) = A(0, \theta^n \varphi).$$

Obviously, $\mathcal{S} = \{A(n, \varphi): n \in \mathbb{Z}, \varphi \in \{\varphi_1, \varphi_2\}\}$ is an SGS. It is not difficult to prove that the measures $\mathsf{Q}_{\{1,2\}}$, $\mathsf{Q}_{\{3,4,5\}}$ defined by

$$\mathsf{Q}^+_{\{1,2\}}(\cdot) = \frac{1}{2} \sum_{j=1}^{2} \mathsf{P}\Big(\Big([[A_n, S_n], x_n(x(j, \Phi), [A_0, S_0], \ldots,$$
$$[A_{n-1}, S_{n-1}]]), n \geqq 0\Big) \in (\cdot)\Big)$$

and

$$\mathsf{Q}^+_{\{3,4,5\}}(\cdot) = \frac{1}{3} \sum_{j=3}^{5} \mathsf{P}\Big(\Big([[A_n, S_n], x_n(x(j, \Phi), [A_0, S_0], \ldots,$$
$$[A_{n-1}, S_{n-1}]]), n \geqq 0\Big) \in (\cdot)\Big)$$

are stationary and ergodic.

Example 1.5.2 suggests that we should consider probability measures Q^+_{J}, $J \subseteqq \{1, \ldots, q\}$ of the form

$$\mathsf{Q}^+_J(\cdot) = \frac{1}{\# J} \sum_{j \in J} \mathsf{P}\Big(\Big([U_n, x_n(x(j, \Phi), U_0, \ldots, U_{n-1})], n \geqq 0\Big) \in (\cdot)\Big). \tag{1.5.5}$$

Assume that for a fixed $J \subseteqq \{1, \ldots, q\}$ the probability measure Q^+_J is stationary. It is easy to see that

$$A_J(0, \varphi) = \{x(j, \varphi): j \in J\},$$

$$A_J(n, \varphi) = A_J(0, \theta^n \varphi),$$

$\varphi \in M$ with $i(\varphi) = q$, defines an SGS $\mathcal{S}_J = \{A_J(n, \varphi), n \in \mathbb{Z}, \varphi \in M, i(\varphi) = q\}$. For the corresponding sets

$$B_J(n, \varphi) \bigcap_{r \leqq n} x_{n-r}(A_J(r, \varphi), u_r, \ldots, u_{n-1}),$$

cf. proof of Theorem 1.4.3, it holds that

$$\# B_J(n, \varphi) = \# J \qquad \mathsf{P}\text{-a.s.}$$

Applying Remark 1.4.4 to the particular SGS $\mathcal{S}_J$ we obtain just the extension Q_J of $\mathsf{Q}_J{}^+$. According to Lemma 1.5.1 Q_J is ergodic iff there is no other stationary weak solution concentrated on $\mathcal{A}(\mathcal{S}_J)$, cf. (1.5.1). In view of this fact one would presume that for sufficiently small sets J (1.5.5) yields ergodic weak solutions. However, the family of probability measures $\{\mathsf{Q}_J: J \subseteqq \{1,\ldots,q\}$ and Q_J is stationary$\}$ depends on the special choice of the functions $x(j, \varphi)$,

$j = 1, \ldots, q$, in the proof of Corollary 1.4.6. For some unfavourable choice of functions $x(j, \varphi)$ this family can be empty.

1.5.3. Example. Consider the single server loss system with the input (Φ, P), given in Example 1.5.2. In contrast to Example 1.5.2 we define

$$x(1, \varphi_1) = 0, \quad x(2, \varphi_1) = 1, \quad x(3, \varphi_1) = 2, \quad x(4, \varphi_1) = 3, \quad x(5, \varphi_1) = 5,$$

$$x(1, \varphi_2) = 0, \quad x(2, \varphi_2) = 1, \quad x(3, \varphi_2) = 2, \quad x(4, \varphi_2) = 3, \quad x(5, \varphi_2) = 4.$$

Then, for each proper subset $J \subset \{1, \ldots, 5\}$, the probability measure Q_J determined by (1.5.5) is non-stationary.

As the following theorem shows, there is always such a choice of functions $x(j, \varphi)$, $j = 1, \ldots, q$, such that all ergodic solutions concentrated on $\mathcal{A}(\mathcal{S})$ have the structure (1.5.5).

1.5.4. Theorem. *Let $\mathcal{S}$ be an SGS for a stationary and ergodic input (Φ, P). Then there are measurable functions $x(j, \varphi)$, $j = 1, \ldots, q$, and integers $0 = k_0 < k_1 < \cdots < k_r = q$ such that Q_m, $m = 1, \ldots, r$,*

$$Q_m^{+}(\cdot) = \frac{1}{k_m - k_{m-1}} \sum_{j=k_{m-1}+1}^{k_m} \mathsf{P}\Big(\big([U_n, x_n(x(j, \Phi), U_0, \ldots, U_{n-1})], n \geqq 0\big) \in (\cdot)\Big), \tag{1.5.6}$$

are just the distributions of all stationary ergodic weak solutions concentrated on $\mathcal{A}(\mathcal{S})$ and

$$\mathsf{Q}^{+}(\cdot) = \frac{1}{q} \sum_{j=1}^{q} \mathsf{P}\Big(\big([U_n, x_n(x(j, \Phi), U_0, \ldots, U_{n-1})], n \geqq 0\big) \in (\cdot)\Big)$$

$$= \sum_{m=1}^{r} \frac{k_m - k_{m-1}}{q} \mathsf{Q}_m^{+}(\cdot).$$

Moreover, for each stationary weak solution (Φ', P') concentrated on $\mathcal{A}(\mathcal{S})$ the probability distribution P' is a mixture of the ergodic distributions $\mathsf{Q}_1, \ldots, \mathsf{Q}_r$, i.e.

$$\mathsf{P}' = \sum_{m=1}^{r} c_m \mathsf{Q}_m, \qquad \sum_{m=1}^{r} c_m = 1.$$

In the proof of Theorem 1.5.4 and later we shall use the following statement.

1.5.5. Lemma. *Let (Φ, P), $\Phi = (U_n)$, be a stationary and ergodic input. Assume (Φ', P'), $\Phi' = ([U_n', X_n'])$, to be a stationary weak solution and $D \subseteq (\mathbf{U} \times \mathbf{X})^{\mathbf{Z}}$ an invariant set with the property $\mathsf{P}'(D) > 0$. Then $\mathsf{P}'(\cdot \mid D)$ is the probability distribution of a stationary weak solution for (Φ, P).*

Proof. Since D is invariant, i.e. $\theta D = D$, the stationarity of P' implies the stationarity of $\mathsf{P}'(\cdot \mid D)$. $\mathsf{P}'(X_{n+1}' = f(X_n', U_n'), n \in \mathbf{Z} \mid D) = 1$ follows immediately from $\mathsf{P}'\big(X_{n+1}' = f(X_n', U_n'), n \in \mathbf{Z}\big) = 1$. It remains to show that $\mathsf{P}'\big((U_n') \in (\cdot) \mid D\big) = \mathsf{P}(\cdot)$. Since this is obviously true if $\mathsf{P}'(D) = 1$, we can assume $0 < \mathsf{P}'(D) < 1$. Then

$$\mathsf{P}(\cdot) = \mathsf{P}'\big((U_n') \in (\cdot) \mid D\big) \mathsf{P}'(D) + \mathsf{P}'\big((U_n') \in (\cdot) \mid D^c\big) \mathsf{P}'(D^c),$$

i.e. P is a mixture of stationary probability distributions. Since P is ergodic, $\mathsf{P}(\cdot) = \mathsf{P}'\big((U_n') \in (\cdot) \mid D\big)$. ∎

Proof of Theorem 1.5.4.

1. First we introduce some notation. According to Corollary 1.4.6 there are measurable mappings $y(j, \varphi)$, $j = 1, \ldots, q$, satisfying

$$\mathsf{Q}^+(\cdot) = \frac{1}{q} \sum_{j=1}^{q} \mathsf{P}\Big(\big([U_n, x_n(y(j, \Phi), U_0, \ldots, U_{n-1})], n \geqq 0\big) \in (\cdot)\Big).$$

In the following the mappings $y(j, \varphi)$ are fixed. Let $\boldsymbol{H}$ be the family of all functions h of the form

$$h\big(([u_n, x_n], n \geqq 0)\big) = \mathbf{1}_{C_0 \times \cdots \times C_{k-1}}([u_0, x_0], \ldots, [u_{k-1}, x_{k-1}]),$$

$$C_j \in \mathcal{U} \otimes \mathcal{X}, \qquad j = 0, \ldots, k-1; \qquad k = 1, 2, \ldots,$$

defined on $(\mathbb{U} \times \mathbb{X})^{\mathbb{Z}_+}$. Define

$$h_j(\varphi) = \lim_{n\to\infty} \frac{1}{n} \sum_{k=1}^{n} h\Big(\big([u_{k+l}, x_{k+l}(y(j, \varphi), u_0, \ldots, u_{k+l-1})], l \geqq 0\big)\Big),$$

$j = 1, \ldots, q$, $h \in \boldsymbol{H}$, $\varphi = (u_k)$. Due to the stationarity of Q, it is easy, by means of the individual ergodic theorem, to prove that $h_j(\varphi)$, $j = 1, \ldots, q$, exists for almost all φ. Furthermore

$$\frac{1}{q}\, \mathsf{E}_{\mathsf{P}} \sum_{j=1}^{q} h_j(\Phi) = \mathsf{E}_{\mathsf{Q}} h(\Psi). \tag{1.5.7}$$

2. The weak solution (Ψ, Q) is concentrated on

$$\Big\{\big([u_n, x_n(y(j, \varphi), u_0, \ldots, u_{n-1})], n \geqq 0\big) : \varphi = (u_k) \in \boldsymbol{M}, j = 1, \ldots, q\Big\}.$$

Thus, in view of the ergodic Theorem A 1.2.2 (Ψ, Q) is ergodic if and only if

$$h_j(\Phi) = \mathsf{E}_{\mathsf{Q}} h(\Psi), \qquad j = 1, \ldots, q \qquad \mathsf{P}\text{-a.s. for all } h \in \boldsymbol{H}.$$

3. First assume that Q is ergodic. From Lemma 1.5.1 we conclude that there is exactly one stationary ergodic weak solution concentrated on $\mathcal{A}(\mathcal{S})$. Thus the statements of the theorem are valid for $r = 1$ and $\mathsf{Q}_1 = \mathsf{Q}$.

4. Now assume Q is not ergodic. Then there is a function $h^* \in \boldsymbol{H}$ such that $\mathsf{P}\left(\bigcup_{i=1}^{q} \{\varphi : h_i^*(\varphi) \neq c^*\}\right) > 0$ with $c^* = \mathsf{E}_{\mathsf{Q}} h^*(\Psi)$. From this and (1.5.7) we get that the events

$$\boldsymbol{M}_- = \bigcup_{i=1}^{q} \{\varphi : h_i^*(\varphi) < c^*\} \quad \text{and} \quad \boldsymbol{M}_+ = \bigcup_{i=1}^{q} \{\varphi : h_i^*(\varphi) > c^*\}$$

have positive probabilities. Since these events are invariant and (Φ, P) is ergodic, it follows that $\mathsf{P}(\boldsymbol{M}_+) = \mathsf{P}(\boldsymbol{M}_-) = 1$ and thus

$$\mathsf{P}(\varphi : \text{There are } k \text{ and } l \text{ with } h_k^*(\varphi) < h_l^*(\varphi),\ i(\varphi) = q) = 1. \tag{1.5.8}$$

Define

$$\underline{I}(\varphi) = \{k : k \in \{1, \ldots, q\}, h_k^*(\varphi) < h_l^*(\varphi) \text{ for all } l \in \{1, \ldots, q\}\},$$
$$\bar{I}(\varphi) = \{1, \ldots, q\} \setminus \underline{I}(\varphi).$$

We have $\underline{I}(\varphi) \neq \emptyset$ and $\bar{I}(\varphi) \neq \emptyset$ (The latter follows from (1.5.8).). For almost all φ a permutation π_φ on $\{1, \ldots, q\}$ is defined by

$$\pi_\varphi(1) = \min \{k : k \in \underline{I}(\varphi)\},$$

$$\pi_\varphi(l+1) = \begin{cases} \min \{m : m \in \underline{I}(\varphi), m > \pi_\varphi(l)\} & \text{for } l < \# \underline{I}(\varphi), \\ \min \{m : m \notin \{\pi_\varphi(1), \ldots, \pi_\varphi(l)\}\} & \text{for } l \geqq \# \underline{I}(\varphi). \end{cases}$$

The permutation π_φ arranges the elements of $\underline{I}(\varphi)$ in front of the elements of $\bar{I}(\varphi)$. We have

$$\underline{I}(\varphi) = \{\pi_\varphi(1), \ldots, \pi_\varphi(\# \underline{I}(\varphi))\},$$
$$\bar{I}(\varphi) = \{\pi_\varphi(\# I(\varphi) + 1), \ldots, \pi_\varphi(q)\}.$$

For $k \in \underline{I}(\varphi)$,

$$x_n(y(k, \varphi), u_0, \ldots, u_{n-1}) \in \{y(l, \theta^n\varphi) : l \in \underline{I}(\theta^n\varphi)\}, \qquad n \geqq 1,$$

since

$$h_k(\varphi) = h_l(\theta^n\varphi) \text{ if } x_n(y(k, \varphi), u_0, \ldots, u_{n-1}) = y(l, \theta^n\varphi).$$

Thus the number $\# \underline{I}(\varphi)$ is invariant with respect to all shifts θ^n, $n \in \mathbb{Z}$, for a.e. φ. In view of the ergodicity of (Φ, P), $\# \underline{I}(\Phi)$ is P-a.s. constant. We denote this number by $\# \underline{I}(\Phi) = \underline{k}$. Obviously $1 \leq \underline{k} < q$. Now we can express the distribution Q in the form

$$\mathsf{Q} = \frac{\underline{k}}{q} \underline{\mathsf{Q}} + \frac{q - \underline{k}}{q} \bar{\mathsf{Q}},$$

where

$$\underline{\mathsf{Q}}^+(\cdot) = \frac{1}{\underline{k}} \sum_{j=1}^{\underline{k}} \mathsf{P}\left(\left(\left[U_n, x_n\left(y(\pi_\Phi(j), \Phi), U_0, \ldots, U_{n-1}\right)\right], n \geqq 0\right) \in (\cdot)\right)$$

and

$$\bar{\mathsf{Q}}^+(\cdot) = \frac{1}{q - \underline{k}} \sum_{j=\underline{k}+1}^{q} \mathsf{P}\left(\left(\left[U_n, x_n\left(y(\pi_\Phi(j), \Phi), U_0, \ldots, U_{n-1}\right)\right], n \geqq 0\right) \in (\cdot)\right).$$

Analogously to the proof of Theorem 1.4.3 one can show the stationarity of $\underline{\mathsf{Q}}$ and $\bar{\mathsf{Q}}$.

5. Define

$$y'(j, \varphi) = y(\pi_\varphi(j), \varphi), \qquad j = 1, \ldots, q.$$

If $\underline{\mathsf{Q}}$ is ergodic, then we put

$$\nu_\varphi(j) = \pi_\varphi(j) \quad \text{for} \quad j \in \{1, \ldots, \underline{k}\}.$$

Analogously, if $\bar{\mathsf{Q}}$ is ergodic, then

$$\nu_\varphi(j) = \pi_\varphi(j) \quad \text{for} \quad j \in \{\underline{k} + 1, \ldots, q\}.$$

If both $\underline{\mathsf{Q}}$ and $\bar{\mathsf{Q}}$ are ergodic, then the proof will be completed in step 7 (with $r = 2$, $k_1 = \underline{k}$). In the other case we continue as follows.

6. Assume $\underline{\mathsf{Q}}$ is not ergodic. (If $\underline{\mathsf{Q}}$ is ergodic, but not $\bar{\mathsf{Q}}$, we can proceed analogously.) Then apply step 4 and 5 anew with $\underline{\mathsf{Q}}$, $\underline{k}$ and $y'(1, \varphi), \ldots, y'(\underline{k}, \varphi)$ instead of Q, q, $y(1, \varphi), \ldots, y(q, \varphi)$, respectively. In this way we obtain a subdivision of the index set $\{1, \ldots, q\}$ and possibly (in step 5) some further values of $\nu_\varphi(\cdot)$. This procedure is continued recursively as long as there are components of the decomposition of Q which are not ergodic. In every recursion step a subset of $\{1, \ldots, q\}$ is devided into two nonempty parts. Hence the recursion terminates after a finite number of steps. (Notice, if one subset consists of a single element, then the "corresponding" measure $\underline{\mathsf{Q}}$ or $\bar{\mathsf{Q}}$, respectively, is ergodic.) The result of this procedure is a subdivision

$$\{1, \ldots, q\} = \{1, \ldots, k_1\} \cup \{k_1 + 1, \ldots, k_2\} \cup \cdots \cup \{k_{r-1} + 1, \ldots, q\}$$

and a permutation $\nu_\varphi(\cdot)$ for a.e. φ.

7. Define

$$x(j, \varphi) = y\big(\nu_\varphi(j), \varphi\big), \qquad j = 1, \ldots, q.$$

Obviously, $B(0, \varphi) = \{x(1, \varphi), \ldots, x(q, \varphi)\}$. According to the termination rule of the recursive procedure described above, $(\Psi_1, \mathsf{Q}_1), \ldots, (\Psi_r, \mathsf{Q}_r)$ defined by

$$\mathsf{Q}_m{}^+(\cdot) = \frac{1}{k_m - k_{m-1}} \sum_{j=k_{m-1}+1}^{k_m} \mathsf{P}\Big(\Big(\big[U_n, x_n\big(x(j, \Phi), U_0, \ldots, U_{n-1}\big)\big], n \geqq 0\Big) \in (\cdot)\Big)$$

$m = 1, \ldots, r$, $k_0 = 0$, $k_r = q$, are stationary ergodic weak solutions. From the remark at the end of the proof of Corollary 1.4.6,

$$\sum_{m=1}^{r} \frac{k_m - k_{m-1}}{q} \mathsf{Q}_m{}^+(\cdot)$$
$$= \frac{1}{q} \sum_{j=1}^{q} \mathsf{P}\Big(\Big(\big[U_n, x_n\big(x(j, \Phi), U_0, \ldots, U_{n-1}\big)\big], n \geqq 0\Big) \in (\cdot)\Big) = \mathsf{Q}^+(\cdot).$$

Define

$$\mathcal{S}_m = \{B_m(n, \varphi) : n \in \mathbb{Z}, \varphi \in M, i(\varphi) = q\},$$

where

$$B_m(n, \varphi) = \{x(k_{m-1} + 1, \theta^n\varphi), \ldots, x(k_m, \theta^n\varphi)\}, \qquad n \in \mathbb{Z}, \qquad \varphi \in M.$$

Obviously $\mathcal{S}_m$ is an SGS for $m = 1, \ldots, r$. Due to Lemma 1.5.1 (Ψ_m, Q_m) is the only weak solution concentrated on $\mathcal{A}(\mathcal{S}_m)$. It remains to show that every stationary ergodic weak solution (Φ', P') which is concentrated on $\mathcal{A}(\mathcal{S})$, belongs to $\{(\Psi_1, \mathsf{Q}_1), \ldots, (\Psi_r, \mathsf{Q}_r)\}$. If (Φ', P') is concentrated on $\mathcal{A}(\mathcal{S})$, then it is concentrated on $\mathcal{A}(\{B(n, \varphi) : n \in \mathbb{Z}, \varphi \in M, i(\varphi) = q\}) = \bigcup_{m=1}^{r} \mathcal{A}(\mathcal{S}_m)$. Since the sets $\mathcal{A}(\mathcal{S}_m)$ are disjoint and invariant, (Φ', P') has to be concentrated on one of them. Thus, (Ψ_i, Q_i), $i = 1, \ldots, r$, are exactly the stationary ergodic weak solutions concentrated on $\mathcal{A}(\mathcal{S})$.

8. Let (Φ', P') be an arbitrary stationary weak solution concentrated on $\mathcal{A}(\mathcal{S})$. We have

$$\mathsf{P}'(\cdot) = \sum_{m=1}^{r} \mathsf{P}'\big(\cdot \mid \mathcal{A}(\mathcal{S}_m)\big)\, \mathsf{P}'\big(\mathcal{A}(\mathcal{S}_m)\big).$$

Since the set $\mathcal{A}(\mathcal{S}_m)$ is invariant,

$$\mathsf{P}'\big(\cdot \mid \mathcal{A}(\mathcal{S}_m)\big)$$

is the distribution of a stationary weak solution concentrated on $\mathcal{A}(\mathcal{S}_m)$, cf. Lemma 1.5.5. From the ergodicity of Q_m and Lemma 1.5.1 we obtain

$$\mathsf{P}'\big(\cdot \mid \mathcal{A}(\mathcal{S}_m)\big) = \mathsf{Q}_m(\cdot) \quad \text{if} \quad \mathsf{P}'\big(\mathcal{A}(\mathcal{S}_m)\big) > 0. \quad \blacksquare$$

Now we are able to prove that the uniqueness of the stationary weak solution implies the uniqueness of the stationary state process.

1.5.6. Theorem. *Let $\big(([U_n', X_n']), \mathsf{P}'\big)$ be a weak solution with (X_n') stationary for the stationary ergodic input (Φ, P). Assume that there is an SGS $\mathcal{S}$ such that $\Phi' = ([U_n', X_n'])$ is concentrated on $\mathcal{A}(\mathcal{S})$, and (Ψ, Q) is the only stationary weak solution concentrated on $\mathcal{A}(\mathcal{S})$. Then $(X_n') \stackrel{\mathcal{D}}{=} (\hat{X}_n)$, and (X_n') is ergodic.*

Proof. For the functions h, h_j used in the proof of Theorem 1.5.4 we have

$$\begin{aligned} h_j(\varphi) &= \lim_{n\to\infty} \frac{1}{n} \sum_{k=1}^{n} h\Big(\big[u_{k+l}, x_{k+l}\big(y(j, \varphi), u_0, \ldots, u_{k+l-1}\big)\big], l \geqq 0\Big) \\ &= \mathsf{E}_{\mathsf{Q}} h([\hat{U}_l, \hat{X}_l], l \geqq 0), \qquad j = 1, \ldots, q, \end{aligned} \tag{1.5.9}$$

for P-a.e. φ, since Q is ergodic. It follows from the construction of $B(n, \varphi)$ (cf. the proof of Theorem 1.4.3) that (Φ', P') is concentrated on $\{B(n, \varphi): n \in \mathbb{Z}, \varphi \in M\}$. From

$$\begin{aligned} &X_0' \in B\big(0, (U_l')\big) = \big\{y\big(1, (U_l')\big), \ldots, y\big(q, (U_l')\big)\big\}, \\ &X_{k+l}' = x_{k+l}(X_0', U_0', \ldots, U_{k+l-1}'), \qquad l \geqq 0, \end{aligned}$$

and from (1.5.9) we obtain

$$\lim_{n\to\infty} \frac{1}{n} \sum_{k=1}^{n} g\big((X_{k+l}', l \geqq 0)\big) = \mathsf{E}_{\mathsf{Q}} g\big((\hat{X}_l, l \geqq 0)\big) \quad \mathsf{P}'\text{-a.s.} \tag{1.5.10}$$

for all functions g of the form

$$\begin{aligned} &g\big((x_n, n \geqq 0)\big) = \mathbf{1}_{C_0 \times \cdots \times C_{k-1}}(x_0, x_1, \ldots, x_{k-1}), \\ &C_j \in \mathcal{X}, \qquad j = 0, \ldots, k-1; \qquad k = 1, 2, \ldots \end{aligned}$$

By the individual ergodic theorem A 1.2.1,

$$\lim_{n\to\infty} \frac{1}{n} \sum_{k=1}^{n} g((X'_{k+1}, l \geqq 0)) = \mathsf{E}_{\mathsf{P}'}\left(g((X_l', l \geqq 0)) \mid \mathcal{J}\right) \qquad \mathsf{P}'\text{-a.s.},$$

where $\mathcal{J}$ is the σ-field of the invariant events. Thus

$$\mathsf{E}_{\mathsf{P}'}\, g((X_l', l \geqq 0)) = \mathsf{E}_{\mathsf{Q}} g((\hat{X}_l, l \geqq 0)),$$

in view of (1.5.10). From this we get $(X_n') \overset{\mathcal{D}}{=} (\hat{X}_n)$. The ergodicity statement follows from Lemma 1.5.1. ∎

In case of a denumerable state space we have

1.5.7. Theorem. *Let $\mathbb{X}$ be a denumerable state space and (Φ, P), $\Phi = (U_n)$, a stationary and ergodic input. Assume (Φ', P'), $\Phi' = ([U_n', X_n'])$, to be a stationary weak solution. Then there are (a finite or infinite number of) positive integers q_i, $i \geqq 1$, positive real numbers p_i, $i \geqq 1$, measurable functions $x(i, \varphi)$, $i \geqq 1$, from $\mathbb{U}^{\mathbb{Z}}$ into $\mathbb{X}$ and sets M_i, respectively, such that:*

(i) *The functions $x(i, \varphi)$ depend only on $(u_n, n < 0)$, i.e. $x(i, \varphi) = x(i, \tilde{\varphi})$ if $(u_n, n < 0) = (\tilde{u}_n, n < 0)$, and $x(i, \varphi) \neq x(j, \varphi)$ for $i \neq j$.*
(ii) *The sets $A_k(n, \varphi) = \{x(i, \theta^n\varphi) : q_1 + \cdots + q_{k-1} + 1 \leqq i \leqq q_1 + \cdots + q_k\}$ define SGS's $\mathcal{S}_k = \{A_k(n, \varphi) : n \in \mathbb{Z}, \varphi \in M_k\}$ with the property* (A 1.4.5).
(iii) *For each of the $k \geqq 1$*

$$\mathsf{P}_k'^{+}(\cdot) = \frac{1}{q_k} \sum_{j=q_1+\cdots+q_{k-1}+1}^{q_1+\cdots+q_k} \mathsf{P}\left(\left([U_n, x_n(x(j, \Phi), U_0, \ldots, U_{n-1})],\ n \geqq 0\right) \in (\cdot)\right)$$

determines the distribution P_k' of a stationary weak solution (Φ_k', P_k') of (1.2.1) which is concentrated on $\mathcal{A}(\mathcal{S}_k)$.

(iv) $\mathsf{P}' = \sum_{k\geqq 1} p_k \mathsf{P}_k'$, $\sum_{k \geqq 1} p_k = 1$.

Proof. 1. For convenience let $\mathbb{X} = \{0, 1, 2, \ldots\}$. Let

$$p_n(\varphi, i) = \mathsf{P}'(X_n' = i \mid (U_k') = \varphi), \qquad i \in \mathbb{X}, \qquad \varphi \in \mathbb{U}^{\mathbb{Z}}$$

be a regular version of the conditional distribution of X_n', where $(U_k') = \varphi$. We have

$$\sum_{i=0}^{\infty} p_n(\varphi, i) = 1. \tag{1.5.11}$$

Using the stationarity of P' we obtain, for $A \in \mathcal{U}_{\mathbb{Z}}$ and $i \in \mathbb{X}$:

$$\int_A p_{n+1}(\varphi, i)\, \mathsf{P}(d\varphi) = \mathsf{P}'(X'_{n+1} = i, (U_k') \in A)$$
$$= \mathsf{P}'(X_n' = i, \theta^{-1}(U_k') \in A) = \int_A p_n(\theta\varphi, i)\, \mathsf{P}(d\varphi),$$

which yields

$$p_{n+1}(\varphi, i) = p_n(\theta\varphi, i) \quad \text{for} \quad \mathsf{P}\text{-a.e. } \varphi. \tag{1.5.12}$$

Since (X_n') satisfies the recurrence equation (1.2.1) we have

$$\mathbf{1}\{X_n' = i\} = \sum_{j=0}^{\infty} \mathbf{1}\{X_k' = j\}\, \mathbf{1}\{x_{n-k}(j, U_k', \ldots, U_{n-1}') = i\}, \quad k < n,$$

and by conditioning we obtain

$$p_n(\varphi, i) = \sum_{j=0}^{\infty} \mathbf{1}\{x_{n-k}(j, u_k, \ldots, u_{n-1}) = i\}\, p_k(\varphi, j) \tag{1.5.13}$$

for P-a.e. $\varphi = (u_n)$.

2. Now we shall show that the $p_n(\cdot, \cdot)$ can be choosen as measurable maps of φ which depend only on $(u_k, k < n)$, i.e. there exist stochastic kernels $p_n^-(\cdot, \cdot)$ from $\{(u_k, k < n): u_k \in \mathbb{U}\}$ into $\mathbb{X}$ such that

$$p_n(\varphi, i) = p_n^-(\varphi_n^-, i) \quad \text{for P-a.e. } \varphi, \tag{1.5.14}$$

where $\varphi_n^- = (u_k, k < n)$ for $\varphi = (u_k)$. Without loss of generality let $n = 0$. Define

$$p_0^{(m)}(\varphi, i) = \mathsf{P}'\big(X_0' = i \mid (U_n', n < m) = (u_n, n < m)\big).$$

(Note, $p_0^{(m)}(\varphi, i)$ depends only on $(u_n, n < m)$ and is a Martingale with time parameter m.) The Martingale Convergence Theorem yields (cf. e.g. BAUER (1974))

$$\int |p_0^{(m)}(\varphi, i) - p_0(\varphi, i)|\, \mathsf{P}(d\varphi) \xrightarrow[m\to\infty]{} 0, \qquad i \in \mathbb{X}.$$

From this and since $\sum_{i=0}^{\infty} p_0^{(m)}(\varphi, i) = \sum_{i=0}^{\infty} p_0(\varphi, i) = 1$ we get

$$\int \sum_{i=0}^{\infty} |p_0^{(m)}(\varphi, i) - p_0(\varphi, i)|\, \mathsf{P}(d\varphi) \xrightarrow[m\to\infty]{} 0. \tag{1.5.15}$$

For $\varphi = (u_n) \in \mathbb{U}^{\mathbb{Z}}$ define

$$g^{(m)}(\varphi^-, i) = \sum_{j=0}^{\infty} p_0^{(m)}(\theta^{-m}\varphi, j)\, \mathbf{1}\{x_m(j, u_{-m}, \ldots, u_{-1}) = i\}$$

with the convention $\varphi^- = \varphi_0^- = (u_n, n < 0)$. From (1.5.12) and (1.5.13) we get

$$\begin{aligned} p_0(\varphi, i) &= \sum_{j=0}^{\infty} p_{-m}(\varphi, j)\, \mathbf{1}\{x_m(j, u_{-m}, \ldots, u_{-1}) = i\} \\ &= \sum_{j=0}^{\infty} p_0(\theta^{-m}\varphi, j)\, \mathbf{1}\{x_m(j, u_{-m}, \ldots, u_{-1}) = i\}. \end{aligned} \tag{1.5.16}$$

Now, using (1.5.15), (1.5.16) and the stationarity of P we find

$$\begin{aligned} &\int \sum_{i=0}^{\infty} |p_0(\varphi, i) - g^{(m)}(\varphi^-, i)|\, \mathsf{P}(d\varphi) \\ &= \int \sum_{i=0}^{\infty} \left| \sum_{j=0}^{\infty} \big(p_0(\theta^{-m}\varphi, j) - p_0^{(m)}(\theta^{-m}\varphi, j)\big)\, \mathbf{1}\{x_m(j, u_{-m}, \ldots, u_{-1}) = i\} \right| \mathsf{P}(d\varphi) \\ &\leqq \int \sum_{j=0}^{\infty} |p_0(\theta^{-m}\varphi, j) - p_0^{(m)}(\theta^{-m}\varphi, j)|\, \mathsf{P}(d\varphi) \\ &= \int \sum_{j=0}^{\infty} |p_0(\varphi, j) - p_0^{(m)}(\varphi, j)|\, \mathsf{P}(d\varphi) \to 0 \end{aligned}$$

which implies

$$g^{(m)}(\Phi^-, i) \xrightarrow[m\to\infty]{\mathsf{P}} p_0(\Phi, i), \qquad i \in \mathbf{X}$$

(convergence in probability). Then, for each $i \in \mathbf{X}$ there is a subsequence $(m_k, k \geqq 0)$ such that

$$\lim_{k\to\infty} g^{(m_k)}(\Phi^-, i) = p_0(\Phi, i) \quad \mathsf{P}\text{-a.s.}$$

(cf. e.g. BAUER (1974, Satz 19.6) and (1.5.14) follows for

$$p_0^-(\varphi^-, i) = \lim_{k\to\infty} g^{(m_k)}(\varphi^-, i).$$

In the following we choose the conditional probabilities $p_n(\varphi, i)$ according to (1.5.14), i.e. the $p_n(\varphi, i)$ depend on $(u_k, k < n)$ only.

3. For $\varphi \in \mathbf{U}^{\mathbf{Z}}$ it holds (1.5.11) and thus $c_n^1(\varphi) = \max\{p_n(\varphi, i): i \in \mathbf{X}\}$ is well-defined and $c_n^1(\varphi) > 0$. From (1.5.13) we obtain $c_n^1(\varphi) \leqq c_{n+1}^1(\varphi)$ and from (1.5.12) we get $c_{n+1}^1(\varphi) = \max_i p_{n+1}(\varphi, i) = \max_i p_n(\theta\varphi, i) = c_n^1(\theta\varphi)$. Thus the sequence $\left(c_n^1(\Phi), n \in \mathbf{Z}\right)$ is a stationary and monotonically one. Hence for a.e. φ (cf. Theorem A 1.1.4) the sequence is constant, i.e. $c_1(\Phi) = c_n^1(\Phi)$, $n \in \mathbf{Z}$, a.s. and by means of the Ergodic theorem A 1.2.2 and $c_n^1(\Phi) = c_0^1(\theta^n\Phi) = c_1(\theta^n\Phi)$ we find

$$\mathsf{E}c_1(\Phi) = \lim_{n\to\infty} \frac{1}{n} \sum_{i=1}^{n} c_1(\theta^i\Phi) = \lim_{n\to\infty} \frac{1}{n} \sum_{i=1}^{n} c_i^1(\Phi) = c_1(\Phi) \quad \text{a.s., i.e.}$$

$$c_1 = c_1(\Phi) = c_n^1(\Phi) \quad \text{a.s.,} \tag{1.5.17}$$

where $c_1 = \mathsf{E}c_1(\Phi)$. Let $q_n^1(\varphi) = \#\{i: p_n(\varphi, i) = c_1\}$. From (1.5.13), (1.5.17) and since c_1 is the maximum of the probabilities $p_n(\varphi, i)$, $i \in \mathbf{X}$, $n \in \mathbf{Z}$, we conclude $q_n^1(\varphi) \leqq q_{n+1}^1(\varphi)$. Further, the invariance property (1.5.12) yields $q_{n+1}^1(\varphi) = q_n^1(\theta\varphi)$. Hence $\left(q_n^1(\Phi)\right)$ is a stationary monotonically increasing sequence, and thus (cf. Theorem A 1.1.4) almost surely constant, i.e.

$$q_1(\Phi) = q_n^1(\Phi), \qquad n \in \mathbf{Z}, \qquad \text{a.s.}$$

Since the events $\{\varphi: q^1(\varphi) = q\}$, $q = 1, 2, \ldots$, are invariant there is a q_1 such that

$$q_1 = q_n^1(\Phi) \quad \text{a.s.} \tag{1.5.18}$$

in view of the ergodicity of Φ.

For a fixed $\varphi = (u_n)$ with $q_n^1(\varphi) = q_1$, $n \in \mathbf{Z}$, the elements of the sets $A_1(n, \varphi) = \{i: p_n(\varphi, i) = c_1\}$ can be numbered in a natural order: $x(1, n, \varphi) < \cdots < x(q_1, n, \varphi)$. Notice, since the $p_n(\varphi, i)$ have the property (1.5.14) the states $x(i, n, \varphi)$ are measurable maps of $(u_k, k < n)$. (1.5.17) and (1.5.18) imply that the right-hand side of (1.5.13) consists of a single summand for $i \in A_1(n, \varphi)$ and

$$f\left(A_1(n, \varphi), u_n\right) = A_1(n + 1, \varphi). \tag{1.5.19}$$

Further, from (1.5.12) it follows that

$$A_1(n, \varphi) = A_1(0, \theta^n \varphi). \tag{1.5.20}$$

Thus $\mathcal{S}_1 = \{A_1(n, \varphi)\colon\ n \in \mathbf{Z},\ \varphi \in M_1\}$ with $M_1 = \{\varphi\colon\ \varphi$ satisfies (1.5.13), (1.5.14) and $c_1 = c_n^1(\varphi),\ q_1 = q_n^1(\varphi),\ n \in \mathbf{Z}\}$ is an SGS with the property (A 1.4.5). Define $x(i, \varphi) = x(i, 0, \varphi),\ i = 1, \ldots, q_1$. From the invariance properties (1.5.19) and (1.5.20) it follows that the distribution P_1' given by

$$\mathsf{P}_1'^{+}(\cdot) = \frac{1}{q_1} \sum_{i=1}^{q_1} \mathsf{P}\left(\left([U_n, x_n(x(i, \Phi), U_0, \ldots, U_{n-1})], n \geqq 0\right) \in (\cdot)\right)$$

defines the distribution of a stationary weak solution $(\mathsf{P}_1', \Phi_1'), \Phi_1' = ([U_n^1, X_n^1])$, which is concentrated on $\mathcal{A}(\mathcal{S}_1)$, (cf. proof of Theorem 1.4.3).

4. Now consider the quantities $c_n^2(\varphi) = \max\{p_n(\varphi, i)\colon i \in \mathbf{X},\ p_n(\varphi, i) < c_1\}$ which are well-defined. By the same arguments as in step 3 and taking into account (1.5.18) and (1.5.19) we get a number c_2 $(< c_1)$ such that $c_2 = c_n^2(\Phi)$ P-a.s. If $c_2 = 0$ then the proof will be completed in step 5. Hence we assume $c_2 > 0$ and, proceeding analogously like in step 3 by taking into account (1.5.18) and (1.5.19) we obtain a number q_2 $(= q_n^2(\varphi) = \#\{i\colon p_n(\varphi, i) = c_2\}$ for a.e. φ), measurable functions $x(j, \varphi),\ j = q_1 + 1, \ldots, q_1 + q_2$, which depend only on $(u_k, k < 0)$, and a set M_2 such that: $x(i, \varphi) \neq x(j, \varphi),\ i \neq j$; the sets $A_2(n, \varphi) = \{i\colon p_n(\varphi, i) = c_2,\ q_n^2(\varphi) = q_2\}$ fulfill $f(A_2(n, \varphi), u_n) = A_2(n + 1, \varphi)$ and $A_2(n, \varphi) = \{x(j, \theta^n \varphi)\colon\ \ q_1 + 1 \leqq j \leqq q_1 + q_2\}$. The family of sets $\mathcal{S}_2 = \{A_2(n, \varphi)\colon n \in \mathbf{Z},\ \varphi \in M_2\}$ form an SGS with the property (A 1.4.5);

$$\mathsf{P}_2'^{+}(\cdot) = \frac{1}{q_2} \sum_{j=q_1+1}^{q_1+q_2} \mathsf{P}\left(\left([U_n, x_n(x(j, \Phi), U_0, \ldots, U_{n-1})], n \geqq 0\right) \in (\cdot)\right)$$

determines the distribution P_2' of a stationary weak solution (Φ_2', P_2'), $\Phi_2' = ([U_n^2, X_n^2])$, concentrated on $\mathcal{A}(\mathcal{S}_2)$.

Now we apply step 4 recursively, i.e. in the $(k + 1)$-st step we consider the quantities $c_n^{k+1}(\varphi) = \max\{p_n(\varphi, i)\colon\ i \in \mathbf{X},\ p_n(\varphi, i) < c_k\}$. The arguments of step 3 together with $f(A_i(n, \varphi), u_n) = A_i(n + 1, \varphi)$ and $q_i = q_n^i(\varphi)$ for $i = 1, \ldots, k$ yield a c_{k+1} such that $c_{k+1} = c_n^{k+1}(\Phi),\ n \in \mathbf{Z}$, a.s. If $c_{k+1} = 0$ then the proof will be completed in step 5 $(N = k)$. Otherwise, by applying the arguments given in step 3, we obtain a number q_{k+1}, measurable functions $x(j, \varphi),\ j = q_1 + \cdots + q_k + 1, \ldots, q_1 + \cdots + q_{k+1}$, and a set M_{k+1} such that $x(i, \varphi) \neq x(j, \varphi)$, for $i \neq j,\ i, j = 1, \ldots, q_1 + \cdots + q_{k+1}$; the sets $A_{k+1}(n, \varphi) = \{x(j, \theta^n \varphi)\colon\ \ q_1 + \cdots + q_k + 1 \leqq j \leqq q_1 + \cdots + q_{k+1}\}$ form an SGS $\mathcal{S}_{k+1} = \{A_{k+1}(n, \varphi)\colon\ n \in \mathbf{Z},\ \varphi \in M_{k+1}\}$ with the property (A 1.4.5); and

$$\mathsf{P}_{k+1}'^{+}(\cdot) = \frac{1}{q_{k+1}} \sum_{j=q_1+\cdots+q_k+1}^{q_1+\cdots+q_{k+1}} \mathsf{P}\left(\left([U_n, x_n(x(j, \Phi), U_0, \ldots, U_{n-1})], n \geqq 0\right) \in (\cdot)\right)$$

determines the distribution P_{k+1}' of a stationary weak solution $(\Phi_{k+1}', \mathsf{P}_{k+1}')$, $\Phi_{k+1}' = ([U_n^{k+1}, X_n^{k+1}])$, concentrated on $\mathcal{A}(\mathcal{S}_{k+1})$. We proceed with the recursion as long as $c_k > 0$.

5. In this way we have obtained a sequence (possibly finite) of numbers q_k, $k = 1, \ldots, N$ ($N \leqq \infty$; if $N < \infty$ the recursive construction terminates at the N-th step), functions $x(j, \varphi)$, $j = 1, \ldots, q_1 + \cdots + q_N$, SGS's $\mathcal{S}_k$, $k = 1, \ldots, N$, and stationary weak solutions (Φ_k', P_k') satisfying (i)–(iii). It remains to show (iv), which can be obtained as follows. By construction of q_k, c_k and (1.5.11) it follows

$$\sum_{k \geqq 1} q_k c_k = \sum_{i=0}^{\infty} p_n(\varphi, i) = 1. \tag{1.5.21}$$

Further, for $A \in \mathbb{U}^{\mathbb{Z}_+}$ we obtain

$$\begin{aligned}
\mathsf{P}'^{+}(A) &= \int \sum_{i=0}^{\infty} \mathsf{P}'\Big(([U_n', X_n'], n \geqq 0) \in A, X_0' = i \mid (U_k') = \varphi\Big)\, \mathsf{P}(\mathrm{d}\varphi) \\
&= \int \sum_{k \geqq 1} \sum_{j = q_1 + \cdots + q_{k-1} + 1}^{q_1 + \cdots + q_k} c_k \mathbf{1}\Big\{\Big(\big[u_n, x_n\big(x(j, \varphi), u_0, \ldots, u_{n-1}\big)\big], n \geqq 0\Big) \in A\Big\}\, \mathsf{P}(\mathrm{d}\varphi) \\
&= \sum_{k \geqq 1} c_k q_k \mathsf{P}_k'^{+}(A)
\end{aligned}$$

which yields (iv) with $p_k = c_k q_k$, $k \geqq 1$, in view of the stationarity of P', P_k' and (1.5.21). ∎

The construction given in the proof of Theorem 1.5.7 yields $B_n(\varphi) = \{j : p_n(\varphi, j) > 0\} = \bigcup_{k \geqq 1} A_k(n, \varphi)$, and since $q_k = \# A_k(n, \varphi)$, $n \in \mathbb{Z}$, $\varphi \in \bigcap_{k \geqq 1} M_k$, it follows that in (1.5.13) the right-hand side consists of precisely one summand provided that $p_n(\varphi, i) > 0$. Hence the map $f(\cdot, u_n)$: $B_n(\varphi) \to B_{n+1}(\varphi)$ is a one-to-one map. Thus the following map is well-defined:

$$h_n(i, \varphi) = \begin{cases} x_n\big(x(i, \varphi), u_0, \ldots, u_{n-1}\big) & \text{for } n \geqq 0, \\ j \text{ if } x_{-n}(j, u_n, \ldots, u_{-1}) = x(i, \varphi) \text{ and } p_n(j, \varphi) > 0 \text{ for } n < 0. \end{cases}$$

It is easy to see that under the assumptions of Theorem 1.5.7 the stationary solution P' has the structure:

$$\mathsf{P}'(\cdot) = \int \sum_{i \geqq 0} \mathbf{1}\Big\{\big([u_n, h_n(i, \varphi)]\big) \in (\cdot)\Big\}\, p_0\big(x(i, \varphi), \varphi\big)\, \mathsf{P}(\mathrm{d}\varphi).$$

From Theorem 1.5.4 and Theorem 1.5.7 we obtain the following representation theorem for stationary weak solutions in case of a denumerable state space.

1.5.8. Theorem. *Let $\mathbb{X}$ be a denumerable state space and (Φ, P) a stationary and ergodic input. Then every stationary weak solution (Φ', P') is a mixture of at most countably many ergodic weak solutions Q_i, $i \geqq 0$, i.e.*

$$\mathsf{P}' = \sum_{i \geqq 1} a_i \mathsf{Q}_i,$$

where $\sum_{i \geqq 1} a_i = 1$ and $a_i > 0$.

1.6. An example: Wald's identity for dependent random variables

In this section we shall apply the method of solution generating systems to the investigation of a sum $\sum_{n=1}^{\tau} U_n$, where (U_n) is a stationary sequence of real-valued random variables with the distribution P and $\tau \geqq 1$ is a stopping time with respect to (U_n), i.e., the event $\{\tau \leqq n\}$ belongs to the σ-field generated by $(U_k, k \leqq n)$, $n = 1, 2, \ldots$

The classical result by WALD (1947) is as follows:

If U_n, $n \geqq 1$, are i.i.d., τ is a stopping time with respect to $(U_n, n \geqq 1)$, $\mathsf{E}\tau < \infty$ and $\mathsf{E}U_1$ does exist, then

$$\mathsf{E}_{\mathsf{P}} \sum_{n=1}^{\tau} U_n = \mathsf{E}_{\mathsf{P}}\tau \cdot \mathsf{E}_{\mathsf{P}} U_1. \tag{1.6.1}$$

We are looking for a suitable generalization of (1.6.1) in the case of dependent random variables. However, for an arbitrary stationary sequence (U_n) we cannot expect (1.6.1), as the following example shows:

1.6.1. Example. Consider a single server loss system, cf. equation (1.2.2), with constant inter-arrival times $A_n \equiv 1$. Let the stationary sequence (S_n) of service times be a Markov chain alternating 1/2 and 3/2, cf. Example 1.3.5. Then $\tau = S_1 + 1/2$ is a stopping time with respect to the sequence (S_n). We obtain

$$\mathsf{E}_{\mathsf{P}}\tau = \frac{3}{2}, \quad \mathsf{E}_{\mathsf{P}} S_1 = 1, \quad \text{but } \mathsf{E}_{\mathsf{P}} \sum_{n=1}^{\tau} S_n = \frac{5}{4} \neq \frac{3}{2} \cdot 1 .$$

A formula for the difference

$$\mathsf{E}_{\mathsf{P}} \sum_{n=1}^{\tau} U_n - (\mathsf{E}_{\mathsf{P}}\tau)\, \mathsf{E}_{\mathsf{P}} U_1$$

was given for the case of a stopped irreducible Markov chain with finite state space by KÜCHLER and SEMJONOV (1979).

In the following we shall present a distinct idea for generalizing (1.6.1). Consider a stationary sequence $\chi' = ([U_n', Y_n'])$, $U_n' \in \mathbb{R}$, $Y_n' \in \{0, 1\}$, with the distribution R'. Define the stopping time

$$\nu(\chi') = \inf \{n : n > 0,\ Y_n' = 1\} .$$

Assuming $\mathsf{R}'(\#\{n : Y_n' = 1\} > 0) = 1$, from Theorem A 1.3.1 we obtain

$$\mathsf{E}_{\mathsf{R}'}\big(\nu(\chi') \mid Y_0' = 1\big) = \frac{1}{\mathsf{R}'(Y_0' = 1)} < \infty$$

and

$$\mathsf{E}_{\mathsf{R}'}\big(\nu(\chi') \mid Y_0' = 1\big)\, \mathsf{R}'(\Phi' \in (\cdot)) = \sum_{n \geqq 0} \mathsf{R}'\big(\theta^n \chi' \in (\cdot), \nu(\chi') > n \mid Y_0' = 1\big)$$

$$= \sum_{n > 0} \mathsf{R}'\big(\theta^n \chi' \in (\cdot), \nu(\chi') \geqq n \mid Y_0' = 1\big).$$

These identities can be rewritten in the form

$$\mathsf{E}_{\mathsf{R}'}\big(\nu(\chi') \mid Y_0' = 1\big) \int \mathbf{1}_{(\cdot)}(\chi)\, \mathsf{R}'(\mathrm{d}\chi) = \int \sum_{n=0}^{\nu(\chi)-1} \mathbf{1}_{(\cdot)}(\theta^n \chi)\, \mathsf{R}'(\mathrm{d}\chi \mid Y_0' = 1).$$

$$\mathsf{E}_{\mathsf{R}'}\big(\nu(\chi') \mid Y_0' = 1\big) \int \mathbf{1}_{(\cdot)}(\chi)\, \mathsf{R}'(\mathrm{d}\chi) = \int \sum_{n=1}^{\nu(\chi)} \mathbf{1}_{(\cdot)}(\theta^n \chi)\, \mathsf{R}'(\mathrm{d}\chi \mid Y_0' = 1).$$

Using standard techniques from integration theory we obtain

$$\mathsf{E}_{\mathsf{R}'}\big(\nu(\chi') \mid Y_0' = 1\big) \int h(\chi)\, \mathsf{R}'(\mathrm{d}\chi) = \int \sum_{n=0}^{\nu(\chi)-1} h(\theta^n \chi)\, \mathsf{R}'(\mathrm{d}\chi \mid Y_0' = 1)$$

$$= \int \sum_{n=1}^{\nu(\chi)} h(\theta^n \chi)\, \mathsf{R}'(\mathrm{d}\chi \mid Y_0' = 1)$$

for all R' integrable functionals h. In particular, for $h(\chi) = u_0$, $\chi = ([u_n, y_n])$:

$$\mathsf{E}_{\mathsf{R}'}\big(\nu(\chi') \mid Y_0' = 1\big)\, \mathsf{E}_{\mathsf{R}'} U_0' = \mathsf{E}_{\mathsf{R}'}\left(\sum_{n=0}^{\nu(\chi')-1} U_n' \mid Y_0' = 1\right) \tag{1.6.2a}$$

$$= \mathsf{E}_{\mathsf{R}'}\left(\sum_{n=1}^{\nu(\chi')} U_n' \mid Y_0' = 1\right). \tag{1.6.2b}$$

(The previous conclusion is allowed if $\mathsf{E}_{\mathsf{R}'} U_0'$ exists.)

Comparing Formula (1.6.2b) with WALD's identity (1.6.1) we observe a formal analogy. Later we shall show that in fact (1.6.1) follows from (1.6.2b) if U_n' are i.i.d., cf. Example 2.2.6. Formula (1.6.2a) is useful for the discussion of busy periods for queueing systems, cf. Section 6.5.

It seems to be a disadvantage of formula (1.6.2a, b) that $\nu(\chi')$ is quite a special stopping time. Therefore we shall investigate the relationship between an arbitrary sequence (U_n) with stopping time $\tau \geqq 1$ and an appropriate sequence $\chi' = ([U_n', Y_n'])$ with stopping time $\nu(\chi')$.

Let (Φ, P), $\Phi = (U_n)$, be a stationary sequence of real valued r.v.'s and $\tau \geqq 1$ a stopping time with respect to $(U_n, n \geqq 1)$. This is equivalent to the existence of a sequence of measurable functions $h_n : \mathbb{R}^n \to \{0, 1\}$, $n = 1, 2, \ldots$, with

$$\mathbf{1}\{\tau \leqq n\} = h_n(U_1, \ldots, U_n) \quad \mathsf{P}\text{-a.s.},$$

i.e.,

$$\tau = \inf\{n : n \geqq 1, h_n(U_1, \ldots, U_n) = 1\} \quad \mathsf{P}\text{-a.s.}$$

In the following we use the notation

$$\tau(\varphi) = \inf\{n : n \geqq 1, h_n(u_1, \ldots, u_n) = 1\},$$

$\varphi = (u_n)$. We are looking for a stationary sequence (χ', R'), $\chi' = ([U_n', Y_n'])$, $Y_n' \in \{0, 1\}$, satisfying

$$\mathsf{R}'\big((U_n') \in (\cdot)\big) = \mathsf{P}\big((U_n) \in (\cdot)\big), \tag{1.6.3}$$

$$\mathsf{R}'(\#\{n : Y_n' = 1\} > 0) = 1, \tag{1.6.4}$$

$$\mathsf{R}'\big(\nu(\chi') = \tau((U_n')) \mid Y_0' = 1\big) = 1. \tag{1.6.5}$$

The following lemma shows the close relation between this problem and the solution of a certain recursive stochastic equation.

1.6.2. Lemma. *Let* (Φ, P), $\Phi = (U_n)$, *be a stationary sequence and* $\tau(\Phi) \geqq 1$ *a stopping time with respect to* $(U_n, n \geqq 1)$. *Then*

$$Y_n' = 1\{X_n' = 0\}$$

resp.

$$X_n' = \inf\{n - k : k \leqq n, Y_k' = 1\},$$

defines a one-to-one mapping between the set of all stationary sequences $\left(([U_n', Y_n']), \mathsf{R}'\right)$, *satisfying* (1.6.3)–(1.6.5) *and the set of all stationary sequences* $\left(([U_n', X_n']), \mathsf{P}'\right)$, $X_n' \in \mathbb{Z}_+$, *satisfying*

$$\mathsf{P}'\left((U_n') \in (\cdot)\right) = \mathsf{P}\left((U_n) \in (\cdot)\right), \tag{1.6.6}$$

$$\mathsf{P}'\left(X_{n+1}' = (X_n' + 1)\, 1\left\{\tau\left(\theta^{n-X_n'}(U_n')\right) \neq X_n' + 1\right\}, n \in \mathbb{Z}\right) = 1, \tag{1.6.7}$$

Proof. 1. Let (χ', R'), $\chi' = ([U_n', Y_n'])$, be a stationary sequence with the properties (1.6.3)–(1.6.5). Define

$$X_n' = \inf\{n - k : k \leqq n, Y_k' = 1\},$$

i.e., X_n' is the distance between n and the last index $k \leqq n$ with $Y_k' = 1$. Let $([U_{n_j}', Y_{n_j}']) = ([U_{n_j}', 1)]$ be the subsequence of χ' with $Y_{n_j}' = 1$. Then $n_{j+1} - n_j = \nu(\theta^{n_j}\chi')$, $j \in \mathbb{Z}$. From (1.6.4) and (1.6.5) it follows that

$$\mathsf{R}'\left(n_{j+1} - n_j = \nu(\theta^{n_j}\chi') = \tau\left(\theta^{n_j}(U_k')\right), j \in \mathbb{Z}\right) = 1 \tag{1.6.8}$$

by means of Theorem A 1.3.1. According to the definition of X_n'

$$X_{n+1}' = (X_n + 1)\, 1\{\nu(\theta^{n-X_n'}\chi') \neq X_n' + 1\}, \quad n \in \mathbb{Z}, \quad \mathsf{R}'\text{-a.s.}$$

Since $\theta^{n-X_n'}\chi' \in \{([U_k, Y_k]) : Y_0 = 1\}$, i.e., $n - X_n' \in \{n_j : j \in \mathbb{Z}\}$, (1.6.7) follows immediately from (1.6.8), where P' is the distribution of the sequence $([U_n' X_n'])$ induced by R'. Condition (1.6.6) coincides with (1.6.3). Thus every probability law R' with (1.6.3)–(1.6.5) induces a probability law R' with (1.6.6) and (1.6.7).

2. Conversely, for every stationary sequence (Φ', P'), $\Phi' = ([U_n', X_n'])$, satisfying (1.6.6), (1.6.7) we have

$$\mathsf{P}'(\#\{n : X_n' = 0\} = \infty) = 1$$

and

$$\nu(\theta^{n_j}\Phi') = \tau\left(\theta^{n_j}(U_k')\right) \quad \mathsf{P}'\text{-a.s.},$$

where $([U_{n_j}', X_{n_j}']) = ([U_{n_j}', 0])$ is the subsequence of Φ' with $X_{n_j}' = 0$; $\chi' = ([U_n', Y_n'])$, $Y_n' = 1\{X_n' = 0\}$. Thus conditions (1.6.3)–(1.6.5) are valid for the sequence $([U_n', Y_n'])$. ∎

Lemma 1.6.2 states that the problem of existence of a stationary sequence $([U_n', Y_n'])$ satisfying (1.6.3)–(1.6.5) reduces to the problem of existence of stationary weak solutions of the equation

$$X_{n+1} = (X_n + 1)\, 1\{\tau(\theta^{n-X_n}(U_k)) \neq X_n + 1\}, \qquad n \in \mathbb{Z}. \tag{1.6.9}$$

Now we shall construct an SGS for this equation. For fixed n and $\varphi = (u_k)$ define

$$l(n, \varphi) = \sum_{k \leq n} 1\{k + \tau(\theta^k \varphi) > n\}.$$

Assume

$$\mathsf{P}(l(0, \Phi) < \infty) = 1. \tag{1.6.10}$$

The indices $k \leq n$ for which $k + \tau(\theta^k \varphi) > n$ will be ordered and denoted by $m(1, n, \varphi), \ldots, m(l(n, \varphi), n, \varphi)$. The sets

$$A(n, \varphi) = \{n - m(1, n, \varphi), \ldots, n - m(l(n, \varphi), n, \varphi)\}, \qquad n \in \mathbb{Z},$$

are finite for P-a.e. φ because of (1.6.10). It is easy to see that $\{A(n, \varphi) : n \in \mathbb{Z}, \varphi \in \mathbb{U}^{\mathbb{Z}}, l(k, \varphi) < \infty, k \in \mathbb{Z}\}$ is an SGS. Applying Theorem 1.4.3 we obtain a stationary weak solution of equation (1.6.9) under the assumption (1.6.10). Notice, in view of Lemma A 1.1.5, $\mathsf{E}_{\mathsf{P}}\tau(\Phi) < \infty$ is sufficient for (1.6.10). Thus we have

1.6.3. Theorem. *Let (Φ, P), $\Phi = (U_n)$, be a stationary sequence of real valued r.v.'s, and $\mathsf{E}_{\mathsf{P}} U_0$ may exist. Furthermore, let $\tau(\Phi) \geq 1$ be a stopping time with respect to $(U_n, n \geq 1)$. Assume $\mathsf{E}_{\mathsf{P}}\tau(\Phi) < \infty$. Then there exists a stationary sequence (χ', R'), $\chi' = ([U_n', Y_n'])$, $Y_n' \in \{0, 1\}$, such that*

$$\mathsf{R}'((U_n') \in (\cdot)) = \mathsf{P}((U_n) \in (\cdot)),$$

$$\mathsf{R}'(\#\{n : Y_n' = 1\} > 0) = 1,$$

$$\mathsf{R}'\left(\nu(\chi') = \tau((U_n')) \mid Y_0' = 1\right) = 1,$$

$$\mathsf{E}_{\mathsf{R}'}\left(\sum_{n=0}^{\tau((U_k'))-1} U_n' \mid Y_0' = 1\right) = \mathsf{E}_{\mathsf{R}'}\left(\tau((U_k')) \mid Y_0' = 1\right) \mathsf{E}_{\mathsf{P}} U_1, \tag{1.6.11a}$$

$$\mathsf{E}_{\mathsf{R}'}\left(\sum_{n=1}^{\tau((U_k'))} U_n' \mid Y_0' = 1\right) = \mathsf{E}_{\mathsf{R}'}\left(\tau((U_k')) \mid Y_0' = 1\right) \mathsf{E}_{\mathsf{P}} U_1. \tag{1.6.11b}$$

Formulae (1.6.11a, b) can be considered as a generalization of WALD's identity. In fact, we shall show later that (1.6.11b) reduces to (1.6.1) if (U_n) are i.i.d., cf. Example 2.2.6.

We emphasize that for a given (Φ, P) and a stopping time $\tau \geq 1$, $\mathsf{E}_{\mathsf{P}}\tau < \infty$, there may exist several labelled sequences (χ', R'), $\chi' = ([U_n', Y_n'])$, satisfying (1.6.3)–(1.6.5) and (1.6.11).

1.6.4. Example. Let $(U_n, n \in \mathbb{Z})$ be a homogeneous Markov chain taking values $1, 2, 3, 4$ with transition probabilities $p_{12} = p_{23} = p_{34} = p_{41} = 1$, $p_{ij} = 0$ otherwise, and unique stationary initial distribution $\mathsf{P}(U_1 = i) = 1/4$, $i = 1, \ldots, 4$. Consider the stopping time

$$\tau(\Phi) = \begin{cases} 1 & \text{if } U_1 = 3, \\ 2 & \text{if } U_1 \in \{2, 4\}, \\ 4 & \text{if } U_1 = 1. \end{cases}$$

Then $\mathsf{E}_{\mathsf{P}}\tau = \frac{9}{4}$, $\mathsf{E}_{\mathsf{P}}U_1 = \frac{5}{2}$ but $\sum_{n=1}^{\tau} U_n = \frac{23}{4} \neq \frac{9}{4} \cdot \frac{5}{2}$, that means Wald's identity (1.6.1) fails.

Consider the Markov chain (Ψ_1, Q_1), $\Psi_1 = ([U_n^1, Y_n^1])$ taking values $[1, 0], [2, 0], [3, 0], [4, 1]$ with the transition probability matrix

$$\begin{pmatrix} 0 & 1 & 0 & 0 \\ 0 & 0 & 1 & 0 \\ 0 & 0 & 0 & 1 \\ 1 & 0 & 0 & 0 \end{pmatrix} \tag{1.6.12}$$

and stationary initial probabilities $[1/4, 1/4, 1/4, 1/4]$. The sequence Ψ_1 satisfies (1.6.3)–(1.6.5). The same is true for the Markov chain (Ψ_2, Q_2), $\Psi_2 = ([U_n^2, Y_n^2])$ taking values $[1, 1], [2, 0], [3, 1], [4, 0]$, with transition probability matrix (1.6.12) and stationary initial probabilities $[1/4, 1/4, 1/4, 1/4]$. We have

$$\mathsf{E}_{\mathsf{Q}_1}\left(\tau((U_k^1)) \mid Y_0^1 = 1\right) = 4, \qquad \mathsf{E}_{\mathsf{Q}_1}\left(\sum_{n=1}^{\tau((U_k^1))} U_n^1 \mid Y_0^1 = 1\right) = 10,$$

$$\mathsf{E}_{\mathsf{Q}_2}\left(\tau((U_k^2)) \mid Y_0^2 = 1\right) = 2, \qquad \mathsf{E}_{\mathsf{Q}_2}\left(\sum_{n=1}^{\tau((U_k^2))} U_n^2 \mid Y_0^2 = 1\right) = 5.$$

1.7. Model continuity

In this section we shall investigate continuity properties of the model

$$X_{n+1} = f(X_n, U_n), \qquad n \in \mathbb{Z},$$

in steady state, i.e., we shall show that under some conditions small perturbations of the input (Φ, P) lead to small deviations of the considered stationary (weak or strong) solution of equation (1.2.1). This problem is of theoretical interest as well as of great practical importance. It arises if the probability distribution of the input is approximated by distributions which are more appropriate for system analysis or if the input distribution is statistically estimated from observed data.

Now we turn towards the rigorous formulation of the problem. Consider a fixed method for constructing weak (or strong) solutions of (1.2.1), and denote by $h(\mathsf{P})$ the probability distribution of the corresponding weak (or strong) solution for the input (Φ, P), for instance $h(\mathsf{P})(\cdot) = \mathsf{P}\left(([U_n, X_n^{\min}]) \in (\cdot)\right)$ or $h(\mathsf{P}) = \mathsf{Q}$, cf. Sections 1.3 and 1.4. Let (Φ, P) and (Φ_k, P_k), $k \geq 1$, be inputs for which $h(\mathsf{P})$ and $h(\mathsf{P}_k)$ are defined. The problem is: Under which additional assumptions does the weak convergence of input distributions

$$\mathsf{P}_k \underset{k\to\infty}{\Longrightarrow} \mathsf{P}$$

imply

$$h(\mathsf{P}_k) \underset{k\to\infty}{\Longrightarrow} h(\mathsf{P})\,?$$

Here we shall present some general ideas for proving model continuity. These ideas will be used for proving continuity properties of special systems in Chapter 8.

A general and often used scheme for proving

$$h(\mathsf{P}_k) \underset{k\to\infty}{\Longrightarrow} h(\mathsf{P})$$

consists in the following steps.

Step 1. One shows the tightness of the sequence $\big(h(\mathsf{P}_k), k \geqq 1\big)$: i.e., for each $\varepsilon > 0$ there is a compact set K_ε such that

$$h(\mathsf{P}_k)\,(K_\varepsilon) \geqq 1 - \varepsilon.$$

From Prokhorov's Theorem, cf. Theorem A 2.5, it follows that $\big(h(\mathsf{P}_k), k \geqq 1\big)$ is relatively compact, i.e., for every subsequence $\big(h(\mathsf{P}_{k_j}), j \geqq 1\big)$ there is a convergent subsequence $\big(h(\mathsf{P}_{k_{j_l}}), l \geqq 1\big)$.

Step 2. One shows that the limit of an arbitrary convergent subsequence $\big(h(\mathsf{P}_{k_j}), j \geqq 1\big)$ coincides with $h(\mathsf{P})$, which ensures the desired property

$$h(\mathsf{P}_k) \underset{k\to\infty}{\Longrightarrow} h(\mathsf{P}).$$

In this procedure the crucial point is the proof of tightness, cf. e.g. Section 8.6. In most cases step 2 is comparably simple. For instance, if $h(P)$ is the probability distribution of the unique stationary weak solution of (1.2.1) for (Φ, P), then it suffices to show that the limit R of an arbitrary convergent subsequence $\big(h(\mathsf{P}_{k_j})\big)$ is the probability distribution of a stationary weak solution of (1.2.1). The stationarity of R is trivial; R is the probability distribution of a weak solution e.g. if f is R-a.s. continuous.

1.7.1. Model continuity in the presence of (x, l)-renewing epochs

Now we shall deal with the continuity of strong solutions constructed by means of (x, l)-renewing epochs, cf. Theorem 1.3.6. Borovkov (1978), (1980) proved

1.7.1. Theorem. *Let (Φ, P), (Φ_k, P_k), $k \geqq 1$, be arbitrary ergodic inputs. Assume:*

(A 1.7.1) $\mathsf{P}_k \Rightarrow \mathsf{P}$.

(A 1.7.2) *For some $x \in \mathbb{X}$ and $l \in \mathbb{Z}_+$ the input Φ has infinitely many (x, l)-renewing epochs, i.e.,* $\mathsf{P}(B_0) = \mathsf{P}(0$ *is an (x, l)-renewing epoch*$) > 0$, *cf. Definition* 1.3.4.

(A 1.7.3) $\liminf\limits_{k\to\infty} \mathsf{P}_k\left(\bigcup\limits_{n=r}^{-1} B_n\right) \xrightarrow[r\to-\infty]{} 1.$

(A 1.7.4) $\liminf\limits_{k\to\infty} \mathsf{P}_k(B_0) > 0.$

(A 1.7.5) $x_r(x, u_{-r}, \ldots, u_{-1})$ *is a* P-*a.s. continuous function of* $u_{-r}, \ldots, u_{-1}$, $r = 1, 2, \ldots$

Then

$$\big([U_{k,n}, \hat{x}_n(x, \Phi_k)]\big) \xrightarrow[k\to\infty]{\mathscr{D}} \big([U_n, \hat{x}_n(x, \Phi)]\big).$$

Proof. From (A 1.7.4) we have $\mathsf{P}_k(B_0) > 0$ for almost all k. Thus, for almost all k, (Φ_k, P_k) satisfies the assumptions of Theorem 1.3.6. Let $\nu(\varphi)$ be the last (x, l)-renewing epoch of φ prior to zero, i.e., $\nu(\varphi) = \sup\{k < 0 : \varphi \in B_k\}$. Then $\nu(\Phi_k)$ is a proper r.v. for almost all k. Obviously

$$\mathsf{P}_k\big(\nu(\Phi_k) < r\big) = 1 - \mathsf{P}_k\left(\bigcup_{n=r}^{-1} B_n\right), \qquad r < 0. \tag{1.7.1}$$

In view of (1.7.1), (A 1.7.3), for a given $\varepsilon > 0$ we can choose an $r_0 < 0$ such that

$$\left.\begin{aligned} &\mathsf{P}_k\big(\nu(\Phi_k) < r_0\big) < \varepsilon \quad \text{for almost all } k \\ \text{and} \\ &\mathsf{P}\big(\nu(\Phi) < r_0\big) < \varepsilon. \end{aligned}\right\} \tag{1.7.2}$$

Define $\bar{x}(\varphi) = x_{l-r_0}(x, u_{r_0-l}, \ldots, u_{-1})$. Taking into account the Definition of (x, l)-renewing epochs we have $\bar{x}(\varphi) = \hat{x}_0(x, \varphi)$ for $\nu(\varphi) \geqq r_0$. From (A 1.7.1) and (A 1.7.5),

$$\bar{x}(\Phi_k) \xrightarrow{\mathscr{D}} \bar{x}(\Phi). \tag{1.7.3}$$

Consider an event $E \subseteqq \mathbf{X}$ whose boundary has the property $\mathsf{P}(\partial E) = 0$. Then

$$\begin{aligned}
&\big|\mathsf{P}_k\big(\hat{x}_0(x, \Phi_k) \in E\big) - \mathsf{P}_k(\bar{x}(\Phi_k) \in E)\big| \\
&= \big|\mathsf{P}_k\big(\hat{x}_0(x, \Phi_k) \in E, \hat{x}_0(x, \Phi_k) \neq \bar{x}(\Phi_k)\big) + \mathsf{P}_k\big(\hat{x}_0(x, \Phi_k) \in E, \hat{x}_0(x, \Phi_k) = \bar{x}(\Phi_k)\big) \\
&\quad - \mathsf{P}_k\big(\bar{x}(\Phi_k) \in E, \hat{x}_0(x, \Phi_k) \neq \bar{x}(\Phi_k)\big) - \mathsf{P}_k\big(\bar{x}(\Phi_k) \in E, \hat{x}_0(x, \Phi_k) = \bar{x}(\Phi_k)\big)\big| \\
&\leqq 2\mathsf{P}_k\big(\hat{x}_0(x, \Phi_k) \neq \bar{x}(\Phi_k)\big) \leqq 2\mathsf{P}_k\big(\nu(\Phi_k) < r_0\big) < 2\varepsilon
\end{aligned}$$

for almost all k in view of (1.7.2). Analogously,

$$\big|\mathsf{P}\big(\hat{x}_0(x, \Phi) \in E\big) - \mathsf{P}\big(\bar{x}(\Phi) \in E\big)\big| \leqq 2\varepsilon.$$

Thus,

$$\begin{aligned}
&\big|\mathsf{P}_k\big(\hat{x}_0(x, \Phi_k) \in E\big) - \mathsf{P}\big(\hat{x}_0(x, \Phi) \in E\big)\big| \\
&\leqq \big|\mathsf{P}_k\big(\hat{x}_0(x, \Phi) \in E\big) - \mathsf{P}_k\big(\bar{x}(\Phi_k) \in E\big)\big| \\
&\quad + \big|\mathsf{P}_k\big(\bar{x}(\Phi_k) \in E\big) - \mathsf{P}\big(\bar{x}(\Phi) \in E\big)\big| + \mathsf{P}\big|\big(\bar{x}(\Phi) \in E\big) - \mathsf{P}\big(\hat{x}_0(x, \Phi) \in E\big)\big| \\
&\leqq \big|\mathsf{P}_k\big(\bar{x}(\Phi_k) \in E\big) - \mathsf{P}\big(\bar{x}(\Phi) \in E\big)\big| + 4\varepsilon
\end{aligned}$$

for all $\varepsilon > 0$ and almost all k. From this and (1.7.3) we obtain

$$\hat{x}_0(x, \Phi_k) \xrightarrow{\mathscr{D}} \hat{x}_0(x, \Phi).$$

The convergence of all finite-dimensional distributions of $\big([U_{k,n}, \hat{x}_n(x, \Phi_k)]\big)$ can be proved in the same way. ∎

1.7.2. Method of metric modification

Another method for proving the continuity of strong solutions consists in modifying the metrization of the set of input sample paths. The idea can be described as follows: Let d be the given metric on $\mathbb{U}$ and $\bar{d}$ the standard metric on $\mathbb{U}^{\mathbb{Z}}$, i.e.,

$$\bar{d}(\varphi_1, \varphi_2) = \sum_{n=-\infty}^{\infty} \frac{d(u_{1,n}, u_{2,n})}{1 + d(u_{1,n}, u_{2,n})} \cdot \frac{1}{2^{|n|}}, \tag{1.7.4}$$

$\varphi_1 = (u_{1,n})$, $\varphi_2 = (u_{2,n})$. Denote by π the corresponding Prokhorov distance in the space of all probability measures on $(\mathbb{U}^{\mathbb{Z}}, \mathcal{U}_{\mathbb{Z}})$, where $\mathcal{U}_{\mathbb{Z}}$ is the Borel σ-field generated by $\bar{d}$. Consider inputs (Φ, P), (Φ_k, P_k), $k \geqq 1$, satisfying $\pi(\mathsf{P}_k, \mathsf{P}) \to 0$. Let $\left(h_n(\cdot), n \in \mathbb{Z}\right)$ be a sequence of measurable functions such that $\left([U_n, h_n(\Phi)]\right)$ and $\left([U_{k,n}, h_n(\Phi_k)]\right)$ are strong solutions of (1.2.1) for (Φ, P) and (Φ_k, P_k), $k \geqq 1$, respectively. We choose a subset $\mathbb{U}_1 \subseteqq \mathbb{U}^{\mathbb{Z}}$ and a suitable metric $\bar{d}_1$ on $\mathbb{U}_1$ with the following properties:

(A 1.7.6) $\mathbb{U}_1 \in \mathcal{U}_{\mathbb{Z}}$ and $\mathsf{P}(\mathbb{U}_1) = 1$, $\mathsf{P}_k(\mathbb{U}_1) = 1$,

(A 1.7.7) $\mathbb{U}_1$ endowed with $\bar{d}_1$ is a Polish space.

(A 1.7.8) The Borel σ-field of subsets of $\mathbb{U}_1$ generated by $\bar{d}_1$ coincides with $\{E : E \in \mathcal{U}_{\mathbb{Z}}, E \subseteqq \mathbb{U}_1\}$.

(A 1.7.9) The mappings $h_n(\varphi)$, $n \in \mathbb{Z}$, are continuous with respect to the metric $\bar{d}_1$.

(A 1.7.10) $\pi_1(\mathsf{P}_k, \mathsf{P}) \to 0$,
where π_1 denotes the Prokhorov distance in the space of probability measures on the Borel σ-field (generated by $\bar{d}_1$) of $\mathbb{U}_1$.

Applying the continuous mapping theorem, cf. Theorem A 2.4, we obtain

$$[h_{-n}(\Phi_k), \ldots, h_n(\Phi_k)] \xrightarrow[k\to\infty]{\mathcal{D}} [h_{-n}(\Phi), \ldots, h_n(\Phi)].$$

The advantage of the method is its constructiveness. By means of a more detailed analysis of the mappings $\varphi \to h_n(\varphi)$, which are enforced to be continuous, one can estimate the rate of convergence:

$$\pi_0\left(\mathsf{P}_k\left(h_0(\Phi_k) \in (\cdot)\right), \mathsf{P}\left(h_0(\Phi) \in (\cdot)\right)\right) \leqq g\left(\pi_1(\mathsf{P}_k, \mathsf{P})\right),$$

where π_0 is the Prokhorov distance in the space of probability measures on $(\mathbb{X}, \mathcal{H})$. This will be illustrated for queues with infinitely many servers in Section 8.2.

1.7.3. Continuity of weak solutions (Ψ, Q)

In this section we shall present a method for proving the continuous dependence of the particular weak solution (Ψ, Q) (cf. Theorem 1.4.3) on the input (Φ, P). This method generalizes the method of metric modification given in the previous section.

Consider the inputs (Φ, P), (Φ_k, P_k), $k \geqq 1$, which are assumed to be stationary and ergodic. Furthermore, let $\mathcal{S} = \{A(n, \varphi): n \in \mathbb{Z}, \varphi \in M\}$ be an SGS for (Φ, P) and (Φ_k, P_k), $k \geqq 1$, as well. Below we use a fixed system of functions $x(i, n, \varphi)$ such that

$$A(n, \varphi) = \{x(1, n, \varphi), \ldots, x(\# A(n, \varphi), n, \varphi)\},$$

cf. Definition 1.4.2. We use the notation

$$\vec{A}(n, \varphi) = \left(x(1, n, \varphi), \ldots, x(\# A(n, \varphi), n, \varphi), e, e, \ldots\right) \in \mathbf{X}^{\mathbb{Z}_+}, \quad n \in \mathbb{Z},$$

$\varphi \in M$, where e is an arbitrary but fixed state. The following assumptions are in some sense similar to (A 1.7.6)–(A 1.7.10).

We assume that there exists a metric d_M on M with the properties:

(A 1.7.11) M endowed with d_M is a Polish space.

(A 1.7.12) The Borel σ-field of subsets of M generated by d_M coincides with the σ-field $E: E \in \mathcal{U}_{\mathbb{Z}}, E \subseteq M$.

(A 1.7.13) The shift operator θ is continuous with respect to d_M.

(A 1.7.14) The function $\varphi \to \# A(0, \varphi)$ is continuous with respect to the metric d_M for P-a.e. φ.

(A 1.7.15) The mappings $\varphi \to x_n(x(j, 0, \varphi), u_0, \ldots, u_{n-1})$, $j \in \{1, \ldots, \# A(0, \varphi)\}$, $n \geqq 0$, are continuous with respect to d_M for P-a.e. φ.

(A 1.7.16) $\pi_M(\mathsf{P}_k, \mathsf{P}) \to 0$,
where π_M denotes the Prokhorov distance in the space of all probability measures on the Borel σ-field (generated by d_M) of M.

Assumptions (A 1.7.11), (A 1.7.12), (A 1.7.16) allow to consider P, P_k, $k \geqq 1$, as probability measures on the Borel σ-field on M generated by d_M and to apply the continuous mapping theorem to continuous mappings defined on M. For $n = 0$, from (A 1.7.15) we obtain that $\varphi \to \vec{A}(0, \varphi)$ is a continuous function on M with respect to d_M.

It should be mentioned that the metric d_M has to be constructed in an appropriate way depending on the problem considered. In general, d_M differs from the standard metric (1.7.4) on $\mathbf{U}^{\mathbb{Z}}$, cf. Section 8.2.

1.7.2. Theorem. *Let (Φ, P), (Φ_k, P_k), $k \geqq 1$, be stationary ergodic inputs and $\mathcal{S} = \{A(n, \varphi), n \in \mathbb{Z}, \varphi \in M\}$ a joint SGS for all these inputs. Assume there is a metric d_M on M satisfying* (A 1.7.11)–(A 1.7.16). *Moreover we assume $q = q_k$ for almost all k, where q, q_k, $k \geqq 1$, are the numbers determined by Corollary* 1.4.6 *for the inputs (Φ, P), (Φ_k, P_k), respectively. Then*

$$\Psi_k \xrightarrow[k \to \infty]{\mathcal{D}} \Psi, \tag{1.7.5}$$

where (Ψ, Q), (Ψ_k, Q_k) are the particular weak solutions given by Corollary 1.4.6 *for the inputs (Φ, P), (Φ_k, P_k), respectively.*

Proof. For a given sample path consider the number $i(\varphi)$ defined by (1.4.6). For every $\varphi \in M$ with $i(\varphi) \leqq q$ define

$$m(\varphi) = \sup\left\{n: n \leqq 0, \# x_{-n}(A(n, \varphi), u_n, \ldots, u_{-1}) \leqq q\right\}.$$

From the proof of Theorem 1.4.3 $-\infty < m(\varphi) \leqq 0$ follows for P-a.e. φ. Let $\varphi \in M$ be a fixed sample path with the property $i(\varphi) = q$. Since M is an invariant set, $\theta^{m(\varphi)}\varphi \in M$. If we choose $\eta = (v_j) \in M$ such that $d_M(\varphi, \eta)$ is sufficiently small, then, in view of (A 1.7.13), $d_M(\theta^{m(\varphi)}\varphi, \theta^{m(\varphi)}\eta)$ also becomes small. Moreover, for P-a.e. φ

$$\begin{aligned} &\# \left\{x_{-m(\varphi)}\Big(A\big(m(\varphi), \eta\big), v_{m(\varphi)}, \ldots, v_{-1}\Big)\right\} \\ &= \# \left\{x_{-m(\varphi)}\Big(A\big(m(\varphi), \varphi\big), u_{m(\varphi)}, \ldots, u_{-1}\Big)\right\} = q \end{aligned}$$

due to (A 1.7.14) and (A 1.7.15). Thus $i(\eta) \leqq q$ for $d_M(\varphi, \eta)$ sufficiently small. Consequently, $m(\eta)$ is finite for all $\eta \in M$ for $d_M(\varphi, \eta)$ sufficiently small, and $m(\eta) \geqq m(\varphi)$. Furthermore, $\# x_{-m(\eta)}\Big(A\big(m(\eta), \eta\big), v_{m(\eta)}, \ldots, v_{-1}\Big) = q$ for $d_M(\varphi, \eta)$ sufficiently small. Let $\tilde{x}(1, \eta), \ldots, \tilde{x}(q, \eta)$ be the elements of $x_{-m(\eta)}\Big(A\big(m(\eta), \eta\big), v_{m(\eta)}, \ldots, v_{-1}\Big)$ arranged in the same order as they occur in $A(0, \eta)$. Note, if $i(\eta) = q$, then $[\tilde{x}(1, \eta), \ldots, \tilde{x}(q, \eta)] = [x(1, \eta), \ldots, x(q, \eta)]$, where $x(i, \eta)$ are defined by (1.4.11) in the proof of Corollary 1.4.6. In particular, $[\tilde{x}(1, \varphi), \ldots, \tilde{x}(q, \varphi)] = [x(1, \varphi), \ldots, x(q, \varphi)]$ for the chosen sample path φ. If $d_M(\varphi, \eta)$ is sufficiently small, then

$$\{\tilde{x}(1, \eta), \ldots, \tilde{x}(q, \eta)\} = x_{-m(\varphi)}\Big(A\big(m(\varphi), \eta\big), v_{m(\varphi)}, \ldots, v_{-1}\Big)$$

because $m(\eta) \geqq m(\varphi)$. In view of (A 1.7.15)

$$\varphi \to [\tilde{x}(1, \varphi), \ldots, \tilde{x}(q, \varphi)]$$

is continuous with respect to d_M for all $\varphi \in M$ with $i(\varphi) = q$. From the continuous mapping theorem we have

$$[\tilde{x}(1, \Phi_k), \ldots, \tilde{x}(q, \Phi_k)] \xrightarrow{\mathfrak{D}} [\tilde{x}(1, \Phi), \ldots, \tilde{x}(q, \Phi)]. \tag{1.7.6}$$

Since $q_k = q$ for almost all k,

$$[x(1, \Phi_k), \ldots, x(q_k, \Phi_k)] = [\tilde{x}(1, \Phi_k), \ldots, \tilde{x}(q_k, \Phi_k)] \qquad \mathsf{P}_k\text{-a.s.}$$

for almost all k, and

$$[x(1, \Phi), \ldots, x(q, \Phi)] = [\tilde{x}(1, \Phi), \ldots, \tilde{x}(q, \Phi)] \qquad \mathsf{P}\text{-a.s.}$$

Thus, according to (1.7.6),

$$[x(1, \Phi_k), \ldots, x(q_k, \Phi_k)] \xrightarrow[k\to\infty]{\mathfrak{D}} [x(1, \Phi), \ldots, x(q, \Phi)].$$

Taking into consideration the definition of Q_k and Q, cf. Corollary 1.4.6, we obtain (1.7.5) immediately. ∎

1.7.3. Corollary. *Let $\mathbb{X}$ be a finite state space and (Φ, P), (Φ_k, P_k), $k \geqq 1$, be stationary ergodic inputs satisfying*

$$\Phi_k \xrightarrow{\mathfrak{D}} \Phi.$$

Assume that $f(x, u)$ is a continuous function of u for all $x \in \mathbb{X}$ and $q_k = q$ for almost all k. Then

$$\Psi_k \xrightarrow{\mathfrak{D}} \Psi.$$

Corollary 1.7.3 follows immediately from Theorem 1.7.2 by using the standard metric (1.7.4) on $\mathbb{U}^{\mathbb{Z}}$ for d_M.

While the conditions (A 1.7.11)–(A 1.7.16) seem to be natural, the condition $q_k = q$, $k \geqq 1$, looks quite artificial. The following example shows that this condition is not necessary for $\Psi_k \xrightarrow{\mathcal{D}} \Psi$.

1.7.4. Example. Consider a single server loss system and a sequence (Φ, P), (Φ_1, P_1), (Φ_2, P_2), ... of inputs defined by $\mathsf{P}(A_n = 1, S_n = 3/2, n \in \mathbb{Z}) = 1$,

$$\mathsf{P}_k\left(A_n = 1, S_n = \begin{Bmatrix} 1/2 \text{ if } n = j \bmod k \\ 3/2 \text{ if } n \neq j \bmod k \end{Bmatrix}, n \in \mathbb{Z}\right) = 1/k,$$

$$j = 0, \ldots, k-1, k \geqq 1.$$

The SGS $\mathscr{S}$ is given in Example 1.4.1, 1.4.7. Obviously, we have $q = 2$, cf. Example 1.2.2, and $q_k = 1$, $k \geqq 1$ (note, (Φ_k, P_k) have renewing epochs). We obtain

$$\mathsf{Q}_k\left(\hat{A}_{k,n} = 1, \hat{S}_{k,n} = \begin{Bmatrix} 1/2 \text{ if } n = j \bmod k \\ 3/2 \text{ if } n \neq j \bmod k \end{Bmatrix},\right.$$

$$\left.\hat{X}_{k,n} = \begin{Bmatrix} 1/2 \text{ if } n = j \bmod 2 \\ 0 \quad \text{ if } n \neq j \bmod 2 \end{Bmatrix}, n \in \mathbb{Z}\right) = \frac{1}{k}, \quad j = 0, \ldots, k-1,$$

if k is even, and

$$\mathsf{Q}_k\left(\hat{A}_{k,n} = 1, \hat{S}_{k,n} = \begin{Bmatrix} 1/2 \text{ if } n = j \bmod k \\ 3/2 \text{ if } n \neq j \bmod k \end{Bmatrix},\right.$$

$$\left.\hat{X}_{k,n} = \begin{Bmatrix} 0 \text{ if } n = j + 1 \bmod k \\ \quad \text{or } n = j + 3 \bmod k \\ \quad \text{or } n = j + 5 \bmod k \\ \quad \vdots \\ \quad \text{or } n = j + k \bmod k \\ \frac{1}{2} \text{ otherwise} \end{Bmatrix}\right) = \frac{1}{k}, \quad j = 0, \ldots, k-1,$$

if k is odd;

$$\mathsf{Q}\left(\hat{A}_n = 1, \hat{S}_n = 3/2, \hat{X}_n = \begin{Bmatrix} 0 \quad \text{if } n \text{ is even} \\ 1/2 \text{ if } n \text{ is odd} \end{Bmatrix}, n \in \mathbb{Z}\right)$$

$$= \mathsf{Q}\left(\hat{A}_n = 1, \hat{S}_n = 3/2, \hat{X}_n = \begin{Bmatrix} 0 \quad \text{if } n \text{ is odd} \\ 1/2 \text{ if } n \text{ is even} \end{Bmatrix}, n \in \mathbb{Z}\right) = \frac{1}{2}.$$

Now it is easy to verify that

$$[\hat{A}_{k,n}, \hat{S}_{k,n}, \hat{X}_{k,n}] \xrightarrow[k\to\infty]{\mathcal{D}} [\hat{A}_n, \hat{S}_n, \hat{X}_n].$$

However, the assumption $q_k = q$ cannot be omitted in general.

1.7.5. Example. Consider a single server loss system and a sequence (Φ, P), (Φ_1, P_1), (Φ_2, P_2), ... of inputs defined by

$$\mathsf{P}\left(A_n = 1, S_n = \begin{Bmatrix} 1.1 \text{ if } n \text{ is even} \\ 1.9 \text{ if } n \text{ is odd} \end{Bmatrix}, n \in \mathbb{Z}\right)$$

$$= \mathsf{P}\left(A_n = 1, S_n = \begin{Bmatrix} 1.1 \text{ if } n \text{ is odd} \\ 1.9 \text{ if } n \text{ is even} \end{Bmatrix}, n \in \mathbb{Z}\right) = \frac{1}{2},$$

$$\mathsf{P}_k(\Phi_k = \theta^j \varphi_k) = \frac{1}{2k}, \qquad j = 0, \ldots, 2k - 1,$$

where

$$\varphi_k = ([a_{k,n}, s_{k,n}]), \quad a_{k,n} \equiv 1,$$

$$s_{k,n} = \begin{cases} 0.5 \text{ if } n = 0 \bmod (2k), \\ 1.9 \text{ if } n \text{ is odd}, \\ 1.1 \text{ if } n \text{ is even}, n \neq 0 \bmod (2k). \end{cases}$$

The SGS $\mathscr{S}$ is given in Example 1.4.1, 1.4.7.
In analogy to Example 1.7.4, we obtain $q = 2$ and $q_k = 1$, $k \geqq 1$. A metric d_M satisfying (A 1.7.11)–(A 1.7.16) will be constructed in Sections 8.2, 8.3. Obviously

$$\mathsf{Q}_k(\hat{X}_{k,0} = 0.9) \xrightarrow[k\to\infty]{} \frac{1}{2},$$

but

$$\mathsf{Q}(\hat{X}_0 = 0.9) = 0.25.$$

Hence Ψ_k does not converge in distribution to Ψ.

If the assumptions of Theorem 1.7.2 excepting $q_k = q$, are fulfilled, we have

$$q_k \leqq q \quad \text{for almost all } k. \tag{1.7.7}$$

This can be shown as follows: Consider the set

$$E = \{\varphi : \varphi \in M, i(\varphi) \leqq q\}.$$

From the first part of the proof of Theorem 1.7.2 we observe that E is an open set with respect to d_M. Moreover, $\mathsf{P}(E) = 1$. Thus the boundary of E has the P measure zero. Hence $\mathsf{P}_k(E) \to \mathsf{P}(E)$ in view of (A 1.7.16). Since E is an invariant set and P_k, $k \geqq 1$, are assumed to be ergodic, $\mathsf{P}_k(E) = 1$ for almost all k. This is equivalent to (1.7.7).

1.7.6. Remark. Now it is easy to see that Theorem 1.7.2 is indeed a generalization of the method of metric modification described in Section 1.7.2: Assume that the assumptions of Theorem 1.7.2 are fulfilled and additionally $q = q_k = 1$. (Notice that $q_k = 1$ follows automatically from $q = 1$ and (1.7.7).) Then the proof of Theorem 1.7.2 shows that the assumptions (A 1.7.6) to (A 1.7.10) are satisfied for $\mathbb{U}_1 = \{\varphi : \varphi \in M, i(\varphi) = 1\}$ and $h_n(\varphi) = \tilde{x}(1, \theta^n \varphi)$.

Finally we point out that sometimes it is easier to check (A 1.7.11) to (A 1.7.16) instead of (A 1.7.6)–(A 1.7.10). In Section 8.3 we shall apply Theorem 1.7.2 in order to prove continuity properties of loss systems.

1.8. Remarks and references

In this book we restrict ourselves to Polish spaces, for convenience. However, some of the results are valid for more general measurable spaces.

1.2. Strong solutions arise in the framework of queueing models in LOYNES (1962), FRANKEN and KERSTAN (1968), FRANKEN (1969, 1970), BOROVKOV (1972a, b). The notion of a weak solution was introduced in FRANKEN (1969, 1970) in a somewhat different form. In the present form the definition is given in LISEK (1979a, b), LISEK (1982), where Φ' is called an extension of Φ.

There are many papers in the field of queueing theory in which the limiting distribution of certain queueing models with i.i.d. input is discussed, cf. e.g. LINDLEY (1952), KIEFER and WOLFOWITZ (1955). Notice that each such limiting distribution determines a stationary (Markovian) weak solution of the corresponding stochastic equation.

A weak solution Φ' is, in general, not a measurable function on the probability space $(\mathbb{U}^{\mathbb{Z}}, \mathcal{U}_{\mathbb{Z}}, \mathsf{P})$. This suggest the following question: is there a "minimal" extension $(\Omega, \mathcal{F}, \mathsf{Pr})$ of $(\mathbb{U}^{\mathbb{Z}}, \mathcal{U}_{\mathbb{Z}}, \mathsf{P})$ such that all stationary weak solutions of equation (1.1.1) for the input (Φ, P) are measurable functions on $(\Omega, \mathcal{F}, \mathsf{Pr})$? This question was answered by NEVEU (1984) for some particular cases of equation (1.1.1) arising in queueing theory.

1.3. The minimal solution $x_n{}^{\min}(\Phi)$ is due to LOYNES (1962). The construction of the maximal stationary solution and Theorem 1.3.3 are given by BRANDT (1983, 1985a) for a special case (many server queues). The proof of Theorem 1.3.3 given here is simpler than the original one.

Stochastic equations of the type

$$X_{i+1} = \mathfrak{f}(\mathbf{X}^{(i)}, U_i), \qquad i \in \mathbb{Z}, \tag{1.8.1}$$

where $\mathbf{X}^{(i)} = (X_i, X_{i-1}, X_{i-2}, \ldots)$, are treated in FICHTNER (1979a). Notice that the system of equations (1.8.1) can be transformed to an equation of type (1.1.1) by choosing an suitable state space. Under the assumption of some strong contractivity and boundeness properties on the function $\mathfrak{f}$ a unique solution of (1.8.1) is constructed by using fixpoint arguments. Furthermore it is proved that the iterations

$$\mathfrak{f}(\mathbf{X}^{(0)}, U_0), \mathfrak{f}\big(\mathfrak{f}(\mathbf{X}^{(0)}, U_0), U_1\big), \ldots$$

converge for every initial state-sequence $\mathbf{X}^{(0)}$ to this unique solution as $n \to \infty$.

Equations of the type (1.8.1) arise for example if investigating motions of collections of points located in an arbitrary space, where the motion depends on the location of the collection of points. Under some monotonicity, continuity and contractivity assumptions on $\mathfrak{f}$ for such a model a (unique) solution is constructed in FICHTNER (1979b).

Certain types of continuous time motions of particle systems can also be described by recursive stochastic equations of the type (1.1.1) by choosing a suitable state space $\mathbf{X}$. For such models it is also possible to apply monotonicity ideas for describing the time evolution. For deterministic motions given by a systems of differential equations see e.g. Fichtner (1979c), Fichtner and Freudenberg (1980), Lang (1979). For stochastic motions given by a system of differential equations we refer e.g. to Manthey (1982).

The method of "renewing epochs" or "renewing events" was developed by Franken (1969, 1970), Borovkov (1972a, b) and others for analyzing special queueing systems. A systematic treatment is given by Borovkov (1978, 1980), cf. also Foss (1986). Concerning results for the rate of convergence (ergodicity) cf. Foss (1986), Borovkov (1988) and Rachev and Kalashnikov (1988). The presentation here is somewhat different from that of Borovkov. It enables us to explain more clearly the relationship between this method and others.

1.4., 1.5. This material is essentially due to Lisek (1979a, b, 1982, 1985a, b). The proof of Theorem 1.5.4 given here is a modified one. (Corresponding arguments, given in Lisek (1985a, b) are not conclusive). Results concerning weak solutions for more general stochastic models, which include recursive stochastic equations of the type (1.2.1), but with a discrete state space $\mathbb{X}$ and input space $\mathbb{U}$, are given in Nawrotzki (1981a). The Theorems 1.5.7 and 1.5.8 are new. The second step in the proof of Theorem 1.5.7 bases on arguments given in Nawrotzki (1981a).

1.6. This presentation is based on Franken and Lisek (1982). Results including (1.6.2) were derived in the framework of point processes by Neveu (1976), and Franken and Streller, cf. [FKAS].

The following particular case was treated by Mogulski and Trofimov (1977): Let $(U_i, i \geqq 0)$ be a stationary irreducible and aperiodic Markov chain taking values in $\{1, \dots, k\}$, and P its distribution. Furthermore, let $(a_{ij}; i, j = 1, \dots, k)$ be a matrix of real numbers. Consider the sums

$$S_n = \sum_{i=1}^{n} a_{U_i U_{i+1}}, \qquad n = 1, 2, \dots$$

(In the special case of $a_{i,j} = i$ we get $S_n = \sum_{i=1}^{n} U_i$).

Let τ be a stopping time of the chain (U_i). Consider a new transition probability matrix $(\tilde{p}_{ij})$ defined by

$$\tilde{p}_{ij} = \mathsf{P}(U_\tau = j \mid U_0 = i).$$

This matrix describes the transition to the time of the next stopping. We remark that there are cases in which $(\tilde{p}_{ij})$ is not irreducible and aperiodic. Mogulski and Trofimov (1977) assume that there is exactly one stationary initial distribution with respect to $(\tilde{p}_{ij})$. Then

$$\mathsf{E}_{\tilde{\mathsf{P}}} S_\tau = (\mathsf{E}_{\tilde{\mathsf{P}}} \tau)\, \mathsf{E}_{P} S_1$$

holds, where $\tilde{\mathsf{P}}$ is the stationary distribution of the Markov chain with transition probabilities $\tilde{p}_{ij}$. The proof is very simple, using the individual ergodic theorem. Comparing this result with (1.6.11b) we see that $\tilde{\mathsf{P}}$ plays the role of $\mathsf{R}'(\cdot \mid Y_0' = 1)$ in our context.

1.7. Instead of "model continuity" the terms "stability" (cf. e.g. Borovkov (1978, 1980) and "robustness" (cf. e.g. Stoyan (1983)) are often used.

The presentation in subsection 1.7.2 is based on Lisek (1981); the results in subsection 1.7.3 are new.

In this book we restrict ourselves to the investigation of model continuity with respect to the Levy-Prokhorov metric and related metrics (when using the method of metric modification). Some other metrics were used e.g. by Zolotarev (1976a, 1977a) and Borovkov (1977, 1980) for proving qualitative as well as quantitative continuity theorems for stochastic models, see also Zolotarev (1976b, 1977b, 1979), Rachev (1981, 1982). There is a good survey in Stoyan (1983). Quantitative model continuity results in the setting of renewing epochs can be found in Borovkov (1988) and Foss (1986).

Concerning special methods for proving model continuity in case of i.i.d. U_n (continuity of Markov chains) we refer also to Stoyan (1983) and references therein.

2. The case of i.i.d. U_n: stationary Markov chains

2.1. Introduction

In this chapter we shall investigate properties of stationary solutions of the equation

$$X_{n+1} = f(X_n, U_n), \qquad n \in \mathbb{Z}, \tag{2.1.1}$$

if U_n are i.i.d. r.v.'s.

In Section 2.2 we show by means of examples that this independence assumption does not ensure the Markov property of stationary weak solutions of (2.1.1) in general. However, the Markov property is valid for all stationary weak solutions concentrated on $\mathcal{A}(\mathcal{S})$ if the underlying SGS $\mathcal{S} = \{A(n, \varphi): n \in \mathbb{Z}, \varphi \in M\}$ has the property (A 1.4.5), cf. Theorem 1.4.11, as shown in Theorem 2.2.3. (The assumption (A 1.4.5) virtually means that

$$A(0, \varphi') = A(0, \varphi'') \quad \text{for} \quad \varphi' = (u_n'), \varphi'' = (u_n'') \in M$$
$$\text{with} \quad u_n' = u_n'', \quad n < 0.)$$

In particular, a stationary weak solution $([U_n', X_n'])$ has the Markov property if the recurrence properties $\mathsf{P}'(\#\{n: X_n' = x\} = \infty) = 1$ and $\mathsf{E} \inf \{n > 0: x_n(x, U_0, \ldots, U_{n-1}) = x\} < \infty$ are fulfilled for some $x \in \mathbb{X}$ (Corollary 2.2.4). For a denumerable state space $\mathbb{X}$, each stationary weak solution $([U_n', X_n'])$ of (2.1.1) has the Markov property (Theorem 2.2.5).

If the assumption (A 1.4.5) is valid and if there is an $x_{\mathrm{P}} \in \mathbb{X}$ such that for every stationary weak solution (Φ', P'), $\Phi' = ([U_n', X_n'])$, concentrated on $\mathcal{A}(\mathcal{S})$ it holds that $\mathsf{P}'(\#\{n: X_n' = x_{\mathrm{P}}\} = \infty) = 1$, then there is exactly one stationary weak solution concentrated on $\mathcal{A}(\mathcal{S})$ (Corollary 2.2.8).

In Section 2.3 we consider time-homogeneous real valued Markov chains as solutions of recursive equations. For a given stochastic kernel $p(y, B)$, $y \in \mathbb{X} \subseteqq \mathbb{R}$, $B \in \mathcal{H}$, and a sequence (U_n) of i.i.d. r.v.'s uniformly distributed on $(0, 1)$ let $f: \mathbb{X} \times (0, 1) \to \mathbb{X}$ be a measurable function such that

$$X_{n+1} = f(X_n, U_n), \qquad n \geqq 0, \quad X_0 = x, \tag{2.1.2}$$

is a time-homogeneous Markov chain with initial state x and transition kernel $p(x, B)$. For example $f(x, u) = \inf \{y: p(x, (-\infty, y]) \geqq u\}$. (Notice that there may exist several such functions, cf. Example 2.3.1) There is a close relationship between the stationary initial distributions for the given transition probabilities $p(y, B)$ and the stationary weak solution of (2.1.1) for an appropriately chosen function f and (U_n) uniformly distributed on $(0, 1)$. This is established in Corollary 2.3.2 and Lemma 2.3.3.

The particular case of denumerable Markov chains is treated in Section 2.4. The main result is as follows (Theorem 2.4.1): There exists a stationary initial distribution with respect to the given transition probability matrix p_{jk} iff there exist an SGS $\mathscr{S}$ with property (A 1.4.5) for equation (2.1.1), where U_n are uniformly distributed on $(0, 1)$. Thus the existence of an SGS is necessary and sufficient for the ergodicity of an irreducible and aperiodic Markov chain (Corollary 2.4.8). As an application of Theorem 2.4.1, we give a new proof of Rosenblatt's representation theorem: A stationary, irreducible, aperiodic Markov chain (X_n) has a representation

$$X_n \stackrel{\mathfrak{D}}{=} g(U_{n-1}, U_{n-2}, \ldots), \qquad n \in \mathbb{Z},$$

where g is a measurable function, and (U_n) is a sequence of i.i.d. r.v.'s uniformly distributed on $(0, 1)$, cf. Theorem 2.4.5.

Finally we give some useful sufficient criteria for the existence of stationary initial distributions for Markov chains (Theorem 2.4.7).

For an arbitrary equation $X_{n+1} = f(X_n, U_n)$ with non-independent U_n in Section 2.5 we develop an idea for finding an SGS by means of investigating a Markov version of the model considered.

2.2. Markov property of stationary weak solutions

In the following we assume that the input $\Phi = (U_n)$ is a sequence of i.i.d. r.v.'s. First we consider two examples showing that equation (2.1.1) may have Markov as well as non-Markov solutions.

2.2.1. Example. Let (U_n) be a sequence of i.i.d. Bernoulli r.v.'s. with $\mathsf{P}(U_n = 0) = \mathsf{P}(U_n = 1) = 1/2$ and $\mathbf{X} = \{0, 1\}^{\mathbb{Z}_+}$. Consider the equation

$$X_{n+1} = f(X_n, U_n) = (U_n, X_{n2}, X_{n3}, \ldots), \tag{2.2.1}$$

where $X_n = (X_{n0}, X_{n1}, X_{n2}, \ldots)$.
Obviously

$$X_n^* = (U_{n-1}, U_n, U_{n+1}, \ldots)$$

defines a stationary strong solution $([U_n, X_n^*])$. It is easy to see that both (X_n^*) and $([U_n, X_n^*])$ possess the Markov property. Another stationary strong solution $([U_n, X_n])$ is given by

$$X_n = (U_{n-1}, U_{n+1}, U_{n+2}, \ldots), \qquad n \in \mathbb{Z}.$$

For $x_{-1} = (x_{-10}, x_{-11}, \ldots) \in \mathbf{X}$ and $x_{-2} = (x_{-20}, 0, x_{-11}, x_{-12}, x_{-13}, \ldots)$ we have

$$\mathsf{P}(X_{00} = 0 \mid X_{-1} = x_{-1}, X_{-2} = x_{-2}) = \mathsf{P}(U_{-1} = 0 \mid U_{-1} = 0) = 1,$$

but

$$\mathsf{P}(X_{00} = 0 \mid X_{-1} = x_{-1})$$
$$= \mathsf{P}(U_{-1} = 0 \mid U_{-2} = x_{-10}, U_0 = x_{-11}, U_1 = x_{-12}) = \frac{1}{2}.$$

Thus (X_n) does not possess the Markov property. However the solution $([U_n, X_n])$ has the Markov property:

$$\begin{aligned}
&\mathsf{P}\big(([U_0, X_0]) \in (\cdot) \mid [U_{-1}, X_{-1}] \\
&= [u_{-1}, x_{-1}], [U_{-2}, X_{-2}] = [u_{-2}, x_{-2}], \ldots\big) \\
&= \mathsf{P}\big([U_0, (U_{-1}, U_1, U_2, \ldots)] \in (\cdot) \mid [U_{-1}, (U_{-2}, U_0, U_1, \ldots)] \\
&= [u_{-1}, (x_{-10}, x_{-11}, \ldots)], [U_{-2}, (U_{-3}, U_{-1}, U_0, \ldots)] \\
&= [u_{-2}, (x_{-20}, x_{-21}, \ldots)], \ldots\big) \\
&= 1\{[x_{-11}, (u_{-1}, x_{-12}; x_{-13}, \ldots)] \in (\cdot)\} = \mathsf{P}\big([U_0, X_0] \in (\cdot) \mid [U_{-1}, X_{-1}] \\
&= [u_{-1}, x_{-1}]\big).
\end{aligned}$$

It can easily be checked that

$$X_n^{**} = (U_{n-1}, U_{n+2}, U_{n+3}, \ldots)$$

defines a stationary strong solution $([U_n, X_n^{**}])$ of (2.2.1) for which neither (X_n^{**}) nor $([U_n, X_n^{**}])$ are Markovian sequences.

2.2.2. Example. Consider the input $\Phi = (U_n)$ given in Example 2.2.1 and the equation

$$X_{n+1} = f(X_n, U_n) = (X_{n1}, X_{n2}, \ldots),$$

where $X_n = (X_{n0}, X_{n1}, X_{n2}, \ldots)$. Define a stationary strong solution $([U_n, X_n])$ by

$$X_n = (U_{n+2}, U_{n+3}, \ldots).$$

Then (X_n) possesses the Markov property, but not $([U_n, X_n])$.

Fortunately, non-Markovian solutions do not occur in the presence of an SGS with the property (A 1.4.5), cf. Theorem 1.4.11.

2.2.3. Theorem. *For a sequence $\Phi = (U_n)$ of i.i.d. r.v.'s. let $\mathcal{S}$ be an SGS with the property* (A 1.4.5) *(cf. Theorem* 1.4.11) *and $\Phi' = ([U_n', X_n'])$ a stationary weak solution concentrated on $\mathcal{A}(\mathcal{S})$. Then X_0' and $(U_n', n \geqq 0)$ are independent. In particular, (X_n') and $([U_n', X_n'])$ are Markov chains.*

Proof. Using the measurable mappings $x(i, n, \varphi)$ given by (A 1.4.5) we obtain mappings $x(k_j, 0, \varphi)$ according to the proof of Theorem 1.4.3. In view of (A 1.4.5) we have

$$\begin{aligned}
&x(k_j, 0, \varphi) = x(k_j, 0, \varphi') \quad \text{for all} \quad \varphi = (u_n), \varphi' = (u_n') \in M \\
&\text{with } u_n = u_n' \text{ for } n < 0.
\end{aligned} \tag{2.2.2}$$

Using the notation from the proof of Theorem 1.4.3 and Corollary 1.4.6 we get

$$\{x(k_1, 0, \theta^n\varphi), \ldots, x(k_q, 0, \theta^n\varphi)\} = B(n, \varphi), \quad n \in \mathbb{Z} \text{ for almost all } \varphi$$

and thus

$$\begin{aligned}
&\{x(k_1, 0, \varphi), \ldots, x(k_q, 0, \varphi)\} \\
&= x_{-m}\big(\{x(k_1, 0, \theta^m\varphi), \ldots, x(k_q, 0, \theta^m\varphi)\}, u_m, \ldots, u_{-1}\big)
\end{aligned}$$

for all $m < 0$. This implies that for every $m < 0$ and $i = 1, \dots, q$ there is exactly one solution $y(i, m, \varphi)$ of

$$x_{-m}\big(y(i, m, \varphi), u_m, \dots, u_{-1}\big) = x(k_i, 0, \varphi).$$

For $m \geqq 0$ we define

$$y(i, m, \varphi) = x_m\big(x(k_i, 0, \varphi), u_0, \dots, u_{m-1}\big), \qquad i = 1, \dots, q.$$

From (2.2.2) it follows

$$\begin{aligned} &y(j, m, \varphi) = y(j, m, \varphi'), \quad m \leqq 0 \text{ for all } \varphi, \varphi' \in \boldsymbol{M} \\ &\text{with } u_n = u_n' \text{ for } n < 0. \end{aligned} \tag{2.2.3}$$

For the functions $h_j(\varphi)$ considered in the proof of Theorem 1.5.4 it is easy to prove by means of the individual ergodic theorem that

$$\begin{aligned} h_j(\varphi) &= \lim_{n\to\infty} \frac{1}{n} \sum_{k=1}^{n} h\Big(\big([u_{k+l}, x_{k+l}\big(y(j, 0, \varphi), u_0, \dots, u_{k+l-1}\big)], l \geqq 0\big)\Big) \\ &= \lim_{n\to\infty} \frac{1}{n} \sum_{k=1}^{n} h\Big(\big([u_{k+l}, y(j, k+l, \varphi)], l \geqq 0\big)\Big) \\ &= \lim_{n\to\infty} \frac{1}{n} \sum_{k=-n}^{-1} h\Big(\big([u_{k+l}, y(j, k+l, \varphi)], l \geqq 0\big)\Big), \end{aligned}$$

$$j = 1, \dots, q, \quad h \in \boldsymbol{H}.$$

From this and (2.2.3),

$$h_j(\varphi) = h_j(\varphi') \quad \text{for all } \varphi, \varphi' \in \boldsymbol{M} \text{ with } u_n = u_n' \text{ for } n < 0.$$

Analogously we obtain

$$\nu_\varphi(j) = \nu_{\varphi'}(j) \quad \text{for all } \varphi, \varphi' \in \boldsymbol{M} \text{ with } u_n = u_n' \text{ for } n < 0,$$
$$j = 1, \dots, q,$$

cf. proof of Theorem 1.5.4. Taking this into account, we obtain the independence of $(U_n^m, n \geqq 0)$ and X_0^m for every stationary and ergodic weak solution (Ψ_m, Q_m), $\Psi_m = \{[U_n^m, X_n^m]\}$ given by Theorem 1.5.4 (i.e. every stationary ergodic weak solution concentrated on $\mathcal{A}(\mathcal{S})$). Now consider an arbitrary weak solution (Φ', P'), $\Phi' = ([U_n', X_n'])$, concentrated on $\mathcal{A}(\mathcal{S})$. In view of the last statement of Theorem 1.5.4,

$$\begin{aligned} &\mathsf{P}'\big(X_0' \in B, (U_n, n \geqq 0) \in C\big) \\ &= \sum_{m=1}^{r} c_m Q_m\big(X_0^m \in B, (U_n^m, n \geqq 0) \in C\big) \\ &= \sum_{m=1}^{r} c_m \cdot \mathsf{Q}_m(X_0^m \in B) \cdot \mathsf{P}\big((U_n, n \geqq 0) \in C\big) \\ &= \mathsf{P}'(X_0' \in B) \cdot \mathsf{P}\big((U_n, n \geqq 0) \in C\big) \end{aligned}$$

for all measurable subsets $B \subseteq \mathbb{X}$ and $C \subseteq \mathbb{U}^{\mathbb{Z}_+}$. ∎

From Theorem 2.2.3 and Theorem 1.4.11 we conclude immediately

2.2.4. Corollary. *For a sequence* $\Phi = (U_n)$ *of i.i.d. r.v.'s let* (Φ', P'), $\Phi' = ([U_n', X_n'])$, *be a stationary weak solution of* (2.1.1). *Assume there is an* $x \in \mathbb{X}$ *which occurs infinitely often in the sequence* (X_n') *and which has a finite mean recurrence time, i.e.*

$$\mathsf{P}'(\#\{n: X_n' = x\} = \infty) = 1$$

and

$$\mathsf{E} \inf \{n > 0: x_n(x, U_0, \ldots, U_{n-1}) = x\} < \infty.$$

Then X_0' *and* $(U_n', n \geqq 0)$ *are independent. In particular,* (X_n') *and* $([U_n', X_n'])$ *are Markov chains.*

In the case of a denumerable state space $\mathbb{X}$, equation (1.2.1) has Markov solutions only.

2.2.5. Theorem. *Let* $\mathbb{X}$ *be a denumerable state space,* (Φ, P), $\Phi = (U_n)$, *a sequence of i.i.d. r.v.'s. and* (Φ', P'), $\Phi' = ([U_n', X_n'])$, *a stationary weak solution. Then* X_0' *and* $(U_n', n \geqq 0)$ *are independent. In particular,* (X_n') *and* $([U_n', X_n'])$ *are Markov chains.*

Proof. From Theorem 1.5.7 we have that (Φ', P') is a mixture of at most countably many stationary weak solutions (Φ_k', P_k'), $\Phi_k' = ([U_n^k, X_n^k])$, which satisfy the assumptions of Theorem 2.2.3. Hence X_0^k and $(U_n^k, n \geqq 0)$ are independent. From this and since P' is a mixture of the P_k' (cf. Theorem 1.5.7, (iv)) we get the assertions of the theorem. ∎

Using Theorem 2.2.3 we are able to show that Theorem 1.6.3 is indeed a generalization of Wald's identity, as announced in Section 1.6.

2.2.6. Example. Let (Φ, P), $\Phi = (U_n)$, be a sequence of i.i.d. real valued r.v.'s with existing expectation $\mathsf{E_P} U_0$. Assume $\tau = \tau(\Phi) \geqq 1$ is a stopping time with respect to $(U_n, n \geqq 1)$ and $\mathsf{E_P}\tau < \infty$. From Theorem 1.6.3 we have a stationary sequence (χ', R'), $\chi' = ([U_n', Y_n'])$, satisfying (1.6.3)–(1.6.5), (1.6.11a, b). Lemma 1.6.2 provides a stationary weak solution (Φ', P'), $\Phi' = ([U_n', X_n'])$, of (1.6.9) with

$$\mathsf{E_{R'}}\left(\sum_{n=1}^{\tau((U_k'))} U_n' \mid Y_0' = 1\right) = \mathsf{E_{P'}}\left(\sum_{n=1}^{\tau((U_k'))} U_n' \mid X_0' = 0\right)$$

and

$$\mathsf{E_{R'}}\big(\tau((U_k')) \mid Y_0' = 1\big) = \mathsf{E_{P'}}\big(\tau((U_k')) \mid X_0' = 0\big).$$

From Theorem 2.2.3

$$\mathsf{E_{P'}}\left(\sum_{n=1}^{\tau((U_k'))} U_n' \mid X_0' = 0\right) = \mathsf{E_{P'}} \sum_{n=1}^{\tau((U_k'))} U_n' = \mathsf{E_P} \sum_{n=1}^{\tau} U_n,$$

$$\mathsf{E_{P'}}\big(\tau((U_k')) \mid X_0' = x\big) = \mathsf{E_{P'}}\tau((U_k')) = \mathsf{E_P}\tau.$$

Thus formula (1.6.11b) reduces to

$$\mathsf{E_P} \sum_{n=1}^{\tau} U_n = \mathsf{E_P} U_1 \mathsf{E_P} \tau.$$

Now our aim is to give a uniqueness criterion for stationary weak solutions. First we shall prove

2.2.7. Theorem. *For a sequence $\Phi = (U_n)$ of i.i.d. r.v.'s let $\mathcal{S}$ be an SGS with the property* (A 1.4.5), *and* (Φ', P'), $\Phi' = ([U_n', X_n'])$, *and* (Φ'', P''), $\Phi'' = ([U_n'', X_n''])$, *be two stationary weak solutions concentrated on $\mathcal{A}(\mathcal{S})$. If there is an $x \in \mathbf{X}$ satisfying*

$$\mathsf{P}'(\#\{n: X_n' = x\} = \infty) = \mathsf{P}''(\#\{n: X_n'' = x\} = \infty) = 1, \qquad (2.2.4)$$

then

$$\mathsf{P}' = \mathsf{P}''.$$

Proof. In view of (2.2.4) the conditional distributions $\mathsf{P}'(\cdot \mid X_0' = x)$ and $\mathsf{P}''(\cdot \mid X_0'' = x)$ are well defined. By means of Theorem 2.2.3 we obtain

$$\begin{aligned}
&\mathsf{P}'\big(([U_n', X_n'], n \geqq 0) \in (\cdot) \mid X_0' = x\big) \\
&= \mathsf{P}'\Big(\big([U_n', x_n(x, U_0', \ldots, U_{n-1}')], n \geqq 0\big) \in (\cdot) \mid X_0' = x\Big) \\
&= \mathsf{P}'\Big(\big([U_n', x_n(x, U_0', \ldots, U_{n-1}')], n \geqq 0\big) \in (\cdot)\Big) \\
&= \mathsf{P}\Big(\big([U_n, x_n(x, U_0, \ldots, U_{n-1})], n \geqq 0\big) \in (\cdot)\Big) \\
&= \mathsf{P}''\big(([U_n'', X_n''], n \geqq 0) \in (\cdot) \mid X_0'' = x\big).
\end{aligned}$$

The desired equality $\mathsf{P}' = \mathsf{P}''$ follows from Theorem A 1.3.2. ∎

2.2.8. Corollary. (Uniqueness criterion). *For a sequence $\Phi = (U_n)$ of i.i.d. r.v.'s let $\mathcal{S}$ be an SGS with property* (A 1.4.5). *Assume there is an $x \in \mathbf{X}$ such that for every stationary weak solution* (Φ', P'), $\Phi' = ([U_n', X_n'])$, *concentrated on $\mathcal{A}(\mathcal{S})$ it holds that*

$$\mathsf{P}'(\#\{n: X_n' = x_{\mathsf{P}}\} = \infty) = 1. \qquad (2.2.5)$$

Then there is exactly one stationary weak solution concentrated on $\mathcal{A}(\mathcal{S})$.

Proof. For each stationary ergodic weak solution (Ψ_m, Q_m), $\Psi_m = ([U_n^m, X_n^m])$, $m = 1, \ldots, r$, cf. Theorem 1.5.4., we have

$$\mathsf{Q}_m(\#\{n: X_n^m = x_{\mathsf{P}}\} = \infty) = 0$$

or

$$\mathsf{Q}_m(\#\{n: X_n^m = x_{\mathsf{P}}\} = \infty) = 1, \qquad (2.2.6)$$

since the event considered is invariant. Theorem 2.2.7 implies that there is one index $m \in \{1, \ldots, r\}$ for which (2.2.6) is true. Since every stationary weak solution (Φ', P') concentrated on $\mathcal{A}(\mathcal{S})$ is a mixture of the ergodic distributions $\mathsf{Q}_1, \ldots, \mathsf{Q}_r$, cf. Theorem 1.5.4, we have $\mathsf{P}' = \mathsf{Q}_m$ in view of (2.2.5). ∎

The usefulness of the preceding corollary will be demonstrated by the following examples.

2.2.9. Example. Consider the single server loss system, cf. Example 1.2.2, under the assumption that the pairs $[A_n, S_n]$, $n \in \mathbb{Z}$, are i.i.d. and $\mathsf{E}A_0 > 0$, $0 < \mathsf{E}S_0 < \infty$. Then, for every stationary weak solution (Φ', P'), $\Phi' = \big([[A_n', S_n'], X_n']\big)$, of (1.2.3) we have

$$\mathsf{P}'(\#\{n: X_n' = 0\} = \infty) = 1$$

(cf. Example 1.4.1). Furthermore, every stationary weak solution is concentrated on the SGS given in Example 1.4.1, 1.4.7. This and Corollary 2.2.8 imply the uniqueness of the stationary weak solution of (1.2.2).

2.2.10. Example. Consider equation (1.6.9) arising in the derivation of the generalized Wald's identity. As mentioned in Example 1.6.4 this equation may have several stationary weak solutions, in general. However in the case of i.i.d. r.v.'s U_n there is only one stationary weak solution. Indeed, we can apply Corollary 2.2.8 by taking $x_P = 0$.

2.3. Markov chains as solutions of recursive equations

In this section (Φ, P), $\Phi = (U_n)$, denotes a sequence of i.i.d. r.v.'s uniformly distributed on the interval $(0, 1)$. Let $p(x, B)$, $x \in \mathbb{X} \subseteq \mathbb{R}$, $B \in \mathcal{H}$, be a stochastic kernel. Define

$$f_1(x, u) = \inf \{y : p(x, (-\infty, y] \cap \mathbb{X}) \geqq u\}, \quad x \in \mathbb{X}, \quad 0 < u < 1. \tag{2.3.1}$$

Then

$$p(x, B) = \mathsf{P}(f_1(x, U_0) \in B), \qquad x \in \mathbb{X}, \quad B \in \mathcal{H}. \tag{2.3.2}$$

Thus, for each fixed $x \in \mathbb{X}$ the sequence $(X_n, n \geqq 0)$, defined by

$$X_{n+1} = f_1(X_n, U_n), \qquad n \geqq 0, \quad X_0 = x, \tag{2.3.3}$$

is a time-homogeneous Markov chain with initial state $X_0 = x$ and transition probabilities $p(x, B)$. The representation (2.3.3) is commonly used for simulating Markov chains.

Notice that the function (2.3.1) is not the only possible one which will generate a Markov chain with the given transition probabilities by means of an equation of type (2.3.3).

2.3.1. Example. Let $\mathbb{X} = \{0, 1\}$. Consider the transition probability matrix

$$\begin{pmatrix} 0.5 & 0.5 \\ 0.5 & 0.5 \end{pmatrix}.$$

Obviously (2.3.1) becomes

$$f_1(x, u) = 1\left\{u \geqq \frac{1}{2}\right\}.$$

Consequently $X_0 = 0$, $X_{n+1} = 1\left\{U_n \geqq \frac{1}{2}\right\}$, $n \geqq 0$, is a Markov chain with transition matrix (2.3.4) and initial state 0. On the other hand the sequence

$$X_0^* = 0, \qquad X_{n+1}^* = f_2(X_n^*, U_n), \qquad n \geqq 0,$$

where

$$f_2(x, u) = \begin{cases} 1\left\{u \geqq \frac{1}{2}\right\} & \text{if } x = 0, \\ 1\left\{u < \frac{1}{2}\right\} & \text{if } x = 1, \end{cases} \tag{2.3.4}$$

is a Markov chain stochastically equivalent to $(X_n, n \geqq 0)$.

In the following we consider an arbitrary function f such that for each fixed $x \in \mathbf{X}$ the sequence $(X_n, n \geqq 0)$ defined by

$$X_{n+1} = f(X_n, U_n), \qquad n \geqq 0, \qquad X_0 = x,$$

is a time homogeneous Markov chain with the given transition probabilities $p(x, B)$. The k step transition probabilities of $(X_n, n \geqq 0)$ are given by

$$p^k(y, B) = \mathsf{P}(X_k \in B \mid X_0 = y) = \mathsf{P}\big(x_k(y, U_0, \ldots, U_{k-1}) \in B\big) \tag{2.3.5}$$

$y \in \mathbf{X}$, $B \in \mathcal{H}$.

It seems to be reasonable to try to find out results about the existence and uniqueness of the stationary initial distribution with respect to the transition probabilities $p(y, B)$ by applying the methods developed in Chapter 1 to equation

$$X_{n+1} = f(X_n, U_n), \qquad n \in \mathbb{Z}. \tag{2.3.6}$$

The following result is an immediate consequence of Theorem 2.2.3.

2.3.2. Corollary. *Assume there is an SGS $\mathcal{S}$ for equation* (2.3.6) *with the property* (A 1.4.5). *Then, for every stationary weak solution* $([U_n', X_n'])$ *concentrated on $\mathcal{A}(\mathcal{S})$ the sequence (X_n') is a stationary Markov chain with the given transition probabilities $p(x, B)$.*

Proof. In view of Theorem 2.2.3, (X_n') is a stationary Markov chain. Furthermore we have

$$\begin{aligned} \mathsf{P}'(X_1' \in B \mid X_0' = y) &= \mathsf{P}'\big(f(X_0', U_0') \in B \mid X_0' = y\big) \\ &= \mathsf{P}'\big(f(y, U_0') \in B\big) = p(y, B). \quad \blacksquare \end{aligned}$$

The converse of Corollary 2.3.2 is true without any assumption.

2.3.3. Lemma. *For every stationary Markov chain (X_n) with the given transition probabilities $p(y, B)$ there is a stationary weak solution $([U_n', X_n'])$ of equation* (2.3.6) *satisfying*

$$(X_n) \overset{\mathcal{D}}{=} (X_n').$$

Proof. Define

$$\begin{aligned} &\mathsf{P}'\big(([U_n', X_n'], n \geqq 0) \in (\cdot)\big) \\ &= \int_{\mathbf{X}} \mathsf{P}\Big(\big([U_n, x_n(x, U_0, \ldots, U_{n-1})], n \geqq 0\big) \in (\cdot)\Big) \, \mathsf{P}(X_0 \in dx). \end{aligned} \tag{2.3.7}$$

The extension $([U_n', X_n'], n \in \mathbb{Z})$ has the desired property. $\blacksquare$

2.4. Countable Markov chains

Assume $\mathbf{X} \subseteq \{0, 1, 2, \ldots\}$, and let $(p_{jk}, j, k \in \mathbf{X})$ be a fixed transition probability matrix. The following theorem, which is based on the relationships between stationary Markov chains and recursive stochastic equations discussed above, gives a necessary and sufficient condition for the existence of a stationary initial distribution.

2.4.1. Theorem. *Let (p_{jk}) be a transition probability matrix and $f: \mathbf{X} \times (0, 1) \to \mathbf{X}$ a fixed function satisfying the following condition. For each $x \in \mathbf{X}$ the sequence (X_n) defined by*

$$X_{n+1} = f(X_n, U_n), \qquad n \geqq 0, \qquad X_0 = x,$$

where $\Phi = (U_n)$ is a sequence of i.i.d. r.v.'s uniformly distributed on (0, 1), is a Markov chain with the transition matrix (p_{jk}) (cf. Section 2.3.). Then the following statements are equivalent:

(i) *There exists a stationary initial distribution with respect to p_{jk}.*
(ii) *There exists an SGS $\mathcal{S}$ with the property (A 1.4.5) for the equation*

$$X_{n+1} = f(X_n, U_n), \qquad n \in \mathbb{Z}. \tag{2.4.1}$$

(iii) *There is a finite number $q \geqq 1$ and measurable mappings $x(i, \varphi)$, $i = 1, \ldots, q$, defined for almost all $\varphi \in (0, 1)^{\mathbb{Z}}$ such that*

$$p_j = \frac{1}{q} \sum_{i=1}^{q} \mathsf{P}\big(x(i, \Phi) = j\big), \qquad j \in \mathbf{X}, \tag{2.4.2}$$

is a stationary initial distribution with respect to p_{jk}.

Proof. $\big((i) \to (ii)\big)$. Let $\big((X_n), \mathsf{P}_X\big)$ be a stationary Markov chain with the transition probabilities p_{jk} and n-step transition probabilities $p_{jk}^{(n)}$. Then there is a state $i^* \in \mathbf{X}$ satisfying

$$\mathsf{P}_X(\# \{n: X_n = i^*\} = \infty) > 0,$$

cf. Theorem A 1.1.2, and i^* is a positively recurrent state. Consider the class of all states communicating with i^*

$$K^* = \{j: j \in \mathbf{X}, p_{i^*j}^{(n)} > 0 \text{ for some } n\}.$$

Then

$$\mathsf{P}^*(\cdot) = \mathsf{P}_X(\cdot \mid X_0 \in K^*)$$

is the probability distribution of a stationary Markov chain (X_n^*) with the transition probabilities (p_{kl}) and the initial distribution

$$p_j^* = \mathsf{P}^*(X_0^* = j), \qquad j \in \mathbf{X}.$$

We have

$$\mathsf{P}^*(\# \{n: X_n^* = i^*\} = \infty) = 1$$

and

$$\mathsf{E} \inf \{n > 0: x_n(i^*, U_0, \ldots, U_{n-1}) = i^*)\} < \infty.$$

In view of Lemma 2.3.3, there is a stationary weak solution (Φ', P'), $\Phi' = ([U_n', X_n'])$, of equation (2.4.1) which satisfies $\mathsf{P}'(\# \{n: X_n' = i^*\} = \infty) = 1$. Now (ii) follows from Theorem 1.4.11.

$\big((\text{ii}) \to (\text{iii})\big)$. Let an SGS $\mathscr{S}$ with property (A 1.4.5) be given. From Corollary 1.4.6 there is a finite number $q \geqq 1$ and measurable mappings $x(i, \varphi)$ such that

$$\mathsf{Q}\big([\hat{U}_0, \hat{X}_0] \in (\cdot)\big) = \frac{1}{q} \sum_{i=1}^{q} \mathsf{P}\big([U_0, x(i, \Phi)] \in (\cdot)\big)$$

is the one-dimensional distribution of a stationary weak solution (Ψ, Q), $\Psi = ([\hat{U}_n, \hat{X}_n])$. This solution is concentrated on $\mathscr{A}(\mathscr{S})$. Then, in view of Theorem 2.2.3 $(\hat{X}_n)$ is a stationary Markov chain, and thus

$$p_j = \mathsf{Q}(\hat{X}_0 = j) = \frac{1}{q} \sum_{i=1}^{q} \mathsf{P}\big(x(i, \Phi) = j\big), \qquad j \in \mathbf{X},$$

is a stationary initial distribution with respect to p_{jk}.

$\big((\text{iii}) \to (\text{i})\big)$. This is trivial. ▮

For given transition probabilities the number q in (2.4.2) may depend on the choice of the function f, as the following example shows.

2.4.2. Remark. The step (i) $\to$ (ii) in the proof of Theorem 2.4.1 can be obtained also from Theorem 1.5.7 directly: In view of Lemma 2.3.3 there is a stationary weak solution (Φ', P'), $\Phi' = ([U_n', X_n'])$, of equation (2.4.1) and from Theorem 1.5.7 it follows that there is an SGS with the property (A 1.4.5).

2.4.3. Example (continuation of Example 2.3.1). Choose $A(n, \Phi) = \{0, 1\}$. Then we obtain $q = 1$ for equation $X_{n+1} = f_1(X_n, U_n)$ and $q = 2$ for equation $X_{n+1} = f_2(X_n, U_n)$, cf. (2.3.4).

Now the question arises whether, for given transition probabilities, there is always a function f such that $q = 1$. The following theorem gives a positive answer.

2.4.4. Theorem *Let (p_{jk}) be an irreducible, aperiodic transition probability matrix possessing a stationary initial distribution. Then there is a measurable function $f: \mathbf{X} \times (0, 1) \to \mathbf{X}$ such that:*

(i) *For each $x \in \mathbf{X}$ the sequence (X_n) defined by*

$$X_{n+1} = f(X_n, U_n), \qquad n \geqq 0, \qquad X_0 = x,$$

where $\Phi = (U_n)$ is a sequence of i.i.d. r.v.'s uniformly distributed on $(0, 1)$, *is a Markov chain with transition matrix (p_{jk}).*

(ii) *There is a measurable mapping $x(\varphi) = x(1, \varphi)$ defined for almost all $\varphi \in (0, 1)^{\mathbf{Z}}$ such that*

$$p_j = \mathsf{P}\big(x(\Phi) = j\big), \qquad j \in \mathbf{X},$$

is a stationary initial distribution with respect to (p_{jk}), i.e. the number q given by Theorem 2.4.1, (iii) equals 1.

Proof. For an arbitrary real number $u \in (0, 1)$ consider the decimal notation

$$u = \sum_{j=1}^{\infty} 10^{-j} \beta_j^{(0)} .$$

Define

$$\beta_j^{(k+1)} = \beta_{2j-1}^{(k)}, \qquad j = 1, 2, \ldots; \qquad k = 0, 1, 2, \ldots,$$

and

$$v^{(k)}(u) = \sum_{j=1}^{\infty} 10^{-j} \beta_{2j}^{(k)}, \qquad k = 0, 1, 2, \ldots$$

Let $\Phi = (U_n, n \in \mathbb{Z})$ be a sequence of independent random variables uniformly distributed on (0, 1). Obviously, for each n $\left(v^{(k)}(U_n), k \in \mathbf{X}\right)$ is also a sequence of i.i.d. r.v.'s uniformly distributed on (0, 1). Define

$$f(k, u) = \min \left\{ l : l \in \mathbf{X}, \sum_{i \leq l} p_{ki} \geqq v^{(k)}(u) \right\}, \qquad k \in \mathbf{X}, \qquad u \in (0, 1).$$

Then condition (i) is satisfied. Since condition (i) in Theorem 2.4.1 is satisfied, we can apply Theorem 2.4.1 to the function f defined above. Thus, there is a finite number $q \geqq 1$ and measurable mappings $x(i, \varphi)$, $i = 1, \ldots, q$, such that (2.4.2) defines the stationary initial distribution. Observe that, according to property (A 1.4.5), the mappings $x(i, \varphi)$ can be chosen in such a way that

$$x(i, \varphi) = x(i, \varphi') \text{ if } \varphi = (u_n),\ \varphi' = (u_n') \text{ and } u_n = u_n' \text{ for } n < 0,$$

cf. the proofs of Theorem 2.4.1 and Corollary 1.4.6. Assume $q > 1$. Then $x(1, \Phi) \neq x(2, \Phi)$ a.s. Consider the sequence of pairs

$$\left[x_n\left(x(1, \Phi), U_0, \ldots, U_{n-1}\right), x_n\left(x(2, \Phi), U_0, \ldots, U_{n-1}\right)\right], \qquad n \geqq 0 .$$

As long as the two components do not coincide, they behave like independent Markov chains with transition probabilities p_{jk} and initial state $x(1, \Phi)$ and $x(2, \varphi)$, respectively. In view of the assumed irreducibility, aperiodicity and positive recurrence, there is an $n_0(\varphi)$ for almost every sample path $\varphi = (u_n)$ of (U_n) such that

$$x_n\left(x(1, \varphi), u_0, \ldots, u_{n-1}\right) = x_n\left(x(2, \varphi), u_0, \ldots, u_{n-1}\right) \quad \text{for } n \geqq n_0(\varphi).$$

(Note, a sequence of pairs $[Y_n, Z_n]$, $n \geqq 0$, where $(Y_n, n \geqq 0)$ and $(Z_n, n \geqq 0)$ are independent, irreducible, aperiodic, positively recurrent Markov chains, is irreducible, aperiodic and positively recurrent itself.) Thus, $i(\varphi) \leqq q - 1$ for almost all φ. However, $i(\Phi) = q$ a.s. Hence the assumption that $q > 1$ is not true. ∎

There is a close relationship between Theorem 2.4.4 and the following well-known representation theorem for stationary Markov chains:

2.4.5. Theorem (Rosenblatt (1960)). *Let $(X_n, n \in \mathbb{Z})$ be a stationary, irreducible, aperiodic Markov chain with the state space $\mathbf{X} = \{0, 1, 2, \ldots\}$. Then there is a measurable function g: $(0, 1)^{\mathbb{Z}_+} \to \mathbf{X}$ such that the sequence $\left(g(U_{n-1}, U_{n-2}, \ldots), n \in \mathbb{Z}\right)$, where $(U_n, n \in \mathbb{Z})$ is a sequence of i.i.d. r.v.'s uniformly distributed on (0, 1), is stochastically equivalent to (X_n).*

Proof. Consider a function f given by Theorem 2.4.4 and the SGS $\mathcal{S}$ generating the mapping $x(\varphi) = x(1, \varphi)$, cf. Theorem 2.4.1. From $q = 1$ and relationship (1.4.9) we obtain $B(n, \varphi) = \{x(\theta^n\varphi)\}$, $n \in \mathbb{Z}$, $\varphi \in M$. From this and $f(B(n, \varphi), u_n) \subseteqq B(n + 1, \varphi)$, $n \in \mathbb{Z}$, $\varphi \in M$, cf. proof of Theorem 1.4.3,

$$x(\theta \Phi) = f(x(\Phi), U_0) \quad \text{P-a.s.}$$

Since the SGS $\mathcal{S}$ possesses the property (A 1.4.5), $x(\varphi)$ is a measurable function of $(u_{-1}, u_{-2}, \ldots)$. Thus, $(x(\theta^n\Phi),\ n \in \mathbb{Z})$ is a stationary Markov chain stochastically equivalent to (X_n). ∎

2.4.6. Remark. Let (X_n) be a stationary, irreducible, aperiodic Markov chain with the state space $\mathbb{X} = \{0, 1, 2, \ldots\}$ and let f be a function satisfying (i), (ii) in Theorem 2.4.4. The first step in the proof of Theorem 2.4.1 shows that we can choose an SGS $\mathcal{S} = \{A(n, \varphi) : n \in \mathbb{Z}, \varphi \in M\}$ such that

$$i^* \in A(n, \Phi), \qquad n \in \mathbb{Z}, \qquad \text{P-a.s.}$$

for some $i^* \in \mathbb{X}$.

Theorem 2.4.1 provides the well-known fact that for Markov chains with finite state space $\mathbb{X}$ there always exists a stationary initial distribution. (Namely, take $A(n, \Phi) = \mathbb{X}$.)

Now we formulate some sufficient conditions for the existence of a stationary initial distribution.

In the following we assume $\mathbb{X} = \{0, 1, 2, \ldots\}$, for convenience. For a given transition probability matrix (p_{jk}) and a fixed function $f(x, u)$ (see Section 2.3) define

$$\nu(j, k, \varphi) = \inf\{n : n \geqq 1, x_n(j, u_0, \ldots, u_{n-1}) \leqq k\},$$

$$\varphi \in (0, 1)^{\mathbb{Z}}, \qquad 0 \leqq j \leqq k,$$

and

$$p(j, k, n) = \frac{\mathsf{P}(\nu(j, k, \Phi) \geqq n + 1)}{\mathsf{P}(\nu(j, k, \Phi) \geqq n)}, \qquad n \geqq 1.$$

Consider the following conditions:

(A 2.4.1) There is a $k \in \mathbb{X}$ such that

$$\inf\{p_{jk} : j \in \mathbb{X}\} > 0.$$

(A 2.4.2) There is an $\varepsilon > 0$ and a state $k \in \mathbb{X}$ such that

$$\sum_{l \leqq k} p_{jl} > \varepsilon \quad \text{for } j > k.$$

(A 2.4.3) There is a $k \in \mathbb{X}$ and δ satisfying

$$p(j, k, n) < \delta < 1 \quad \text{for all } j \in \{0, \ldots, k\}$$

and all n for which $p(j, k, n)$ is defined.

(A 2.4.4) There is a $k \in \mathbb{X}$ such that $\mathsf{E}_{\mathsf{P}}\nu(j, k, \Phi) < \infty$ for all $j \in \{0, \ldots, k\}$.

2.4.7. Theorem. (i) *Condition* (A 2.4.4) *is sufficient for the existence of a stationary initial distribution.*

(ii) *The following implications are true:*

$$(\text{A } 2.4.1) \to (\text{A } 2.4.2) \to (\text{A } 2.4.3) \to (\text{A } 2.4.4).$$

Proof. (i) The proof of statement (i) is similar to that of Theorem 1.4.11. Consider

$$L_n = \sum_{m \leqq n} 1\{\max\{m + \nu(j, k, \theta^m \Phi) : j \leqq k\} > n\}, \qquad n \in \mathbb{Z}.$$

We obtain

$$\begin{aligned} \mathsf{E}_{\mathsf{P}} L_n = \mathsf{E}_{\mathsf{P}} L_0 &= \sum_{m \leqq 0} \mathsf{P}(\max\{m + \nu(j, k, \theta^m \Phi) : j \leqq k\} > 0) \\ &= \sum_{m=0}^{\infty} \mathsf{P}(\max\{\nu(j, k, \theta^m \Phi) : j \leqq k\} > m) \\ &= \mathsf{E}_{\mathsf{P}} \max\{\nu(j, k, \Phi) : j \leqq k\} < \infty \end{aligned}$$

in view of (A 2.4.4). In particular

$$L_n < \infty, \qquad n \in \mathbb{Z}, \qquad \mathsf{P}\text{-a.s.} \tag{2.4.3}$$

Now define $I(n, \varphi) = \{m : m \leqq n;\ \max\{m + \nu(j, k, \theta^m \Phi) : j \leqq k\} > n\}$. On account of (2.3.4), $I(n, \Phi)$ is P-a.s. finite for each $n \in \mathbb{Z}$. Analogously to the proof of Theorem 1.4.11,

$$A(n, \varphi) = \bigcup_{j=0}^{k} \{x_{n-m}(j, u_m, \ldots, u_{n-1}) : m \in I(n, \varphi)\}$$

defines an SGS:

$$\mathcal{S} = \{A(n, \varphi) : n \in \mathbb{Z}, \varphi \in (0, 1)^{\mathbb{Z}}, \# I(k, \varphi) < \infty, k \in \mathbb{Z}\}$$

with the property (A 1.4.5). Now assertion (i) follows from Theorem 2.4.1.

(ii) a) (A 2.4.2) follows immediately from (A 2.4.1).

b) From (A 2.4.2) we have

$$\sum_{l > k} p_{jl} = \mathsf{P}\big(f(j, U_0) > k\big) < \delta < 1 \quad \text{for all } j > k.$$

Then

$$\begin{aligned} p(j, k, n) = \mathsf{P}\big(x_n(j, U_0, \ldots, U_{n-1}) > k \mid x_i(j, U_0, \ldots, U_{i-1}) \\ > k, i = 1, \ldots, n-1\big) < \delta < 1, \qquad j \leqq k, \end{aligned}$$

i.e. (A 2.4.2) implies (A 2.4.3).

c) Condition (A 2.4.3) is the well-known d'Alembert criterion for the convergence of the series

$$\mathsf{E}_{\mathsf{P}} \nu(j, k, \Phi) = \sum_{n=1}^{\infty} \mathsf{P}\big(\nu(j, k, \Phi) \geqq n\big) < \infty, \qquad j \in \{0, \ldots, k\}. \quad \blacksquare$$

If the transition probability matrix (p_{jk}) is irreducible and aperiodic, then there is at most one stationary initial distribution. In this case the existence of the stationary initial distribution (p_j) is equivalent to the ergodicity of the

Markov chain, i.e. to

$$p_{ij}^{(n)} \xrightarrow[n\to\infty]{} p_j, \qquad i, j \in \mathbb{X},$$

cf. e.g. Feller (1971). Theorem 2.4.1 yields:

2.4.8. Corollary. *An irreducible aperiodic Markov chain is ergodic iff there is an SGS with the property* (A 1.4.5) *for the equation* (2.4.1).

2.5. Markovian techniques for non-Markovian systems

Consider a stochastic system described by means of the equation

$$X_{n+1} = f(X_n, U_n), \qquad n \in \mathbb{Z}, \tag{2.5.1}$$

where $\Phi = (U_n)$ is an arbitrary stationary input.

One method for proving the existence of stationary weak solutions consists in constructing an SGS, cf. Theorem 1.4.3. The question is, how to find an SGS? In the following we give an idea for constructing an SGS via the investigation of a Markovian version of the system. For this purpose consider an appropriately chosen sequence $(\tilde{\Phi}, \tilde{\mathsf{P}})$, $\tilde{\Phi} = (\tilde{U}_n)$, of $\mathbb{U}$-valued i.i.d. r.v.'s. Then the equation

$$\tilde{X}_{n+1} = f(\tilde{X}_n, \tilde{U}_n), \qquad n \in \mathbb{Z}, \tag{2.5.2}$$

describes the given system with the i.i.d. input $\tilde{\Phi}$. First we search for a stationary initial distribution of a Markov chain governed by the transition probabilities

$$p(x, B) = \tilde{\mathsf{P}}\big(f(x, \tilde{U}_0) \in B\big), \qquad x \in \mathbb{X}, \qquad B \in \mathcal{H}. \tag{2.5.3}$$

Notice that (2.5.2) may be interpreted as a Markov version of the model (2.5.1). There are some useful criteria for proving the existence of stationary initial distributions with respect to the transition probabilities (2.5.3). One powerful criterion is stated in the following theorem:

2.5.1. Theorem. *Let $\mathbb{X}$ be a closed subset of $\mathbb{R}$. Assume:*

(A 2.5.1) *The kernel $p(x, B)$ is ϕ-irreducible, i.e. there exists a non-zero measure ϕ on $\mathcal{H}$ such that for any $x \in \mathbb{X}$ and $B \in \mathcal{H}$ with $\phi(B) > 0$ there is an n for which $p^n(x, B) = \tilde{\mathsf{P}}\big(x_n(x, \tilde{U}_0, \ldots, \tilde{U}_{n-1}) \in B\big) > 0$.*

(A 2.5.2) $\sup_{x\in A} \mathsf{E}_{\tilde{\mathsf{P}}} \inf \{n > 0 \colon x_n(x, \tilde{U}_0, \ldots, \tilde{U}_{n-1}) \in A\} < \infty$, *where A satisfies one of the following conditions:*

a) $A = \{x\}$ *for some* $x \in \mathbb{X}$;

b) $\phi(A) > 0$, *A is bounded, and $g(x) = \int h(y)\, \mathsf{P}(x, dy)$ is continuous for every bounded continuous function h.*

Then there is exactly one stationary initial distribution with respect to the kernel $p(x, B)$.

In case of $\mathbb{X} = [0, \infty)$, *a sufficient condition for* (A 2.5.2) *is: There exist* $\varepsilon > 0$, $M < \infty$ *and a bounded set* $A \in \mathcal{H}$ *such that*

$$\mathsf{E}_{\tilde{\mathsf{P}}} f(x, \tilde{U}_0) \leqq x - \varepsilon \quad \textit{for } x \notin A,$$

$$\mathsf{E}_{\tilde{\mathsf{P}}} f(x, \tilde{U}_0) \leqq M \quad \textit{for } x \in A.$$

For a proof of Theorem 2.5.1 see Laslett, Pollard and Tweedie (1978), Tweedie (1975).

Now assume that we have already shown the existence of a stationary initial distribution p with respect to the transition probabilities (2.5.3). Then a stationary weak solution $\big(([\tilde{U}_n', \tilde{X}_n']), \tilde{\mathsf{P}}'\big)$ of equation (2.5.2) is defined by

$$\tilde{\mathsf{P}}'\big(([\tilde{U}_n', \tilde{X}_n'], n \geqq 0) \in (\cdot)\big)$$
$$= \int_{\mathbb{X}} \tilde{\mathsf{P}}\Big(([\tilde{U}_n, x_n(x, \tilde{U}_0, \ldots, \tilde{U}_{n-1})], n \geqq 0\Big) \in (\cdot)\Big)\, p(\mathrm{d}x).$$

Next we assume that there exists an SGS $\mathcal{S} = \{A(n, \varphi) : n \in \mathbb{Z}, \varphi \in \tilde{M}\}$ for (2.5.2). This assumption is fulfilled, e.g. if

$$\begin{aligned} &p(\{x\}) > 0 \text{ for some } x \in \mathbb{X} \\ \text{and} \qquad & \\ &\mathsf{E} \inf \{n > 0 : x_n(x, U_0, \ldots, U_{n-1}) = x\} < \infty. \end{aligned} \tag{2.5.4}$$

Namely, since $D = \big\{[u_n, x_n] : \# \{n : x_n = x\} = \infty\big\}$ is invariant and $\tilde{\mathsf{P}}'(D) > 0$, we obtain from Lemma 1.5.5 that $\tilde{\mathsf{P}}'(\cdot \mid \# \{n : \tilde{X}_n' = x\} = \infty)$ is the distribution of a stationary weak solution of (2.5.2) for $(\tilde{\Phi}, \tilde{\mathsf{P}})$ satisfying the conditions of Theorem 1.4.11. This ensures the existence of an SGS. Thus $\tilde{\mathsf{P}}'(\cdot \mid \# \{n : \tilde{X}_n' = x\} = \infty)$ is the distribution of a stationary weak solution of (2.5.2) for $(\tilde{\Phi}, \tilde{\mathsf{P}})$ satisfying the conditions of Theorem 1.4.11, which ensures the existence of an SGS.

Sometimes the constructive proof of Theorem 1.4.11 allows us to find $A(n, \varphi)$ explicitly. Let $A(n, \varphi) = \{x(i, n, \varphi),\ i = 1, \ldots, \# A(n, \varphi)\}$, $\varphi \in \tilde{M}$. Finally assume that $(\tilde{\Phi}, \tilde{\mathsf{P}})$ and $\tilde{M}$ can be chosen in such a way that

$$\mathsf{P}(\tilde{M}) = 1.$$

(The latter assumption means that the sample paths of Φ and $\tilde{\Phi}$ respectively have similar properties with respect to the problems considered.) Then $\mathcal{S}$ is an SGS for equation (2.5.1), too. Thus, in view of Theorem 1.4.3 there is a stationary weak solution of (2.5.1).

2.6. Remarks and references

The material in Sections 2.2–2.5 is essentially due to Lisek (1985a, b).

2.2. As far as we are aware, the fact of the existence of non-Markovian solutions of equation (2.1.1) for i.i.d. input seems not to be commonly known.

Theorem 2.2.5 is a special case of the following result due to Fassler (1982).

2.6.1. Theorem. *Let* $\mathbb{X}$ *be an arbitrary Polish state space,* (Φ, P), $\Phi = (U_n)$, *a sequence of i.i.d. r.v.'s, and* (Φ', P'), $\Phi' = ([U_n', X_n'])$, *a stationary weak* *solution of* (2.1.1). *If there exists a σ-finite measure μ such that*

$$\mathsf{P}'\big([X_0', (U_n', n \geqq 0)] \in (\cdot)\big) \ll (\mu \otimes \mathsf{P}^+)(\cdot), \tag{2.6.1}$$

then X_0' *and* $(U_n', n \geqq 0)$ *are independent.*

Taking $\mu(\cdot) = \sum_{x \in \mathbf{X}} \mathbf{1}_{\{x\}}(\cdot)$, one obtains Theorem 2.2.5 immediately from Theorem 2.6.1. Fassler's proof of Theorem 2.6.1 differs from our approach; it uses martingale arguments.

Using Theorem 2.6.1 we are able to give another proof of Theorem 2.2.3, avoiding the complicated details of the construction of ergodic solutions that were used in our proof given in Section 2.2. In fact, every stationary weak solution (Φ', P') concentrated on $\mathcal{A}(\mathcal{S})$ for some SGS $\mathcal{S}$ with the property (A 1.4.5) is absolutely continuous with respect to the probability measure Q given in Theorem 1.4.3. By the construction of Q (proof of Theorem 1.4.3) we have

$$\mathsf{Q}\big(\hat{X}_0 \in C_1, (\hat{U}_n, n \geqq 0) \in C_2\big) = \mathsf{Q}(\hat{X}_0 \in C_1) \cdot \mathsf{P}\big((U_n, n \geqq 0) \in C_2\big).$$

Hence (2.6.1) is satisfied with $\mu(\cdot) = \mathsf{Q}\big(\hat{X}_0 \in (\cdot)\big)$.

It should be mentioned that Fassler (1982) investigated a more general stochastic model which includes recursive stochastic equations of the type (1.1.1), see also Nawrotzki (1981a).

The interpretation of Theorem 1.6.3 as a generalization of Wald's identity was verified in a somewhat different way in Franken and Lisek (1982).

2.4. Theorem 2.4.5 was formulated in Rosenblatt (1960), cf. Rosenblatt (1962) and Rosenblatt (1971). For a generalization of the result see Hanson (1963). Brandt, Lisek and Nerman (1986) contains three different proofs of Theorem 2.4.5.

In Isaacson and Tweedie (1978) it is shown that the condition (A 2.4.2) is sufficient for the so-called strong ergodicity of an irreducible Markow chain.

The necessary and sufficient condition for the ergodicity of an irreducible and aperiodic Markov chain given in Corollary 2.4.7 seems to be useless in practical situations. A sufficient and "almost necessary" criterion is due to Tweedie (1975): Let (X_n) be an irreducible aperiodic Markov chain with transition probability matrix p_{jk} and $g(j) \geqq 0$, $j \in \mathbf{X} \subseteq \mathbf{Z}$, a function satisfying

$$\sum_{m \in \mathbf{X}} \big(g(m) - g(j)\big)\, p_{jm} < \infty, \qquad j \in \mathbf{X}.$$

(i) If there is an $l > 0$ and a $k \in \mathbf{X}$ such that

$$\liminf_{n \to \infty} \sum_{j \geqq 1} p_{kj}^{(n)} \sum_{m \in l} \big(g(m) - g(j)\big)\, p_{jm} < 0, \tag{2.6.2}$$

then (X_n) is ergodic.

(ii) If g is a strictly increasing function and there is a $b < \infty$ such that $\left|\sum_{m \in \mathbf{X}} \big(g(m) - g(j)\big)\, p_{jm}\right| \leqq b$, $j \in \mathbf{X}$, then (2.6.2) is necessary for the ergodicity.

From this criterion one obtains several well-known sufficient criteria for ergodicity in terms of the one-step probability matrix, cf. Moustafa (1957), Crabill (1968), and Lamperti (1963).

3. Stationary sequences and stationary marked point processes

3.1. Introduction

In previous chapters we have interpreted the equation

$$X_{n+1} = f(X_n, U_n), \qquad n \in \mathbb{Z}, \tag{3.1.1}$$

as a mathematical model of the temporal behaviour of a given stochastic system working in discrete time. For systems working on the real time axis, (3.1.1) may describe their behaviour at time epochs at which certain events occur, e.g. at arrival epochs of customers in a queueing system. This means that (3.1.1) only yields an event-oriented description. However, in general one is also interested in the continuous time behaviour of such systems.

In this chapter we provide some mathematical tools which are useful for the treatment of these questions.

In Section 3.2 we define a *marked point process* (*abbr. MPP*) with the mark space $\mathbb{K}$ as a sequence (Ψ, R), $\Psi = ([T_n, K_n])$ of $\mathbb{R} \times \mathbb{K}$-valued r.v.'s with the properties

$$T_n \leqq T_{n+1}, \qquad n \in \mathbb{Z}, \qquad \lim_{n \to \pm\infty} T_n = \pm\infty \quad \mathsf{R}\text{-a.s.} \tag{3.1.2}$$

(A sequence (T_n) of real valued r.v.'s satisfying (3.1.2) is called a *point process.*) This definition differs somewhat from the usual definition of a marked point process (point process) as a counting measure on $\mathbb{R} \times \mathbb{K}$ (on $\mathbb{R}$), cf. Sections 3.2 and 3.8.

The *intensity measure* ν_R of an MPP (Ψ, R) is defined by

$$\nu_\mathsf{R}(C) = \mathsf{E}_\mathsf{R} \,\#\, \{n : [T_n, K_n] \in C\}, \qquad C \in \mathcal{R} \otimes \mathcal{K},$$

cf. Definition 3.2.3. The well-known Campbell's theorem (Theorem 3.2.4) reads as follows: For all $h \in F_+(\mathbb{R} \times \mathbb{K})$

$$\mathsf{E}_\mathsf{R} \left(\sum_{n \in \mathbb{Z}} h(T_n, K_n) \right) = \int_{\mathbb{R} \times \mathbb{K}} h(a)\, \nu_\mathsf{R}(\mathrm{d}a).$$

For an MPP (Ψ, R), $\Psi = ([T_n, K_n])$, and any $t \in \mathbb{R}$ we define the shifted MPP

$$\theta_t \Psi = ([T_n(t), K_n(t)],\ n \in \mathbb{Z}),$$

where $T_n(t) = T_{n+N(t)} - t$, $K_n(t) = K_{n+N(t)}$, $N(t) = \max\{n : T_n \leqq t\}$. An MPP (Ψ, R) is called *stationary* if $\theta_t \Psi$ has the same probability distribution R

for all t, cf. Definition 3.2.5. We always use the notation $(\overline{\Psi}, \overline{\mathsf{P}})$ for a stationary MPP and its distribution and equip all symbols related with a stationary MPP $(\overline{\Psi}, \overline{\mathsf{P}})$ with the sign "–", e.g. $\overline{\Psi} = ([\overline{T}_n, \overline{K}_n])$. Stationarity implies $\cdots \leqq \overline{T}_0 < 0 < \overline{T}_1 \leqq \cdots$. The intensity measure of a stationary MPP $(\overline{\Psi}, \overline{\mathsf{P}})$ is of the form, cf. Theorem 3.2.7,

$$\nu_{\overline{\mathsf{P}}}(B \times L) = |B| \, \nu_{\overline{\mathsf{P}}}\big((0, 1] \times L\big), \quad B \in \mathcal{R}, \quad L \in \mathcal{K},$$

where $|B|$ is the Lebesgue measure of B.

In Section 3.3 we give a brief outline of some well-known results concerning renewal and Markov-renewal processes. Our goal is to introduce them as a special class of point processes and MPP's, respectively. This enables us to interpret the key result of Section 3.4 concerning the one-to-one correspondence between probability distributions of stationary sequences of $\mathbb{R}_+ \times \mathbb{K}$-valued r.v.'s and stationary MPP's with mark space $\mathbb{K}$ as a straightforward generalization of the classical theorems about the structure of stationary renewal and stationary Markov renewal processes.

For a stationary sequence (Φ, P), $\Phi = ([A_n, K_n])$, with the property $A_n \geqq 0$, set

$$a_{\mathsf{P}} = \mathsf{E}_{\mathsf{P}} A_0, \qquad \psi(\Phi) = ([T_n, K_n]),$$

where $T_0 = 0$, $T_n = \sum_{j=0}^{n-1} A_j$, $n \geqq 1$, $T_n = -\sum_{j=-1}^{n} A_j$, $n < 0$. (The MPP $\psi(\Phi)$ is called the *synchronous MPP* generated by the sequence Φ.) For a stationary MPP $(\overline{\Psi}, \overline{\mathsf{P}})$, $\overline{\Psi} = ([\overline{T}_n, \overline{K}_n])$, we set

$$\lambda_{\overline{\mathsf{P}}} = \nu_{\overline{\mathsf{P}}}\big((0, 1] \times \mathbb{K}\big), \qquad \varphi(\overline{\Psi}) = ([\overline{T}_{n+1} - \overline{T}_n, \overline{K}_n]).$$

($\lambda_{\overline{\mathsf{P}}}$ is called the *intensity* of $(\overline{\Psi}, \overline{\mathsf{P}})$.)

The main result of Section 3.4 is as follows (Theorem 3.4.1): There exists a one-to-one mapping between the families

$\mathbf{P} = \{\mathsf{P}$: P is the probability distribution of a stationary sequence $\Phi = ([A_n, K_n])$ of $\mathbb{R}_+ \times \mathbb{K}$-valued r.v.'s with the properties $a_{\mathsf{P}} < \infty$ and $\sum_{n=-\infty}^{-1} A_n = \sum_{n=0}^{\infty} A_n = \infty\}$,

$\overline{\mathbf{P}} = \{\overline{\mathsf{P}}$: $\overline{\mathsf{P}}$ is the probability distribution of a stationary MPP $\overline{\Psi} = ([\overline{T}_n, \overline{K}_n])$ with the intensity $\lambda_{\overline{\mathsf{P}}} < \infty\}$.

For a given (Φ, P), $\Phi = ([A_n, K_n])$, with $\mathsf{P} \in \mathbf{P}$ the corresponding stationary $(\overline{\Psi}, \overline{\mathsf{P}})$ is determined by

$$\overline{\mathsf{P}}\big(\overline{\Psi} \in (\cdot)\big) = a_{\mathsf{P}}^{-1} \int_0^\infty \mathsf{P}\big(A_0 > t, \theta_t \psi(\Phi) \in (\cdot)\big) \, dt.$$

Conversely, we have

$$\mathsf{P}\big(\Phi \in (\cdot)\big) = (\lambda_{\overline{\mathsf{P}}})^{-1} \sum_{j=1}^{\infty} \overline{\mathsf{P}}\big(\overline{T}_j \leqq 1, \theta^j \varphi(\overline{\Psi}) \in (\cdot)\big).$$

Furthermore,

$$\lambda_{\overline{\mathsf{P}}} = a_{\mathsf{P}}^{-1}.$$

In Section 3.5 some conclusions from this theorem are given. In particular, for a stationary MPP $(\overline{\Psi}, \overline{\mathsf{P}})$ without multiple points (i.e. $\overline{T}_n < \overline{T}_{n+1}$, $n \in \mathbb{Z}$) and with the property $\lambda_{\overline{\mathsf{P}}} < \infty$ we have

$$\overline{\mathsf{P}}(\overline{T}_2 \leqq t) = o(t) \quad \text{as} \quad t \to 0 \quad \text{(Dobrushin's theorem)}$$

and

$$\overline{\mathsf{P}}(\overline{T}_1 \leqq t < \overline{T}_2) = \lambda_{\overline{\mathsf{P}}} \cdot t + o(t) \quad \text{as} \quad t \to 0 \quad \text{(Korolyuk's theorem)}.$$

Section 3.6 deals with ergodic marked point processes. A stationary MPP $(\overline{\Psi}, \overline{\mathsf{P}})$ is *ergodic* if and only if

$$\frac{1}{t}\int_0^t h(\theta_u\overline{\Psi})\,du \xrightarrow[t\to\infty]{} \mathsf{E}_{\overline{\mathsf{P}}}h(\overline{\Psi}) \qquad \overline{\mathsf{P}}\text{-a.s.}$$

for all real-valued non-negative measurable functions h with $\mathsf{E}_{\overline{\mathsf{P}}}h(\overline{\Psi}) < \infty$ (Theorem 3.6.2).

Let $(\overline{\Psi}, \overline{\mathsf{P}})$ be a stationary, ergodic MPP with $\lambda_{\overline{\mathsf{P}}} < \infty$ and (Φ, P) the corresponding stationary sequence. Then

$$\frac{1}{t}\int_0^t \mathsf{P}\big(\theta_u\psi(\Phi) \in (\cdot)\big)\,du \xrightarrow[t\to\infty]{} \overline{\mathsf{P}}\big(\overline{\Psi} \in (\cdot)\big)$$

and

$$\frac{1}{n}\sum_{j=1}^{n} \overline{\mathsf{P}}\big(\theta^j\varphi(\overline{\Psi}) \in (\cdot)\big) \xrightarrow[n\to\infty]{} \mathsf{P}\big(\Phi \in (\cdot)\big),$$

cf. Theorem 3.6.6. This leads to the interpretation of $\overline{\mathsf{P}}$ as the distribution of the synchronous MPP $\psi(\Phi)$ seen by an observer located at a "randomly chosen" point of the time axis; P appears as the distribution of the sequence $\varphi(\overline{\Psi}) = ([\overline{T}_{n+1} - \overline{T}_n, \overline{K}_n])$ seen by an observer located at a "randomly chosen" point of $\overline{\Psi}$.

Finally, in Section 3.7 we show the following continuity property of the mapping $\overline{\mathsf{P}} \to \mathsf{P}$ (Theorem 3.7.1):

Let $(\overline{\Psi}, \overline{\mathsf{P}})$, $(\overline{\Psi}_1, \overline{\mathsf{P}}_1)$, $(\overline{\Psi}_2, \overline{\mathsf{P}}_2)$, ... be stationary MPP's with $\lambda_{\overline{\mathsf{P}}} < \infty$, $\lambda_{\overline{\mathsf{P}}_m} < \infty$, $m = 1, 2, \ldots$, and (Φ, P), (Φ_1, P_1), (Φ_2, P_2), ... the corresponding stationary sequences. Under the assumption $\lambda_{\overline{\mathsf{P}}_m} \xrightarrow[m\to\infty]{} \lambda_{\overline{\mathsf{P}}}$ we have

$$\overline{\Psi}_m \xrightarrow[m\to\infty]{\mathcal{D}} \overline{\Psi} \quad \text{iff} \quad \Phi_m \xrightarrow[m\to\infty]{\mathcal{D}} \Phi.$$

3.2. Marked point processes. Basic notations

Let $[\mathbb{K}, \mathcal{K}]$ be an arbitrary Polish space with the metric d_K, the so-called *mark space*. Denote by $M_K{}^*$ the set of all sequences $\psi = ([t_n, k_n])$, $t_n \leqq t_{n+1}$, $k_n \in \mathbb{K}$ or all n. Then, cf. e.g. BILLINGSLEY (1968, Appendix 1),

$$d(\psi, \psi') = \sum_{n=-\infty}^{\infty} \left(\frac{|t_n - t_n'|}{1 + |t_n - t_n'|} + \frac{d_K(k_n, k_n')}{1 + d_K(k_n, k_n')} \right) 2^{-|n|}, \tag{3.2.1}$$

$\psi = ([t_n, k_n])$, $\psi' = ([t_n', k_n'])$ defines a metric on $M_K{}^*$. Endowed with the metric d the space $M_K{}^*$ becomes Polish; by $\mathcal{M}_K{}^*$ we denote the σ-field of Borel subsets of $M_K{}^*$. Consider the subset

$$M_K = \left\{ \psi : \psi = ([t_n, k_n)] \in M_K{}^*, \lim_{n \to \pm\infty} t_n = \pm\infty \right\}$$

and the σ-field $\mathcal{M}_K = \mathcal{M}_X{}^* \cap M_K$.

3.2.1. Definition. A random element (Ψ, R) of the measurable space $[M_K{}^*, \mathcal{M}_K{}^*]$ with $\mathsf{R}(M_K) = 1$ is called a *random marked point process*.

In the following we drop the word "random" and use the abbreviation MPP for "marked point process".

It should be noted that Definition 3.2.1 and all results of this chapter, except those of Section 3.7, make use only of the fact that the mark space $[\mathbb{K}, \mathcal{K}]$ is a measurable one.

$\mathcal{M}_K{}^*$ is the smallest σ-field of subsets of $M_K{}^*$ such that all projections

$$\psi \to [t_n, k_n] = [t_n(\psi), k_n(\psi)], \qquad n \in \mathbb{Z},$$

are measurable, cf. BILLINGSLEY (1968). Thus we can define an MPP also as a sequence $\Psi = ([T_n, K_n], n \in \mathbb{Z})$ of $\mathbb{R} \times \mathbb{K}$-valued r.v.'s satisfying

$$T_n \leqq T_{n+1}, \qquad n \in \mathbb{Z}; \qquad \lim_{n \to \pm\infty}, T_n = \pm\infty \tag{3.2.2}$$

with probability one.

We call the T_n the *points*, the K_n the *marks* and the $[T_n, K_n]$ the *marked points of the MPP* Ψ. A sequence (T_n) of real valued r.v.'s satisfying (3.2.2) is called a *(random) point process*. A (marked) point process Ψ is called *simple* (or without multiple points) if $T_n < T_{n+1}$ for all n almost surely.

For example, the input of a queueing system can be described by an MPP $\Psi = ([T_n, K_n])$, where T_n is the n-th arrival epoch. The marks can be chosen to carry whatever information about the customers we need. For example, K_n could be a vector recording the priority class and service time of the n-th customer. As we do not assume that T_n is strictly less than T_{n+1} for every n, batch arrivals are included. However, since the marks are indexed, it follows that the customers in a batch are also numbered.

If one refrains from numbering the marks associated with the same point (e.g. from numbering customers within a batch), one obtains another definition of an MPP, namely as a random counting measure on $\mathbb{R} \times \mathbb{K}$, cf. e.g. [KMM], [FKAS]. Both definitions coincide if multiple points are excluded, cf. Section 3.8.

In the following we use the space S_K^*,

$$S_K^* = \{\varphi: \varphi = ([a_n, k_n]), a_n \in \mathbb{R}_+; k_n \in \mathbb{K}\}.$$

Endowed with the metric d given by (3.2.1), S_K^* is Polish. Let $\mathscr{S}_K^*$ be the σ-field of Borel subsets of S_K^*. Consider the subset

$$S_K = \left\{\varphi: \varphi = ([a_n, k_n]) \in S_K^*, \sum_{n=-\infty}^{-1} a_n = \sum_{n=0}^{\infty} a_n = \infty\right\}$$

and the σ-field $\mathscr{S}_K = \mathscr{S}_K^* \cap S_K$.

Obviously, the projections

$$\varphi \to [a_n, k_n] = [a_n(\varphi), k_n(\varphi)], \qquad n \in \mathbb{Z},$$

are measurable.

For $\psi = ([t_n, k_n]) \in M_K$ define

$$\varphi(\psi) = ([t_{n+1} - t_n, k_n]).$$

Conversely, for $\varphi = ([a_n, k_n]) \in S_K$ define

$$\psi(\varphi) = \big([t_n(\varphi), k_n(\varphi)]\big),$$

where

$$\begin{aligned} &k_n(\varphi) = k_n; \quad t_0(\varphi) = 0; \quad t_n(\varphi) = \sum_{j=0}^{n-1} a_j \text{ for } n \geqq 1; \\ &t_n(\varphi) = -\sum_{j=n}^{-1} a_j \text{ for } n < 0. \end{aligned} \tag{3.2.3}$$

An important class of MPP's are the so-called synchronous MPP's.

3.2.2. Definition. Let (Φ, P), $\Phi = ([A_n, K_n])$, be a stationary sequence of $\mathbb{R}_+ \times \mathbb{K}$-valued r.v.'s satisfying the condition

$$\sum_{n=-\infty}^{-1} A_n = \sum_{n=0}^{\infty} A_n = \infty \qquad \mathsf{P}\text{-a.s.}, \tag{3.2.4}$$

i.e. $\mathsf{P}(S_K) = 1$. The MPP $\psi(\Phi) = ([T_n, K_n]) = \big([t_n(\Phi), k_n(\Phi)]\big)$, where

$$\begin{aligned} &T_0 = t_0(\Phi) = 0; \qquad T_n = t_n(\Phi) = \sum_{j=0}^{n-1} A_j \quad \text{for } n \geqq 1; \\ &T_n = t_n(\Phi) = -\sum_{j=n}^{-1} A_j \quad \text{for } n < 0, \end{aligned} \tag{3.2.5}$$

is called a *synchronous MPP generated by* Φ. The corresponding point process (T_n) is called a *synchronous point process generated by* (A_n).

Notice that the non-negativity of A_n and the property (3.2.4) ensure the validity of (3.2.2). The MPP $\psi(\Phi)$ is simple iff $\mathsf{P}(A_0 > 0) = 1$.

In accordance with Definition 3.2.1 we call a sequence $([T_n, K_n], n \geqq 1)$ with the property $T_1 \geqq 0$, $\lim_{n\to\infty} T_n = \infty$, an *MPP with mark space* $\mathbb{K}$ *on* $\mathbb{R}_+$ ($(T_n, n \geqq 1)$ is called a point process on $\mathbb{R}_+$)

For a given stationary sequence $\Phi = ([A_n, K_n], n \geqq 0)$, $A_n \geqq 0$, the MPP $\psi(\Phi) = ([T_n, K_n], n \geqq 1)$,

$$T_n = \sum_{i=0}^{n-1} A_i,$$

is called a *synchronous MPP on* $\mathbb{R}_+$ generated by Φ ($(T_n, n \geqq 1)$ is called a synchronous point process on $\mathbb{R}_+$ generated by $(A_n, n \geqq 0)$).

A very important characteristic of an MPP is the so-called intensity measure.

3.2.3. Definition. The *intensity measure* ν_{R} of an MPP (Ψ, R), $\Psi = ([T_n, K_n])$, is defined by

$$\nu_{\mathsf{R}}(C) = \mathsf{E}_{\mathsf{R}} \# \{n: [T_n, K_n] \in C\}, \qquad C \in \mathcal{R} \otimes \mathcal{K}. \tag{3.2.6}$$

The intensity measure ν_{R} of a point process (Ψ, R), $\Psi = (T_n)$, is defined by

$$\nu_{\mathsf{R}}(B) = \mathsf{E}_{\mathsf{R}} \# \{n: T_n \in B\}, \qquad B \in \mathcal{R}.$$

In the following we always assume

$$\nu_{\mathsf{R}}(B \times \mathbb{K}) < \infty \quad \text{for all bounded } B \in \mathcal{R}.$$

For a given $h \in F_+(\mathbb{R} \times \mathbb{K})$ we can regard $h(t, k)$ as the effect (reward) associated with the marked point $[t, k]$. Thus

$$\sum_{n=-\infty}^{\infty} h(T_n, K_n)$$

may be interpreted as the total effect related with the given MPP (Ψ, R), $\Psi = ([T_n, K_n])$. The following theorem shows that the expected total effect can be calculated by means of the intensity measure ν_{R} of (Ψ, R); for some applications cf. Chapters 6 and 7.

3.2.4. Theorem. (Campbell theorem). *For an arbitrary MPP* (Ψ, R), $\Psi = ([T_n, K_n])$, *and an arbitrary function* $h \in F_+(\mathbb{R} \times \mathbb{K})$ *the mapping*

$$\psi = ([t_n, k_n]) \to \sum_{n=-\infty}^{\infty} h(t_n, k_n)$$

from M_K *into* $[0, \infty]$ *is measurable and it holds that*

$$\mathsf{E}_{\mathsf{R}} \left(\sum_{n=-\infty}^{\infty} h(T_n, K_n) \right) = \int_{\mathbb{R} \times \mathbb{K}} h(a)\, \nu_{\mathsf{R}}(\mathrm{d}a). \tag{3.2.7}$$

Proof. For $h = \mathbf{1}_C$, $C \in \mathbb{R} \times \mathbb{K}$, the equation (3.2.7) coincides with (3.2.6). The rest of the proof follows by standard techniques of integration theory (monotone approximation). ∎

For a sample path $\psi = ([t_n, k_n]) \in M_K$ and $t \in \mathbb{R}$ we denote

$$n(t, \psi) = \max\{n: t_n \leqq t\}.$$

The shift operator θ_t, $t \in \mathbb{R}$, on M_K is defined by

$$\theta_t \psi = \big([t_n(t), k_n(t)]\big),$$

where $\psi = ([t_n, k_n])$, $t_n(t) = t_{n+n(t,\psi)} - t$, $k_n(t) = k_{n+n(t,\psi)}$, $n \in \mathbb{Z}$. In this way the points of ψ are shifted in the usual way; the marks are shifted along with the corresponding points, and both are renumbered. Since an MPP Ψ takes values from M_K a.s., $\theta_t\Psi$ is well-defined. Notice, $t_0(\theta_t\Psi) \leqq 0 < t_1(\Psi)$ for any t. For a sample path $\psi = ([t_n, k_n], n \geqq 1)$, $t_1 \geqq 0$, of an MPP on $\mathbb{R}_+$ we define the shifted sample path $\theta_t\psi = \big([t_n(t), k_n(t)], n \geqq 1\big)$, $t \geqq 0$, in the same way as for a $\psi \in M_K$.

3.2.5. Definition. An MPP (Ψ, R), $\Psi = ([T_n, K_n], n \in \mathbb{Z})$, is called *stationary* if for every $t \in \mathbb{R}$ the shifted MPP $\theta_t\Psi$ has the same probability distribution R. The stationarity of a point process is defined similarly. The stationarity of an MPP or of a point process on $\mathbb{R}_+$ is defined in the same way, where only the shifts θ_t, $t \geqq 0$, are considered. We always use the notation $(\overline{\Psi}, \overline{\mathsf{P}})$ for a stationary MPP and its distribution and equip all symbols related with a stationary MPP $(\overline{\Psi}, \overline{\mathsf{P}})$ with the sign "$-$", e.g. $\overline{\Psi} = ([\overline{T}_n, \overline{K}_n])$.

It follows immediately from Definition 3.2.5 that

$$\overline{\mathsf{P}}(\overline{T}_n \neq 0, n \in \mathbb{Z}) = 1 \tag{3.2.8}$$

for a stationary MPP $\overline{\Psi} = ([\overline{T}_n, \overline{K}_n])$. Moreover, we have

$$\cdots \leqq \overline{T}_0 < 0 < \overline{T}_1 \leqq \cdots,$$

since $\overline{\mathsf{P}}\big(n(0, \overline{\Psi}) = 0\big) = \overline{\mathsf{P}}\big(n(0, \theta_0\overline{\Psi}) = 0\big)$.

Notice that in view of (3.2.5) and (3.2.8) a synchronous MPP on $\mathbb{R}$ is never stationary.

For an arbitrary MPP $\Psi = ([T_n, K_n])$ and $L \in \mathcal{K}$, $L \neq \emptyset$, we define the jump process $\big(N_L(t), t \geqq 0\big)$, $N_L(t) = \# \{n : 0 < T_n \leqq t, K_n \in L\}$. Obviously, for $0 < u < t$

$$N_L(t) - N_L(u) = \# \{n : u < T_n \leqq t, K_n \in L\}.$$

For a stationary MPP $\overline{\Psi} = ([\overline{T}_n, \overline{K}_n])$ the process $\big(\overline{N}_L(t), t \geqq 0\big)$,

$$\overline{N}_L(t) = \# \{n : 0 < \overline{T}_n \leqq t, \overline{K}_n \in L\},$$

possesses stationary non-negative integer-valued increments, $L \in \mathcal{K}$. Thus, cf. Theorem A 1.1.2,

$$\overline{\mathsf{P}}\left(\overline{N}_L(t) \xrightarrow[t\to\infty]{} \infty\right) + \overline{\mathsf{P}}\big(\overline{N}_L(t) \equiv 0\big) = 1, \qquad L \in \mathcal{K}. \tag{3.2.9}$$

In the special case $L = \mathbb{K}$ the processes $\big(N(t), t \geqq 0\big)$, $N(t) = N_{\mathbb{K}}(t)$, and $\big(\overline{N}(t), t \geqq 0\big)$, $\overline{N}(t) = \overline{N}_{\mathbb{K}}(t)$, are usually called counting processes associated with $\Psi = ([T_n, K_n])$ and $\overline{\Psi} = ([\overline{T}_n, \overline{K}_n])$, respectively.

3.2.6. Definition. For an arbitrary stationary MPP $(\overline{\Psi}, \overline{\mathsf{P}})$ we define the measure $\lambda_{\overline{\mathsf{P}}}(\cdot)$ on $\mathcal{K}$ by

$$\lambda_{\overline{\mathsf{P}}}(L) = \nu_{\overline{\mathsf{P}}}\big((0, 1] \times L\big), \qquad L \in \mathcal{K}.$$

The number $\lambda_{\overline{\mathsf{P}}} = \lambda_{\overline{\mathsf{P}}}(\mathbb{K})$ is called the *intensity* of the MPP $(\overline{\Psi}, \overline{\mathsf{P}})$.

3.2.7. Theorem. *For an arbitrary stationary MPP* $(\overline{\Psi}, \overline{\mathsf{P}})$ *with finite intensity* $\lambda_{\overline{\mathsf{P}}}$ *the equation*

$$\nu_{\overline{\mathsf{P}}}(B \times L) = |B|\, \lambda_{\overline{\mathsf{P}}}(L), \qquad B \in \mathcal{R}, \qquad L \in \mathcal{K},$$

holds, where $|B|$ *is the Lebesgue measure of* B. (*We put* $\infty \cdot 0 = 0$.)

Proof. For every $L \in \mathcal{K}$ it is obvious that $\nu_{\overline{\mathsf{P}}}\big((\cdot) \times L\big)$ is a θ_t-invariant measure on $\mathcal{R}$ with $\nu_{\overline{\mathsf{P}}}\big((0, 1] \times L\big) < \infty$. Therefore, $\nu_{\overline{\mathsf{P}}}\big((\cdot) \times L\big)$ is a multiple of Lebesgue measure. The corresponding factor is equal to $\lambda_{\overline{\mathsf{P}}}(L)$ because of $\nu_{\overline{\mathsf{P}}}\big((0, 1] \times L\big) = \lambda_{\overline{\mathsf{P}}}(L)$. Thus we have

$$\nu_{\overline{\mathsf{P}}}(B \times L) = |B|\, \lambda_{\overline{\mathsf{P}}}(L), \qquad B \in \mathcal{R}, \qquad L \in \mathcal{K}. \quad \blacksquare$$

Obviously, for a set $L \in \mathcal{K}$ with the property $\overline{\mathsf{P}}\left(\overline{N}_L(t) \xrightarrow[t\to\infty]{} \infty\right) = 1$ the value $\lambda_{\overline{\mathsf{P}}}(L)$ is the intensity of the stationary MPP with mark space L which arises from $\overline{\Psi}$ by deleting the points with marks belonging to $\mathbb{K} \setminus L$.

For a stationary MPP $(\overline{\Psi}, \overline{\mathsf{P}})$ with $\lambda_{\overline{\mathsf{P}}} < \infty$ the ratio $\lambda_{\overline{\mathsf{P}}}(\cdot)/\lambda_{\overline{\mathsf{P}}}$ defines a probability measure on $\mathcal{K}$. If in addition

$$\frac{\overline{N}_L(t)}{t} \xrightarrow[t\to\infty]{} \lambda_{\overline{\mathsf{P}}}(L) \qquad \overline{\mathsf{P}}\text{-a.s.}$$

for every $L \in \mathcal{K}$, cf. Corollary 3.6.3, the ratio

$$\frac{\lambda_{\overline{\mathsf{P}}}(L)}{\lambda_{\overline{\mathsf{P}}}} = \lim_{t\to\infty} \frac{\overline{N}_L(t)}{\overline{N}(t)}$$

can be interpreted as the probability that the mark of a "randomly chosen" marked point of $\overline{\Psi}$ belongs to L, cf. also formula (3.5.6).

3.2.8. Remark. Most of the definitions and results in this chapter are formulated for (marked) point processes on the whole line $\mathbb{R}$. We dispense with explicit formulations for (marked) point processes on $\mathbb{R}_+$.

3.3. Renewal and Markov renewal processes

In this section we give a short survey of standard renewal and Markow renewal theory. We consider renewal and Markov renewal processes to be point processes and marked point processes on $\mathbb{R}_+$, respectively. Some well-known results for these processes will be formulated in quite an unusual way. However, this enables us to recognize general statements in the next sections as straightforward generalizations of just these well-known results.

Let $(A_n, n \geqq 0)$ be a sequence of independent non-negative r.v.'s with distribution functions

$$\mathsf{F}_0(x) = \mathsf{P}(A_0 \leqq x), \qquad \mathsf{F}(x) = \mathsf{P}(A_n \leqq x), \qquad n \geqq 1,$$

where $\mathsf{F}(0) < 1$. The point process $\Psi = (T_n, n \geqq 1)$ with

$$T_n = \sum_{j=0}^{n-1} A_j$$

is called *renewal process on* $\mathbb{R}_+$ with delay distribution function $\mathsf{F}_0(x)$ and underlying (lifetime) distribution function $\mathsf{F}(x)$. The points T_n are called renewal epochs. Obviously, Ψ has no multiple points iff $\mathsf{F}(0) = 0$. If $\mathsf{F}_0(x) = \mathsf{F}(x)$, then the renewal process Ψ is called ordinary (or synchronous, or zero-delayed).

For the associated counting process $N(t) = \max\{n : T_n \leqq t\}$, $t \geqq 0$, we have

$$\mathsf{P}\big(N(t) \geqq k\big) = \mathsf{P}(T_k \leqq t) = \mathsf{F}_0 * \mathsf{F}^{*(k-1)}(t), \tag{3.3.1}$$

where $*$ denotes convolution, $\mathsf{F}^{*0}(x) \equiv 1$, $x \geqq 0$, and $\mathsf{F}^{*j}(x) = \mathsf{F} * \mathsf{F}^{*(j-1)}(x)$, $j \geqq 1$.

The function

$$H(t) = \mathsf{E}N(t) = \sum_{k=1}^{\infty} \mathsf{F}_0 * \mathsf{F}^{*(k-1)}(t) \tag{3.3.2}$$

is called the *renewal function* of Ψ. (Notice, the renewal function determines the intensity measure of Ψ.) $H(t)$ is the unique finite solution of the renewal equation

$$H(t) = \mathsf{F}_0(t) + \int_0^t \mathsf{F}(t-u)\, \mathrm{d}H(u) \tag{3.3.3}$$

which in terms of Laplace-Stieltjes-transforms becomes

$$\mathcal{H}(s) = \frac{\mathcal{F}_0(s)}{1 - \mathcal{F}(s)} \tag{3.3.4}$$

(where $\mathcal{F}(s) = \int_0^\infty \mathrm{e}^{-sx}\, \mathrm{d}\mathsf{F}(x)$, etc.).

For an arbitrary $t \geqq 0$ we consider the so-called forward recurrence time of $\Psi = (T_n)$ at the instant t

$$V_t = T_{N(t)+1} - t.$$

(V_t is the time distance between t and the next renewal epoch after t.) For the distribution function $\mathsf{V}_t(x)$ of V_t we have

$$\mathsf{V}_t(x) = \mathsf{F}_0(t+x) - \mathsf{F}_0(t) + \int_0^t \big(\mathsf{F}(t+x-u) - \mathsf{F}(t-u)\big)\, \mathrm{d}H(u), \quad t \geqq 0. \tag{3.3.5}$$

Applying Theorem A 3.4.1 to the stopping time $N(t) + 1$ we conclude that for every $t > 0$ the shifted renewal process $\theta_t\Psi$ is also a renewal process (with the same underlying distribution function $\mathsf{F}(x)$ and delay distribution function $\mathsf{V}_t(x)$ given by (3.3.5)). In particular,

$$\mathsf{P}\big(N(t+u) - N(t) \geqq k\big) = \mathsf{V}_t * \mathsf{F}^{*(k-1)}(u). \tag{3.3.6}$$

Moreover, a renewal process is stationary if and only if the distribution function $\mathsf{V}_t(x)$ does not depend on $t \geqq 0$.

Now we consider renewal processes under the additional assumption

$$a = \mathsf{E}A_1 = \int_0^\infty (1 - \mathsf{F}(u))\, du < \infty. \tag{3.3.7}$$

Then we obtain the following necessary and sufficient criterion for the stationarity of a renewal process.

3.3.1. Theorem. *A renewal process with the underlying distribution function* $\mathsf{F}(x)$ *satisfying* (3.3.7) *and the delay distribution* $\mathsf{F}_0(x)$ *is stationary iff*

(a) $\mathsf{F}_0(x) = \dfrac{1}{a} \displaystyle\int_0^x (1 - \mathsf{F}(u))\, du$

or equivalently

(b) $H(x) = \dfrac{1}{a}\, x.$

Proof. The equivalence of (a) and (b) follows immediately from (3.3.4), because the Laplace-Stieltjes transforms of $\frac{1}{a}\, x$ and

$$\frac{1}{a} \int_0^x (1 - \mathsf{F}(u))\, du$$

are equal to $\frac{1}{as}$ and $\frac{1}{as}\big(1 - \mathcal{F}(s)\big)$, respectively.

Assume the considered renewal process to be stationary. Then, in view of Theorem 3.2.7 (cf. Remark 3.2.8), its renewal function has the form

$$H(x) = \lambda x, \qquad 0 < \lambda < \infty.$$

From this and (3.3.3),

$$\mathsf{F}_0(x) = \lambda \int_0^x (1 - \mathsf{F}(u))\, du. \tag{3.3.8}$$

Taking $x \to \infty$ in (3.3.8) we obtain $\lambda = 1/a$ and thus

$$H(x) = \frac{1}{a}\, x.$$

Conversely from $H(x) = \frac{1}{a}\, x$ and

$$\mathsf{F}_0(x) = \frac{1}{a} \int_0^x (1 - \mathsf{F}(u))\, du$$

by using (3.3.5), we obtain

$$\mathsf{V}_t(x) = \frac{1}{a} \int_0^x \big(1 - \mathsf{F}(u)\big)\, du, \qquad t \geqq 0,$$

and thus the stationarity of the renewal process. ∎

Theorem 3.3.1 can be reformulated in the following way.

3.3.2. Theorem. *There exists a one-to-one mapping between the families*

$$\{\mathsf{F}: \mathsf{F} \textit{ is a distribution function with } \mathsf{F}(0-0) = 0 \textit{ and } a = \int_0^\infty \big(1 - \mathsf{F}(u)\big)\, du < \infty\}$$

and

$$\{\overline{\mathsf{P}}: \overline{\mathsf{P}} \textit{ is the probability distribution of a stationary renewal process on } \mathbb{R}_+ \textit{ with finite intensity } \lambda_{\overline{\mathsf{P}}}\}.$$

For a given F the corresponding $\overline{\mathsf{P}}$ is the distribution of the renewal process with underlying distribution function F and delay distribution

$$\frac{1}{a} \int_0^x \big(1 - \mathsf{F}(u)\big)\, du.$$

The intensity $\lambda_{\overline{\mathsf{P}}}$ is equal to $1/a$. Conversely, for a given $\overline{\mathsf{P}}$ the corresponding F is the underlying distribution function of $\overline{\mathsf{P}}$.

3.3.3. Remark. Theorem 3.3.2 states a one-to-one mapping between the family of probability distributions of sequences $(A_n, n \geqq 0)$ of i.i.d. r.v.'s with the properties $A_0 \geqq 0$, $0 < a = \mathsf{E}A_0 < \infty$ and the family of probability distributions of stationary renewal processes with finite intensity.

3.3.4. Example. A renewal process with $\mathsf{F}_0(x) = \mathsf{F}(x) = 1 - e^{-\lambda x}$ is called a *Poisson process*. From (3.3.1) we obtain for Poisson processes

$$\mathsf{P}\big(N(t) = k\big) = \frac{(\lambda t)^k}{k!}\, e^{-\lambda t}.$$

Since condition (a) in Theorem 3.3.1 is satisfied, the Poisson process is a stationary point process. Moreover, the counting process $N(t)$ has independent increments. Note that

$$\mathsf{F}(x) = \frac{1}{a} \int_0^x \big(1 - \mathsf{F}(u)\big)\, du$$

iff $\mathsf{F}(x) = 1 - e^{-\lambda x}$ for some $\lambda > 0$. Thus, $1 - e^{-\lambda x}$, $\lambda > 0$, are the only distribution functions for which the corresponding ordinary renewal process is stationary.

Now we introduce a particular class of MPP's, known by the name of Markov renewal processes. For simplicity, we assume here that the mark space $\mathbb{K}$ is at most countable, say $\mathbb{K} = \{1, \ldots, N\}$ or $\{1, 2, \ldots\}$; the general case will be considered in Section 4.3. Let $(p_{ij},\ i, j \in \mathbb{K})$ be a stochastic matrix, $(q_i, i \in \mathbb{K})$ a probability distribution on $\mathbb{K}$, $\{\mathsf{F}^0_{ij}(x),\ i, j \in \mathbb{K}\}$ and $\{\mathsf{F}_{ij}(x),\ i, j \in \mathbb{K}\}$ families of distribution functions on $\mathbb{R}_+$.

Consider a sequence $([A_n, K_n], n \geqq 0)$ of $\mathbb{R}_+ \times \mathbb{K}$-valued r.v.'s with the following properties:

(A 3.3.1) $\mathsf{P}(A_n \leqq x, K_{n+1} = j \mid K_0 = i_0, \ldots, K_{n-1} = i_{n-1}, K_n = i, A_0 = x_0, \ldots, A_{n-1} = x_{n-1}) = \mathsf{P}(A_n \leqq x, K_{n+1} = j \mid K_n = i)$, for all $i, j, i_0, \ldots, i_{n-1} \in \mathbb{K}$, $x, x_0, \ldots, x_{n-1} \in \mathbb{R}_+$, $n \geqq 1$.

(A 3.3.2) $\mathsf{P}(K_0 = i) = q_i$, $\mathsf{P}(K_{n+1} = j \mid K_n = i) = p_{ij}$, $n \geqq 0$;

$\mathsf{P}(A_0 \leqq x \mid K_0 = i, K_1 = j) = \mathsf{F}^0_{ij}(x)$,

$\mathsf{P}(A_n \leqq x \mid K_n = i, K_{n+1} = j) = \mathsf{F}_{ij}(x)$, $i, j \in \mathbb{K}$, $x \in \mathbb{R}_+$, $n \geqq 1$.

The MPP

$$([T_n, K_n], n \geqq 1) \quad \text{with} \quad T_n = \sum_{j=0}^{n-1} A_j, \qquad n \geqq 1, \tag{3.3.9}$$

is called a *Markov renewal process* determined by the parameters q_i, p_{ij}, F^0_{ij} and F_{ij}, $i, j \in \mathbb{K}$. The properties (A 3.3.1) and (A 3.3.2) imply in particular that $(K_n, n \geqq 0)$ is a homogeneous Markov chain with the initial distribution $(q_i, i \in \mathbb{K})$ and the transition matrix (p_{ij}). Property (A 3.3.1) implies that the r.v.'s $A_0, A_1, \ldots$ are conditionally independent given the Markov chain K_0, $K_1, \ldots$ with the distribution of A_n depending only on K_n and K_{n+1}. In particular, if the state space $\mathbb{K}$ consists of a single point, by deleting the marks we obtain a renewal process $(T_n, n \geqq 1)$. This and the Markov property of $(K_n, n \geqq 0)$ justify the term "Markov renewal process" somewhat. Note that (A 3.3.1) and (A 3.3.2) imply that $([A_n, K_{n+1}], n \geqq 0)$ and also $([A_n, K_n], n \geqq 0)$ have the Markov property.

Assume there is a unique stationary initial distribution $(p_i, i \in \mathbb{K})$ with respect to the stochastic matrix (p_{ij}). Then the sequence $([A_n, K_n], n \geqq 0)$ is stationary if and only if $q_i = p_i$ and $\mathsf{F}^0_{ij}(x) \equiv \mathsf{F}_{ij}(x)$, $i, j \in \mathbb{K}$. The one-dimensional distribution of this stationary sequence is given by

$$\mathsf{P}(A_0 \leqq x, K_0 = i) = \sum_{j \in \mathbb{K}} p_i p_{ij} \mathsf{F}_{ij}(x).$$

In fact, the stationarity of $([A_n, K_n], n \geqq 0)$ implies the stationarity of $(K_n, n \geqq 0)$ and thus $p_i = q_i$, $i \in \mathbb{K}$. Moreover $[A_0, K_0, K_1] \overset{\mathfrak{D}}{=} [A_1, K_1, K_2]$. Thus, $p_i p_{ij} \mathsf{F}^0_{ij}(x) = p_i p_{ij} \mathsf{F}_{ij}(x)$ for all $i, j \in \mathbb{K}$. On the other hand, if $q_i = p_i$ and $\mathsf{F}^0_{ij}(x) \equiv \mathsf{F}_{ij}(x)$, $i, j \in \mathbb{K}$, then

$$\mathsf{P}(K_n = i_0, \ldots, K_{n+r} = i_r, A_n \leqq x_0, \ldots, A_{n+r-1} \leqq x_{r-1})$$

$$= p_{i_0} \prod_{j=0}^{r-1} p_{i_j i_{j+1}} \mathsf{F}_{i_j i_{j+1}}(x_j) \quad \text{for all } n \geqq 0.$$

Thus $([A_n, K_n], n \geqq 0)$ is stationary.

Using the above mentioned stationary sequence $([A_n, K_n], n \geqq 0)$ in (3.3.9) we obtain the so-called synchronous Markov renewal process $([T_n, K_n], n \geqq 1)$ determined by p_{ij} and $\mathsf{F}_{ij}(x)$ (in other words: the Markov renewal process determined by q_i, p_{ij}, F^0_{ij}, F_{ij} is a synchronous MPP if and only if $q_i = p_i$, and $\mathsf{F}^0_{ij}(x) \equiv \mathsf{F}_{ij}(x)$, $i, j \in \mathbb{K}$).

Obviously, for fixed $j \in \mathbb{K}$ the points T_n with $K_n = j$ constitute a renewal process (the point process of j-renewals). The corresponding renewal functions $H_j(x)$ satisfy the following Markov renewal equation (cf. Theorem A 3.3.1):

$$H_j(x) = \sum_i q_i p_{ij} \mathsf{F}^0_{ij}(x) + \sum_l p_{lj} \int_0^x \mathsf{F}_{lj}(x-u)\, \mathrm{d}H_l(u), \qquad j \in \mathbb{K}. \tag{3.3.10}$$

Using (3.3.10) we obtain the following generalization of Theorem 3.3.1.

3.3.5. Theorem. *Assume that* (p_{ij}) *and* $\{\mathsf{F}_{ij}(x)\}$ *have the properties:*

(A 3.3.3) *There is a unique stationary initial distribution* (p_i) *with respect to* (p_{ij}).

(A 3.3.4) $a_{ij} = \int_0^\infty \big(1 - F_{ij}(u)\big)\, \mathrm{d}u < \infty, \qquad i, j \in \mathbb{K},$

$$a = \sum_{i,j} p_i p_{ij} a_{ij} < \infty.$$

Then the Markov renewal process determined by (q_i), (p_{ij}), $\{\mathsf{F}^0_{ij}(x)\}$, $\{\mathsf{F}_{ij}(x)\}$ *is stationary if and only if*

(a) $$q_i = \frac{p_i a_i}{a} \quad \textit{with} \quad a_i = \sum_j p_{ij} a_{ij};$$

$$\mathsf{F}^0_{ij}(x) = \frac{1}{a_{ij}} \int_0^x \big(1 - \mathsf{F}_{ij}(u)\big)\, \mathrm{d}u, \qquad i, j \in \mathbb{K},$$

or, equivalently, if and only if

(b) $$H_i(x) = \frac{p_i x}{a} \quad \textit{for all} \quad i \in \mathbb{K}.$$

3.3.6. Remark. We can reformulate Theorem 3.3.5 in a manner more appropriate for further generalizations: There is a one-to-one mapping between the families

$\{\mathsf{P}$: P is the probability distribution of a stationary sequence $([A_n, K_n], n \geqq 0)$ satisfying (A 3.3.1)–(A 3.3.4) with $\mathsf{F}_{ij}(x) = \mathsf{F}^0_{ij}(x)$, $q_i = p_i\}$

and

$\{\bar{\mathsf{P}}$: $\bar{\mathsf{P}}$ is the probability distribution of a stationary Markov renewal process satisfying (A 3.3.3) and (A 3.3.4), and (q_i) and $\{\mathsf{F}^0_{ij}\}$ are given in (a) of Theorem 3.3.5$\}$.

This mapping is given by (a) of Theorem 3.3.5.

3.4. Relationships between stationary sequences and stationary MPP's. I: The basic theorem

In this section we show that there is a close relation between stationary sequences (Φ, P), $\Phi = ([A_n, K_n])$, $A_n \geqq 0$, and stationary MPP's $(\overline{\Psi}, \overline{\mathsf{P}})$, $\overline{\Psi} = ([\overline{T}_n, \overline{K}_n])$. This relation is a straightforward generalization of Theorem 3.3.2 and 3.3.5 for renewal and Markov renewal processes, respectively (cf. Remarks 3.3.3 and 3.3.6). In this section we give only the formal result. Some useful interpretations will be given later.

3.4.1. Theorem. *There exists a one-to-one mapping between the families* $\mathbf{P} = \{\mathsf{P}$: P *is the probability distribution of a stationary sequence* $\Phi = ([A_n, K_n])$ *of* $\mathbb{R}_+ \times \mathbb{K}$*-valued r.v.'s with the properties* $\sum_{n=-\infty}^{-1} A_n = \sum_{n=0}^{\infty} A_n = \infty$ P*-a.s. and* $\mathsf{E}_\mathsf{P} A_0 < \infty\}$

and

$\overline{\mathbf{P}} = \{\overline{\mathsf{P}}$: $\overline{\mathsf{P}}$ *is the probability distribution of a stationary MPP* $\overline{\Psi} = ([\overline{T}_n, \overline{K}_n])$ *with mark space* $\mathbb{K}$ *and intensity* $\lambda_{\overline{\mathsf{P}}} < \infty\}$.

For given (Φ, P) *with* $\mathsf{P} \in \mathbf{P}$ *the corresponding stationary MPP* $(\overline{\Psi}, \overline{\mathsf{P}})$ *is determined by*

$$\overline{\mathsf{P}}(\overline{\Psi} \in (\cdot)) = a_\mathsf{P}^{-1} \int_0^\infty \mathsf{P}(A_0 > t, \theta_t \psi(\Phi) \in (\cdot))\, dt, \tag{3.4.1}$$

where $a_\mathsf{P} = \mathsf{E}_\mathsf{P} A_0$, *and* $\psi(\Phi)$ *is the synchronous MPP generated by* Φ, *cf. Definition* 3.2.2.

Conversely, we have

$$\mathsf{P}(\Phi \in (\cdot)) = \lambda_{\overline{\mathsf{P}}}^{-1} \sum_{j=1}^\infty \overline{\mathsf{P}}(\overline{T}_j \leqq 1, \theta^j \varphi(\overline{\Psi}) \in (\cdot)), \tag{3.4.2}$$

where $\varphi(\overline{\Psi}) = ([\overline{T}_{n+1} - \overline{T}_n, \overline{K}_n])$.

Moreover,

$$\lambda_{\overline{\mathsf{P}}} = a_\mathsf{P}^{-1}.$$

3.4.2. Remark. The following useful formulae are equivalent to (3.4.1):

$$\overline{\mathsf{P}}(\overline{\Psi} \in (\cdot)) = a_\mathsf{P}^{-1} \mathsf{E}_\mathsf{P} \int_0^{A_0} \mathbf{1}_{(\cdot)}(\theta_u \psi(\Phi))\, du, \tag{3.4.3}$$

$$\overline{\mathsf{P}}(\overline{\Psi} \in (\cdot)) = a_\mathsf{P}^{-1} \int_0^\infty \mathsf{P}(A_{-1} > u, \theta_{-u} \psi(\Phi) \in (\cdot))\, du, \tag{3.4.4}$$

$$\overline{\mathsf{P}}(\overline{\Psi} \in (\cdot)) = a_\mathsf{P}^{-1} \mathsf{E}_\mathsf{P} \int_0^{A_{-1}} \mathbf{1}_{(\cdot)}(\theta_{-u} \psi(\Phi))\, du. \tag{3.4.5}$$

For every $n \in \mathbb{Z}$,

$$\overline{\mathsf{P}}(\overline{\Psi} \in (\cdot)) = a_\mathsf{P}^{-1} \mathsf{E}_\mathsf{P} \int_0^{A_n} \mathbf{1}_{(\cdot)}(\theta_{T_n + u} \psi(\Phi))\, du, \tag{3.4.6}$$

where $T_0 = 0$, $T_n = A_0 + \cdots + A_{n-1}$ if $n \geqq 1$ and $T_n = -(A_n + \cdots + A_{-1})$ if $n < 0$.

For every $h \in F_+(M_K)$,

$$\mathsf{E}_{\bar{\mathsf{P}}} h(\overline{\Psi}) = a_{\mathsf{P}}{}^{-1} \mathsf{E}_{\mathsf{P}} \int_0^{A_0} h\big(\theta_u \psi(\Phi)\big)\, du, \tag{3.4.7}$$

$$\mathsf{E}_{\bar{\mathsf{P}}} h(\overline{\Psi}) = a_{\mathsf{P}}{}^{-1} \mathsf{E}_{\mathsf{P}} \int_0^{A_{-1}} h\big(\theta_{-u} \psi(\Phi)\big)\, du. \tag{3.4.8}$$

The equivalence of (3.4.3) and (3.4.6) follows from the stationarity of (Φ, P). The equivalence of (3.4.3) and (3.4.5) follows from

$$\int_0^{A_{-1}} \mathbf{1}_{(\cdot)}\big(\theta_u \psi(\theta^{-1}\Phi)\big)\, du = \int_0^{A_{-1}} \mathbf{1}_{(\cdot)}\big(\theta_{u-A_{-1}} \psi(\Phi)\big)\, du$$

$$= \int_0^{A_{-1}} \mathbf{1}_{(\cdot)}\big(\theta_{-u} \psi(\Phi)\big)\, du \qquad \mathsf{P}\text{-a.s.}$$

and the stationarity of (Φ, P).

By means of Fubini's theorem we immediately obtain the equivalence of (3.4.1), (3.4.3) and (3.4.4), (3.4.5), respectively. The equivalence of (3.4.3), (3.4.7) and (3.4.5), (3.4.8), respectively, follows by standard techniques of integration theory.

We shall show the truth of Theorem 3.4.1 by proving four lemmas. First we introduce some notation.

For a sample path $\psi = ([t_n, k_n]) \in M_k$ let

$$\varphi(\psi) = ([a_n, k_n]), \qquad a_n = t_{n+1} - t_n. \tag{3.4.9}$$

Consider the sequence $\theta^i \varphi(\psi) = ([a_{n+i}, k_{n+i}], n \in \mathbb{Z})$ as a new mark associated with the point t_i. ($\theta^i \varphi(\psi)$ can be interpreted as "past, present and future" of ψ observed by the i-th point of ψ.) In this way we obtain a measurable one-to-one mapping

$$\psi \to \psi^* = \big([t_i, \theta^i \varphi(\psi)]\big)$$

from M_K into M_{S_K}.

Let $(\overline{\Psi}, \bar{\mathsf{P}})$, $\overline{\Psi} = ([\overline{T}_n, \overline{K}_n])$, be a stationary MPP with finite intensity $\lambda_{\bar{\mathsf{P}}}$. The mapping $\psi \to \psi^*$ forms $(\overline{\Psi}, \bar{\mathsf{P}})$ into an MPP (Ψ^*, P^*), $\Psi^* = \big([\overline{T}_i, \theta^i \varphi(\overline{\Psi})]\big)$ with the mark space S_K. It is easy to see that (Ψ^*, P^*) is stationary too. For the intensities we have the relationship

$$\lambda_{\mathsf{P}^*} = \lambda_{\bar{\mathsf{P}}}. \tag{3.4.10}$$

Consider the ratio

$$\frac{\nu_{\mathsf{P}^*}\big((0, t] \times (\cdot)\big)}{\lambda_{\mathsf{P}^*} t} = \frac{1}{\lambda_{\bar{\mathsf{P}}} t} \int_{M_K} \sum_{j=1}^{\infty} \mathbf{1}\{0 < t_j(\psi) \leqq t, \theta^j \varphi(\psi) \in (\cdot)\}\, \bar{\mathsf{P}}(d\psi), \; t > 0. \tag{3.4.11}$$

Changing integral and sum, we obtain from (3.4.11)

$$\frac{\nu_{\mathsf{P}*}\big((0,t]\times(\cdot)\big)}{\lambda_{\mathsf{P}*}t} = \frac{1}{\lambda_{\overline{\mathsf{P}}}t}\sum_{j=1}^{\infty}\overline{\mathsf{P}}\big(\overline{T}_j \leqq t, \theta^j\varphi(\overline{\Psi}) \in (\cdot)\big). \tag{3.4.12}$$

Using Theorem 3.2.7 and (3.4.12) we get

$$\frac{\nu_{\mathsf{P}*}\big((0,t]\times(\cdot)\big)}{\lambda_{\mathsf{P}*}t} = \frac{\nu_{\mathsf{P}*}\big((0,1]\times(\cdot)\big)}{\lambda_{\overline{\mathsf{P}}}}$$

$$= \frac{1}{\lambda_{\overline{\mathsf{P}}}}\sum_{j=1}^{\infty}\overline{\mathsf{P}}\big(\overline{T}_j \leqq 1, \theta^j\varphi(\overline{\Psi}) \in (\cdot)\big).$$

Thus, for the probability distribution P given by (3.4.2), we have

$$\mathsf{P}(\cdot) = \frac{\nu_{\mathsf{P}*}\big((0,t]\times(\cdot)\big)}{\lambda_{\mathsf{P}*}t} \tag{3.4.13}$$

for each $t > 0$.

Formulae (3.4.11) and (3.4.13) yield the following useful expression for the distribution P given by (3.4.2)

$$\mathsf{P}(\cdot) = \frac{1}{\lambda_{\overline{\mathsf{P}}}t}\,\mathsf{E}_{\overline{\mathsf{P}}}\sum_{j=1}^{\overline{N}(t)}\mathbf{1}_{(\cdot)}\big(\theta^j\varphi(\overline{\Psi})\big), \quad t > 0, \quad \overline{N}(t) = \max\{n : \overline{T}_n \leqq t\}. \tag{3.4.14}$$

By standard techniques of integration theory formula (3.4.14) leads to

$$\lambda_{\overline{\mathsf{P}}}t\int_{S_K} h(\varphi)\,\mathsf{P}(\mathrm{d}\varphi) = \int_{M_K}\left(\sum_{j=1}^{n(t)} h\big(\theta^j\varphi(\psi)\big)\right)\overline{\mathsf{P}}(\mathrm{d}\psi) \tag{3.4.15}$$

for all $h \in F_+(S_K)$, where $n(t) = n(t,\psi) = \max\{i : t_i \leqq t\}$ for $\psi = ([t_i, k_i]) \in M_K$.

3.4.3. Lemma. *Let* $(\overline{\Psi}, \overline{\mathsf{P}})$ *be a stationary MPP with finite intensity* $\lambda_{\overline{\mathsf{P}}}$. *Then the sequence* (Φ, P), *where* P *is defined by* (3.4.2), *is stationary, and*

$$\sum_{n=-\infty}^{-1} A_n = \sum_{n=0}^{\infty} A_n = \infty \qquad \mathsf{P}\text{-}a.s.$$

Proof. Using formula (3.4.14) we get for an arbitrary measurable subset $E \subseteq S_K$ and all $t > 0$

$$|\mathsf{P}(E) - \mathsf{P}(\theta E)| \leqq \frac{1}{\lambda_{\overline{\mathsf{P}}}t}\,\mathsf{E}_{\overline{\mathsf{P}}}\left|\sum_{j=1}^{\overline{N}(t)}\Big(\mathbf{1}_E\big(\theta^j\varphi(\overline{\Psi})\big) - \mathbf{1}_E\big(\theta^{j-1}\varphi(\overline{\Psi})\big)\Big)\right|$$

$$\leqq \frac{1}{\lambda_{\overline{\mathsf{P}}}t}\,\mathsf{E}_{\overline{\mathsf{P}}}\Big(\mathbf{1}_E\big(\theta^{\overline{N}(t)}\varphi(\overline{\Psi})\big) + \mathbf{1}_E\big(\varphi(\overline{\Psi})\big)\Big) \leqq \frac{2}{\lambda_{\overline{\mathsf{P}}}t}.$$

Hence, with $t \to \infty$ the stationarity follows. In view of (3.4.14) the second statement is obvious. ∎

3.4.4. Lemma. *Let* $(\overline{\Psi}, \overline{\mathsf{P}})$ *be a stationary MPP with finite intensity* $\lambda_{\overline{\mathsf{P}}}$ *and* (Φ, P), $\Phi = ([A_n, K_n])$, *a stationary sequence satisfying* (3.4.2). *Then* $\mathsf{E}_{\mathsf{P}} A_0 < \infty$,

$$\overline{\mathsf{P}}(\overline{\Psi} \in (\cdot)) = (\mathsf{E}_{\mathsf{P}} A_0)^{-1} \int_0^\infty \mathsf{P}(A_0 > t, \theta_t \psi(\Phi) \in (\cdot)) \, \mathrm{d}t,$$

and $a_{\mathsf{P}} = \mathsf{E}_{\mathsf{P}} A_0 = \lambda_{\overline{\mathsf{P}}}^{-1}$.

Proof. For a sequence $\varphi = ([a_i, k_i]) \in S_K$ and $n \geq 1$ let $t_n(\varphi) = a_0 + \cdots + a_{n-1}$. From Fubini's theorem

$$\lambda_{\overline{\mathsf{P}}} \int_0^\infty \mathsf{P}(A_0 > t, \theta_t \psi(\Phi) \in (\cdot)) \, \mathrm{d}t = \lambda_{\overline{\mathsf{P}}} \int_{S_K} \int_0^{t_1(\varphi)} \mathbf{1}_{(\cdot)}(\theta_t \psi(\varphi)) \, \mathrm{d}t \mathsf{P}(\mathrm{d}\varphi).$$

Thus, using (3.4.15) for $t = 1$ and

$$h(\varphi) = \int_0^{t_1(\varphi)} \mathbf{1}_{(\cdot)}(\theta_t \psi(\varphi)) \, \mathrm{d}t,$$

we obtain

$$\lambda_{\overline{\mathsf{P}}} \int_0^\infty \mathsf{P}(A_0 > t, \theta_t \psi(\Phi) \in (\cdot)) \, \mathrm{d}t$$

$$= \int_{M_K} \sum_{j=1}^{n(1,\eta)} \int_0^{t_1(\theta^j \varphi(\eta))} \mathbf{1}_{(\cdot)}(\theta_t \psi(\theta^j \varphi(\eta))) \, \mathrm{d}t \, \overline{\mathsf{P}}(\mathrm{d}\eta)$$

$$= \int_{M_K} \sum_{j=1}^{n(1,\eta)} \int_{t_j(\eta)}^{t_{j+1}(\eta)} \mathbf{1}_{(\cdot)}(\theta_t \eta) \, \mathrm{d}t \overline{\mathsf{P}}(\mathrm{d}\eta).$$

By rearranging the domain of integration we obtain $(n(1) = n(1, \psi))$

$$\lambda_{\overline{\mathsf{P}}} \int_0^\infty \mathsf{P}(A_0 > t, \theta_t \psi(\Phi) \in (\cdot)) \, \mathrm{d}t$$

$$= \int_{M_K} \left(\int_0^1 \mathbf{1}_{(\cdot)}(\theta_t \psi) \, \mathrm{d}t + \int_1^{t_{n(1)+1}(\psi)} \mathbf{1}_{(\cdot)}(\theta_t \psi) \, \mathrm{d}t - \int_0^{t_1(\psi)} \mathbf{1}_{(\cdot)}(\theta_t \psi) \, \mathrm{d}t \right) \overline{\mathsf{P}}(\mathrm{d}\psi). \tag{3.4.16}$$

In view of

$$\int_1^{t_{n(1)+1}(\psi)} \mathbf{1}_{(\cdot)}(\theta_t \psi) \, \mathrm{d}t = \int_0^{t_1(\theta_1 \psi)} \mathbf{1}_{(\cdot)}(\theta_t(\theta_1 \psi)) \, \mathrm{d}t,$$

the stationarity of $\overline{\mathsf{P}}$ implies

$$\int_{M_K} \int_1^{t_{n(1)+1}(\psi)} \mathbf{1}_{(\cdot)}(\theta_t \psi) \, \mathrm{d}t \overline{\mathsf{P}}(\mathrm{d}\psi) = \int_{M_K} \int_0^{t_1(\psi)} \mathbf{1}_{(\cdot)}(\theta_t \psi) \, \mathrm{d}t \overline{\mathsf{P}}(\mathrm{d}\psi). \tag{3.4.17}$$

(Note that both sides in (3.4.17) can be infinite. Thus their difference cannot automatically be neglected in (3.4.16).)

Now consider an event D satisfying

$$D \subseteq \{\psi : t_1(\psi) - t_0(\psi) < c\}, \tag{3.4.18}$$

where c is an arbitrary but fixed constant. Since

$$t_1(\theta_t\psi) - t_0(\theta_t\psi) = t_1(\psi) - t_0(\psi) \quad \text{for} \quad t_0(\psi) \leqq 0 < t < t_1(\psi)$$

we have

$$\int_{M_K} \int_0^{t_1(\psi)} 1_D(\theta_t\psi) \, dt\overline{\mathsf{P}}(d\psi) \leqq \int_{M_K} \int_0^{t_1(\psi)} 1_{\{\eta: t_1(\eta) - t_0(\eta) < c\}}(\theta_t\psi) \, dt\overline{\mathsf{P}}(d\psi) \leqq c.$$

Thus, in view of (3.4.17), equation (3.4.16) applied to D reduces to

$$\lambda_{\overline{\mathsf{P}}} \int_0^\infty \mathsf{P}(A_0 > t, \theta_t\psi(\Phi) \in D) \, dt$$

$$= \int_{M_K} \int_0^1 1_D(\theta_t\psi) \, dt\overline{\mathsf{P}}(d\psi) = \int_0^1 \int_{M_K} 1_D(\theta_t\psi) \, \overline{\mathsf{P}}(d\psi) \, dt = \overline{\mathsf{P}}(\overline{\Psi} \in D). \tag{3.4.19}$$

Formula (3.4.19) can be extended to all events $D \in \mathcal{M}_K$ by using the monotone convergence theorem:

$$\overline{\mathsf{P}}(\overline{\Psi} \in (\cdot)) = \lambda_{\overline{\mathsf{P}}} \int_0^\infty \mathsf{P}(A_0 > t, \theta_t\psi(\Phi) \in (\cdot)) \, dt. \tag{3.4.20}$$

Applying (3.4.20) to M_K we get

$$1 = \lambda_{\overline{\mathsf{P}}} \int_0^\infty \mathsf{P}(A_0 > t) \, dt = \lambda_{\overline{\mathsf{P}}} \mathsf{E}_{\mathsf{P}} A_0. \tag{3.4.21}$$

The assertion of Lemma 3.4.4 follows from (3.4.20), (3.4.21). ∎

Lemma 3.4.4 shows that the mapping $\overline{\mathsf{P}} \to \mathsf{P}$ defined by (3.4.2) is one-to-one onto its range. In order to complete the proof of Theorem 3.4.1 it remains to show that the range of this mapping is $\mathbf{P}$ and (3.4.1) defines the inversion of (3.4.2). This will be done by the following two lemmas.

3.4.5. Lemma. *Consider a stationary sequence* (Φ, P), $\Phi = ([A_n, K_n])$, *with* $\mathsf{P} \in \mathbf{P}$. *Then* (3.4.1) *defines the distribution* $\overline{\mathsf{P}}$ *of a stationary MPP* $\overline{\Psi}$ *with finite intensity* $\lambda_{\overline{\mathsf{P}}} = a_{\mathsf{P}}^{-1}$.

Proof. Obviously, $\overline{\mathsf{P}}$ is a probability measure on $(M_K, \mathcal{M}_K)$. From (3.4.3) and the stationarity of (Φ, P) we obtain

$$\overline{\mathsf{P}}(\overline{\Psi} \in D) = a_{\mathsf{P}}^{-1} \int_{S_K} \int_0^{a_0(\varphi)} 1_D(\theta_u\psi(\varphi)) \, du\mathsf{P}(d\varphi)$$

$$= a_{\mathsf{P}}^{-1} \int_{S_K} \int_0^{a_n(\varphi)} 1_D(\theta^n\psi(\theta/\varphi)) \, du\mathsf{P}(d\varphi), \qquad D \in \mathcal{M}_K, \quad n \geqq 0.$$

Thus,

$$\overline{\mathsf{P}}(\overline{\Psi} \in D) = \frac{a_{\mathsf{P}}^{-1}}{n+1} \sum_{m=0}^{n} \int_{S_K} \int_0^{a_m(\varphi)} \mathbf{1}_D\big(\theta_u \psi(\theta^m \varphi)\big)\, \mathrm{d}u \mathsf{P}(\mathrm{d}\varphi)$$

$$= \frac{a_{\mathsf{P}}^{-1}}{n+1} \int_{S_K} \int_0^{t_{n+1}(\varphi)} \mathbf{1}_D\big(\theta_u \psi(\varphi)\big)\, \mathrm{d}u \mathsf{P}(\mathrm{d}\varphi), \quad D \in \mathcal{M}_K, \quad n \geqq 0,$$

where $t_0(\varphi) = 0$, $t_n(\varphi) = a_0 + \cdots + a_{n-1}$ if $n \geqq 1$, $t_n(\varphi) = -(a_n + \cdots + a_{-1})$ if $n < 0$ for $\varphi \in S_K$. Analogously, for every fixed $t \in \mathbb{R}$,

$$\overline{\mathsf{P}}(\theta_t \overline{\Psi} \in D) = a_{\mathsf{P}}^{-1} \int_{S_K} \int_0^{a_0(\varphi)} \mathbf{1}_D\big(\theta_{u+t} \psi(\varphi)\big)\, \mathrm{d}u \mathsf{P}(\mathrm{d}\varphi)$$

$$= \frac{a_{\mathsf{P}}^{-1}}{n+1} \int_{S_K} \int_0^{t_{n+1}(\varphi)} \mathbf{1}_D\big(\theta_{u+t} \psi(\varphi)\big)\, \mathrm{d}u \mathsf{P}(\mathrm{d}\varphi)$$

$$= \frac{a_{\mathsf{P}}^{-1}}{n+1} \int_{S_K} \int_t^{t_{n+1}(\varphi)+t} \mathbf{1}_D\big(\theta_u \psi(\varphi)\big)\, \mathrm{d}u \mathsf{P}(\mathrm{d}\varphi), \quad D \in \mathcal{M}_K, \quad n \geqq 0.$$

Thus,

$$|\overline{\mathsf{P}}(\overline{\Psi} \in D) - \overline{\mathsf{P}}(\theta_t \overline{\Psi} \in D)|$$

$$\leqq \frac{a_{\mathsf{P}}^{-1}}{n+1} \int_{S_K} \left(\int_0^t \mathbf{1}_D\big(\theta_u \psi(\varphi)\big)\, \mathrm{d}u + \int_{t_{n+1}(\varphi)}^{t_{n+1}(\varphi)+t} \mathbf{1}_D\big(\theta_u \psi(\varphi)\big)\, \mathrm{d}u \right) \mathsf{P}(\mathrm{d}\varphi)$$

$$\leqq \frac{2t a_{\mathsf{P}}^{-1}}{n+1} \xrightarrow[n\to\infty]{} 0.$$

Hence $(\overline{\Psi}, \overline{\mathsf{P}})$ is stationary.

It remains to show that $\lambda_{\overline{\mathsf{P}}}$ is finite. Using formula (3.4.3) we obtain

$$\overline{\mathsf{P}}(\overline{T}_i \leqq 1) = a_{\mathsf{P}}^{-1} \int_{S_K} \int_0^{a_0(\varphi)} \mathbf{1}_{\{\psi : t_i \leqq 1\}}\big(\theta_u \psi(\varphi)\big)\, \mathrm{d}u \mathsf{P}(\mathrm{d}\varphi)$$

$$= a_{\mathsf{P}}^{-1} \int_{S_K} \int_{t_i(\varphi)-t_1(\varphi)}^{t_i(\varphi)} \mathbf{1}_{(0,1]}(s)\, \mathrm{d}s\, \mathsf{P}(\mathrm{d}\varphi), \quad i \geqq 1.$$

In view of $t_i(\varphi) - t_1(\varphi) = -t_{1-i}(\theta^i \varphi)$ and $t_i(\varphi) = -t_{-i}(\theta^i \varphi)$, we can continue as follows:

$$\overline{\mathsf{P}}(\overline{T}_i \leqq 1) = a_{\mathsf{P}}^{-1} \int_{S_K} \int_{-t_{1-i}(\theta^i \varphi)}^{-t_{-i}(\theta^i \varphi)} \mathbf{1}_{(0,1]}(s)\, \mathrm{d}s \mathsf{P}(\mathrm{d}\varphi)$$

$$= a_{\mathsf{P}}^{-1} \int_{S_K} \int_{-t_{1-i}(\varphi)}^{-t_{-i}(\varphi)} \mathbf{1}_{(0,1]}(s)\, \mathrm{d}s \mathsf{P}(\mathrm{d}\varphi).$$

Summation over i yields

$$\lambda_{\bar{\mathsf{P}}} = \sum_{i=1}^{\infty} \bar{\mathsf{P}}(\bar{T}_i \leqq 1) = a_{\mathsf{P}}^{-1} \int_{S_K} \int_0^{\infty} \mathbf{1}_{(0,1]}(s)\, ds \mathsf{P}(d\varphi) = a_{\mathsf{P}}^{-1} < \infty. \quad ∎$$

3.4.6. Lemma. *Let* (Φ_1, P_1), $\Phi_1 = ([A_{1,n}, K_{1,n}])$, *be a stationary sequence with* $\mathsf{P}_1 \in \mathbf{P}$. *Consider the stationary MPP* $(\bar{\Psi}, \bar{\mathsf{P}})$ *satisfying*

$$\bar{\mathsf{P}}(\bar{\Psi} \in (\cdot)) = a_{\mathsf{P}_1}^{-1} \int_0^{\infty} \mathsf{P}_1(A_{1,0} > t, \theta_t \psi(\Phi_1) \in (\cdot))\, dt, \tag{3.4.22}$$

cf. Lemma 3.4.5. *Then the distribution* P *defined by* (3.4.2) *coincides with* P_1.

Proof. From (3.4.22) and (3.4.6), for P_1 instead of P, we obtain for $n \geqq 0$

$$\bar{\mathsf{P}}(\bar{\Psi} \in (\cdot)) = a_{\mathsf{P}_1}^{-1} \int_{S_K} \int_0^{a_n(\varphi)} \mathbf{1}_{(\cdot)}(\theta_{t_n(\varphi)+u} \psi(\varphi))\, du \mathsf{P}_1(d\varphi).$$

Using Lemma 3.4.4, Lemma 3.4.5 and (3.4.6) we get

$$\bar{\mathsf{P}}(\bar{\Psi} \in (\cdot)) = a_{\mathsf{P}}^{-1} \int_{S_K} \int_0^{a_n(\varphi)} \mathbf{1}_{(\cdot)}(\theta_{t_n(\varphi)+u} \psi(\varphi))\, du \mathsf{P}(d\varphi), \qquad n \geqq 0.$$

Taking into account Lemma 3.4.4 and Lemma 3.4.5 we find

$$a_{\mathsf{P}} = \lambda_{\bar{\mathsf{P}}}^{-1} = a_{\mathsf{P}_1}.$$

Thus

$$\int_{S_K} \int_0^{a_n(\varphi)} \mathbf{1}_{(\cdot)}(\theta_{t_n(\varphi)+u} \psi(\varphi))\, du \mathsf{P}_1(d\varphi)$$

$$= \int_{S_K} \int_0^{a_n(\varphi)} \mathbf{1}_{(\cdot)}(\theta_{t_n(\varphi)+u} \psi(\varphi))\, du \mathsf{P}(d\varphi), \quad n \geqq 0.$$

Applying this equation to the events $\{\eta : \psi(\theta^{-n} \varphi(\theta_{t_0(\eta)} \eta)) \in D\}$, $D \in \mathcal{M}_K$, we get

$$\int_{S_K} a_n(\varphi)\, \mathbf{1}_D(\psi(\varphi))\, \mathsf{P}_1(d\varphi) = \int_{S_K} a_n(\varphi)\, \mathbf{1}_D(\psi(\varphi))\, \mathsf{P}(d\varphi), \qquad n \geqq 0,$$

and thus

$$\int_{S_K} \frac{1}{n} \sum_{i=0}^{n-1} a_i(\varphi)\, \mathbf{1}_D(\psi(\varphi))\, \mathsf{P}_1(d\varphi) = \int_{S_K} \frac{1}{n} \sum_{i=0}^{n-1} a_i(\varphi)\, \mathbf{1}_D(\psi(\varphi))\, \mathsf{P}(d\varphi). \tag{3.4.23}$$

In view of the individual ergodic theorem, cf. Theorem A 1.2.1,

$$\bar{a}(\Phi_1) = \lim_{n\to\infty} \frac{1}{n} \sum_{i=0}^{n-1} A_{1,i} \quad \text{and} \quad \bar{a}(\Phi) = \lim_{n\to\infty} \frac{1}{n} \sum_{i=0}^{n-1} A_i$$

exists P_1-a.s. and P-a.s., respectively, and

$$\mathsf{E}_{\mathsf{P}_1} \bar{a}(\Phi_1) = a_{\mathsf{P}_1} < \infty, \qquad \mathsf{E}_{\mathsf{P}} \bar{a}(\Phi) = a_{\mathsf{P}} < \infty. \tag{3.2.24}$$

Now we show

$$\bar{a}(\Phi_1) > 0 \quad \mathsf{P}_1\text{-a.s.}, \qquad \bar{a}(\Phi) > 0 \quad \mathsf{P}\text{-a.s.} \tag{3.4.25}$$

Namely, if $\mathsf{P}(\bar{a}(\Phi) = 0) > 0$, then $\mathsf{P}(\cdot \mid \bar{a}(\Phi) = 0)$ is the distribution of a stationary sequence, because the event $\{\varphi : \bar{a}(\varphi) = 0\}$ is invariant. Applying the individual ergodic theorem A 1.2.1 again, we obtain

$$0 = \mathsf{E}_\mathsf{P}(\bar{a}(\Phi) \mid \bar{a}(\Phi) = 0) = \mathsf{E}_\mathsf{P}(A_1 \mid \bar{a}(\Phi) = 0).$$

Hence, $\mathsf{P}(A_i = 0, i \in \mathbb{Z} \mid \bar{a}(\Phi) = 0) = 1$, which contradicts $\mathsf{P}(S_k) = 1$. Analogously it follows $\bar{a}(\Phi_1) > 0$ P_1-a.s. In view of (3.4.24), we can take the limit as $n \to \infty$ in (3.4.23):

$$\int_{S_K} \bar{a}(\varphi)\, 1_D(\psi(\varphi))\, \mathsf{P}_1(d\varphi) = \int_{S_K} \bar{a}(\varphi)\, 1_D(\psi(\varphi))\, \mathsf{P}(d\varphi). \tag{3.4.26}$$

By standard techniques of integration theory (3.4.26) can be extended to all $h \in F_+(M_k)$:

$$\int_{S_K} \bar{a}(\varphi)\, h(\psi(\varphi))\, \mathsf{P}_1(d\varphi) = \int_{S_K} \bar{a}(\varphi)\, h(\psi(\varphi))\, \mathsf{P}(d\varphi).$$

Consider an arbitrary function $g \in F_+(S_k)$. Setting

$$h(\psi) = \frac{g(\varphi(\psi))}{\bar{a}(\varphi(\psi))},$$

we get

$$\int_{S_K} \bar{a}(\varphi) \frac{g\big(\varphi(\psi(\varphi))\big)}{\bar{a}\big(\varphi(\psi(\varphi))\big)} \mathsf{P}_1(d\varphi) = \int_{S_K} \bar{a}(\varphi) \frac{g\big(\varphi(\psi(\varphi))\big)}{\bar{a}\big(\varphi(\psi(\varphi))\big)} \mathsf{P}(d\varphi),$$

i.e.

$$\int_{S_K} g(\varphi)\, \mathsf{P}_1(d\varphi) = \int_{S_K} g(\varphi)\, \mathsf{P}(d\varphi), \qquad g \in F_+(S_k).$$

This implies $\mathsf{P}_1(\cdot) = \mathsf{P}(\cdot)$. ∎

Lemmas 3.4.3–3.4.6 show that Theorem 3.4.1 is true. ∎

3.5. Relationships between stationary sequences and stationary MPP's. II: Conclusions

In this section we derive some useful results for stationary MPP's and their associated stationary sequences. The proofs of these results are based on Theorem 3.4.1. Throughout this section let $(\bar{\Psi}, \bar{\mathsf{P}})$, $\bar{\Psi} = ([\bar{T}_n, \bar{K}_n])$, be a stationary MPP with $\lambda_{\bar{\mathsf{P}}} < \infty$ and (Φ, P), $\Phi = ([A_n, K_n])$, the stationary sequence associated with $(\bar{\Psi}, \bar{\mathsf{P}})$ according to (3.4.1) and (3.4.2). (Notice, $0 < a_\mathsf{P} < \infty$.) By $([T_n, K_n])$ we denote the corresponding synchronous MPP $\psi(\Phi)$.

First we consider some relationships between the sequences $(\bar{K}_n)$ and (K_n).

3.5.1. Theorem.

(i) *If A_0 and K_0 are independent, then* $\bar{K}_0 \overset{\mathcal{D}}{=} K_0$.

(ii) *If A_0 and (K_n) are independent, then* $(\bar{K}_n) \overset{\mathcal{D}}{=} (K_n)$.

(iii) *If (A_n) and (K_n) are independent, then $(\bar{T}_n)$ and $(\bar{K}_n)$ are independent, too.*

(iv) *If $(\bar{T}_n)$ and $(\bar{K}_n)$ are independent and $(\bar{K}_n)$ is stationary, then* $(\bar{K}_n) \overset{\mathcal{D}}{=} (K_n)$, *and (A_n) and (K_n) are independent sequences.*

(v) *If $(\bar{T}_n)$ and $(\bar{K}_n)$ are independent and $(\bar{\Psi}, \bar{\mathsf{P}})$ is a simple MPP, then $(\bar{K}_n)$ is stationary.*

Proof. (i) If A_0 and K_0 are independent, then

$$\bar{\mathsf{P}}(\bar{K}_0 \in (\cdot)) = a_{\mathsf{P}}^{-1} \int_0^\infty \mathsf{P}(A_0 > t, K_0 \in (\cdot))\, dt$$
$$= \mathsf{P}(K_0 \in (\cdot))\, a_{\mathsf{P}}^{-1} \int_0^\infty \mathsf{P}(A_0 > t)\, dt = \mathsf{P}(K_0 \in (\cdot)),$$

in view of (3.4.1).

(ii) The proof of assertion (ii) is analogous to (i).

(iii) Assume that (A_n) and (K_n) are independent. Together with (ii) formula (3.4.1) yields

$$\bar{\mathsf{P}}((\bar{T}_n) \in B, (\bar{K}_n) \in C) = a_{\mathsf{P}}^{-1} \int_0^\infty \mathsf{P}(A_0 > t, (T_n - t) \in B, (K_n) \in C)\, dt$$
$$= a_{\mathsf{P}}^{-1}\mathsf{P}((K_n) \in C) \int_0^\infty \mathsf{P}(A_0 > t, (T_n - t) \in B)\, dt$$
$$= \bar{\mathsf{P}}((\bar{K}_n) \in C)\, \bar{\mathsf{P}}((\bar{T}_n) \in B)$$

for all measurable sets B and C.

(iv) Assume that $(\bar{T}_n)$ and $(\bar{K}_n)$ are independent, and let $(\bar{K}_n)$ be stationary. Using formula (3.4.2) we obtain

$$\mathsf{P}((K_n) \in D) = \lambda_{\mathsf{P}}^{-1} \sum_{j=1}^\infty \bar{\mathsf{P}}(\bar{T}_j \leqq 1, (\bar{K}_{n+j}) \in D)$$
$$= \lambda_{\mathsf{P}}^{-1} \sum_{j=1}^\infty \bar{\mathsf{P}}((\bar{K}_n) \in D)\, \bar{\mathsf{P}}(\bar{T}_j \leqq 1)$$
$$= \bar{\mathsf{P}}((\bar{K}_n) \in D)$$

and

$$\mathsf{P}((K_n) \in D, (A_n) \in E)$$
$$= \lambda_{\bar{\mathsf{P}}}^{-1} \sum_{j=1}^\infty \bar{\mathsf{P}}(\bar{T}_j \leqq 1, (\bar{T}_{n+j+1} - \bar{T}_{n+j}) \in E, (\bar{K}_{n+j}) \in D)$$
$$= \lambda\ ^{-1}\bar{\mathsf{P}}((\bar{K}_n) \in D) \sum_{j=1}^\infty \bar{\mathsf{P}}(\bar{T}_j \leqq 1, (\bar{T}_{n+j+1} - \bar{T}_{n+j}) \in E)$$
$$= \bar{\mathsf{P}}((\bar{K}_n) \in D)\, \mathsf{P}((A_n) \in E) = \mathsf{P}((K_n) \in D)\, \mathsf{P}((A_n) \in E).$$

(v) Let $p(s) = \bar{\mathsf{P}}(\bar{T}_1 < s < \bar{T}_2)$. From (3.4.1) it follows that

$$p(s) = a_{\mathsf{P}}^{-1} \int_0^\infty \mathsf{P}(u < A_0 < u + s < A_0 + A_1) \, \mathrm{d}u$$

and thus (since $\bar{\Psi}$ is a simple MPP, i.e. $\mathsf{P}(A_1 > 0) = 1$), there is an $s_0 > 0$ such that

$$p(s_0) > 0. \tag{3.5.1}$$

The stationarity of $\bar{\Psi}$ implies

$$p(s_0) = \bar{\mathsf{P}}(\bar{T}_{-1} < -t < \bar{T}_0). \tag{3.5.2}$$

We use the notation $\bar{K}_n = k_n(\bar{\Psi})$, $n \in \mathbb{Z}$. From (3.5.1), (3.5.2), and from the independence of $(\bar{T}_n)$ and $(\bar{K}_n)$ we get

$$\begin{aligned}
&\bar{\mathsf{P}}\left(\left(k_n(\bar{\Psi})\right) \in (\cdot)\right) \\
&= \bar{\mathsf{P}}\left(\left(k_n(\bar{\Psi})\right) \in (\cdot), \bar{T}_{-1} < -s_0 < \bar{T}_0\right) / p(s_0) \\
&= \bar{\mathsf{P}}\left(\left(k_{n+1}(\theta_{-s_0}\bar{\Psi})\right) \in (\cdot), t_1(\theta_{-s_0}\bar{\Psi}) < s_0 < t_2(\theta_{-s_0}\bar{\Psi})\right) / p(s_0) \\
&= \bar{\mathsf{P}}\left(\left(k_{n+1}(\bar{\Psi})\right) \in (\cdot), t_1(\bar{\Psi}) < s_0 < t_2(\bar{\Psi})\right) / p(s_0) \\
&= \bar{\mathsf{P}}\left(\left(k_{n+1}(\bar{\Psi})\right) \in (\cdot)\right). \quad \blacksquare
\end{aligned}$$

3.5.2. Remark. From Theorem 3.5.1, (ii) it follows that $(\bar{K}_n)$ is stationary if A_0 and (K_n) are independent. This is not true in general, since the distributions

$$\bar{\mathsf{P}}(\bar{K}_0 \in (\cdot)) = a_{\mathsf{P}}^{-1} \int_0^\infty \mathsf{P}(A_0 > t, K_0 \in (\cdot)) \, \mathrm{d}t$$

and

$$\bar{\mathsf{P}}(\bar{K}_1 \in (\cdot)) = a_{\mathsf{P}}^{-1} \int_0^\infty \mathsf{P}(A_0 > t, K_1 \in (\cdot)) \, \mathrm{d}t$$

differ e.g. for suitably chosen Markov renewal processes.

At first glance one would expect that (A_n) and (K_n) are independent if and only if $(\bar{T}_n)$ and $(\bar{K}_n)$ are independent. The following example shows that this is not true.

3.5.3. Example. Define an MPP $(\bar{\Psi}, \bar{\mathsf{P}})$, $\bar{\Psi} = ([\bar{T}_n, \bar{K}_n])$, by

$$\bar{K}_n = \begin{cases} 0 & \text{if } n \text{ is even}, \\ 1 & \text{if } n \text{ is odd}, \end{cases}$$

$$\bar{\mathsf{P}}(\bar{T}_1 \leqq t) = t \quad \text{for } 0 \leqq t \leqq 1,$$

$$\bar{T}_n = \begin{cases} \bar{T}_{n-1} & \text{if } n \text{ is even}, \\ \bar{T}_{n-1} + 1 & \text{if } n \text{ is odd}, \end{cases} \qquad n \in \mathbb{Z}.$$

It is easy to see that $(\overline{\Psi}, \overline{\mathsf{P}})$ is a stationary MPP, but $(\overline{K}_n)$ is a non-stationary sequence. Obviously, $(\overline{T}_n)$ and $(\overline{K}_n)$ are independent. Formula (3.4.2) yields

$$\mathsf{P}(A_0 = 1, K_0 = 0) = \frac{1}{2} \sum_{j=1}^{\infty} \overline{\mathsf{P}}(\overline{T}_j \leqq 1, \overline{T}_{j+1} - \overline{T}_j = 1, \overline{K}_j = 0) = \frac{1}{2},$$

$$\mathsf{P}(A_0 = 1) = \frac{1}{2} \sum_{j=1}^{\infty} \overline{\mathsf{P}}(\overline{T}_j \leqq 1, \overline{T}_{j+1} - \overline{T}_j = 1) = \frac{1}{2},$$

$$\mathsf{P}(K_0 = 0) = \frac{1}{2} \sum_{j=1}^{\infty} \overline{\mathsf{P}}(\overline{T}_j \leqq 1, \overline{K}_j = 0) = \frac{1}{2}.$$

Thus, A_0 and K_0 are not independent.

For the distribution function of $\overline{T}_1$ of a stationary MPP $(\overline{\Psi}, \overline{\mathsf{P}})$ we obtain the same formula as for stationary renewal processes, cf. Theorem 3.3.1:

3.5.4. Theorem. *Let* $\mathsf{F}(t) = \mathsf{P}(A_0 \leqq t)$. *Then*

$$\overline{\mathsf{P}}(-\overline{T}_0 > u, \overline{T}_1 > v) = a_{\mathsf{P}}^{-1} \int_{u+v}^{\infty} (1 - \mathsf{F}(t))\, dt, \qquad u, v \geqq 0, \tag{3.5.3}$$

$$\overline{\mathsf{P}}(\overline{T}_1 > v) = \overline{\mathsf{P}}(-\overline{T}_0 > v) = a_{\mathsf{P}}^{-1} \int_{v}^{\infty} (1 - \mathsf{F}(t))\, dt, \qquad v \geqq 0. \tag{3.5.4}$$

Proof. From (3.4.1)

$$\overline{\mathsf{P}}(-\overline{T}_0 > u, \overline{T}_1 > v) = a_{\mathsf{P}}^{-1} \int_0^{\infty} \mathsf{P}(A_0 > t, t > u, t + v < A_0)\, dt$$

$$= a_{\mathsf{P}}^{-1} \int_{u+v}^{\infty} \mathsf{P}(A_0 > t)\, dt.$$

Formula (3.5.4) follows immediately from (3.5.3), because $-\overline{T}_0 > 0$ and $\overline{T}_1 > 0$ $\overline{\mathsf{P}}$-a.s. ∎

Now we introduce synchronous and stationary renewal and Markov renewal processes on the whole axis. Consider a sequence (Φ, P), $\Phi = (A_n, n \in \mathbb{Z})$ of i.i.d. non-negative r.v.'s; $\mathsf{F}(x) = \mathsf{P}(A_0 \leqq x)$. Then $\psi(\Phi) = (T_n)$, where T_n is defined according to (3.2.5), is called an ordinary (synchronous) renewal process on $\mathbb{R}$ with the underlying distribution function $\mathsf{F}(x)$. The associated stationary point process $(\overline{\Psi}, \overline{\mathsf{P}})$ is called a stationary renewal process on $\mathbb{R}$ with the underlying distribution function $\mathsf{F}(x)$.

3.5.5. Theorem. *The stationary renewal process* $(\overline{\Psi}, \overline{\mathsf{P}})$ *with the underlying distribution function* $\mathsf{F}(x)$ *is determined by the following properties:* $\overline{T}_{n+1} - \overline{T}_n$, $n \neq 0$, *are i.i.d. r.v.'s, and* $\overline{\mathsf{P}}(\overline{T}_{n+1} - \overline{T}_n \leqq x) = \mathsf{F}(x)$, $n \neq 0$; *the pair* $[-\overline{T}_0, \overline{T}_1]$ *is independent of* $(\overline{T}_{n+1} - T_n, n \neq 0)$; *the distribution of the pair* $[-\overline{T}_0, \overline{T}_1]$ *is given by* (3.5.3).

Proof. From formula (3.4.1) we have

$$\overline{\mathsf{P}}(\overline{T}_{n+1} - \overline{T}_n \leqq x_n, n = \pm 1, \ldots, \pm k)$$
$$= a_{\mathsf{P}}^{-1} \int_0^\infty \mathsf{P}(A_0 > t, A_n \leqq x_n, n = \pm 1, \ldots, \pm k)\,\mathrm{d}t = \prod_{\substack{n=-k\\ n\neq 0}}^{k} \mathsf{F}(x_n),$$

$$\overline{\mathsf{P}}(-\overline{T}_0 > u, \overline{T}_1 > v, \overline{T}_{n+1} - \overline{T}_n \leqq x_n, n = \pm 1, \ldots, \pm k)$$
$$= a_{\mathsf{P}}^{-1} \int_0^\infty \mathsf{P}(A_0 > t, t > u, A_0 - t > v, A_n \leqq x_n, n = \pm 1, \ldots, \pm k)\,\mathrm{d}t$$
$$= a_{\mathsf{P}}^{-1} \int_0^\infty \mathsf{P}(A_0 > t, t > u, A_0 - t > v)\,\mathrm{d}t \mathsf{P}(A_n \leqq x_n, n = \pm 1, \ldots, \pm k)$$
$$= a_{\mathsf{P}}^{-1} \int_{u+v}^\infty \big(1 - \mathsf{F}(t)\big)\,\mathrm{d}t \prod_{\substack{n=-k\\ n\neq 0}}^{k} \overline{\mathsf{P}}(\overline{T}_{n+1} - \overline{T}_n \leqq x_n).$$

Since the distribution $\overline{\mathsf{P}}$ of $\overline{\Psi}$ is determined by the joint distribution of $[-\overline{T}_0, \overline{T}_1]$ and $(\overline{T}_{n+1} - \overline{T}_n, n \neq 0)$, the assertion of Theorem 3.5.5 is proved. ∎

In case of $\mathsf{F}(x) = 1 - \mathrm{e}^{-\lambda x}$, $\lambda > 0$, we use the symbol Π_λ instead of $\overline{\mathsf{P}}$ and call $(\overline{\Psi}, \Pi_\lambda)$ a stationary (homogeneous) Poisson process on $\mathbb{R}$ with the intensity λ. From Theorem 3.5.5 we obtain the following useful property of Poisson processes:

$$(\overline{T}_n, n > 0) \overset{\mathcal{D}}{=} (T_n, n > 0), \qquad (\overline{T}_{n+1}, n < 0) \overset{\mathcal{D}}{=} (T_n, n < 0). \qquad (3.5.5)$$

This means that the stationary Poisson process $(\overline{\Psi}, \Pi_\lambda)$ can be obtained from the associated synchronous point process by deleting the point $T_0 = 0$. In fact this property characterizes stationary Poisson processes within the class of stationary point processes, cf. [KMM]. For MPP's we can state the following result.

3.5.6. Theorem. *Consider the stationary sequence $\Phi = ([A_n, K_n])$, where A_n are i.i.d. r.v.'s with $\mathsf{P}(A_n \leqq x) = 1 - \mathrm{e}^{-\lambda x}$, and the sequences (A_n) and (K_n) are independent. Then the synchronous MPP $\psi(\Phi) = ([T_n, K_n])$ and the associated stationary MPP $\overline{\Psi} = ([\overline{T}_n, \overline{K}_n])$ have the property*

$$([T_n, K_n], n < 0) \overset{\mathcal{D}}{=} ([\overline{T}_{n+1}, \overline{K}_{n+1}], n < 0),$$
$$([T_n, K_n], n > 0) \overset{\mathcal{D}}{=} ([\overline{T}_n, \overline{K}_n], n > 0).$$

Proof. According to (3.5.5) we have $(T_n, n < 0) \overset{\mathcal{D}}{=} (\overline{T}_{n+1}, n < 0)$. On the other hand, from Theorem 3.5.1 (ii), (iii) it follows that $(\overline{K}_n)$ and $(\overline{T}_n)$ are independent, and $(\overline{K}_n) \overset{\mathcal{D}}{=} (K_n)$. Thus, $(K_n, n < 0) \overset{\mathcal{D}}{=} (\overline{K}_{n+1}, n < 0)$ in view of the stationarity of (K_n). This proves the first part of the theorem. The second can be proved in the same way. ∎

Consider a stochastic matrix (p_{ij}) having the unique stationary initial distribution (p_i) and a family $\{\mathsf{F}_{ij}(x)\}$ of distribution functions on $\mathbb{R}_+$. As mentioned in Section 3.3, there exists a unique stationary sequence $([A_n, K_n], n \geqq 0)$ satisfying (A 3.3.1) and (A 3.3.2) with $q_i = p_i$, $\mathsf{F}^0_{ij}(x) = \mathsf{F}_{ij}(x)$, $i, j \in \mathbb{K}$. This sequence can be extended in the usual way to the stationary sequence $\Phi = ([A_n, K_n], n \in \mathbb{Z})$ on the whole axis with the same transition mechanism. The synchronous MPP $\psi(\Phi) = ([T_n, K_n])$, with T_n giv enby (3.2.5), is called a *synchronous Markov renewal process* on the whole line governed by (p_{ij}) and $\{\mathsf{F}_{ij}(x)\}$. If

$$a = \sum_{i,j} p_i p_{ij} \int_0^\infty \left(1 - \mathsf{F}_{ij}(x)\right) \mathrm{d}x < \infty,$$

then the associated stationary Markov renewal process $(\overline{\Psi}, \overline{\mathsf{P}})$ on $\mathbb{R}$ defined by formula (3.4.1) exists. In particular we obtain

$$\begin{aligned}&\overline{\mathsf{P}}(\overline{K}_1 = k_1, \overline{T}_1 \leqq v, \overline{K}_2 = k_2, \overline{T}_2 - \overline{T}_1 \leqq x_1, \ldots,\\ &\overline{K}_n = k_n, \overline{T}_{n+1} - \overline{T}_n \leqq x_n, \overline{K}_{n+1} = k_{n+1})\\ &= \frac{1}{a} \sum_i p_i p_{ik_1} \int_0^v \left(1 - \mathsf{F}_{ik_1}(x)\right) \mathrm{d}x \cdot \prod_{j=1}^n p_{k_j k_{j+1}} \mathsf{F}_{k_j k_{j+1}}(x_j).\end{aligned}$$

Now we formulate some well-known useful results.

3.5.7. Theorem. *For a stationary simple MPP $(\overline{\Psi}, \overline{\mathsf{P}})$ with $\lambda_{\mathsf{P}} < \infty$ we have*

(i) (*Dobrushin's theorem*)

$$\overline{\mathsf{P}}\left(\overline{N}(t) > 1\right) = o(t),$$

(ii) $$\overline{\mathsf{P}}\left(\overline{N}(t) > 0\right) = \lambda_{\overline{\mathsf{P}}} t + o(t),$$

and, as a consequence,

(iii) (*Korolyuk's theorem*)

$$\overline{\mathsf{P}}\left(\overline{N}(t) = 1\right) = \lambda_{\overline{\mathsf{P}}} t + o(t).$$

Proof. In accordance with (3.4.4) we have

$$\begin{aligned}\overline{\mathsf{P}}\left(\overline{N}(t) > 1\right) &= \overline{\mathsf{P}}(\overline{T}_2 \leqq t)\\ &= a_{\mathsf{P}}^{-1} \int_0^\infty \mathsf{P}(A_{-1} > u, A_0 \leqq t - u)\, \mathrm{d}u\\ &\leqq a_{\mathsf{P}}^{-1} \int_0^t \mathsf{P}(A_0 \leqq t)\, \mathrm{d}u = o(t),\end{aligned}$$

because of $\mathsf{P}(A_0 \leqq t) \xrightarrow[t \to 0+0]{} 0$ (in view of simplicity). Analogously, from (3.4.4) we get

$$\begin{aligned}\overline{\mathsf{P}}\left(\overline{N}(t) > 0\right) &= \overline{\mathsf{P}}(\overline{T}_1 \leqq t)\\ &= a_{\mathsf{P}}^{-1} \int_0^\infty \mathsf{P}(A_{-1} > u, u \leqq t)\, \mathrm{d}u\end{aligned}$$

$$= a_{\mathsf{P}}^{-1} \int_0^t \mathsf{P}(A_{-1} > u)\,\mathrm{d}u$$

$$= a_{\mathsf{P}}^{-1} t + a_{\mathsf{P}}^{-1} \int_0^t \big(\mathsf{P}(A_{-1} > u) - \mathsf{P}(A_{-1} > 0)\big)\,\mathrm{d}u$$

$$= \lambda_{\bar{\mathsf{P}}} t + o(t). \quad \blacksquare$$

In the following we shall give some interpretations of the mapping $\bar{\mathsf{P}} \to \mathsf{P}$. Formula (3.4.2) yields

$$\mathsf{P}\big(K_0 \in (\cdot)\big) = \frac{\lambda_{\bar{\mathsf{P}}}(\cdot)}{\lambda_{\bar{\mathsf{P}}}}. \tag{3.5.6}$$

Thus, if

$$\frac{\bar{N}_L(t)}{t} \xrightarrow[t\to\infty]{} \lambda_{\bar{\mathsf{P}}}(L) \qquad \bar{\mathsf{P}}\text{-a.s.}, \quad L \in \mathcal{K},$$

then the distribution of K_0 can be interpreted as the distribution of the mark of a "randomly chosen" marked point of $\bar{\Psi}$, cf. the interpretation of $\lambda_{\bar{\mathsf{P}}}(\cdot)/\lambda_{\bar{\mathsf{P}}}$ in Section 3.2.

In case of a simple MPP $(\bar{\Psi}, \bar{\mathsf{P}})$ the following statement gives a clear interpretation of the probability distribution $\mathsf{P}\big(\psi(\Phi) \in (\cdot)\big)$ corresponding to $\bar{\mathsf{P}}$ as the conditional distribution of $\bar{\Psi}$ under the condition that there is a point of the point process at the origin.

3.5.8. Theorem. *Let $(\bar{\Psi}, \bar{\mathsf{P}})$ be a simple MPP satisfying $\lambda_{\bar{\mathsf{P}}} < \infty$. Then*

$$\sup_{D \in \mathcal{S}_K} \big|\mathsf{P}(D) - \bar{\mathsf{P}}\big(\varphi(\theta_{\bar{T}_1}\bar{\Psi}) \in D \mid \bar{T}_1 \leq t\big)\big| \xrightarrow[t\to 0]{} 0.$$

Proof. All conditional probabilities occurring in the assertion exist, because $\bar{\mathsf{P}}(\bar{T}_1 \leq t) > 0$, $t > 0$, cf. Theorem 3.5.4.
For every $D \in \mathcal{S}_K$ and $t > 0$

$$\bar{\mathsf{P}}\big(\varphi(\theta_{\bar{T}_1}\bar{\Psi}) \in D \mid \bar{T}_1 \leq t\big) = \frac{\lambda_{\bar{\mathsf{P}}} t}{\bar{\mathsf{P}}(\bar{T}_1 \leq t)} \cdot \frac{\bar{\mathsf{P}}\big(\bar{T}_1 \leq t, \varphi(\theta_{\bar{T}_1}\bar{\Psi}) \in D\big)}{\lambda_{\bar{\mathsf{P}}} t}.$$

In view of Theorem 3.5.7, (ii) the first factor converges to 1 as $t \to 0$. For the second factor we obtain by (3.4.4)

$$\bar{\mathsf{P}}\big(\bar{T}_1 \leq t, \varphi(\theta_{\bar{T}_1}\bar{\Psi}) \in D\big)$$

$$= a_{\mathsf{P}}^{-1} \int_0^\infty \mathsf{P}(A_{-1} > u, u < t, \Phi \in D)\,\mathrm{d}u$$

$$= a_{\mathsf{P}}^{-1} \int_0^t \mathsf{P}(A_{-1} > u, \Phi \in D)\,\mathrm{d}u$$

for every $t > 0$ and $D \in \mathcal{S}_K$. On the other hand,

$$\lambda_{\bar{\mathsf{P}}} t \mathsf{P}(D) = a_{\mathsf{P}}^{-1} \int_0^t \mathsf{P}(D)\,\mathrm{d}u.$$

Thus

$$\left|\overline{\mathsf{P}}\left(\overline{T}_1 \leqq t, \varphi(\theta_{\overline{T}_1}\overline{\Psi}) \in D\right) - \lambda_{\overline{\mathsf{P}}} t \mathsf{P}(D)\right|$$
$$= \lambda_{\overline{\mathsf{P}}} \left| \int_0^t \left(\mathsf{P}(A_{-1} > u, \Phi \in D) - \mathsf{P}(D)\right) \mathrm{d}u \right|$$
$$= \lambda_{\overline{\mathsf{P}}} \int_0^t \mathsf{P}(A_{-1} \leqq u, \Phi \in D) \, \mathrm{d}u \leqq \lambda_{\overline{\mathsf{P}}} t \mathsf{P}(A_{-1} \leqq t) = o(t)$$

as $t \to 0$, uniformly in D. This finishes the proof. ∎

Formula (3.5.6) and Theorem 3.2.7 allow us to write down Campbell's theorem (3.2.7) for a stationary MPP $(\overline{\Psi}, \overline{\mathsf{P}})$ with $\lambda_{\overline{\mathsf{P}}} < \infty$ in the following form:

$$\mathsf{E}_{\overline{\mathsf{P}}} \sum_{n=-\infty}^{+\infty} h(\overline{T}_n, \overline{K}_n) = \int_{\mathbb{R}\times\mathbb{K}} h(t, k) \, \nu_{\overline{\mathsf{P}}}(\mathrm{d}[t, k])$$
$$= \int_{-\infty}^{+\infty} \int_{\mathbb{K}} h(t, k) \, \lambda_{\overline{\mathsf{P}}}(\mathrm{d}k) \, \mathrm{d}t$$
$$= \lambda_{\overline{\mathsf{P}}} \int_{-\infty}^{+\infty} \mathsf{E}_{\mathsf{P}} h(t, K_0) \, \mathrm{d}t \tag{3.5.7}$$

for all $h \in F_+(\mathbb{R} \times \mathbb{K})$.

Finally we prove the following useful result:

3.5.9. Theorem. *Let $(\overline{\Psi}, \overline{\mathsf{P}})$ be a stationary MPP with $0 < \lambda_{\overline{\mathsf{P}}} < \infty$ and (Φ, P) the corresponding stationary sequence with $0 < \mathsf{E}_{\mathsf{P}} A_0 < \infty$. Consider an invariant event $D \in \mathcal{M}_K$, i.e. $\theta_t D = D$ for all $t \in \mathbb{R}$. Then*

$$\overline{\mathsf{P}}(\overline{\Psi} \in D) = 1 \quad \textit{implies} \quad \mathsf{P}(\theta_0 \psi(\Phi) \in D) = 1,$$
$$\mathsf{P}\left(\psi(\Phi) \in D\right) = 1 \quad \textit{implies} \quad \overline{\mathsf{P}}(\overline{\Psi} \in D) = 1.$$

Proof. Since D is invariant with respect to θ_t,

$$\mathbf{1}_D\left(\psi(\varphi)\right) \leqq \mathbf{1}_D\left(\theta_t \psi(\varphi)\right) \leqq \mathbf{1}_D\left(\theta_0 \psi(\varphi)\right)$$

for all $t \in \mathbb{R}$ and $\varphi \in S_K$. Now (3.4.6) yields

$$a_{\mathsf{P}}^{-1} \mathsf{E}_{\mathsf{P}} \left(A_n \mathbf{1}_D\left(\psi(\Phi)\right)\right) \leqq \overline{\mathsf{P}}(\overline{\Psi} \in D) \leqq a_{\mathsf{P}}^{-1} \mathsf{E}_{\mathsf{P}} \left(A_n \mathbf{1}_D\left(\theta_0 \psi(\Phi)\right)\right), \qquad n \in \mathbb{Z}.$$

Thus,

$$a_{\mathsf{P}}^{-1} \mathsf{E}_{\mathsf{P}} \left(\frac{1}{n} \sum_{i=0}^{n-1} A_i \mathbf{1}_D\left(\psi(\Phi)\right)\right)$$
$$\leqq \overline{\mathsf{P}}(\overline{\Psi} \in D) \leqq a_{\mathsf{P}}^{-1} \mathsf{E}_{\mathsf{P}} \left(\frac{1}{n} \sum_{i=0}^{n-1} A_i \mathbf{1}_D\left(\theta_0 \psi(\Phi)\right)\right), \qquad n \in \mathbb{Z}.$$

Analogously to the proof of Lemma 3.4.6, we obtain

$$a_{\mathsf{P}}^{-1} \mathsf{E}_{\mathsf{P}} \left(\bar{a}(\Phi) \, \mathbf{1}_D\left(\psi(\Phi)\right)\right)$$
$$\leqq \overline{\mathsf{P}}(\overline{\Psi} \in D) \leqq a_{\mathsf{P}}^{-1} \mathsf{E}_{\mathsf{P}} \left(\bar{a}(\Phi) \, \mathbf{1}_D\left(\theta_0 \psi(\Phi)\right)\right), \tag{3.5.8}$$

where

$$\bar{a}(\Phi) = \lim_{n\to\infty} \frac{1}{n} \sum_{i=0}^{n-1} A_i > 0 \qquad \mathsf{P}\text{-a.s.} \tag{3.5.9}$$

Let $\bar{\mathsf{P}}(\bar{\Psi} \in D) = 1$. Since $\mathsf{E}_{\mathsf{P}}\bar{a}(\Phi) = a_{\mathsf{P}}$, we have

$$1_D\big(\theta_0\psi(\Phi)\big) = 1 \qquad \mathsf{P}\text{-a.s.}$$

in view of (3.5.8), (3.5.9).

Conversely, if $1_D\big(\psi(\Phi)\big) = 1$ P-a.s., then the left-hand inequality of (3.5.8) yields $\bar{\mathsf{P}}(\bar{\Psi} \in D) = 1$. ∎

3.5.10. Corollary. *Let $(\bar{\Psi}, \bar{\mathsf{P}})$ be a simple, stationary MPP with $0 < \lambda_{\bar{\mathsf{P}}} < \infty$ and (Φ, P) the corresponding stationary sequence with $0 < \mathsf{E}_{\mathsf{P}}A_0 < \infty$. Consider an invariant event $D \in \mathcal{M}_K$. Then $\bar{\mathsf{P}}(\bar{\Psi} \in D) = 1$ iff $\mathsf{P}\big(\psi(\Phi) \in D\big) = 1$.*

Proof. Applying Theorem 3.5.9 to the event $D = \{\psi : t_{n+1} - t_n > 0, n \in \mathbb{Z}\}$, we obtain $A_n > 0$, $n \in \mathbb{Z}$, P-a.s. and thus $\theta_0\psi(\Phi) = \psi(\Phi)$ P-a.s. ∎

3.6. Ergodic marked point processes

We define the ergodicity of an MPP in the usual way.

3.6.1. Definition. A stationary MPP $(\bar{\Psi}, \bar{\mathsf{P}})$ is said to be *ergodic* iff the probability $\bar{\mathsf{P}}(D)$ of each invariant event D (i.e. $\theta_t D = D$ for all $t \in \mathbb{R}$) equals 0 or 1.

First we prove the following theorem.

3.6.2. Theorem. *A stationary MPP $(\bar{\Psi}, \bar{\mathsf{P}})$ is ergodic iff*

$$\frac{1}{t}\int_0^t h(\theta_u\bar{\Psi})\,du \xrightarrow[t\to\infty]{} \mathsf{E}_{\bar{\mathsf{P}}}h(\bar{\Psi}) \qquad \bar{\mathsf{P}}\text{-a.s.} \tag{3.6.1}$$

for all $h \in F_+(M_K)$ with $\mathsf{E}_{\bar{\mathsf{P}}}h(\bar{\Psi}) < \infty$.

Proof. 1. Observe that for a stationary MPP $(\bar{\Psi}, \bar{\mathsf{P}})$ and $h \in F_+(M_K)$ the sequence

$$\left(\int_n^{n+1} h(\theta_u\bar{\Psi})\,du,\ n \in \mathbb{Z}\right) \tag{3.6.2}$$

is also stationary.

Assume $\mathsf{E}_{\bar{\mathsf{P}}}h(\bar{\Psi}) < \infty$. Then

$$\mathsf{E}_{\bar{\mathsf{P}}}\int_0^1 h(\theta_u\bar{\Psi})\,du = \int_0^1 \mathsf{E}_{\bar{\mathsf{P}}}h(\theta_u\bar{\Psi})\,du = \mathsf{E}_{\bar{\mathsf{P}}}h(\bar{\Psi}) < \infty. \tag{3.6.3}$$

For an arbitrary $t > 0$ we have

$$\frac{1}{[t]+1}\int_0^{[t]} h(\theta_u\bar{\Psi})\,du \le \frac{1}{t}\int_0^t h(\theta_u\bar{\Psi})\,du \le \frac{1}{[t]}\int_0^{[t]+1} h(\theta_u\bar{\Psi})\,du. \tag{3.6.4}$$

Applying the individual ergodic Theorem A 1.2.1 to the sequence (3.6.2), we obtain

$$\lim_{n\to\infty} \frac{1}{n} \sum_{j=0}^{n-1} \int_j^{j+1} h(\theta_u \overline{\Psi})\, du < \infty \qquad \overline{\mathsf{P}}\text{-a.s.} \tag{3.6.5}$$

and in view of (3.6.3)

$$\mathsf{E}_{\overline{\mathsf{P}}} \left(\lim_{n\to\infty} \frac{1}{n} \sum_{j=0}^{n-1} \int_j^{j+1} h(\theta_u \overline{\Psi})\, du \right) = \mathsf{E}_{\overline{\mathsf{P}}} h(\overline{\Psi}).$$

From this and (3.6.4) we obtain

$$\mathsf{E}_{\overline{\mathsf{P}}} \left(\lim_{t\to\infty} \frac{1}{t} \int_0^t h(\theta_u \overline{\Psi})\, du \right) = \mathsf{E}_{\overline{\mathsf{P}}} h(\overline{\Psi}). \tag{3.6.6}$$

2. Let $(\overline{\Psi}, \overline{\mathsf{P}})$ be ergodic. The events

$$\left\{ \psi : \lim_{t\to\infty} \frac{1}{t} \int_0^t h(\theta_u \psi)\, du \leqq x \right\}, \qquad 0 \leqq x < \infty,$$

are invariant and consequently their probabilities are either 0 or 1. Thus, the r.v.

$$\lim_{t\to\infty} \frac{1}{t} \int_0^t h(\theta_u \overline{\Psi})\, du$$

is almost surely constant and (3.6.6) yields (3.6.1).

3. Now assume that (3.6.1) is true for all $h \in F_+(M_K)$ with $\mathsf{E}_{\overline{\mathsf{P}}} h(\overline{\Psi}) < \infty$. Let $D \in \mathcal{M}_K$ be invariant. Applying (3.6.1) to $h(\cdot) = 1_D(\cdot)$ we obtain

$$\lim_{t\to\infty} \frac{1}{t} \int_0^t 1_D(\theta_u \overline{\Psi})\, du = \lim_{t\to\infty} \frac{1}{t} \int_0^t 1_D(\overline{\Psi})\, du = \mathsf{E}_{\overline{\mathsf{P}}} 1_D(\overline{\Psi}) = \overline{\mathsf{P}}(D) \quad \overline{\mathsf{P}}\text{-a.s.}$$

Thus, $\overline{\mathsf{P}}(D)$ equals either 0 or 1. ∎

3.6.3. Corollary. *For an arbitrary stationary ergodic MPP $(\overline{\Psi}, \overline{\mathsf{P}})$ and an arbitrary $L \in \mathcal{K}$*

$$\lim_{t\to\infty} \frac{\overline{N}_L(t)}{t} = \lambda_{\overline{\mathsf{P}}}(L) \qquad \overline{\mathsf{P}}\text{-}a.s.$$

Proof. For a fixed $L \in \mathcal{K}$ define $h(\overline{\Psi}) = \#\{n : \overline{T}_n \in (0, 1], \overline{K}_n \in L\}$. Taking into account the relationship $h(\theta_u \overline{\Psi}) = \overline{N}_L(u + 1) - \overline{N}_L(u)$, $u > 0$, we obtain for $t > 1$

$$\overline{N}_L(t) - \overline{N}_L(1) \leqq \int_0^t h(\theta_u \overline{\Psi})\, du \leqq \overline{N}_L(t + 1).$$

The desired property follows from this and (3.6.1). ∎

Sometimes we shall use the following criterion for the ergodicity of a stationary MPP.

3.6.4. Theorem. *A stationary MPP* $(\overline{\Psi}, \overline{\mathsf{P}})$ *is ergodic if and only if every representation of* $\overline{\mathsf{P}}$ *as a mixture* $\overline{\mathsf{P}} = \alpha\overline{\mathsf{P}}_1 + (1-\alpha)\,\overline{\mathsf{P}}_2$, $0 \leq \alpha \leq 1$, *of probability distributions* $\overline{\mathsf{P}}_1, \overline{\mathsf{P}}_2$ *of stationary MPP's* $\overline{\Psi}_1, \overline{\Psi}_2$ *is trivial, i.e.* $\alpha = 0$ *or* $\alpha = 1$ *or* $\overline{\mathsf{P}}_1 = \overline{\mathsf{P}}_2$.

Proof. 1. Assume $(\overline{\Psi}, \overline{\mathsf{P}})$ is not ergodic. Then there is an invariant set $D \in \mathcal{M}_K$ with $0 < \overline{\mathsf{P}}(D) < 1$. The conditional probability distributions

$$\overline{\mathsf{P}}(\cdot \mid D) \quad \text{and} \quad \overline{\mathsf{P}}(\cdot \mid M_K \setminus D)$$

on $\mathcal{M}_K$ are also stationary. Hence,

$$\overline{\mathsf{P}}(\cdot) = \overline{\mathsf{P}}(D)\,\overline{\mathsf{P}}(\cdot \mid D) + \big(1 - \overline{\mathsf{P}}(D)\big)\,\overline{\mathsf{P}}(\cdot \mid M_K \setminus D)$$

is a representation of $\overline{\mathsf{P}}$ as a non-trivial mixture of stationary probability distributions on $\mathcal{M}_K$.

2. Assume there is a representation

$$\overline{\mathsf{P}} = \alpha\overline{\mathsf{P}}_1 + (1-\alpha)\,\overline{\mathsf{P}}_2$$

of $\overline{\mathsf{P}}$ as a non-trivial mixture of stationary probability distributions $\overline{\mathsf{P}}_1$ and $\overline{\mathsf{P}}_2$ on $\mathcal{M}_K$. Then $\overline{\mathsf{P}}_1$ is absolutely continuous with respect to $\overline{\mathsf{P}}$, $\overline{\mathsf{P}}_1 \ll \overline{\mathsf{P}}$. Thus there is a version of the Radon-Nikodym derivative $h = \mathrm{d}\overline{\mathsf{P}}_1/\mathrm{d}\overline{\mathsf{P}}$:

$$\overline{\mathsf{P}}_1(D) = \int_D h(\psi)\,\overline{\mathsf{P}}(\mathrm{d}\psi), \qquad D \in \mathcal{M}_K.$$

In view of the stationarity of $\overline{\mathsf{P}}_1$ and $\overline{\mathsf{P}}$, for each $u \in \mathbb{R}$, $h\big(\theta_u(\cdot)\big)$ is also a version of the Radon-Nikodym derivative $\mathrm{d}\overline{\mathsf{P}}_1/\mathrm{d}\overline{\mathsf{P}}$, and thus

$$h(\theta_u\overline{\Psi}) = h(\overline{\Psi}) \quad \overline{\mathsf{P}}\text{-a.s.}, \qquad u \in \mathbb{R}$$

Hence, for all $n > 0$

$$\int_{M_K} \left| \frac{1}{n} \int_0^n h(\theta_u\psi)\,\mathrm{d}u - h(\psi) \right| \overline{\mathsf{P}}(\mathrm{d}\psi)$$

$$\leq \int_{M_K} \frac{1}{n} \int_0^n |h(\theta_u\psi) - h(\psi)|\,\mathrm{d}u\overline{\mathsf{P}}(\mathrm{d}\psi)$$

$$= \frac{1}{n} \int_0^n \int_{M_K} |h(\theta_u\psi) - h(\psi)|\,\overline{\mathsf{P}}(\mathrm{d}\psi)\,\mathrm{d}u = 0$$

From this we get

$$\int_{M_K} \left| \lim_{t\to\infty} \frac{1}{t} \int_0^t h(\theta_u\psi)\, du - h(\psi) \right| \overline{\mathsf{P}}(d\psi)$$

$$= \int_{M_K} \left| \lim_{n\to\infty} \frac{1}{n} \int_0^n h(\theta_u\psi)\, du - h(\psi) \right| \overline{\mathsf{P}}(d\psi) = 0,$$

i.e. $\bar{h}(\psi) = \lim_{t\to\infty} \frac{1}{t} \int_0^t h(\theta_u\psi)\, du$ is also a version of $d\overline{\mathsf{P}}_1/d\overline{\mathsf{P}}$.

If $\bar{h}(\psi)$ were constant almost everywhere, then $\overline{\mathsf{P}}_1$ and $\overline{\mathsf{P}}$ would coincide, which contradicts the assumption. Thus

$$0 < \overline{\mathsf{P}}(\psi : \bar{h}(\psi) > 1) < 1.$$

On the other hand, the event $\{\psi : \bar{h}(\psi) > 1\}$ is invariant. Therefore $\overline{\mathsf{P}}$ is not ergodic. ∎

3.6.5. Theorem. *A stationary MPP* $(\overline{\Psi}, \overline{\mathsf{P}})$ *with* $\lambda_{\overline{\mathsf{P}}} < \infty$ *is ergodic if and only if the associated (in the sense of Theorem* 3.4.1) *stationary sequence* (Φ, P) *is ergodic.*

Proof. Consider the mixture

$$\overline{\mathsf{P}} = \alpha\overline{\mathsf{P}}_1 + (1-\alpha)\,\overline{\mathsf{P}}_2, \qquad 0 \leqq \alpha \leqq 1,$$

where $\overline{\mathsf{P}}, \overline{\mathsf{P}}_1, \overline{\mathsf{P}}_2$ are stationary probability distributions on $\mathcal{M}_K$. For the corresponding stationary distributions $\mathsf{P}, \mathsf{P}_1, \mathsf{P}_2$ on $\mathcal{S}_K$ formula (3.4.2) provides

$$\mathsf{P} = \frac{\lambda_{\overline{\mathsf{P}}_1}}{\lambda_{\overline{\mathsf{P}}}}\,\alpha\mathsf{P}_1 + \frac{\lambda_{\overline{\mathsf{P}}_2}}{\lambda_{\overline{\mathsf{P}}}}\,(1-\alpha)\,\mathsf{P}_2. \tag{3.6.7}$$

Conversely, if P_i with $a_{\mathsf{P}_i} < \infty$, $i = 1, 2$, are stationary distributions on $\mathcal{S}_K$ and

$$\mathsf{P} = \beta\mathsf{P}_1 + (1-\beta)\,\mathsf{P}_2, \qquad 0 \leqq \beta \leqq 1,$$

then formula (3.4.1) provides

$$\overline{\mathsf{P}} = \frac{a_{\mathsf{P}_1}}{a_{\mathsf{P}}}\,\beta\overline{\mathsf{P}}_1 + \frac{a_{\mathsf{P}_2}}{a_{\mathsf{P}}}\,(1-\beta)\,\overline{\mathsf{P}}_2 \tag{3.6.8}$$

for the corresponding stationary distributions on $\mathcal{M}_K$. Now the desired result follows from (3.6.7), (3.6.8) by means of Theorem 3.6.4 and Theorem A 1.2.3. ∎

The following statements show that in the ergodic case $\overline{\mathsf{P}}$-probabilities arise as limiting averages of the corresponding P-probabilities and vice versa.

3.6.6. Theorem. *Consider a stationary ergodic MPP* $(\overline{\Psi}, \overline{\mathsf{P}})$ *with* $\lambda_{\overline{\mathsf{P}}} < \infty$ *and the associated (in sense of Theorem* 3.4.1) *stationary ergodic sequence* (Φ, P).

Then, for every $h \in F_+(M_K)$ *with* $\mathsf{E}_{\bar{\mathsf{P}}} h(\overline{\Psi}) < \infty$

$$\lim_{t\to\infty} \frac{1}{t} \int_0^t h(\theta_u \psi(\Phi))\, du = \mathsf{E}_{\bar{\mathsf{P}}} h(\overline{\Psi}) \qquad \mathsf{P}\text{-}a.s. \tag{3.6.9}$$

Moreover, for every $D \in \mathcal{M}_K$

$$\lim_{t\to\infty} \frac{1}{t} \int_0^t \mathsf{P}(\theta_u \psi(\Phi) \in D)\, du = \bar{\mathsf{P}}(\overline{\Psi} \in D). \tag{3.6.10}$$

For every $g \in F_+(S_K)$ *with* $\mathsf{E}_{\mathsf{P}} g(\Phi) < \infty$

$$\lim_{n\to\infty} \frac{1}{n} \sum_{j=1}^{n} g(\theta^j \varphi(\overline{\Psi})) = \mathsf{E}_{\mathsf{P}} g(\Phi) \qquad \bar{\mathsf{P}}\text{-}a.s. \tag{3.6.11}$$

Moreover, for every $C \in \mathcal{S}_K$

$$\lim_{n\to\infty} \frac{1}{n} \sum_{j=1}^{n} \bar{\mathsf{P}}(\theta^j \varphi(\overline{\Psi}) \in C) = \mathsf{P}(\Phi \in C). \tag{3.6.12}$$

Proof. 1. Let

$$D' = \left\{\psi : \lim_{t\to\infty} \frac{1}{t} \int_0^t h(\theta_u \psi)\, du = \mathsf{E}_{\bar{\mathsf{P}}} h(\overline{\Psi})\right\}.$$

In view of the ergodic Theorem 3.6.2 we have $\bar{\mathsf{P}}(D') = 1$. Since D' is invariant, by Theorem 3.5.9

$$1 = \mathsf{P}(\varphi : \varphi \in S_K, \psi(\varphi) \in D') = \mathsf{P}\left(\lim_{t\to\infty} \frac{1}{t} \int_0^t h(\theta_u \psi(\Phi))\, du = \mathsf{E}_{\bar{\mathsf{P}}} h(\overline{\Psi})\right),$$

i.e. (3.6.9) is true. Applying (3.6.9) to $h = 1_D$ and integrating with respect to P we obtain (3.6.10).

2. Consider

$$C' = \left\{\varphi : \lim_{n\to\infty} \frac{1}{n} \sum_{j=1}^{n} g(\theta^j \varphi) = \mathsf{E}_{\mathsf{P}} g(\Phi)\right\}.$$

By the ergodic Theorem A 1.2.2, $\mathsf{P}(C') = 1$. Let

$$D' = \{\psi : \psi \in M_K, \varphi(\psi) \in C'\}.$$

Obviously, D' is invariant, and $C' = \{\varphi : \psi(\varphi) \in D'\}$. By Theorem 3.5.9,

$$1 = \mathsf{P}(C') = \bar{\mathsf{P}}(D') = \bar{\mathsf{P}}\left(\lim_{n\to\infty} \frac{1}{n} \sum_{j=1}^{n} g(\theta^j \varphi(\overline{\Psi})) = \mathsf{E}_{\mathsf{P}} g(\Phi)\right),$$

i.e. (3.6.11) is true. Applying (3.6.11) to $g = 1_C$ and integrating with respect to $\bar{\mathsf{P}}$, we obtain (3.6.12). ∎

3.6.7. Corollary. *Under the assumptions of Theorem* 3.6.6 *we have*

$$\lim_{t\to\infty} \frac{N_L(t)}{t} = \lambda_{\overline{\mathsf{P}}}(L) \quad \mathsf{P}\text{-}a.s., \qquad L \in \mathcal{K},$$

$$\lim_{n\to\infty} \frac{\overline{T}_n}{n} = a_{\mathsf{P}} \quad \overline{\mathsf{P}}\text{-}a.s.$$

Proof. Set $h(\psi) = \#\{n: 0 < t_n \leq 1, k_n \in L\}$ in (3.6.9) and $g(\varphi) = a_0$ in (3.6.11), respectively. ∎

3.7. Continuity of the mapping $\overline{\mathsf{P}} \to \mathsf{P}$

The continuity properties of the mapping $\overline{\mathsf{P}} \to \mathsf{P}$ are stated in the following theorem.

3.7.1. Theorem. *Let* $(\overline{\Psi}, \overline{\mathsf{P}})$, $(\overline{\Psi}_1, \overline{\mathsf{P}}_1)$, $(\overline{\Psi}_2, \overline{\mathsf{P}}_2)$, ... *be stationary MPP's with* $\lambda_{\overline{\mathsf{P}}} < \infty$, $\lambda_{\overline{\mathsf{P}}_1} < \infty$, $\lambda_{\overline{\mathsf{P}}_2} < \infty$, ..., *and* (Φ, P), (Φ_1, P_1), (Φ_2, P_2), ... *the corresponding stationary sequences, cf. Theorem* 3.4.1. *The following statements are equivalent:*

(i) $\overline{\Psi}_m \xrightarrow[m\to\infty]{\mathcal{D}} \overline{\Psi}$ *and* $\lambda_{\overline{\mathsf{P}}_m} \xrightarrow[m\to\infty]{} \lambda_{\overline{\mathsf{P}}}$.

(ii) $\Phi_m \xrightarrow[m\to\infty]{\mathcal{D}} \Phi$ *and* $a_{\mathsf{P}_m} \xrightarrow[m\to\infty]{} a_{\mathsf{P}}$.

Proof. 1. Let (i) be fulfilled. In view of the stationarity of $(\overline{\Psi}, \overline{\mathsf{P}})$, the mapping $\psi \to n(1, \psi) = \sup\{i: t_i \leq 1\}$, $\psi \in M_K{}^*$, is continuous for $\overline{\mathsf{P}}$-almost all ψ. Thus, by the continuous mapping theorem,

$$\overline{N}^{(m)}(1) \xrightarrow[m\to\infty]{\mathcal{D}} \overline{N}(1),$$

where

$$\overline{N}^{(m)}(1) = \sup\{n: t_n(\overline{\Psi}_m) \leq 1\},$$

$$\overline{N}(1) = \sup\{n: t_n(\overline{\Psi}) \leq 1\}.$$

Lemma A 2.1 yields

$$\lim_{r\to\infty} \sup_m \mathsf{E}_{\overline{\mathsf{P}}_m}\left(\overline{N}^{(m)}(1)\, \mathbf{1}_{[r,\infty)}\left(\overline{N}^{(m)}(1)\right)\right) = 0. \tag{3.7.1}$$

Let $g: S_K{}^* \to \mathbb{R}$ be a bounded, continuous function. Obviously, for each fixed $k \geq 1$, the mapping

$$\psi \to \sum_{n=1}^{k} g\left(\theta^n \varphi(\psi)\right) \mathbf{1}_{(0,1]}\left(t_n(\psi)\right)$$

is continuous for $\bar{\mathsf{P}}$-almost all ψ and bounded. From (3.4.15), (3.7.1) and (i),

$$\int_{S_K} g(\varphi)\, \mathsf{P}_m(\mathrm{d}\varphi) = \frac{1}{\lambda_{\bar{\mathsf{P}}_m}} \int_{M_K} \sum_{n=1}^{\infty} \mathbf{1}_{(0,1]}\big(t_n(\psi)\big)\, g\big(\theta^n \varphi(\psi)\big)\, \bar{\mathsf{P}}_m(\mathrm{d}\psi)$$

$$\xrightarrow[m\to\infty]{} \frac{1}{\lambda_{\bar{\mathsf{P}}}} \int_{M_K} \sum_{n=1}^{\infty} \mathbf{1}_{(0,1]}\big(t_n(\psi)\big)\, g\big(\theta^n \varphi(\psi)\big)\, \bar{\mathsf{P}}(\mathrm{d}\psi)$$

$$= \int_{S_K} g(\varphi)\, \mathsf{P}(d\varphi).$$

Thus, $\Phi_m \xrightarrow[m\to\infty]{\mathcal{D}} \Phi$. Moreover,

$$a_{\mathsf{P}_m} = \frac{1}{\lambda_{\bar{\mathsf{P}}_m}} \xrightarrow[m\to\infty]{} \frac{1}{\lambda_{\bar{\mathsf{P}}}} = a_{\mathsf{P}}.$$

2. Now assume (ii) is true. Since

$$a_0(\Phi_m) \xrightarrow[m\to\infty]{\mathcal{D}} a_0(\Phi) \quad \text{and} \quad a_{\mathsf{P}_m} \to a_{\mathsf{P}},$$

Lemma A 2.1 yields

$$\lim_{a\to\infty} \sup_m \mathsf{E}_{\mathsf{P}_m}\Big(a_0(\Phi_m)\, \mathbf{1}_{(a,\infty)}\big(a_0(\Phi_m)\big)\Big) = 0. \tag{3.7.2}$$

Let $h\colon M_K^* \to \mathbb{R}$ be a bounded, continuous function. Obviously, for each fixed $t \geqq 0$, the mapping

$$\varphi \to \int_0^t \mathbf{1}_{[0,a_0(\varphi))}(u)\, h\big(\theta_u \psi(\varphi)\big)\, \mathrm{d}u$$

is continuous and bounded. From (3.4.7), (3.7.2) and (ii),

$$\int_{M_K} h(\psi)\, \bar{\mathsf{P}}_m(\mathrm{d}\psi) = a_{\mathsf{P}_m}^{-1} \int_{S_K} \int_0^{a_0(\varphi)} h\big(\theta_u \psi(\varphi)\big)\, \mathrm{d}u \mathsf{P}_m(\mathrm{d}\varphi)$$

$$\xrightarrow[m\to\infty]{} a_{\mathsf{P}}^{-1} \int_{S_K} \int_0^{a_0(\varphi)} h\big(\theta_u \psi(\varphi)\big)\, \mathrm{d}u \mathsf{P}(\mathrm{d}\varphi) = \int_{M_K} h(\psi)\, \bar{\mathsf{P}}(\mathrm{d}\psi).$$

Thus, $\bar{\Psi}_m \xrightarrow[m\to\infty]{\mathcal{D}} \bar{\Psi}$. Moreover,

$$\lambda_{\bar{\mathsf{P}}_m} = a_{\mathsf{P}_m}^{-1} \to a_{\mathsf{P}}^{-1} = \lambda_{\bar{\mathsf{P}}}. \quad \blacksquare$$

The conditions $\lambda_{\bar{\mathsf{P}}_m} \to \lambda_{\bar{\mathsf{P}}}$ and $a_{\mathsf{P}_m} \to a_{\mathsf{P}}$, respectively, in Theorem 3.7.1 cannot be neglected in general, as the following examples show.

3.7.2. Example. Consider stationary sequences (Φ, P), (Φ_1, P_1), (Φ_2, P_2), ... $\big(\Phi = ([A_i, K_i], i \in \mathbb{Z})$, $\Phi_m = ([A_i^m, K_i^m], i \in \mathbb{Z})$, $m \geqq 1\big)$ defined by

$$\mathsf{P}_m(A_i^m = m, K_i^m = 0) = 1 - \mathsf{P}_m(A_i^m = 1, K_i^m = 0) = 1/m,$$

$$i \in \mathbb{Z}, \qquad m = 1, 2, \ldots,$$

$$\mathsf{P}(A_i = 1, K_i = 0) = 1, \qquad i \in \mathbb{Z},$$

where A_i^m, $i \in \mathbb{Z}$ are independent r.v.'s for every fixed m. Obviously,

$$\Phi_m \xrightarrow[m\to\infty]{\mathcal{D}} \Phi,$$

but $a_{\mathsf{P}_m} = 2 - 1/m \xrightarrow[m\to\infty]{} 2 \neq 1 = a_{\mathsf{P}}$.

Using Theorem 3.5.4, for the corresponding stationary MPP's $(\overline{\Psi}, \overline{\mathsf{P}})$, $(\overline{\Psi}_1, \overline{\mathsf{P}}_1)$, $(\overline{\Psi}_2, \overline{\mathsf{P}}_2)$, ... we obtain

$$\overline{\mathsf{P}}_m(\overline{T}_1{}^m > v) = a_{\mathsf{P}_m}^{-1} \int_v^\infty \left(1 - P_m(A_0{}^m \leqq t)\right) dt$$

$$= \begin{cases} \dfrac{m - 1 + (1 - v)\, m}{2m - 1} & \text{if } 0 \leqq v < 1, \\ \dfrac{m - v}{2m - 1} & \text{if } 1 \leqq v < m, \\ 0 & \text{if } v \geqq m \end{cases}$$

$$\xrightarrow[m\to\infty]{} \begin{cases} 1 - v/2 & \text{if } 0 \leqq v < 1, \\ 1/2 & \text{if } v \geqq 1. \end{cases}$$

On the other hand,

$$\overline{\mathsf{P}}(\overline{T}_1 > v) = \begin{cases} 1 - v & \text{if } 0 \leqq v < 1, \\ 0 & \text{if } v \geqq 1. \end{cases}$$

Thus $\overline{\Psi}_m \xrightarrow{\mathcal{D}} \overline{\Psi}$ does not hold.

3.7.3. Example. Consider the stationary sequences (Φ, P) and (Φ_m, P_m), $m \geqq 2$, with $\Phi = ([A_n, K_n)]$, $A_n \equiv 1$, $K_n \equiv 0$, and $\Phi_m = ([A_n{}^m, K_n{}^m])$, $A_n{}^m \equiv \dfrac{1}{m}$, $K_n{}^m \equiv 0$. The corresponding stationary MPP's are denoted by $(\overline{\Phi}, \overline{\mathsf{P}})$ and $(\overline{\Phi}_m, \overline{\mathsf{P}}_m)$, $m \geqq 2$, respectively. Further let $(\overline{\Phi}_m{}^*, \overline{\mathsf{P}}_m{}^*)$, $m \geqq 2$, be stationary MPP's with the distribution

$$\overline{\mathsf{P}}_m{}^* = \frac{1}{m}\, \overline{\mathsf{P}}_m + \left(1 - \frac{1}{m}\right) \overline{\mathsf{P}}.$$

Obviously, $\overline{\Phi}_m{}^* \xrightarrow{\mathcal{D}} \overline{\Phi}$. By formula (3.6.7) the distributions of the corresponding stationary sequences $(\Phi_m{}^*, \mathsf{P}_m{}^*)$ are of the form

$$\mathsf{P}_m{}^* = \frac{1}{m} \frac{\lambda_{\overline{\mathsf{P}}_m}}{\lambda_{\overline{\mathsf{P}}_m{}^*}}\, \mathsf{P}_m + \left(1 - \frac{1}{m}\right) \frac{\lambda_{\overline{\mathsf{P}}}}{\lambda_{\overline{\mathsf{P}}_m{}^*}}\, \mathsf{P} = \frac{m}{2m - 1}\, \mathsf{P}_m + \frac{m - 1}{2m - 1}\, \mathsf{P}.$$

Now it is easy to see that $\Phi_m{}^* \xrightarrow{\mathcal{D}} \Phi$ does not hold.

3.8. Remarks and references

The first detailed treatment of point processes was given by KHINCHIN (1955), where earlier investigations, in particular those of C. PALM (1943) were systematized and further developed. Khinchin considered point processes as stochastic processes with stepwise increasing trajectories, i.e. he dealt with counting processes $(N(t), t \in \mathbb{R}_+)$. The notion of a marked point process is due to MATTHES (1963). In the last thirty years the theory of point processes has undergone a stormy development. In order to develop an intuitive feeling for the subject we recommend COX and ISHAM (1980), GRANDELL (1976, 1977), and STOYAN, KENDALL and MECKE (1987), Chapters 2, 4, 5. For a detailed description of some basic aspects of the theory of point processes we refer to MECKE (1967), [KMN], LEWIS (1972), PORT and STONE (1973), [KMM], KALLENBERG (1975), NEVEU (1977), [FKAS], STOYAN, KENDALL and MECKE (1987). Concerning the martingale approach to point processes we refer to BRÉMAUD (1981), JACOBSEN (1982), LIPTSER and SHIRYAYEV (1978), and KARR (1986).

In most of the references above (marked) point processes were treated as random counting measures. It should be noted that our notion of point processes and that based on random measures differ in general. The random measure approach neglects the order of "particles" and marks within a multiple point. However, for stationary MPP's on $\mathbb{R}$ both notions are equivalent if there are no multiple points.

Let Ω be the set of all counting measures ω on $\mathbb{R} \times \mathbb{K}$ satisfying $\omega(\mathbb{R}_+ \times \mathbb{K}) = \omega(\mathbb{R} \setminus \mathbb{R}_+ \times \mathbb{K}) = \infty$, $\omega(\{a\} \times \mathbb{K}) \leqq 1$ for all $a \in \mathbb{R}$, and $\omega(B \times \mathbb{K}) < \infty$ for all bounded $B \in \mathcal{R}$. Obviously, there is a one-to-one mapping between Ω and $\tilde{M}_K = \{\psi : \psi = ([t_n, k_n]) \in M_K, \ldots, < t_{-1} < t_0 \leqq 0 < t_1 < t_2 < \cdots\}$. This correspondence is a homeomorphism with respect to the vague topology on Ω and $M_K \cap \tilde{M}_K$. However, the restriction of this mapping on the subspaces $\tilde{\tilde{\Omega}} = \{\omega : \omega \in \Omega,\ \omega(\{0\} \times \mathbb{K}) = 0\}$ and $\tilde{\tilde{M}}_K = \{\psi : \psi = ([t_n, k_n]),\ \psi \in \tilde{M}_K,\ t_0 < 0 < t_1\}$ is a homeomorphism. Since all simple, stationary MPP's in the sense of Definition 3.2.1 take values in $\tilde{\tilde{M}}_K$, the notions of simple stationary MPP's in this monograph and in [KMM], [KMN] are equivalent.

3.3. The results of this section are well-known and can be found in several textbooks, cf. e.g. FELLER (1971, Vol. II), CINLAR (1975), STÖRMER (1970).

3.4. The relationships between the family $\bar{\mathsf{P}}$ of distributions of stationary ordered MPP's and the family P of distributions of stationary sequences of $\mathbb{R}_+ \times \mathbb{K}$-valued r.v.'s, as stated in Theorem 3.4.1, are suitable modifications of the relationships between probabilities and Palm probabilities of stationary simple MPP's, cf. [KMN], [KMM], [FKAS], NEVEU (1977), and BACCELLI and BRÉMAUD (1978). The notion of Palm probabilities goes back to RYLL-NARDZEWSKI (1961), SLIVNYAK (1962), and MATTHES (1963). The approach to Palm probabilities based on the conditioning of stationary point processes can be found in JAGERS (1973) and PAPANGELOU (1974).

Our approach goes back to PORT and STONE (1973) and STRELLER (1980) (however, the presentation in Streller (1980) seems to be not completely conclusive in some details); without proofs it is sketched already in the Russian edition of [FKAS]. As mentioned above, there is no difference between the Palm distribution and distribution P of the stationary sequences associated with a given $\overline{\mathsf{P}}$ if there are no multiple points. However, in presence of multiplicities, the notion of ordered MPP's and their associated stationary sequences of $\mathbb{R}_+ \times \mathbb{K}$-valued r.v.'s is more convenient for the applications treated in this book than that of stationary counting measures and their Palm probabilities, cf. [KMM]. In particular, the statements of Theorem 3.4.1 and some of their corollaries, cf. Section 3.5, are valid regardless of whether $\mathsf{P}(A_0 = 0) = 0$ or $\mathsf{P}(A_0 = 0) > 0$, $\mathsf{P} \in \mathbf{P}$ (cf. the results of Section 3.3 as the "traditional" forerunners).

Using directly the relationships between probabilities and Palm probabilities of stationary simple MPP's one obtains another proof of Theorem 3.4.1 for MPP's with multiple points. The idea is as follows: The distribution P of a stationary sequence $\Phi = ([A_n, K_n])$ with $a_{\mathsf{P}} < \infty$ is uniquely determined by the conditional distribution $\mathsf{P}(\Phi \in (\cdot) \mid A_{-1} > 0)$, cf. Theorem A 1.3.2. Let $(\tilde{\Phi}, \tilde{\mathsf{P}})$, $\tilde{\Phi} = ([\tilde{A}_n, \tilde{K}_n])$, be a sequence with $\tilde{\mathsf{P}}(\tilde{\Phi} \in (\cdot)) = \mathsf{P}(\Phi \in (\cdot) \mid A_{-1} > 0)$. Denote the subsequence of all strictly positive elements of $(\tilde{A}_n)$ by $(\tilde{A}_{n_k})$. (In order to make this definition unique, we claim $n_{-1} = -1$. This can be done because of $\tilde{\mathsf{P}}(\tilde{A}_{-1} > 0) = 1$.) Define

$$\hat{A}_j = \tilde{A}_{n_j}, \qquad \hat{K}_j = [\tilde{K}_{n_{j-1}+1}, \ldots, \tilde{K}_{n_j}], \qquad j \in \mathbb{Z}.$$

Note that $\hat{A}_j > 0$, $j \in \mathbb{Z}$. Using Theorem A 1.3.2, it is not difficult to prove that $(\hat{\Phi}, \hat{\mathsf{P}})$, $\hat{\Phi} = ([\hat{A}_j, \hat{K}_j], j \in \mathbb{Z})$, is a stationary sequence. Thus $\psi(\hat{\Phi})$ is a simple, synchronous MPP with the mark space

$$\bigcup_{j=1}^{\infty} \mathbb{K}^j.$$

From the theory of Palm distributions, cf. e.g. [KMN], we obtain a one-to-one correspondence between $\hat{\mathsf{P}}$ and the distribution $\bar{\hat{\mathsf{P}}}$ of a simple, stationary MPP $\overline{\Psi} = ([\bar{\hat{T}}_n, \bar{\hat{K}}_n])$,

$$\bar{\hat{K}}_n = [\bar{K}_{n,1}, \bar{K}_{n,2}, \ldots, \bar{K}_{n,j_n}] \in \bigcup_{j=1}^{\infty} \mathbb{K}^j.$$

Define

$$\overline{T}_n = \bar{\hat{T}}_k \quad \text{if } \sum_{i=1}^{k-1} j_i < n \leqq \sum_{i=1}^{k} j_i, \qquad n > 0,$$

$$\overline{T}_n = \bar{\hat{T}}_k \quad \text{if } -\sum_{i=k}^{0} j_i < n \leqq -\sum_{i=k+1}^{0} j_i, \qquad n \leqq 0,$$

$$\overline{K}_n = \overline{K}_{k,1} \quad \text{if } n = \sum_{i=1}^{k-1} j_i + 1, \qquad n > 0,$$

$$\overline{K}_n = \overline{K}_{k,1} \quad \text{if } n = -\sum_{i=k}^{0} j_i + 1, \qquad n \leqq 0,$$

and let $\overline{\mathsf{P}}$ be the probability distribution of $\overline{\Psi} = ([\overline{T}_n, \overline{K}_n])$. Obviously, the MPP $(\overline{\Psi}, \overline{\mathsf{P}})$ is stationary. Thus we obtain a one-to-one correspondence between stationary sequences (Φ, P) and stationary MPP's $(\overline{\Psi}, \overline{\mathsf{P}})$. Formulae (3.4.1), (3.4.2) can also be proved in this way.

Finally, a remark concerning the Palm distribution of a stationary MPP with respect to a given subset L of the mark space $\mathbb{K}$. Let (Φ, P), $\Phi = ([A_n, K_n])$, be a stationary sequence with the property

$$\mathsf{P}(\# \{n : n < 0, K_n \in L\} = \# \{n : n \geqq 0, K_n \in L\} = \infty) = 1.$$

The conditional distribution $\mathsf{P}^L(\cdot) = \mathsf{P}(\cdot \mid K_0 \in L)$ determines uniquely the distribution P, cf. Theorem A 1.3.2. Thus, in view of Theorem 3.4.1 there is a one-to-one correspondence between the families $\{\mathsf{P}_L\}$ and $\{\overline{\mathsf{P}}\}$, L fixed. Let $\Phi^L = ([A_n^L, K_n^L])$ be a sequence with the probability law P_L. If $\mathsf{P}(A_0 > 0) = 1$, the distribution of the ("L-synchronous") MPP $\Psi^L = ([T_n^L, K_n^L])$, where $T_0^L = 0$; $T_n^L = \sum_{j=0}^{n-1} A_n^L$, $n > 0$, $T_n^L = -\sum_{j=n}^{-1} A_j^L$, $n < 0$, is called the Palm distribution of $\overline{\mathsf{P}}$ with respect to L, cf. [KMN], [FKAS].

3.5. The proofs of the Theorems 3.5.4, 3.5.7 and 3.5.8 stem from [KMN].

3.6. Most of the proofs are due to [KMM]. The relationships (3.6.10) and (3.6.12) remain valid if ergodicity is replaced by the weaker assumption

$$\lim_{n\to\infty} \frac{1}{n} \sum_{j=0}^{n-1} A_j = a_{\mathsf{P}} \quad \mathsf{P}\text{-a.s.},$$

cf. Nawrotzki (1978).

4. Continuous time models

4.1. Introduction

In this chapter we shall investigate stochastic systems working in continuous time. For many systems there are particular instants at which certain events occur. In the analysis of stochastic systems such epochs are often called *embedded epochs*. For example, in queueing systems such embedded epochs could be arrival or departure instants. Another typical example is that of level crossings of a state process. In simulation studies one frequently estimates time averages of a stochastic process via randomly located observations. An appropriate mathematical tool for treating the temporal behaviour of systems with embedded epochs is the theory of *processes* containing an *embedded marked point process* (abbr. *PEMP*): A PEMP is defined as a pair $\chi = \left[\left(X(t), t \in \mathbb{R}\right), \Psi\right]$ consisting of an $\mathbb{X}$-valued stochastic process $\left(X(t)\right)$ with sample paths in the Skorokhod-space $D\left((-\infty, \infty), \mathbb{X}\right)$ and a marked point process (MPP) $\Psi = \left([T_n, [X_n, Y_n]]\right)$ with mark space $\mathbb{X} \times \mathbb{Y}$ satisfying the condition $X_n = X(T_n)$ if $T_{n-1} < T_n$, Section 4.2. If we think of $\left(X(t)\right)$ as a stochastic process describing the time evolution of a certain system in continuous time, then the mark X_n indicates the state of the system just prior to T_n if $T_{n-1} < T_n$. If $T_{n-1} < T_n = T_{n+1} = \cdots = T_{n+r-1} < T_{n+r}$, then the system has exactly r jumps at the instant T_n, and $X_{n+1}, \ldots, X_{n+r-1}$ are the instantaneous states entered, ordered according to their occurence. The Y_n can be interpreted as the random influence from outside. For instance, if we think of the T_n as the arrival epochs of customers in a queueing system, then the Y_n are usually the service times, whereas the X_n are the system states seen by the arriving customers.

The stationarity of a PEMP is defined in the usual way, cf. Definition 4.2.3. (We again supply all the symbols related to a stationary PEMP with the sign "$-$", e.g. $(\overline{\chi}, \overline{\mathsf{P}})$, $\overline{X}(t)$, $\overline{T}_n$. The intensity of a stationary PEMP $\overline{\chi} = \left[\left(\overline{X}(t)\right), \left([\overline{T}_n, \overline{X}_n, \overline{Y}_n]\right)\right]$ is defined as that of the embedded stationary point process $(\overline{T}_n)$.)

For a given PEMP χ we define

$$A_n = T_{n+1} - T_n,$$

$$K_n = \begin{cases} \left[\left(X(T_n + u), 0 < u \leqq A_n\right), X_n, Y_n\right] & \text{if } A_n > 0, \\ [X_n, X_n, Y_n] & \text{otherwise}. \end{cases}$$

We call $[A_n, K_n]$, $n \in \mathbb{Z}$, the *cycles* of χ. If we think of K_n as a new mark associated with the point T_n, then the MPP $([T_n, K_n])$ is an equivalent description of the PEMP χ. Using this and Theorem 3.4.1 we obtain a one-to-one correspondence between the family of all probability distributions P of stationary sequences of cycles with finite mean cycle length and the family of all probability distributions $\bar{\mathsf{P}}$ of stationary PEMP's $\bar{\chi} = \left[\left(\bar{X}(t)\right), \bar{\Psi}\right]$ with finite intensity, Theorem 4.2.5. This result is a far-reaching generalization of a well-known result for regenerative processes — the i.i.d.-assumption about cycles is replaced by stationarity. As a particular consequence we obtain the so-called "*stochastic mean value theorem*": For an arbitrary non-negative function h on $\mathbb{X}$

$$\mathsf{E}_{\bar{\mathsf{P}}} h\left(\bar{X}(0)\right) = \frac{1}{\mathsf{E}_{\mathsf{P}} A_0} \mathsf{E}_{\mathsf{P}} \left(\int_0^{A_0} h\left(X(t)\right) \mathrm{d}t\right),$$

formula (4.2.10). Thus, the probabilities $\bar{\mathsf{P}}\left(\bar{X}(0) \in (\cdot)\right)$ can be calculated by means of the probability distribution of the generic cycle $[A_0, K_0]$.

Several well-known classes of stochastic processes appear as particular PEMP's where the cycles $([A_n, K_n])$ form a time-homogeneous Markov chain. This is demonstrated in Section 4.3 for semi-Markov and semi-regenerative processes. Finally, we sketch briefly how the modelling of the time evolution of complex systems with repair by means of PEMP's leads to a unified approach for deriving stationary (long-run) reliability characteristics of such systems. The approach also works well in cases where there are no regeneration points. To illustrate this we finally derive the mean time between failures (MTBF) of a coherent system with independent and separately maintained components.

In Section 4.4 we construct stationary PEMP's $\bar{\chi} = \left[\left(\bar{X}(t), t \in \mathbb{R}\right), \bar{\Psi}\right]$, $\bar{\Psi} = \left(\left[\bar{T}_n, [\bar{X}_n, \bar{Y}_n]\right], n \in \mathbb{Z}\right)$ describing the time evolution of stochastic systems with the following properties:

(i) The consecutive states X_n of the system at the embedded epochs T_n satisfy the recursive equation

$$X_{n+1} = f(X_n, [A_n, Y_n]), \qquad n \in \mathbb{Z}, \tag{4.1.1}$$

where $A_n = T_{n+1} - T_n$.

(ii) Given the states X_k, $k \in \mathbb{Z}$, the state $X(t)$ of the system at an arbitrary time t is determined by

$$X(t) = f(X_n, [t - T_n, Y_n]), \quad \text{if } T_n < t \leqq T_{n+1}. \tag{4.1.2}$$

It is shown that to each stationary weak solution $\left([[A_n', Y_n'], X_n']\right)$ of (4.1.1) there corresponds a stationary PEMP $\left[\left(\bar{X}'(t)\right), \left([\bar{T}_n', [\bar{X}_n', \bar{Y}_n']]\right)\right]$ satisfying (4.1.1) and (4.1.2) with $\bar{X}_n'$, $\bar{X}'(t)$, $\bar{Y}_n'$, $\bar{T}_n'$, and $\bar{A}_n'$ $(= \bar{T}'_{n+1} - \bar{T}_n')$, instead of X_n, $X(t)$, Y_n, T_n, and A_n, respectively. We call (X_n') an *event-stationary* and $\left(\bar{X}'(t)\right)$ the *associated time-stationary state process* of the system under consideration in steady state.

In Section 4.5 we discuss the following continuous time analogue of (4.1.1):

$$X(t) = \bar{f}\left(X(u), t - u, \theta_u \bar{\Psi}\right), \qquad u, t \in \mathbb{R}, \qquad u < t, \tag{4.1.3}$$

where $\bar{\Psi} = ([\bar{T}_n, \bar{Y}_n])$ is a simple stationary MPP and $\bar{f}(x, a, \psi)$ is left-continuous in a and satisfies

$$\bar{f}(x, t, \psi) = \bar{f}(\bar{f}(x, u, \psi), t - u, \theta_u\psi), \quad 0 < u < t, \quad x \in \mathbb{X}.$$

4.2. Stochastic processes with an embedded point process

As mentioned in the previous chapter, MPP's appear as an appropriate mathematical tool for describing the input of a queueing system. Consider an arbitrary system which changes its state only by jumps. The temporal behaviour of such a system can be described by a simple MPP $\Psi = ([T_n, X_n])$ with the mark space $\mathbb{X}$, where the T_n are the epochs at which jumps occur, $\mathbb{X}$ is the state space of the system and X_n is the state of the system just prior to T_n. Another, more common description of the time evolution of the system is given by the $\mathbb{X}$-valued stochastic process

$$X(t) = X_n \quad \text{if} \quad T_{n-1} < t \leqq T_n. \tag{4.2.1}$$

Obviously both descriptions are equivalent. In particular, the process $(X(t))$ is strictly stationary if and only if the MPP Ψ is stationary in the sense of Definition 3.2.5. The assumption "Ψ is simple" means that "the system under consideration has no instantaneous states". Later we shall see that this assumption can be dropped.

Pure jumpwise processes (or equivalent MPP's) are used to describe the dynamics of systems in reliability framework; in this case $\mathbb{X}$ is finite. Processes of the type (4.2.1) also arise in queueing theory if we are interested e.g. in the queue length only. However, for the investigation of queueing quantities such as virtual waiting time which change their value not only by jumps but also between consecutive jump epochs one needs a somewhat more general notion.

Let $\mathbb{X}$, $\mathbb{Y}$ be arbitrary Polish spaces. We denote by $D((-\infty, \infty), \mathbb{X})$ the space of all left-hand continuous, $\mathbb{X}$-valued functions on $\mathbb{R}$ with right-hand limits, cf. Whitt (1980).

4.2.1. Definition. Let $(X(t), t \in \mathbb{R})$ be a stochastic process with sample paths belonging to $D((-\infty, \infty), \mathbb{X})$ and $\Psi = ([T_n, X_n])$ or $\Psi = ([T_n, [X_n, Y_n]])$ an MPP with the mark space $\mathbb{X}$ or $\mathbb{X} \times \mathbb{Y}$, respectively. Assume that $(X(t))$ and Ψ are both defined on a common probability space and that the marks X_n are connected with $(X(t))$ as follows:

$$X_n = X(T_n) \quad \text{if} \quad T_{n-1} < T_n. \tag{4.2.2}$$

Then $\chi = [(X(t)), \Psi]$ is called a *process with an embedded marked point process* (abbr. *PEMP*). The r.v.'s T_n are called the *embedded points* of χ.

Here we have defined a PEMP on the whole line $\mathbb{R}$. Sometimes one needs PEMP's $\chi = [(X(t), t \geqq 0), ([T_n, [X_n, Y_n]], n \geqq 1)]$ on the non-negative half-line only. The necessary modification of the definition is obvious. In this case the process $(X(t), t \geqq 0)$ has sample paths in $D([0, \infty], \mathbb{X})$ (the space of all left-hand continuous, $\mathbb{X}$-valued functions on $[0, \infty)$ with right-hand limits, cf. Lindvall (1973), Whitt (1980), Ivanoff (1980), Kalashnikov (1983)).

If we think of $(X(t))$ as a stochastic process describing the temporal behaviour of a certain system in continuous time, then the mark X_n indicates the system state just prior to T_n if $T_{n-1} < T_n$. If $T_{n-1} < T_n = T_{n+1} = \cdots = T_{n+r-1} < T_{n+r}$, then the system has exactly r jumps at the instant T_n, and $X_{n+1}, \ldots, X_{n+r-1}$ are the instantaneous states entered, ordered according to their occurence. These "intermediate states" cannot be extracted from the process $(X(t))$ and are therefore "put into Ψ" as marks. If there are no multiple points, we can drop the mark components X_n in view of (4.2.2).

Often the MPP Ψ has the form $\Psi = ([T_n, X_n])$ and is defined by the evolution of the considered system itself, i.e. Ψ is determined by the process $(X(t))$. This will happen if the embedded points are epochs at which the process $(X(t))$ crosses some fixed level or has some well-defined jumps, etc. However, the T_n, $n \in \mathbb{Z}$, may also be moments at which some "influence from outside" occurs, e.g. the T_n may be epochs at which the environmental load changes. In the latter case Y_n may denote the value of the load within the interval $(T_n, T_{n+1}]$, whereas $(X(t))$ describes the system behaviour. In some cases it is convenient to denote by Y_n the system state just after the n-th jump; then $Y_n = X(T_n + 0)$ if $T_n < T_{n+1}$, and $Y_n = X_{n+1}$ if $T_n = T_{n+1}$, cf. Section 4.3.

4.2.2. Example. Consider a single server loss system with the input $([A_n, S_n])$. Let $([[A_n, S_n], X_n])$ be a strong solution of the equation

$$X_{n+1} = \max(X_n + S_n \mathbf{1}\{X_n = 0\} - A_n, 0), \qquad n \in \mathbb{Z}. \tag{4.2.3}$$

Customer number 0 is supposed to arrive at time 0. Then the arrival epochs are determined by

$$T_0 = 0; \quad T_n = \sum_{i=0}^{n-1} A_i, \quad n > 0; \quad T_n = \sum_{i=n}^{-1} A_i, \quad n < 0.$$

The state X_n is the residual service time seen by the n-th arriving customer (the residual service time of the idle server is assumed to be equal to 0). Notice that if $T_{n-1} < T_n = \cdots = T_{n+k} < T_{n+k+1}$ then $X_{n+1} = \cdots = X_{n+k}$, i.e. all rejected customers within a batch see the same state. The residual service time $X(t)$ at an arbitrary instant t is given by

$$X(t) = \max\left(X_n + S_n \mathbf{1}\{X_n = 0\} - (t - T_n), 0\right) \text{ if } T_n < t \leqq T_{n+1}.$$

Obviously, $[(X(t)), ([T_n, [X_n, S_n]])]$ is a PEMP.

Using other embedded points, we can describe the same system by means of a different PEMP: The departure epochs of customers are given by

$$T_k' = \begin{cases} T_k & \text{if } X_k > 0, \\ T_k + S_k & \text{if } X_k = 0. \end{cases}$$

Notice that there are two types of departures: If $X_k = 0$, then T_k' is the departure epoch of a served customer. In case of $X_k > 0$ the k-th customer is rejected. By renumbering the sequence (T_k') in ascending order we obtain the sequence (T_n^*) of consecutive departures of customers (here we consider coinciding departures as a single epoch):

$$\cdots < T_{-1}^* < T_0^* \leqq 0 < T_1^* < \cdots.$$

Let $X_n^* = X(T_n^*)$, $n \in \mathbb{Z}$. Then $[(X(t)), ([T_n^*, X_n^*])]$ is a PEMP. Since (X_n^*) is uniquely determined by $(X(t))$ and (T_n^*), we may drop the marks X_n^*. The pair $[(X(t)), (T_n^*)]$ describes completely the system under consideration.

While in the previous chapter we have obtained a generalization of results from renewal theory, our aim here is to extend some facts known for regenerative processes. This extension follows immediately from the results of Chapter 3 if we use the following interpretation of a PEMP as an MPP with a suitable chosen mark space.

For a given PEMP $\chi = [(X(t)), \Psi]$, $\Psi = ([T_n, [X_n, Y_n]])$, define

$$A_n = T_{n+1} - T_n, \quad K_n = \begin{cases} [(X(T_n + u), 0 < u \leqq A_n), X_n, Y_n] & \text{if } A_n > 0, \\ [X_n, X_n, Y_n] & \text{otherwise}. \end{cases}$$

We call the $[A_n, K_n]$, $n \in \mathbb{Z}$, the *cycles* of χ. A cycle $[A_n, K_n]$ is a complete description of the temporal behaviour of the system and the environmental load between the embedded points T_n and T_{n+1}. Since we need not distinguish between K_n and $[(\check{X}_n(u), 0 < u < \infty), X_n, Y_n]$ where

$$\check{X}_n(u) = \begin{cases} X(T_n + u) & \text{if } 0 < u \leqq A_n, \quad A_n > 0, \\ X(T_{n+1}) = X_{n+1} & \text{if } u > A_n, \quad A_n > 0, \\ X_n & \text{if } u > 0, \quad A_n = 0, \end{cases}$$

we can consider K_n as a random element of $D((0, \infty), \mathbb{X}) \times \mathbb{X} \times \mathbb{Y}$.

If we think of K_n as a new mark associated with the point T_n, then the PEMP appears as the MPP $\Psi(\chi) = ([T_n, K_n])$, cf. (3.2.3), with the mark space $D((0, \infty), \mathbb{X}) \times \mathbb{X} \times \mathbb{Y}$.

An important particular case of PEMP's on $[0, \infty)$ are the regenerative processes. We call the PEMP $\chi = [(X(t), t \geqq 0), ([T_n, X_n, Y_n], n \geqq 1)]$ a *regenerative process* if the cycles $[A_n, K_n]$, $n \geqq 1$, are i.i.d. and independent of $[A_0, K_0]$, where

$$A_0 = T_1, \qquad K_0 = (X(u), 0 \leqq u \leqq A_0).$$

(Usually, the process $(X(t), t \geqq 0)$ is called a regenerative process if there is a point process $(T_n, n \geqq 1)$ such that the "cycles" $(X(T_n + u), 0 < u \leqq T_{n+1} - T_n)$, $n \geqq 1$, are i.i.d. and independent of $(X(u), 0 \leqq u \leqq T_1)$, cf. e.g. Brown and Ross (1972), Çinlar (1975), Cohen (1976b).)

For $t \in \mathbb{R}$ and a PEMP $\chi = [(X(u), u \in \mathbb{R}), \Psi]$ define the shifted PEMP $\theta_t\chi$ by

$$\theta_t\chi = [(X(u + t), u \in \mathbb{R}), \theta_t\Psi],$$

where $\theta_t\Psi$ is defined in Section 3.2.

4.2.3. Definition. A PEMP $(\bar{\chi}, \bar{\mathsf{P}})$, $\bar{\chi} = [(\bar{X}(u), u \in \mathbb{R}), \bar{\Psi}]$, is called *stationary* if for every $t \in \mathbb{R}$ the shifted PEMP $\theta_t\bar{\chi}$ has the same probability distribution $\bar{\mathsf{P}}$ as $\bar{\chi}$.

Obviously, $\bar{\chi}$ is stationary if and only if the corresponding MPP $\psi(\bar{\chi})$ is stationary. For the embedded points of a stationary PEMP we have, cf. (3.2.8),

$$\cdots \leqq \overline{T}_0 < 0 < \overline{T}_1 \leqq \cdots \quad \overline{\mathsf{P}}\text{-a.s.}$$

The value

$$\lambda_{\overline{\mathsf{P}}} = \mathsf{E}_{\overline{\mathsf{P}}} \overline{N}(1), \qquad \overline{N}(1) = \max\{n : \overline{T}_n \leqq 1\},$$

is called the intensity of the stationary PEMP $(\bar{\chi}, \overline{\mathsf{P}})$.

Now we want to generalize the well-known relationship between stationary and synchronous regenerative processes, cf. ÇINLAR (1975), COHEN (1976b). For this reason we consider a stationary sequence $\left(\left[A_n, \left[\left(X_n(t), t > 0\right), X_n, Y_n\right]\right]\right)$ of $\mathbb{R}_+ \times \left(D\left((0, \infty), \mathbf{X}\right) \times \mathbf{X} \times \mathbf{Y}\right)$-valued r.v.'s. Assume

$$A_n \geqq 0, \qquad \sum_{n=-\infty}^{-1} A_n = \sum_{n=0}^{\infty} A_n = \infty \tag{4.2.4}$$

and

$$X_n(t) = \begin{cases} X_{n+1} & \text{if } t > A_n > 0, \\ X_n & \text{if } A_n = 0, \quad t > 0. \end{cases}$$

Define

$$K_n = \begin{cases} \left[\left(X_n(t), 0 < t \leqq A_n\right), X_n, Y_n\right] & \text{if } A_n > 0, \\ \left[X_n, X_n, Y_n\right] & \text{if } A_n = 0. \end{cases} \tag{4.2.5}$$

Then $\xi = ([A_n, K_n])$ is stationary; it can be interpreted as the sequence of cycles of a PEMP.

4.2.4. Definition. Let (ξ, P), $\xi = ([A_n, K_n])$, be a stationary sequence of cycles, i.e. (A_n) satisfies (4.2.4) and the r.v.'s K_n, $n \in \mathbb{Z}$, have the form (4.2.5). The PEMP $\chi(\xi) = \left[\left(X(t)\right), \left(\left[T_n, [X_n, Y_n]\right]\right)\right]$, where

$$T_0 = 0; \qquad T_n = \sum_{j=0}^{n-1} A_j, \quad n > 0; \qquad T_n = -\sum_{j=n}^{-1} A_j, \quad n < 0,$$

$$X(t) = X_n(t - T_n) \quad \text{for } T_n < t \leqq T_{n+1}, \tag{4.2.6}$$

and X_n, Y_n are the same as in K_n, is called the *synchronous PEMP generated by* ξ. (Thus, a synchronous PEMP is equivalent to a stationary sequence of cycles.)

As a consequence of Theorem 3.4.1 we obtain:

4.2.5. Theorem. *There exists a one-to-one mapping between the families*

$\mathsf{F} = \{\mathsf{P}$: P *is the probability distribution of a stationary sequence of cycles* $\xi = ([A_n, K_n])$, *where* (A_n) *satisfies* $a_{\mathsf{P}} = \mathsf{E}_{\mathsf{P}} A_0 < \infty$, (4.2.4), *and* K_n *are defined by* (4.2.5)$\}$

and

$\overline{\mathsf{F}} = \{\overline{\mathsf{P}}$: $\overline{\mathsf{P}}$ *is the probability distribution of a stationary PEMP* $\bar{\chi} = \left(\left(\overline{X}(t)\right), \overline{\Psi}\right)$, $\overline{\Psi} = \left(\left[\overline{T}_n, [\overline{X}_n, \overline{Y}_n]\right]\right)$, *with intensity* $\lambda_{\overline{\mathsf{P}}} < \infty\}$.

For a given stationary sequence of cycles (ξ, P) *with* $\mathsf{P} \in \mathbf{F}$ *the corresponding stationary PEMP* $(\bar{\chi}, \bar{\mathsf{P}})$ *is determined by*

$$\bar{\mathsf{P}}\big(\bar{\chi} \in (\cdot)\big) = a_{\mathsf{P}}^{-1} \int_0^\infty \mathsf{P}\big(A_0 > t, \theta_t \chi(\xi) \in (\cdot)\big)\, \mathrm{d}t, \tag{4.2.7}$$

where $\chi(\xi)$ *is the synchronous PEMP generated by* ξ, *cf. Definition* 4.2.4.
Conversely, we have

$$\mathsf{P}\big(\xi \in (\cdot)\big) = \lambda_{\bar{\mathsf{P}}}^{-1} \sum_{j=1}^{\infty} \bar{\mathsf{P}}\big(\bar{T}_j \leqq 1, \theta^j \bar{\xi}(\bar{\chi}) \in (\cdot)\big), \tag{4.2.8}$$

where

$$\bar{\xi}(\bar{\chi}) = \big([\bar{T}_{n+1} - \bar{T}_n, \bar{K}_n]\big),$$

$$\bar{K}_n = \begin{cases} \big[\big(\bar{X}(t), 0 < t \leqq \bar{T}_{n+1} - \bar{T}_n\big), \bar{X}_n, \bar{Y}_n\big] & \text{if } \bar{T}_{n+1} - \bar{T}_n > 0, \\ [\bar{X}_n, \bar{X}_n, \bar{Y}_n] & \text{if } \bar{T}_{n+1} = \bar{T}_n. \end{cases}$$

Moreover,

$$\lambda_{\bar{\mathsf{P}}} = a_{\mathsf{P}}^{-1}.$$

In the same way as for MPP's, cf. Remark 3.4.2, one can state several formulae which are equivalent to (4.2.7). In particular, (4.2.7) is equivalent to

$$\mathsf{E}_{\bar{\mathsf{P}}} h(\bar{\chi}) = a_{\mathsf{P}}^{-1} \mathsf{E}_{\mathsf{P}} \int_0^{A_0} h\big(\theta_t \chi(\xi)\big)\, \mathrm{d}t \tag{4.2.9}$$

for every measurable real-valued non-negative function h. For any non-negative function h on $\mathbf{X}$ we obtain

$$\mathsf{E}_{\bar{\mathsf{P}}} h\big(\bar{X}(0)\big) = a_{\mathsf{P}}^{-1} \mathsf{E}_{\mathsf{P}} \int_0^{A_0} h\big(X_0(t)\big)\, \mathrm{d}t = a_{\mathsf{P}}^{-1} \mathsf{E}_{\mathsf{P}} \int_0^{A_0} h\big(X(t)\big)\, \mathrm{d}t. \tag{4.2.10}$$

Formula (4.2.10) is called the *stochastic mean value theorem*. Taking $h(\cdot) = \mathbf{1}_L(\cdot)$, L a measurable subset of $\mathbf{X}$, from (4.2.10) we obtain

$$\bar{\mathsf{P}}\big(\bar{X}(0) \in L\big) = a_{\mathsf{P}}^{-1} \mathsf{E}_{\mathsf{P}} \int_0^{A_0} \mathbf{1}_L\big(X(t)\big)\, \mathrm{d}t.$$

Thus, the stationary probability $\bar{\mathsf{P}}\big(\bar{X}(0) \in L\big)$ appears as the ratio of the mean sojourn time of $\big(X(t)\big)$ in the set L during a generic cycle to the mean length of a generic cycle.

Theorem 3.6.6 yields the following

4.2.6. Theorem. *Consider a stationary ergodic sequence of cycles* (ξ, P), $\xi = ([A_n, K_n])$, *where* (A_n) *satisfies* $a_{\mathsf{P}} < \infty$, (4.2.4.), *and* K_n *are defined by* (4.2.5.). *Let* $(\bar{\chi}, \bar{\mathsf{P}})$ *be the stationary PEMP associated with* (ξ, P) *in the sense of Theorem* 4.2.5. *Then for every measurable, non-negative function* h *with* $\mathsf{E}_{\bar{\mathsf{P}}} h(\bar{\chi}) < \infty$

$$\lim_{t \to \infty} \frac{1}{t} \int_0^t h\big(\theta_u \chi(\xi)\big)\, \mathrm{d}u = \mathsf{E}_{\bar{\mathsf{P}}} h(\bar{\chi}) \qquad \mathsf{P}\text{-}a.s. \tag{4.2.11}$$

Moreover,

$$\lim_{t\to\infty} \frac{1}{t} \int_0^t \mathsf{P}(\theta_u\chi(\xi) \in (\cdot))\, du = \overline{\mathsf{P}}(\overline{\chi} \in (\cdot)), \tag{4.2.12}$$

$$\lim_{t\to\infty} \frac{1}{t} \int_0^t \mathsf{P}(X(u) \in (\cdot))\, du = \overline{\mathsf{P}}(\overline{X}(0) \in (\cdot)). \tag{4.2.13}$$

For every measurable, non-negative function g with $\mathsf{E}_{\mathsf{P}}g(\xi) < \infty$

$$\lim_{n\to\infty} \frac{1}{n} \sum_{j=1}^{n} g(\theta^j\xi(\overline{\chi})) = \mathsf{E}_{\mathsf{P}}g(\xi), \quad \overline{\mathsf{P}}\text{-a.s.} \tag{4.2.14}$$

Moreover,

$$\lim_{n\to\infty} \frac{1}{n} \sum_{j=1}^{n} \overline{\mathsf{P}}(\theta^j\xi(\overline{\chi}) \in (\cdot)) = \mathsf{P}(\xi \in (\cdot)). \tag{4.2.15}$$

If ξ is ergodic, then, in view of Theorem 4.2.6, we can interpret $\overline{\chi}$ as the model of the temporal behaviour of a certain system in its steady state seen by an observer located in an arbitrary point of the time axis (the time origin is an arbitrarily chosen point of the time axis). On the other hand, $\chi(\xi)$ provides the description of the behaviour of the system seen by an observer located in an arbitrary embedded point.

Assume that there are no marks Y_n and that $\xi = ([A_n, K_n])$ consists of i.i.d. cycles. Then $\chi(\xi)$ and $\overline{\chi}$, where $\overline{\chi}$ is given by Theorem 4.2.5, are the synchronous and stationary regenerative processes, respectively, determined by the probability distribution of the generic cycle $[A_0, K_0]$. The stochastic mean value theorem (4.2.10) and the limiting property (4.2.13) are well-known in this case, cf. e.g. Brown and Ross (1972), Çinlar (1975), Cohen (1976b). In some important cases the regeneration points may be identified with the entries into some fixed state $x \in \mathbf{X}$. For example in queueing theory x could be the idle state; in Markov processes x could be any positively recurrent state.

Let (χ, P), $\chi = [(X(t)), \Psi]$, $\Psi = ([T_n, [X_n, Y_n]])$, $a_{\mathsf{P}} < \infty$, and $(\overline{\chi}, \overline{\mathsf{P}})$, $\overline{\chi} = [(\overline{X}(t)), \overline{\Psi}]$, $\overline{\Psi} = ([\overline{T}_n, [\overline{X}_n, \overline{Y}_n]])$, be a synchronous PEMP and the corresponding stationary, PEMP, respectively. We obtain thef ollowing useful result, cf. Theorem 3.5.1.

4.2.7. Theorem.

(i) *If A_0 $(= T_1)$ and $X(0)$ are independent, then*

$$X(0) \overset{\mathcal{D}}{=} \overline{X}(\overline{T}_0). \tag{4.2.16}$$

(ii) *If A_0 and $[X(0), X_0, Y_0]$ are independent, then*

$$[X(0), X_0, Y_0] \overset{\mathcal{D}}{=} [\overline{X}(\overline{T}_0), \overline{X}_0, \overline{Y}_0]. \tag{4.2.17}$$

(iii) *If A_0 and $(X(t), t \leqq 0)$ are independent, then*

$$(\overline{X}(t), t \leqq \overline{T}_0) \overset{\mathcal{D}}{=} (X(t), t \leqq 0). \tag{4.2.18}$$

(iv) *If* A_0 *and* $[(X(t), t \leqq 0), X_0, Y_0]$ *are independent, then*

$$[(\overline{X}(t), t \leqq \overline{T}_0), \overline{X}_0, \overline{Y}_0] \overset{\mathfrak{D}}{=} [(X(t), t \leqq 0), X_0, Y_0]. \tag{4.2.19}$$

(v) *If* A_0 *and* $[(X(t), t \leqq 0), ([T_n, [X_n, Y_n]], n \leqq 0)]$ *are independent, then*

$$[(\overline{X}(t), t \leqq \overline{T}_0), ([\overline{T}_n, [\overline{X}_n, \overline{Y}_n]], n \leqq 0)]$$
$$\overset{\mathfrak{D}}{=} [(X(t), t \leqq 0), ([T_n, [X_n, Y_n]], n \leqq 0)]. \tag{4.2.20}$$

Proof.

(i) Applying (4.2.7), we get

$$\overline{\mathsf{P}}(\overline{X}(\overline{T}_0) \in (\cdot)) = a_{\mathsf{P}}^{-1} \int_0^\infty \mathsf{P}(A_0 > t, X(0) \in (\cdot)) \, \mathrm{d}t$$
$$= \mathsf{P}(X(0) \in (\cdot)).$$

The proofs of (ii)–(v) are similar. ∎

4.2.8. Remark. For the description of the temporal behaviour of a given system the choice of the sequence of embedded points and thus the sequence ξ of cycles, the synchronous PEMP $\chi(\xi)$ and the stationary PEMP $\overline{\chi}$ are by no means unique, cf. Example 4.2.2. Let $(\overline{X}(t))$ be a stationary stochastic process describing the behaviour of a certain system, and let $\overline{\Psi}_1$ and $\overline{\Psi}_2$ be two different MPP's (defined on the same probability space as $(\overline{X}(t))$) such that $[(\overline{X}(t)), \overline{\Psi}_1]$ and $[(\overline{X}(t)), \overline{\Psi}_2]$ are stationary PEMP's with finite intensities. Then the corresponding stationary sequences ξ_1 and ξ_2 of cycles are of course different. Thus, applying formula (4.2.7) to ξ_1 and ξ_2, one obtains two different expressions for the distribution of $(\overline{X}(t))$. It depends on the choice of embedded points (i.e. on the choice of $\overline{\Psi}_1$ or $\overline{\Psi}_2$), which of these expressions is less complicated. In general one can expect that the calculation of the distribution of $(\overline{X}(t))$ is easier for short cycles. This becomes clear if one thinks of the stochastic mean value theorem (4.2.10). In Chapter 6 we shall discuss situations in which there are regeneration points, but for calculating stationary system characteristics it is much more convenient to use the more frequent embedded points. These embedded points have no regeneration properties. The advantage of the approach presented here is precisely that independence assumptions on the stationary sequence of cycles are not required.

4.3. Semi-Markov and semi-regenerative processes. Reliability analysis of complex systems with repair

Several well-known classes of stochastic processes can be considered as particular PEMP's where the cycles $([A_n, K_n])$ form a homogeneous Markov chain. Here we discuss in detail semi-Markov and semi-regenerative processes with an arbitrary (Polish) state space; cf. Section 4.6 for references to further examples.

For the description of stochastic objects arising in this section by means of PEMP's a somewhat modified notion of PEMP's is more appropriate. One considers a pair $[(X(t)), ([T_n, X_n])]$, where $X(t)$ and T_n are of the same meaning as in Definition 4.2.1. However, X_n denotes the state of the system just after the n-th jump, i.e. $X_n = X(T_n + 0)$ if $T_n < T_{n+1}$. In order to avoid confusion, we will here use PEMP's in the sense of Definition 4.2.1 with the following special form: $[(X(t)), ([T_n, Y_n])]$, where Y_n denotes the state of the system just after the n-th jump, i.e. $Y_n = X(T_n + 0)$ if $T_n < T_{n+1}$; if $T_{n-1} < T_n = T_{n+1} = \cdots = T_{n+r-1} < T_{n+r}$, then we have exactly r jumps at the instant T_n, and $Y_n, \ldots, Y_{n+r-2}$ are the instantaneous states entered, ordered according to their occurence. (We can drop the marks X_n because $X_n = X(T_n)$ if $T_{n-1} < T_n$ and $X_n = Y_{n-1}$ if $T_{n-1} = T_n$.)

Consider a synchronous PEMP (χ, P), $\chi = [(X(t)), ([T_n, Y_n])]$, of the type just mentioned where

$$X(t) = \sum_{n=-\infty}^{\infty} Y_n \mathbf{1}_{(T_n, T_{n+1}]}(t)$$

$(X(t)$ is a pure jumpwise process$)$. Define

$$p(\cdot) = \mathsf{P}(Y_0 \in (\cdot)),$$

$$\mathsf{F}_y(u) = \mathsf{P}(A_0 \leqq u \mid Y_0 = y), \qquad a_y = \mathsf{E}_{\mathsf{P}}(A_0 \mid Y_0 = y).$$

(Note that $\mathsf{F}_y(u)$ and a_y are well-defined since $\mathbf{X}$ is Polish.) Using the general formula

$$\overline{\mathsf{P}}(\overline{X}(0) \in L) = a_{\mathsf{P}}^{-1} \mathsf{E}_{\mathsf{P}} \left(\int_0^{A_0} \mathbf{1}_L(X(u)) \, \mathrm{d}u \right) \tag{4.3.1}$$

(which follows from (4.2.10) for $h(\cdot) = \mathbf{1}_L(\cdot)$, L a measurable subset of $\mathbf{X}$) and taking into account that the state Y_0 cannot change within the interval $(0, A_0)$, we obtain

$$\begin{aligned}\overline{\mathsf{P}}(\overline{X}(0) \in L) &= a_{\mathsf{P}}^{-1} \int_{\mathbf{X}} \mathsf{E} \left(\int_0^{A_0} \mathbf{1}_L(X(u)) \, \mathrm{d}u \mid Y_0 = y \right) p(\mathrm{d}y) \\ &= a_{\mathsf{P}}^{-1} \int_L \mathsf{E}_{\mathsf{P}}(A_0 \mid Y_0 = y) \, p(\mathrm{d}y) \\ &= \left(\int_{\mathbf{X}} a_y p(\mathrm{d}y) \right)^{-1} \int_L a_y p(\mathrm{d}y). \end{aligned} \tag{4.3.2}$$

Finally, by means of (4.2.7) we derive the joint stationary distribution of the system state and the "forward recurrence time"

$$\begin{aligned}\overline{\mathsf{P}}(\overline{X}(0) \in L, \overline{T}_1 \leqq t) &= a_{\mathsf{P}}^{-1} \int_0^{\infty} \mathsf{P}(A_0 > u, A_0 - u \leqq t, X(u) \in L) \, \mathrm{d}u \\ &= a_{\mathsf{P}}^{-1} \int_0^{\infty} \mathsf{P}(A_0 > u, A_0 - u \leqq t, Y_0 \in L) \, \mathrm{d}u\end{aligned}$$

$$= a_{\mathsf{P}}^{-1} \int_L \int_0^\infty \mathsf{P}(A_0 > u, A_0 - u \leqq t \mid Y_0 = y) \, du p(dy)$$

$$= a_{\mathsf{P}}^{-1} \int_L \int_0^\infty \big(F_y(t + u) - \mathsf{F}_y(u)\big) \, du p(dy)$$

$$= a_{\mathsf{P}}^{-1} \int_L \int_0^t \big(1 - \mathsf{F}_y(u)\big) \, du p(dy) . \tag{4.3.3}$$

Using (4.3.2) we can rewrite (4.3.3) as follows:

$$\overline{\mathsf{P}}\big(\overline{X}(0) \in L, \overline{T}_1 \leqq t\big) = \int_L a_y^{-1} \int_0^t \big(1 - \mathsf{F}_y(u)\big) \, du \overline{\mathsf{P}}\big(\overline{X}(0) \in dy\big)$$

(we take $a_y^{-1} \int_0^t \big(1 - \mathsf{F}_y(u)\big) \, du = 0$ if $a_y = 0$). Thus,

$$\overline{\mathsf{P}}\big(\overline{T}_1 \leqq t \mid \overline{X}(0) = y\big) = a_y^{-1} \int_0^t \big(1 - \mathsf{F}_y(u)\big) \, du \qquad \overline{\mathsf{P}}\text{-a.s}$$

4.3.1. Semi-Markov processes

First we introduce Markov renewal processes with an arbitrary state space $\mathbf{X}$, cf. Sections 3.3 and 3.5 for the discrete case. Consider a probability measure $q(\cdot)$ on the Borel σ-field $\mathscr{H}$ of $\mathbf{X}$, a stochastic kernel $p(y, \cdot)$ from $\mathbf{X}$ into $\mathbf{X}$, and families $\{\mathsf{F}^0_{yz}(u), y, z \in \mathbf{X}\}$ and $\{\mathsf{F}_{yz}(u), y, z \in \mathbf{X}\}$ of distribution functions on $\mathbb{R}_+$ which are measurable in y, z for every fixed $u \geqq 0$. Let $([A_n, Y_n], n \geqq 0)$ be a sequence of $\mathbb{R}_+ \times \mathbf{X}$-valued r.v.'s with the properties:

(A 4.3.1) $\mathsf{P}(A_n \leqq u, Y_{n+1} \in (\cdot) \mid Y_0 = y_0, \ldots, Y_n = y_n, A_0 = x_0, \ldots, A_{n-1} = x_{n-1})$

$$= \mathsf{P}\big(A_n \leqq u, Y_{n+1} \in (\cdot) \mid Y_n = y_n\big), \quad n \geqq 1,$$

(A 4.3.2) $\mathsf{P}\big(Y_0 \in (\cdot)\big) = q(\cdot)$;

$$\mathsf{P}\big(Y_{n+1} \in (\cdot) \mid Y_n = y\big) = p(y, \cdot), \qquad n \geqq 0;$$

$$\mathsf{P}(A_0 \leqq u \mid Y_0 = y, Y_1 = z) = \mathsf{F}^0_{yz}(u);$$

$$\mathsf{P}(A_n \leqq u \mid Y_n = y, Y_{n+1} = z) = \mathsf{F}_{yz}(u), \qquad n \geqq 1.$$

The MPP $([T_n, Y_n], n \geqq 1)$ with

$$T_n = \sum_{j=0}^{n-1} A_j, \qquad n \geqq 1,$$

is called a *Markov renewal process* determined by the parameters $q(\cdot)$, $p(y, \cdot)$, $\{\mathsf{F}^0_{yz}(u)\}$, $\{\mathsf{F}_{yz}(u)\}$, cf. Section 3.3. The stochastic process $\big(X(t), \geqq 0\big)$,

$$X(t) = Y_0 \mathbf{1}_{(0,T_1]}(t) + \sum_{n=1}^\infty Y_n \mathbf{1}_{(T_n, T_{n+1}]}(t),$$

is called the semi-Markov process determined by $q(\cdot)$, $p(y, \cdot)$, $\{\mathsf{F}^0_{yz}(u)\}$, and $\{\mathsf{F}_{yz}(u)\}$.

The sequence $([A_n, Y_n], n \geqq 0)$ is a Markov chain. If there exists a unique probability measure $p(\cdot)$, which is invariant with respect to $p(y, \cdot)$, i.e. which satisfies the equation

$$p(\cdot) = \int_{\mathbf{X}} p(y, \cdot)\, p(\mathrm{d}y),$$

then the sequence $([A_n, Y_n], n \geqq 0)$ is stationary if and only if $q = p$ and $\mathsf{F}^0_{yz}(u) = \mathsf{F}_{yz}(u)$ for all $y, z \in \mathbf{X}$. This can be shown by the same arguments as are used in Section 3.3. In this case there exists a unique stationary sequence (Φ, P), $\Phi = ([A_n, Y_n], n \in \mathbb{Z})$, with the transition mechanism given by (A 4.3.1) and (A 4.3.2) and with the one-dimensional distribution

$$\mathsf{P}(A_0 \leqq u, Y_0 \in L) = \int_L \int_{\mathbf{X}} \mathsf{F}_{yz}(u)\, p(y, \mathrm{d}z)\, p(\mathrm{d}y).$$

According to Definition 3.2.2, Φ defines the so-called *synchronous Markov renewal process* $([T_n, Y_n], n \in \mathbb{Z})$ determined by $p(y, \cdot)$, $\{\mathsf{F}_{yz}(u)\}$. The process

$$X(t) = \sum_{n=-\infty}^{\infty} Y_n \mathbf{1}_{(T_n, T_{n+1}]}(t), \qquad t \in \mathbb{R},$$

is the *synchronous semi-Markov process* on $\mathbb{R}$ determined by $p(y, \cdot)$, $\{\mathsf{F}_{yz}(u)\}$.

Obviously (χ, P), $\chi = \left[\big(X(t)\big), \big([T_n, Y_n]\big)\right]$, is a synchronous PEMP. (Since the distribution of χ is generated by that of $\Phi = ([A_n, Y_n])$, we use the symbol P again.) If

$$a_{\mathsf{P}} = \mathsf{E}_{\mathsf{P}} A_0 = \int_{\mathbf{X}} \int_{\mathbf{X}} \int_0^{\infty} \big(1 - \mathsf{F}_{yz}(u)\big)\, \mathrm{d}u\, p(y, \mathrm{d}z)\, p(\mathrm{d}y) < \infty,$$

then the stationary PEMP $(\bar{\chi}, \bar{\mathsf{P}})$, $\bar{\chi} = \left[\big(\bar{X}(t)\big), \big([\bar{T}_n, \bar{Y}_n]\big)\right]$, associated with (χ, P) is well-defined by Theorem 4.2.5. The process $\big(\bar{X}(t)\big)$,

$$\bar{X}(t) = \sum_{n=-\infty}^{\infty} \bar{Y}_n \mathbf{1}_{(\bar{T}_n, \bar{T}_{n+1}]}(t),$$

is the stationary semi-Markov process governed by $p(y, \cdot)$, $\{\mathsf{F}_{yz}(u)\}$. Taking into account

$$\mathsf{F}_y(u) = \mathsf{P}(A_0 \leqq u \mid Y_0 = y) = \int_{\mathbf{X}} \mathsf{F}_{yz}(u)\, p(y, \mathrm{d}z),$$

$$a_y = \mathsf{E}_{\mathsf{P}}(A_0 \mid Y_0 = y) = \int_{\mathbf{X}} \int_0^{\infty} \big(1 - \mathsf{F}_{yz}(u)\big)\, \mathrm{d}u\, p(y, \mathrm{d}z),$$

from (4.3.2) and (4.3.3) we obtain

$$\bar{\mathsf{P}}\big(\bar{X}(0) \in L\big) = \int_L a_y p(\mathrm{d}y) \Big(\int_{\mathbf{X}} a_y p(\mathrm{d}y)\Big)^{-1}$$

$$= a_{\mathsf{P}}^{-1} \int_L \int_{\mathbf{X}} \int_0^{\infty} \big(1 - \mathsf{F}_{yz}(u)\big)\, \mathrm{d}u\, p(y, \mathrm{d}z)\, p(\mathrm{d}y), \qquad (4.3.4)$$

$$\bar{\mathsf{P}}\left(\bar{X}(0) \in L, \bar{T}_1 \leqq t\right) = a_{\mathsf{P}}^{-1} \int_L \int_0^t \left(1 - \mathsf{F}_y(u)\right) du\, p(dy)$$

$$= a_{\mathsf{P}}^{-1} \int_L \int_{\mathbf{X}} \int_0^t \left(1 - \mathsf{F}_{yz}(u)\right) du\, p(y, dz)\, p(dy). \tag{4.3.5}$$

4.3.2. Semi-regenerative processes

A very useful generalization of semi-Markov processes are semi-regenerative processes. Let $\mathbf{Y}$ be a closed subset of the Polish space $[\mathbf{X}, \mathscr{H}]$. Consider a PEMP $\left[\left(X(t), t \geqq 0\right), ([T_n, Y_n], n \geqq 1)\right]$ and a $\mathbf{Y}$-valued r.v. Y_0, both given on a common probability space. Assume that the cycles $([A_n, K_n], n \geqq 0)$ with $A_0 = T_1$, $A_n = T_{n+1} - T_n$, $K_0 = \left[\left(X(t), 0 \leqq t \leqq A_0\right), Y_0\right]$,

$$K_n = \begin{cases} Y_n & \text{if } A_n = 0, \\ \left[\left(X(T_n + u), 0 < u \leqq A_n\right), Y_n\right] & \text{otherwise} \end{cases}$$

satisfy the following conditions, where $q(\cdot)$ is a given probability distribution on $\mathscr{Y} = \mathbf{Y} \cap \mathscr{H}$, $p(y, \cdot)$, $\mathsf{Q}^0_{yz}(\cdot)$, $\mathsf{Q}_{yz}(\cdot)$ are given stochastic kernels (remember that the marks X_n have been dropped):

(A 4.3.3) $\mathsf{P}\left([A_n, K_n] \in (\cdot), Y_{n+1} \in C \mid Y_0 = y_0, \ldots, Y_n = y_n,\right.$

$\left.[A_0, K_0] = [a_0, k_0], \ldots, [A_{n-1}, K_{n-1}] = [a_{n-1}, k_{n-1}]\right)$

$= \mathsf{P}\left([A_n, K_n] \in (\cdot), Y_{n+1} \in C \mid Y_n = y_n\right), \qquad n \geqq 1;$

(A 4.3.4) $\mathsf{P}\left(Y_0 \in (\cdot)\right) = q(\cdot),$

$\mathsf{P}\left(Y_{n+1} \in (\cdot) \mid Y_n = y\right) = p(y, \cdot), \qquad n \geqq 0,$

$\mathsf{P}\left([A_0, K_0] \in (\cdot) \mid Y_0 = y, Y_1 = z\right) = \mathsf{Q}^0_{yz}(\cdot),$

$\mathsf{P}\left([A_{n+1}, K_{n+1}] \in (\cdot) \mid Y_n = y, Y_{n+1} = z\right) = \mathsf{Q}_{yz}(\cdot), \qquad n \geqq 1.$

Then the process $\left(X(t), t \geqq 0\right)$ is called the semi-regenerative process determined by $q(\cdot)$, $p(y, \cdot)$, $\mathsf{Q}^0_{yz}(\cdot)$, and $\mathsf{Q}_{yz}(\cdot)$. Obviously, $([T_n, Y_n], n \geqq 1)$ is the Markov renewal process determined by $q(\cdot)$, $p(y, \cdot)$, $\{\mathsf{F}^0_{yz}(u)\}$, and $\{\mathsf{F}_{yz}(u)\}$, where

$$\mathsf{F}^0_{yz}(u) = \mathsf{P}(A_0 \leqq u \mid Y_0 = y, Y_1 = z),$$

$$\mathsf{F}_{yz}(u) = \mathsf{P}(A_n \leqq u \mid Y_n = y, Y_{n+1} = z), \qquad n \geqq 1,$$

are given by $\mathsf{Q}^0_{yz}(\cdot)$ and $\mathsf{Q}_{yz}(\cdot)$, respectively. The process

$$Y(t) = Y_0 \mathbf{1}_{[0, T_1]}(t) + \sum_{n=1}^{\infty} Y_n \mathbf{1}_{(T_n, T_{n+1}]}(t)$$

is the corresponding semi-Markov process with the state space $\mathbf{Y}$. This fact explains the original name for semi-regenerative processes — semi-Markov processes with auxiliary paths, cf. PYKE and SCHAUFELE (1966). The auxiliary path $\left(X(t), T_n < t \leqq T_{n+1}\right)$ between the consecutive jumps T_n and T_{n+1} starts with the state Y_n; its evolution is completely determined by the kernels $p(y, \cdot)$ and $Q_{yz}(\cdot)$. From (A 4.3.3) and (A 4.3.4) we conclude that $(Y_n, n \geqq 0)$ is a homogeneous Markov chain with the transition probabilities $p(y, \cdot)$. If there is a unique stationary distribution $p(\cdot)$ with respect to the kernel $p(y, \cdot)$, then the sequence $([A_n, K_n], n \geqq 0)$ is stationary if and only if $q(\cdot) = p(\cdot)$ and $Q^0_{yz}(\cdot) = Q_{yz}(\cdot)$, $y, z \in \mathbb{Y}$ (this can be shown by the same arguments as are used in Section 3.3 for Markov renewal processes with discrete state space). In this case there is a unique stationary sequence (Φ, P), $\Phi = ([A_n, K_n], n \in \mathbb{Z})$, with the one-dimensional distribution

$$P\big(Y_0 \in L, [A_0, K_0] \in (\cdot)\big) = \int_L \int_{\mathbb{X}} Q_{yz}(\cdot)\, p(y, dz)\, p(dy). \tag{4.3.6}$$

Using Φ, via Definition 4.2.4 and Theorem 4.2.5 we can define a synchronous PEMP (χ, P) and (if $a_P = E_P A_0 < \infty$) the associated stationary PEMP $\bar{\chi} = \big[\big(\bar{X}(t)\big), ([\bar{T}_n, \bar{Y}_n])\big]$. The process $\big(\bar{X}(t)\big)$ is called the stationary semi-regenerative process determined by $p(y, \cdot)$ and $\{Q_{yz}(\cdot)\}$.

From (4.2.7) we obtain

$$\bar{P}\big(\bar{Y}_0 \in L, \bar{X}(0) \in C, \bar{T}_1 \leqq t\big)$$

$$= a_P^{-1} \int_L \int_{\mathbb{Y}} \int_0^\infty Q_{yz}\big(u < A_0 \leqq u + t, X(u) \in C\big)\, du\, p(y, dz)\, p(dy). \tag{4.3.7}$$

The asymptotic behaviour of semi-Markov processes with an at most countable state space $\mathbb{X}$ as well as of semi-regenerative processes with an at most countable space $\mathbb{Y}$ can be investigated by means of the key renewal theorem, cf. Çinlar (1975, Chapter 10). Now we prove an ergodic theorem for semi-regenerative processes with arbitrary spaces $\mathbb{X}$ and $\mathbb{Y}$, starting with the ergodic statement (4.2.11), Theorem 4.2.6.

The disadvantage of (4.2.11) for practical needs lies in the fact that it ensures the convergence to the steady state distribution $\bar{P}$ if one starts with a very special initial distribution (namely, with the distribution P of the associated synchronous PEMP). One would like to replace P by the distribution of a PEMP which has the same dynamics as the synchronous PEMP but with an "arbitrary initial phase". Unfortunately, in general one is already unable to define precisely what is meant by an "arbitrary initial phase". However, this difficulty does not occur if the cycles constitute a Markov chain, in particular, in case of semi-regenerative processes.

In the following we consider semi-regenerative processes on $\mathbb{R}_+$ with fixed parameters $p(y, \cdot)$ and $\{Q_{yz}(\cdot)\}$ but varying initial conditions. We assume that there is a unique stationary initial distribution $p(\cdot)$ with respect to $p(y, \cdot)$ and that

$$a_P = \int_{\mathbb{Y}} \int_{\mathbb{Y}} \int_0^\infty \big(1 - F_{yz}(u)\big)\, du\, p(y, dz)\, p(dy) < \infty$$

holds. By (χ, P) and $(\bar{\chi}, \bar{P})$ we denote the synchronous and the stationary PEMP determined by $p(y, \cdot)$ and $\{Q_{yz}(\cdot)\}$, respectively. Let P_μ be the distribution of the semi-regenerative process $[(X(t), t \geqq 0), ([T_n, Y_n], n \geqq 1)]$ with the distribution μ of the "initial phase" $[(X(t), 0 \leqq t \leqq T_1), Y_1]$ and with parameters $p(y, \cdot)$ and $\{\mathsf{Q}_{yz}(\cdot)\}$. Note that μ can directly be given or can be calculated by means of the parameters $q(\cdot)$, $\{\mathsf{Q}^0_{yz}(\cdot)\}$ (introduced above) and $p(y, \cdot)$. Notice that the restrictions of (χ, P) and $(\bar{\chi}, \bar{\mathsf{P}})$ to $[0, \infty)$ correspond to particular choices of the initial distribution μ.

Taking into account the structure of P_μ, we obtain

$$\mathsf{P}_\mu\Big([(X(t), T_1 \leqq t < \infty), ([T_n - T_1, Y_n], n \geqq 1)] \in (\cdot)\Big)$$
$$= \int_{\mathbf{Y}} \mathsf{P}_\mu\Big([(X(t), T_1 \leqq t < \infty), ([T_n - T_1, Y_n], n \geqq 1)] \in (\cdot) \mid Y_1 = y\Big) \mu(Y_1 \in dy)$$
$$= \int_{\mathbf{Y}} \mathsf{P}\Big([(X(t), 0 \leqq t < \infty), ([T_n, Y_n], n \geqq 0)] \in (\cdot) \mid Y_0 = y\Big)\, \mu(Y_1 \in dy). \tag{4.3.8}$$

Now we prove the following ergodic theorem.

4.3.1. Theorem. *If there exists a σ-finite non-zero measure ϱ on the Borel σ-field $\mathcal{Y}$ of $\mathbf{Y}$ such that for an arbitrary $B \in \mathcal{Y}$ with $\varrho(B) > 0$ and each $y \in \mathbf{Y}$*

$$\mathsf{P}(\inf\{n: Y_n \in B\} < \infty \mid Y_0 = y) = 1 \tag{4.3.9}$$

(this is the so-called Harris recurrence of the embedded Markov chain $(Y_n, n \geqq 0)$, cf. e.g. Nummelin *(1984)), then for every initial distribution μ we have*

$$\lim_{t\to\infty} t^{-1} \int_0^t h(\theta_u \chi)\, du = \mathsf{E}_{\bar{\mathsf{P}}} h(\bar{\chi}) \qquad \mathsf{P}_\mu\text{-a.s.} \tag{4.3.10}$$

for every measurable, non-negative function h with $\mathsf{E}_{\bar{\mathsf{P}}} h(\bar{\chi}) < \infty$.

Proof. Denote

$$A = \left\{ \lim_{t\to\infty} t^{-1} \int_0^t h(\theta_u \chi)\, du = \mathsf{E}_{\bar{\mathsf{P}}} h(\bar{\chi}) \right\}.$$

In view of (4.3.8) it is sufficient to show $\mathsf{P}(A \mid Y_0 = y) = 1$ for every $y \in \mathbf{Y}$. In other words, we have to show

$$E = \{y: \mathsf{P}(A \mid Y_0 = y) < 1\} = \emptyset.$$

From (4.2.11) and

$$\mathsf{P}(A) = \int_{\mathbf{Y}} \mathsf{P}(A \mid Y_0 = y)\, p(dy)$$

we obtain

$$p(E) = 0. \tag{4.3.11}$$

Assume $E \neq \emptyset$, and let z be an element of $\mathbf{Y}$ with the property $\mathsf{P}(A \mid Y_0 = z) < 1$. Assume

$$\mathsf{P}(\tau = \inf\{n: n \geqq 1, Y_n \in \mathbf{Y} \setminus E\} < \infty \mid Y_0 = z) = 1. \tag{4.3.12}$$

Using the structure of a semi-regenerative process we obtain

$$\begin{aligned}
&\mathsf{P}(A \mid Y_0 = z)\\
&= \sum_{n=1}^{\infty} \int_{\mathbb{Y}\setminus E} \mathsf{P}(A \mid \tau = n, Y_n = y)\, \mathsf{P}(\tau = n, Y_n \in dy \mid Y_0 = z)\\
&= \sum_{n=1}^{\infty} \int_{\mathbb{Y}\setminus E} \mathsf{P}\Big(\lim_{t\to\infty} t^{-1} \int_0^t h(\theta_u \theta_{T_n} \chi)\, du\\
&= \mathsf{E}_{\bar{\mathsf{P}}} h(\bar{\chi}) \mid \tau = n, Y_n = y\Big)\, \mathsf{P}(\tau = n, Y_n \in dy \mid Y_0 = z)\\
&= \sum_{n=1}^{\infty} \int_{\mathbb{Y}\setminus E} \mathsf{P}(A \mid Y_0 = y)\, \mathsf{P}(\tau = n, Y_n \in dy \mid Y_0 = z)\\
&= 1
\end{aligned}$$

in view of (4.3.12) and $\mathsf{P}(A \mid Y_0 = y) = 1$ for each $y \notin E$. This contradicts $z \in E$. Thus we have

$$\mathsf{P}(\inf \{n: Y_n \in \mathbb{Y} \setminus E\} < \infty \mid Y_0 = z) < 1.$$

In view of the assumed Harris recurrence we obtain $\varrho(\mathbb{Y} \setminus E) = 0$, i.e. $\varrho(E) > 0$. From $\varrho(E) > 0$ we obtain $p(E) > 0$, cf. e.g. NUMMELIN (1984), in contradiction to (4.3.11). ∎

4.3.3. Reliability analysis of complex systems with repair

Assume that the time evolution of a complex system with repair can be described by a stochastic process $(X(t), t \geqq 0)$ with a finite state space $\mathbb{X}$ and sample paths belonging to $D([0, \infty), \mathbb{X})$. Furthermore, assume that $\mathbb{X}$ consists of two disjoint sets $\mathbb{X}_+$ and $\mathbb{X}_-$; $\mathbb{X}_+$ being the set of "good" states (where the system is able to function; in other words, the system is up) and $\mathbb{X}_-$ the set of "bad" states (where the system is failed; in other words, it is down); $\mathbb{X} = \mathbb{X}_+ \cup \mathbb{X}_-$. For example, if we consider a coherent binary system consisting of n binary components, then $\mathbb{X} = \{\boldsymbol{x}: \boldsymbol{x} = [x_1, \ldots, x_n], x_i \in \{0, 1\}\}$ and $\mathbb{X}_+ = \{\boldsymbol{x}: \boldsymbol{x} \in \mathbb{X}, \varphi(\boldsymbol{x}) = 1\}$, where $x_i = 1$ $(x_i = 0)$ if the i-th component is up (down) and φ denotes the system function, cf. e.g. BARLOW and PROSCHAN (1975). In the following we extend $(X(t), t \geqq 0)$ to a PEMP of the form $[(X(t), t \geqq 0), ([T_n, Y_n], n \geqq 1)]$, where the embedded epochs T_n are characterized by $X(T_n) \in \mathbb{X}_-$, $X(T_n + 0) \in \mathbb{X}_+$, and $Y_n = X(T_n + 0)$. (Note that $T_n < T_{n+1}$, $n \geqq 1$.) In the following we drop the marks Y_n. The length A_n of the n-th cycle has the form $A_n = U_n + D_n > 0$, where U_n (D_n) is the up-time (down-time) of the system within (T_n, T_{n+1}).

In the following we assume that the system is already in steady state at time 0, i.e. its time evolution can be described by a stationary PEMP $(\bar{\chi}, \bar{\mathsf{P}})$ on $\mathbb{R}$, $\bar{\chi} = [(\bar{X}(t)), (\bar{T}_n)]$, where the sequence $(\bar{T}_n)$ consists of all epochs $\bar{T}$ with $\bar{X}(\bar{T}) \in \mathbb{X}_-$, $\bar{X}(\bar{T} + 0) \in \mathbb{X}_+$. (We prefer to deal here with stationary and synchronous PEMP's on the whole line, as done in the previous sections. Notice, that each stationary (synchronous) PEMP on $\mathbb{R}_+$ can be extended to a stationary (synchronous) PEMP on the whole line.) The most important reliability characteristics such as stationary state probabilities $\bar{p}_i$, stationary availability $\bar{A}$, and stationary interval reliability $\bar{A}_s$ can be defined in terms of $(\bar{\chi}, \bar{\mathsf{P}})$:

$$\bar{p}_i = \bar{\mathsf{P}}\big(\bar{X}(0) = i\big), \qquad i \in \mathbf{X},$$

$$\bar{A} = \bar{\mathsf{P}}\big(\bar{X}(0) \in \mathbf{X}_+\big),$$

$$\bar{A}_s = \bar{\mathsf{P}}\big(\bar{X}(t) \in \mathbf{X}_+, 0 \leqq t \leqq s\big).$$

Other important characteristics such as the mean time between failures (MTBF), the mean up-time m_U, and the mean down-time m_D can be defined in terms of the synchronous PEMP (χ, P), $\chi = \big[\big(X(t)\big), (T_n)\big]$, associated with $(\bar{\chi}, \bar{\mathsf{P}})$:

$$\text{MTBF} = \mathsf{E}_\mathsf{P} A_0 = \mathsf{E}_\mathsf{P}(U_0 + D_0),$$

$$m_U = \mathsf{E}_\mathsf{P} U_0 = \mathsf{E}_\mathsf{P} \inf \{t : X(t) \in \mathbf{X}_-\},$$

$$m_D = \mathsf{E}_\mathsf{P} D_0 = \mathsf{E}_\mathsf{P}(T_1 - U_0).$$

4.3.2. Theorem. *The following relations are valid:*

$$\bar{A} = \frac{\mathsf{E}_\mathsf{P} U_0}{\mathsf{E}_\mathsf{P} A_0} = \frac{\mathsf{E}_\mathsf{P} U_0}{\mathsf{E}_\mathsf{P} U_0 + \mathsf{E}_\mathsf{P} D_0}, \tag{4.3.13}$$

$$\bar{A}_s = (\mathsf{E}_\mathsf{P} A_0)^{-1} \int_s^\infty \mathsf{P}(U_0 > u)\, du, \tag{4.3.14}$$

$$\mathsf{E}_\mathsf{P} A_0 = \lambda_{\bar{\mathsf{P}}}^{-1}, \quad \textit{where} \quad \lambda_{\bar{\mathsf{P}}} = \mathsf{E}_{\bar{\mathsf{P}}}(\# \{n : \bar{T}_n < 1\}). \tag{4.3.15}$$

Proof. From (4.2.7) we obtain

$$\begin{aligned}
\bar{A}_s &= \bar{\mathsf{P}}\big(\bar{X}(t) \in \mathbf{X}_+, 0 \leqq t \leqq s\big) \\
&= (\mathsf{E}_\mathsf{P} A_0)^{-1} \int_0^\infty \mathsf{P}\big(A_0 > u, X(t) \in \mathbf{X}_+, u \leqq t \leqq u + s\big)\, du \\
&= (\mathsf{E}_\mathsf{P} A_0)^{-1} \int_0^\infty \mathsf{P}\big(A_0 > u, U_0 > u + s\big)\, du \\
&= (\mathsf{E}_\mathsf{P} A_0)^{-1} \int_0^\infty \mathsf{P}(U_0 > u + s)\, du \\
&= (\mathsf{E}_\mathsf{P} A_0)^{-1} \int_s^\infty \mathsf{P}(U_0 > u)\, du.
\end{aligned}$$

Thus, formula (4.3.14) is true.
Formula (4.3.13) follows from (4.3.14) in view of

$$A = \lim_{s \to 0} A_s.$$

Formula (4.3.15) is just the last statement of Theorem 4.2.5. ∎

Now we describe the time evolution of the system considered by the stationary PEMP $(\bar{\chi}', \bar{\mathsf{P}}')$, $\bar{\chi}' = [(\bar{X}(t)), (\bar{T}_n')]$, where the $\bar{T}_n'$ are the epochs at which $\bar{X}(t)$ jumps. Denote by (χ', P') the associated synchronous PEMP. Taking $L = \{i\}$ in formula (4.3.2), we obtain the following useful formula for the stationary state probabilities $\bar{p}_i = \bar{\mathsf{P}}'(\bar{X}(0) = i) = \bar{\mathsf{P}}(\bar{X}(0) = i)$:

$$\bar{p}_i = \frac{p_i' a_i'}{\sum\limits_{j \in \mathbf{X}} p_j' a_j'},$$

where $p_k' = \mathsf{P}'(X'(0+0) = k)$, $a_k' = \mathsf{E}_{\mathsf{P}'}(A_0' \mid X'(0+0) = k)$.

Consider a coherent binary system consisting of N independent and separately maintained binary components. Assume that at time 0 the system is already in steady state. Then the time evolution of the i-th component can be described by the stationary PEMP $(\bar{\chi}^{(i)}, \bar{\mathsf{P}}^{(i)})$, $\bar{\chi}^{(i)} = [(\bar{X}^{(i)}(t)), (\bar{T}_n^{(i)})]$, or by the associated synchronous PEMP $(\chi^{(i)}, \mathsf{P}^{(i)})$, $\chi^{(i)} = [(\hat{X}^{(i)}(t)), (T_n^{(i)})]$, where the embedded epochs are the epochs of finishing a repair of the i-th component. The process $(\hat{X}^{(i)}(t))$ is of the form

$$\hat{X}^{(i)}(t) = \begin{cases} 1 & \text{if } T_n^{(i)} < t \leqq T_n^{(i)} + U_n^{(i)}, \\ 0 & \text{if } T_n^{(i)} + U_n^{(i)} < t \leqq T_n^{(i)} + U_n^{(i)} + D_n^{(i)} = T_{n+1}^{(i)}, \end{cases}$$

where $U_n^{(i)}, D_n^{(i)}$ are the successive life and repair times of the i-th component,

$$T_0^{(i)} = 0, \qquad T_n^{(i)} = \sum_{j=0}^{n-1} (U_j^{(i)} + D_j^{(i)}), \qquad n \geqq 1,$$

$$T_n^{(i)} = -\sum_{j=n}^{-1} (U_j^{(i)} + D_j^{(i)}], \qquad n \leqq -1.$$

We assume $U_n^{(i)} > 0$, $D_n^{(i)} > 0$ a.s. The processes $\bar{X}^{(i)}(t)$ are of similar structure. By assumption the PEMP's $\chi^{(i)}$, $i = 1, \ldots, N$, $(\bar{\chi}^{(i)}, i = 1, \ldots, N)$ are independent. However, we need no independence assumptions concerning the life and repair times of the i-th component.

The time evolution of the system will be described by the stationary PEMP $(\bar{\chi}, \bar{\mathsf{P}})$, $\bar{\chi} = [(\underline{\bar{X}}(t)), (\bar{T}_n)]$, $\underline{\bar{X}}(t) = [\bar{X}^{(1)}(t), \ldots, \bar{X}^{(N)}(t)]$, where the $\bar{X}^{(i)}(t)$ are defined above and the $\bar{T}_n$ are epochs defined by the condition

$$\varphi(\underline{X}(\bar{T}_n + 0)) = 1, \qquad \varphi(\underline{X}(\bar{T}_n)) = 0$$

(φ is the system function). The corresponding synchronous PEMP will be denoted by (χ, P), $\chi = [(\underline{\bar{X}}(t)), (T_n)]$, $\underline{\bar{X}}(t) = [X^{(1)}(t), \ldots, X^{(N)}(t)]$. Note that both the binary processes $(X^{(i)}(t))$ and $(\hat{X}^{(i)}(t))$ describe the time evolution of the i-th component, but they do not coincide, in general (in view of different choice of the embedded epochs).

Using the notations

$$m_U^{(i)} = \mathsf{E}_{\mathsf{P}^{(i)}} U_0^{(i)}, \qquad m_D^{(i)} = \mathsf{E}_{\mathsf{P}^{(i)}} D_0^{(i)},$$

from (4.3.13) we obtain the stationary availability of the n-th component

$$\bar{A}_i = \bar{\mathsf{P}}^{(i)}\big(\bar{X}^{(i)}(0) = 1\big) = \frac{m_U{}^{(i)}}{m_U{}^{(i)} + m_D{}^{(i)}}. \tag{4.3.16}$$

The stationary availability of the whole system is given by

$$\bar{A} = \bar{\mathsf{P}}\big(\varphi(\underline{\bar{X}}(0)) = 1\big) = h(\bar{A}_1, \ldots, \bar{A}_N), \tag{4.3.17}$$

where h is the reliability function of the system, cf. e.g. BARLOW and PROSCHAN (1975), BEICHELT and FRANKEN (1983).

4.3.3. Theorem. *Consider a coherent binary system consisting of N independent and separately maintained binary components with the system function φ and the reliability function h. The system is assumed to be in steady state. Then the failure intensity $\lambda_{\bar{\mathsf{P}}}$ of the system (i.e. the intensity of the PEMP $(\bar{\chi}, \bar{\mathsf{P}})$ is given by*

$$\lambda_{\bar{\mathsf{P}}} = \sum_{i=1}^{N} \frac{1}{m_U{}^{(i)} + m_D{}^{(i)}} \big(h(\bar{A}_1, \ldots, \bar{A}_{i-1}, 1, \bar{A}_{i+1}, \ldots, \bar{A}_N) - h(\bar{A}_1, \ldots, \bar{A}_{i-1}, 0, \bar{A}_{i+1}, \ldots, \bar{A}_N)\big), \tag{4.3.18}$$

where $m_U{}^{(i)}$ and $m_D{}^{(i)}$ are the mean generic life and repair time of the i-th component, respectively, and $\bar{A}_i$ is given by (4.3.16).

The MTBF of the system is equal to

$$\mathsf{E}_{\mathsf{P}} A_0 = \mathsf{E}_{\mathsf{P}}(U_0 + D_0) = \lambda_{\bar{\mathsf{P}}}^{-1}. \tag{4.3.19}$$

The mean up time is

$$\mathsf{E}_{\mathsf{P}} U_0 = \frac{h(\bar{A}_1, \ldots, \bar{A}_N)}{\lambda_{\bar{\mathsf{P}}}}. \tag{4.3.20}$$

Proof. Consider the PEMP's $(\bar{\chi}_{(1)}, \bar{\mathsf{P}}_{(1)})$, $\bar{\chi}_{(1)} = \big[\big([\bar{X}^{(1)}(t), \ldots, \bar{X}^{(N)}(t)]\big), (\bar{T}_n{}^{(1)})$ and $(\chi_{(1)}, \mathsf{P}_{(1)})$, $\chi_{(1)} = \big[\big([\hat{X}^{(1)}(t), \bar{X}^{(2)}(t), \ldots, \bar{X}^{(N)}(t)]\big), (T_n{}^{(1)})\big]$ where we assume that $\big[\big(\bar{X}^{(1)}(t)\big), (\bar{T}_n{}^{(1)})\big], \big(\bar{X}^{(2)}(t)\big), \ldots, \big(\bar{X}^{(N)}(t)\big)$ are independent (the same holds for $\chi_{(1)}$). Using this independence assumption, it is easy to show that $(\bar{\chi}_{(1)}, \bar{\mathsf{P}}_{(1)})$ is stationary and $(\chi_{(1)}, \mathsf{P}_{(1)})$ is a synchronous PEMP. Applying (4.2.7) to $\big[\big(\hat{X}^{(1)}(t)\big), (T_n{}^{(1)})\big]$ we obtain

$$(\mathsf{E}_{\mathsf{P}_{(1)}} T_1{}^{(1)})^{-1} \int_0^\infty \mathsf{P}_{(1)}\Big(T_1{}^{(1)} > u, \theta_u\big[\big(\hat{X}^{(1)}(t)\big), (T_n{}^{(1)})\big] \in C_1, \theta_u\big(\bar{X}^{(j)}(t)\big) \in C_j,$$

$$j = 2, \ldots, N\Big)\, \mathrm{d}u = \bar{\mathsf{P}}^{(1)}\Big(\big[\big(\bar{X}^{(1)}(t)\big), (\bar{T}_n{}^{(1)})\big] \in C_1\Big) \prod_{j=2}^{N} \bar{\mathsf{P}}^{(j)}\Big(\big(\bar{X}^{(j)}(t)\big) \in C_j\Big).$$

Thus, $(\bar{\chi}_{(1)}, \bar{\mathsf{P}}_{(1)})$ and $(\chi_{(1)}, \mathsf{P}_{(1)})$ are associated in the sense of Theorem 4.2.5.

Consider the stationary point process $(\bar{T}^{(i)}_{1,n})$ characterized by the properties

$$\bar{X}^{(i)}(\bar{T}^{(i)}_{1,n} - 0) = 0, \qquad \bar{X}^{(i)}(\bar{T}^{(i)}_{1,n} + 0) = 1,$$

$$\varphi\big(\underline{\bar{X}}(\bar{T}^{(i)}_{1,n} - 0)\big) = 0, \qquad \varphi\big(\underline{\bar{X}}(\bar{T}^{(i)}_{1,n} + 0)\big) = 1.$$

($\overline{T}_{1,n}^{(i)}$ are just the epochs of finishing a repair of the system caused by finishing a repair of the i-th component.) The point process ($\overline{T}_n$) defined above is just the superposition of the ($\overline{T}_{1,n}^{(i)}$), $i = 1, \ldots, N$, i.e. ($\overline{T}_n$) consists of the points of

$$\bigcup_{i=1}^{N} \{\overline{T}_{1,n}^{(i)} : i = 1, \ldots, N;\, n \in \mathbb{Z}\}$$

renumbered in ascending order. The intensity $\lambda_{\overline{\mathsf{P}}}$ of $(\overline{\chi}, \overline{\mathsf{P}})$ is equal to

$$\lambda_{\overline{\mathsf{P}}} = \sum_{i=1}^{N} \lambda_{0i}, \tag{4.3.21}$$

where λ_{0i} is the intensity of ($\overline{T}_{1,n}^{(i)}$). (Notice that in view of the assumed simplicity of the point processes ($\overline{T}_n{}^{(i)}$), $i = 1, \ldots, N$, and property (3.2.8) the points of ($\overline{T}_{1,n}^{(i)}$) and ($\overline{T}_{1,n}^{(j)}$), $i \neq j$, never coincide.)

Now we calculate λ_{01}. Let

$$\lambda_1 = \frac{1}{m_U{}^{(1)} + m_D{}^{(1)}} \text{ — the intensity of } (\overline{T}_n{}^{(1)}, n \geqq 1),$$

$$\underline{X}_{(1)}(t) = [\hat{X}^{(1)}(t), \overline{X}^{(2)}(t), \ldots, \overline{X}^{(N)}(t)],$$

$$B_1 = \{\varphi(\underline{X}_{(1)}(T_n{}^{(1)})) = 0,\, \varphi(\underline{X}_{(1)}(T_n{}^{(1)} + 0)) = 1\}$$

(n is an arbitrary but fixed number, $n \geqq 1$). According to formula (3.5.6)

$$\lambda_{01} = \lambda_1 \mathsf{P}_{(1)}(B_1).$$

Obviously,

$$B_1 = \{\varphi(\underline{X}_{(1)}(T_n{}^{(1)} + 0)) = 1\}$$
$$\{\varphi(\underline{X}_{(1)}(T_n{}^{(1)} + 0)) = 1,\, \varphi(\underline{X}_{(1)}(T_n{}^{(1)})) = 1\},$$

and

$$\mathsf{P}_{(1)}(\varphi(\underline{X}_{(1)}(T_n{}^{(1)} + 0)) = 1,\, \varphi(\underline{X}_{(1)}(T_n{}^{(1)})) = 1)$$
$$= \mathsf{P}_{(1)}(\varphi(\underline{X}_{(1)}(T_n{}^{(1)})) = 1)$$

in view of monotonicity of φ. Observe that

$$\mathsf{P}_{(1)}(\hat{X}^{(1)}(T_n{}^{(1)} + 0) = 1) = 1,$$

$$\mathsf{P}_{(1)}(\hat{X}^{(1)}(T_n{}^{(1)} - 0) = 1) = 0,$$

$$\mathsf{P}_{(1)}(\overline{X}^{(j)}(T_n{}^{(1)} - 0) = 1) = \mathsf{P}_{(1)}(\overline{X}^{(j)}(T_n{}^{(1)} + 0) = 1) = \bar{A}_j, \quad j \geqq 2.$$

(The last formula is true in view of the independence of ($T_n{}^{(1)}$) and ($\overline{X}^{(j)}(t)$),

$j \geqq 2$.) Thus, we have

$$\lambda_{01} = \frac{1}{m_U^{(1)} + m_D^{(1)}} \mathsf{P}_{(1)}(B_1)$$

$$= \frac{1}{m_U^{(1)} + m_D^{(1)}} \Big(\mathsf{P}_{(1)}\big(\varphi(\underline{X}_{(1)}(T_n^{(1)} + 0)\big) = 1\Big)$$

$$- \mathsf{P}_{(1)}\big(\varphi(\underline{X}_{(1)}(T_n^{(1)})) = 1\big)\Big)$$

$$= \frac{1}{m_U^{(1)} + m_D^{(1)}} \big(h(1, \bar{A}_2, \ldots, \bar{A}_N) - h(0, \bar{A}_2, \ldots, \bar{A}_N)\big).$$

Analogous formulae can be obtained for $\lambda_{0j}, j = 2, \ldots, N$. Then formula (4.3.18) follows from (4.3.21).

Formulae (4.3.19) and (4.3.20) are particular cases of (4.3.13) and (4.3.15) (cf. (4.3.17)), respectively. ∎

4.4. Continuous time state processes associated with the equation $X_{n+1} = f(X_n, U_n)$

In this section we construct stationary PEMP's describing the temporal behaviour of stochastic systems of the following kind. Let (Φ, P), $\Phi = ([A_n, Y_n])$, be a stationary sequence, where $A_n \geqq 0$ and Y_n are $\mathbb{Y}$-valued r.v.'s ($\mathbb{Y}$ is assumed to be Polish). We think of the A_n as distances between the embedded epochs T_n:

$$T_0 = 0; \quad T_n = \sum_{i=0}^{n-1} A_i, \quad n > 0; \quad T_n = -\sum_{i=n}^{-1} A_i, \quad n < 0.$$

The mark Y_n is a certain characteristic of the embedded epoch T_n. For example, in the queueing framework, we can think of T_n as the arrival epoch of the n-th customer and Y_n as the associated required service time.

Assume that the evolution of the system is described by the recursive equation

$$X_{n+1} = f(X_n, [A_n, Y_n]), \qquad n \in \mathbb{Z}, \tag{4.4.1}$$

where the X_n are interpreted as the consecutive states at the epochs T_n (for instance, X_n could be the state of the queueing system seen by the n-th arriving customer). We suppose that, given the states X_k, $k \in \mathbb{Z}$, the state $X(t)$ of the system at an arbitrary time t is determined by

$$X(t) = f(X_n, [t - T_n, Y_n]), \qquad T_n < t \leqq T_{n+1}. \tag{4.4.2}$$

In most queueing systems, the state of the system (e.g. the virtual waiting time of a single server queue) between the arrival instants of customers is determined by an equation of the type (4.4.2).

Assume there is a stationary weak solution of (4.1.1),

$$(\Phi', \mathsf{P}'), \qquad \Phi' = \big([[A_n', Y_n'], X_n'], n \in \mathbb{Z}\big),$$

say. Define

$$T_0' = 0; \quad T_n' = \sum_{i=0}^{n-1} A_i', \quad n > 0; \quad T_n' = -\sum_{i=n}^{-1} A_i', \quad n < 0, \tag{4.4.3}$$

$$X'(t) = f(X_n', [t - T_n', Y_n']), \qquad T_n' < t \leqq T'_{n+1}. \tag{4.4.4}$$

If, in addition, we assume that the function $f(x, [a, y])$ is left-continuous with respect to a, then $X'(T_n') = X_n'$ for $T'_{n-1} < T_n'$ and hence $\chi' = \left[\left(X'(t), t \in \mathbb{R}\right), \left([T_n', [X_n', Y_n']], n \in \mathbb{Z}\right)\right]$ is a PEMP in the sense of Definition 4.2.1. Notice that χ' is given as a measurable function of Φ'. For this reason we shall use the symbol P' for the probability distribution of χ', too. Obviously, the stochastic process $\left(X'(t), t \in \mathbb{R}\right)$ is non-stationary in general. However, our aim consists in constructing a stationary, continuous time state process. In view of the stationarity of Φ', χ' is a synchronous PEMP. If $\mathsf{E}_{\mathsf{P}'}A_0' = \mathsf{E}_{\mathsf{P}}A_0 < \infty$, then the stationary PEMP $(\bar{\chi}', \bar{\mathsf{P}}')$,

$$\bar{\chi}' = \left[\left(\bar{X}'(t)\right), \left([\bar{T}_n', [\bar{X}_n', \bar{Y}_n']]\right)\right],$$

associated with (χ', P') by formula (4.2.7) exists, cf. Theorem 4.2.5. Since Φ' is a weak solution, from (4.2.7) we get

$$\bar{X}'_{n+1} = f(\bar{X}_n', [\bar{T}'_{n+1} - \bar{T}_n', \bar{Y}_n']), \qquad n \in \mathbb{Z}, \quad \bar{\mathsf{P}}'\text{-a.s.} \tag{4.4.5}$$

and

$$\bar{X}'(t) = f(\bar{X}_n', [t - \bar{T}_n', \bar{Y}_n']) \ \text{ if } \ \bar{T}_n' < t \leqq \bar{T}'_{n+1} \quad \bar{\mathsf{P}}'\text{-a.s.} \tag{4.4.6}$$

in view of (4.4.4).

Summarizing these considerations and taking into account Theorem 4.2.6, we obtain

4.4.1. Theorem. *Let* (Φ', P'), $\Phi' = \left([[A_n', Y_n'], X_n']\right)$, *be a stationary weak solution of equation* (4.4.1) *for the input* (Φ, P), $\Phi = ([A_n, Y_n])$, *satisfying* $\mathsf{E}_{\mathsf{P}}A_0 < \infty$. *Assume* $f(x, [a, y])$ *to be left-continuous in* a. *Let* (T_n') *and* $\left(X'(t)\right)$ *be given by* (4.4.3) *and* (4.4.4), *respectively. Then there exists a stationary PEMP* $(\bar{\chi}', \bar{\mathsf{P}}')$,

$$\bar{\chi}' = \left[\left(\bar{X}'(t)\right), \left([\bar{T}_n', [\bar{X}_n', \bar{Y}_n']]\right)\right],$$

such that (4.4.5), (4.4.6) *are valid. If, in addition,* (Φ', P') *is ergodic, then it holds that*

$$\lim_{t\to\infty} \frac{1}{t} \int_0^t 1_{(\cdot)}\left(X'(u)\right) du = \bar{\mathsf{P}}'\left(\bar{X}'(0) \in (\cdot)\right) \quad \mathsf{P}'\text{-}a.s. \tag{4.4.7}$$

and

$$\lim_{n\to\infty} \frac{1}{n} \sum_{j=1}^{n} 1_{(\cdot)}(\bar{X}_j') = \mathsf{P}'\left(X_0' \in (\cdot)\right) \quad \bar{\mathsf{P}}'\text{-}a.s. \tag{4.4.8}$$

Now we sketch a somewhat different way of constructing a stationary PEMP describing the behaviour of the system considered. We start again with a stationary weak solution (Φ', P'), $\Phi' = \left([[A_n', Y_n'], X_n']\right)$, and assume

$\mathsf{E}A_0' = \mathsf{E}A_0 < \infty$. By Theorem 3.4.1 we get a corresponding stationary MPP $(\tilde{\Phi}', \tilde{\mathsf{P}}')$, $\tilde{\Phi}' = \left([\tilde{T}_n', [\tilde{X}_n', \tilde{Y}_n']]\right)$, satisfying

$$\tilde{X}_{n+1}' = f(\tilde{X}_n', [\tilde{T}_{n+1}' - \tilde{T}_n', \tilde{Y}_n']), \qquad n \in \mathbb{Z}, \quad \tilde{\mathsf{P}}'\text{-a.s.}$$

Define

$$\tilde{X}'(t) = f(\tilde{X}_n', [t - \tilde{T}_n', \tilde{Y}_n']), \qquad \tilde{T}_n' < t \leqq \tilde{T}_{n+1}'. \tag{4.4.9}$$

Obviously, $(\tilde{\chi}', \tilde{\mathsf{P}}')$, $\tilde{\chi}' = \left[\left(\tilde{X}'(t)\right), \left([\tilde{T}_n', [\tilde{X}_n', \tilde{Y}_n']]\right)\right]$, is a stationary PEMP. In view of (4.2.7) and (3.4.1),

$$\left([\tilde{T}_n', [\tilde{X}_n', \tilde{Y}_n']]\right) \overset{\mathcal{D}}{=} \left([\overline{T}_n', [\overline{X}_n', \overline{Y}_n']]\right).$$

Now $\tilde{\mathsf{P}}' = \overline{\mathsf{P}}'$ follows from (4.4.6) and (4.4.9). Thus, the second way provides the same result as the first one.

Theorem 4.4.1 shows that the PEMP's (χ', P') and $(\overline{\chi}', \overline{\mathsf{P}}')$ (and their corresponding stochastic processes $\left(X'(t)\right)$ and $\left(\overline{X}'(t)\right)$) are dual in some sense: The $\overline{\mathsf{P}}'$-probabilities appear as Cesaro limits in the model (χ', P') and vice versa. Together with (4.4.3) and (4.4.4), this duality shows that $(\overline{\chi}', \overline{\mathsf{P}}')$ is indeed the appropriate model for describing the continuous time evolution of a system (governed by (4.4.1) and (4.4.2)) in steady state viewed by an observer located at an arbitrary time instant. For this reason, the PEMP $(\overline{\chi}', \overline{\mathsf{P}}')$ is called a *time-stationary model* and $\left(\overline{X}'(t)\right)$ a *time-stationary state process* of the system described by equations (4.4.1), (4.4.2). The PEMP (χ', P') is an event-oriented description of the system in steady state; we call (χ', P') an *event-stationary model* and (X_n') an *event-stationary state process*. (If the T_n are arrival epochs of customers in a queueing system, then (χ', P') and (X_n') are called an *arrival-stationary model* and an *arrival-stationary state process*, respectively.)

4.5. Recursive stochastic equations in continuous time

Till now we have described the dynamics of a stochastic system by means of a "discrete" equation of the type

$$X_{n+1} = f(X_n, [T_{n+1} - T_n, Y_n]), \qquad n \in \mathbb{Z},$$

where $\Phi = ([A_n, Y_n])$, $A_n = T_{n+1} - T_n$, is the input. For a given weak solution $\left([[A_n', Y_n'], X_n']\right)$ in Section 4.4 we have introduced the time-stationary model $(\overline{\chi}', \overline{\mathsf{P}}')$, $\overline{\chi}' = \left[\left(\overline{X}'(t)\right), \left([\overline{T}_n', [\overline{X}_n', \overline{Y}_n']]\right)\right]$, via the equation

$$\overline{X}'(t) = f(\overline{X}_n', [t - \overline{T}_n', \overline{Y}_n']), \qquad \overline{T}_n' < t \leqq \overline{T}_{n+1}'. \tag{4.5.1}$$

The time-stationary model includes the MPP $\overline{\Psi}' = ([\overline{T}_n', \overline{Y}_n'])$. In view of $([A_n', Y_n']) \overset{\mathcal{D}}{=} ([A_n, Y_n])$ and (4.2.7) we obtain $\overline{\Psi}' \overset{\mathcal{D}}{=} \overline{\Psi}$, where $\overline{\Psi} = ([\overline{T}_n, \overline{Y}_n])$ is the stationary MPP associated (in the sense of Theorem 3.4.1) with the input $\Phi = ([A_n, Y_n])$. Thus we can speak about $\Phi = ([A_n, Y_n])$ and $\overline{\Psi} = ([\overline{T}_n, \overline{Y}_n])$ as the event-stationary (in queueing setup: arrival-stationary) and the time-stationary input, respectively. For defining the event- and time-stationary model of the system under consideration, cf. Chapter 1 and Section 4.4, respectively, we have used the event-stationary input Φ, the dynamic "arrival by arrival" (4.4.1), Theorem 3.4.1, and formula (4.5.1). Nothing was said about the dynamic "from one time instant to the other". This dynamic will be considered now.

As in Section 4.4, let $f(x, [a, y])$ be left continuous in a. Assume the time-stationary input $\overline{\Psi} = ([\overline{T}_n, \overline{Y}_n])$ to be simple. This is equivalent to $\mathsf{P}(A_0 > 0) = 1$. Moreover, assume that there is a measurable $\mathbb{X}$-valued function

$$\bar{f}(x, t, \psi), \quad x \in \mathbb{X}, \quad t \in \mathbb{R}_+, \quad \psi = ([t_n, y_n]) \in M_y,$$

which is left continuous in t and satisfies

$$\left.\begin{aligned} &\bar{f}(x, t, \psi) = f(x, [t, y_0]), \qquad x \in \mathbb{X}, \quad \psi = ([t_n, y_n]), \\ &t_{-1} < 0 = t_0 < t \leqq t_1, \end{aligned}\right\} \tag{4.5.2}$$

and

$$\bar{f}(x, t, \psi) = \bar{f}\big(\bar{f}(x, u, \psi), t - u, \theta_u\psi\big), \quad 0 < u < t, \quad x \in \mathbb{X}. \tag{4.5.3}$$

The function $\bar{f}$ (if it exists) is an extension of f, as (4.5.2) shows. We interpret $\bar{f}$ as follows: If the system is in state x at time 0, then the state at time $t > 0$ is given by $\bar{f}(x, t, \psi)$, where $\psi = ([t_n, y_n])$ is a fixed sample path of the input $\overline{\Psi}$.

4.5.1. Example. Consider the single server loss system, which is characterized by the equation

$$X_{n+1} = f(X_n, [A_n, S_n]) = (X_n + S_n \mathbf{1}\{X_n = 0\} - A_n)_+,$$

$X_n, A_n, S_n \geqq 0$, $n \in \mathbb{Z}$. Then a function $\bar{f}$ following the above interpretation is defined by

$$\bar{f}(x, t, \psi) = \begin{cases} x_{r+1}(x, [t_1, 0], [t_2 - t_1, s_1], \ldots, [t_r - t_{r-1}, s_{r-1}], [t - t_r, s_r]) & \text{if} \\ \qquad t_0 < 0 < t_1 < \cdots < t_r < t \leqq t_{r+1}, & \\ x_{r+1}(x, [t_1 - t_0, s_0], \ldots, [t_r - t_{r-1}, s_{r-1}], [t - t_r, s_r]) & \text{if} \\ \qquad 0 = t_0 < t_1 < \cdots < t_r < t \leqq t_{r+1}, & \end{cases} \tag{4.5.4}$$

$x \in \mathbb{R}_+$, $t > 0$, $\psi = ([t_n, s_n])$.

In Example 4.5.1 the function $\bar{f}$ was obtained in an obvious way from f. However, there are examples in which $\bar{f}$ cannot be defined by (4.5.4).

4.5.2. Example. Consider the single server queue G/G/1/∞ with the following queueing discipline: At each arrival of a customer the order of all (waiting) customers in the queue will be reversed. Between consecutive arrivals the queue works in the usual FCFS queueing discipline. The state $X_n = (X_{n1}, X_{n2}, \ldots)$ consists of the residual service time X_{n1} of the customer just served and the service times of the customers waiting in the queue arranged in their order in the queue. Then there is a function f such that

$$X_{n+1} = f(X_n, [A_n, S_n]).$$

But (4.5.4) does not lead to a function satisfying (4.5.2) and (4.5.3). However, there is a function $\bar{f}$ with these properties.

Consider a time-stationary model $(\bar{\chi}', \bar{\mathsf{P}}')$, $\bar{\chi}' = \left[\left(\bar{X}'(t)\right), \left([\bar{T}_n', [\bar{X}_n', \bar{Y}_n']]\right)\right]$, cf. Theorem 4.4.1. Using (4.5.2), (4.5.3), (4.4.3), (4.4.4), and the left continuity of $\bar{f}$ with respect to the second component, we obtain

$$\bar{X}'(t) = \bar{f}\left(\bar{X}'(u), t - u, \theta_u \bar{\Psi}'\right), \qquad u < t, \quad \bar{\mathsf{P}}'\text{-a.s.} \tag{4.5.5}$$

for the time stationary model $(\bar{\chi}', \bar{\mathsf{P}}')$. (Note that in view of the left continuity of $\bar{f}$, the event $\left\{\left[\left(x(\cdot)\right), \psi\right] : x(t) = \bar{f}\left(x(u), t - u, \theta_u \psi\right), u, t \in \mathbb{R}, u < t\right\}$ is measurable.) Equation (4.5.5) can be considered as a continuous time analogue of equation (4.4.1). This interpretation leads to the following problem which is of intrinsic interest:

Let $(\bar{\Psi}, \bar{\mathsf{P}})$, $\bar{\Psi} = ([\bar{T}_n, \bar{Y}_n])$, be a simple, stationary MPP with the $\mathbb{Y}$-valued marks $\bar{Y}_n$. Assume $\lambda_{\bar{\mathsf{P}}} < \infty$. Consider the equation

$$X(t) = \bar{f}\left(X(u), t - u, \theta_u \bar{\Psi}\right), \qquad u, t \in \mathbb{R}, \quad u < t, \tag{4.5.6}$$

where $\bar{f}(x, a, \psi)$ is an $\mathbb{X}$-valued function, left continuous in a and satisfying (4.5.3). We look for solutions of (4.5.6). First we have to clarify what is meant by a solution of this equation. By analogy with the definition of a weak solution of equation (1.1.1) we define:

4.5.3. Definition. The (stationary) PEMP $(\bar{\chi}', \bar{\mathsf{P}}')$, $\bar{\chi}' = \left[\left(\bar{X}'(t)\right), \left([\bar{T}_n', [\bar{X}_n', \bar{Y}_n']]\right)\right]$, is called a *(stationary) weak solution of equation* (4.5.6) *for the input* $(\bar{\Psi}, \bar{\mathsf{P}})$, $\bar{\Psi} = ([\bar{T}_n, \bar{Y}_n])$, iff

(i) $\bar{\Psi}' = ([\bar{T}_n', \bar{Y}_n']) \overset{\mathcal{D}}{=} \bar{\Psi}$,

(ii) $\bar{X}'(t) = \bar{f}\left(\bar{X}'(u), t - u, \theta_u \bar{\Psi}'\right)$, $u < t$, $\bar{\mathsf{P}}'$-a.s.

(Since a weak solution of (4.5.6) is a PEMP and $\bar{\Psi}'$ is simple, we have

$$\bar{X}'(\bar{T}_n') = \bar{X}_n', \qquad n \in \mathbb{Z}, \quad \bar{\mathsf{P}}'\text{-a.s.}$$

for each weak solution, and thus the component $\bar{X}_n'$ is, in general, not needed.)

In view of (4.5.5) we can interpret a time-stationary model as a stationary weak solution of equation (4.5.6). Remember, this stationary weak solution is defined by means of a stationary weak solution of equation (4.4.1). The methods discussed in Chapter 1 for proving the existence of a stationary weak solution of (4.4.1) use conditions on the function f and the input $\Phi = ([A_n, Y_n])$. However, if a simple stationary MPP $\bar{\Psi}$ and the function $\bar{f}$ are given, then one is interested in conditions on $\bar{f}$ and $\bar{\Psi}$, ensuring the existence of a stationary solution of (4.5.6). Fortunately, the methods presented in Chapter 1 for equation (1.1.1) can be used in a modified form for solving equation (4.5.6). Here we give only a short outline of the method of solution generating systems for equation (4.5.6); the other methods can be transcribed in a similar way.

4.5.4. Definition. Let $(\bar{\Psi}, \bar{\mathsf{P}})$, $\bar{\Psi} = ([\bar{T}_n, \bar{Y}_n])$, be a simple, stationary MPP (time-stationary model of the input) and M an invariant measurable subset of M_Y satisfying $\bar{\mathsf{P}}(M) = 1$. A family of sets $\mathcal{S} = \{A(t, \psi): t \in \mathbb{R}, \psi \in M\}$ is called a *solution generating system* (abbr. *SGS*) *for equation* (4.5.6) if the following conditions are fulfilled.

(A 4.5.1) $A(t, \psi) \subseteqq \mathbb{X}$ and $1 \leqq \# A(t, \psi) < \infty$ for $\psi \in M$, $t \in \mathbb{R}$.

(A 4.5.2) The sets $\{\psi: \psi \in M, i \leqq \# A(t, \psi)\}$, $i = 1, 2, \ldots$, $t \in \mathbb{R}$, are measurable. There exist $\mathbb{X}$-valued measurable mappings $\psi \to x(i, t, \psi)$ defined on $\{\psi: \psi \in M, i \leqq \# A(t, \psi)\}$ satisfying $A(t, \psi) = \{x(1, t, \psi), \ldots, x(\# A(t, \psi), t, \psi)\}$.

(A 4.5.3) $\bar{f}(A(u, \psi), t - u, \theta_u \psi) \subseteqq A(t, \psi)$ for $\psi \in M$, $u < t$.

(A 4.5.4) $A(t, \psi) = A(0, \theta_t \psi)$ for $\psi \in M$, $t \in \mathbb{R}$.

Similarly to Theorem 1.4.3, one can prove

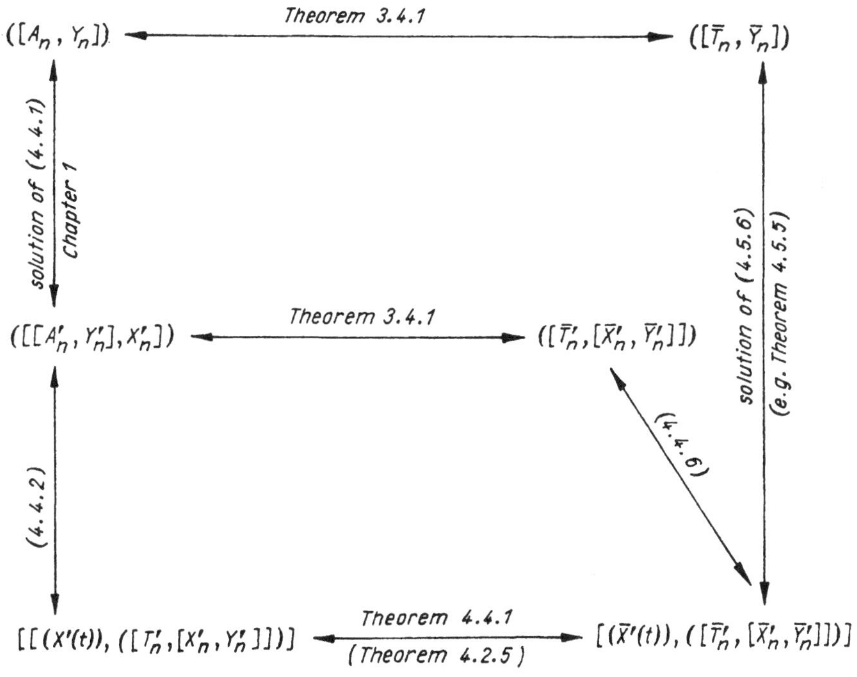

Fig. 2. From the input to the arrival- and time-stationary model.

4.5.5. Theorem. *Let* $(\bar{\Psi}, \bar{\mathsf{P}})$ *be a simple stationary MPP with* $\lambda_{\bar{\mathsf{P}}} < \infty$ *and* $\bar{f}(x, a, \psi)$ *a left continuous function in a, satisfying* (4.5.3). *Assume there is an SGS for equation* (4.5.6). *Then there exists a stationary weak solution of* (4.5.6).

At the beginning of this section we have seen that each stationary weak solution of the event-oriented equation (4.4.1) via (4.4.2) provides a stationary weak solution of the continuous time equation (4.5.6) if f and $\bar{f}$ are related one to the other according to (4.5.2), (4.5.3). The reverse is also true: Let

$\overline{\Psi} = ([\overline{T}_n, \overline{Y}_n])$ be a simple MPP with $\lambda_{\overline{P}} < \infty$ and $\bar{f}(x, t, \psi)$ a function which has the properties (4.5.2), (4.5.3) and is left continuous in t. Consider a stationary weak solution of (4.5.6):

$$\left[(\overline{X}'(t)), ([\overline{T}_n', [\overline{X}_n', \overline{Y}_n']])\right].$$

Let $([A_n', [X_n', Y_n']])$ be a stationary sequence associated with the stationary MPP $([\overline{T}_n', [\overline{X}_n', \overline{Y}_n']])$ according to Theorem 3.4.1. Then $([[A_n', Y_n'], X_n'])$ is a stationary weak solution of (4.4.1).

The considerations of Sections 4.4 and 4.5 can be summarized in the following scheme. (Notice that for solving (4.5.6) one needs, for technical reasons, $([\overline{T}_n, \overline{Y}_n])$ to be simple.)

In order to obtain a time-stationary and an event-stationary model it suffices to solve one of the equations (4.4.1) or (4.5.6), respectively. It depends on the particular system which of the ways shown in Figure 2 from $([A_n, Y_n])$ or $([\overline{T}_n, \overline{Y}_n])$ to the time-stationary or event-stationary model, respectively, is the best one.

4.6. Remarks and references

4.2. Synchronous PEMP's of the type $\left[(X(t)), ([T_n, X_n])\right]$, where $(X(t))$ is a real-valued process, appear first in Serfozo (1972) under the name semi-stationary processes. In the presented form the concept of PEMP's was independently introduced by Franken and co-workers, and by König and co-workers, at the end of the seventies, cf. e.g. König et al. (1978), Arndt and Franken (1979), Franken and Streller (1979). These papers also discuss several well-known classes of stochastic processes in terms of PEMP's. The duality between (χ, P) and $(\bar{\chi}, \overline{\mathsf{P}})$ was also considered by Serfozo and Stidham (1978). For a comprehensive discussion of PEMP's with a simple embedded point process and their applications we refer to [FKAS], Rolski (1981b), Beichelt and Franken (1983), and references therein.

The key point in König et al. (1978) is the so-called "intensity conservation principle" for PEMP's with a simple embedded point process, which is a very appropriate tool for some applications, cf. also [FKAS], Rolski (1981b). A particular case of this principle is derived in Cohen (1977).

The ergodic Theorem 4.2.6 was proved by Franken and Streller (1979); a somewhat different proof was given by Nawrotzki (1981b). For another version (convergence in variation) see Zähle (1980).

PEMP's of the type $\left[(X(t), t \geqq 0), ([T_n, X_n], n \geqq 0)\right]$ arise in simulation framework if one is interested in estimating time averages via randomly located observations (the T_n are the epochs of observations), cf. Fox and Glynn (1987). For a real-valued function h one is interested in asymptotic properties of the averages

$$r(t) = t^{-1} \int_0^t h(X(u))\, \mathrm{d}u$$

and

$$r_n = T_n^{-1} \int_0^{T_n} h(X(u))\,\mathrm{d}u$$

as $t \to \infty$ and $n \to \infty$, respectively.

Consider a synchronous PEMP (χ, P), $\chi = [(X(t), t \geqq 0), ([T_n, Y_n], n \geqq 0)]$, $T_0 = 0$, $Y_n = X(T_n + 0)$, on $\mathbb{R}_+$ and the associated stationary PEMP $(\bar{\chi}, \bar{\mathsf{P}})$, $\bar{\chi} = [(\bar{X}(t), t \geqq 0), ([\bar{T}_n, \bar{Y}_n], n \geqq 1)]$, cf. Section 4.3. Assume:

(A 4.6.1) The cycles of χ form a φ-mixing sequence with φ-mixing coefficients satisfying

$$\sum_{k=1}^{\infty} \varphi_k^{1/2} < \infty,$$

cf. BILLINGSLEY (1968, Section 20).

(A 4.6.2) $\mathsf{E}_{\mathsf{P}}\left(A_0^2 + \left(\int_0^{A_0} |h(X(u))|\,\mathrm{d}u\right)^2\right) < \infty.$

Call

$$\bar{r} = \mathsf{E}_{\bar{\mathsf{P}}} h(\bar{X}(0)) = a_{\mathsf{P}}^{-1} \mathsf{E}_{\mathsf{P}} \int_0^{A_0} h(X(u))\,\mathrm{d}u,$$

cf. (4.2.10). The following statements are established in FOX and GLYNN (1987).

$$r(t) \to \bar{r} \qquad \mathsf{P}\text{-a.s.}, \tag{4.6.1}$$

$$r_n \to \bar{r} \qquad \mathsf{P}\text{-a.s.}, \tag{4.6.2}$$

$$\frac{n^{1/2} a_{\mathsf{P}}}{\sigma} (r_n - \bar{r}) \Rightarrow \mathrm{N}(0, 1), \tag{4.6.3}$$

$$\frac{(t a_{\mathsf{P}})^{1/2}}{\sigma} \left(r(t) - \bar{r}\right) \Rightarrow \mathrm{N}(0, 1), \tag{4.6.4}$$

$$\bar{r}(t) = t^{-1} \int_0^t h(\bar{X}(u))\,\mathrm{d}u \to \bar{r} \qquad \bar{\mathsf{P}}\text{-a.s.}, \tag{4.6.5}$$

$$\bar{r}_n = \bar{T}_n^{-1} \int_0^{\bar{T}_n} h(\bar{X}(u))\,\mathrm{d}u \to \bar{r} \qquad \bar{\mathsf{P}}\text{-a.s.}, \tag{4.6.6}$$

$$\frac{(t a_{\mathsf{P}})^{1/2}}{\sigma} \left(\bar{r}(t) - \bar{r}\right) \Rightarrow \mathrm{N}(0, 1), \tag{4.6.7}$$

where

$$\sigma^2 = \mathsf{E}_{\mathsf{P}} \left(\int_0^{A_0} h(X(u))\,\mathrm{d}u - A_0 \bar{r}\right)^2 + 2 \sum_{k=1}^{\infty} \mathsf{E}_{\mathsf{P}} \left(\int_0^{A_0} h(X(u))\,\mathrm{d}u - A_0 \bar{r}\right) \times \left(\int_{T_k}^{T_{k+1}} h(X(u))\,\mathrm{d}u - (T_{k+1} - T_k)\,\bar{r}\right).$$

Fox and Glynn (1987) provided some interesting applications of these results and some further discussions concerning the weakening of assumptions.

Obviously, (4.6.5) follows from the Individual Ergodic Theorem. Formula (4.6.1) follows as a particular case from (4.2.11). Taking into account $\overline{T}_n/n \to a_{\mathrm{P}}$ $\overline{\mathsf{P}}$-a.s., one obtains (4.6.6) from (4.2.14). From $T_n/n \to a_{\mathrm{P}}$ and the ergodic Theorem A 1.2.2 one gets (4.6.2). The Central Limit Theorem (4.6.4) was proved by Serfozo (1972). Cumulative processes related to a given PEMP, somewhat more general than integrals of the type

$$\int_0^t h(X(u))\, du,$$

were treated by Streller (1980). Theorem 3.2 in Streller (1980) contains (4.6.4) and (4.6.7). It should be noted that in practical situations one is interested in asymptotic relationships for $r(t)$ and r_n not only in steady state but even for an "arbitrary initial phase". Some hints in this direction are given by Fox and Glynn (1987). Streller (1980) has shown that the relations (4.6.1)–(4.6.7) remain valid in the case of an arbitrary initial phase if the underlying PEMP is a semi-regenerative process with an arbitrary state space and with a Harris-recurrent embedded Markov chain, cf. Section 4.3.2.

4.3. Semi-Markov processes with an arbitrary state space were discussed e.g. in Çinlar (1969). Semi-regenerative processes with discrete space $\mathbb{X}$ were introduced by Pyke and Schaufele (1966) under the name semi-Markov processes with auxiliary paths, cf. also Schäl (1970, 1971) and Çinlar (1975). The presentation here goes back to Arndt and Franken (1979).

Formulae (4.3.4) and (4.3.5) are well-known for semi-Markov processes. They are commonly derived as limiting results by means of the key renewal theorem, cf. Çinlar (1975). As a generalization of (4.3.4) and (4.3.5), our approach shows the validity of (4.3.2) and (4.3.3) for arbitrary stationary pure jumpwise processes. However, the right-hand sides in (4.3.2) and (4.3.3) are unknown in general. Similar formulae in a much more general setting are given in Last (1990).

The proof of Theorem 4.3.1 is due to Streller (1982). Convergence in variation of $\mathsf{P}(X(t) \in (\cdot) \mid Y_0 = x)$ to $\overline{\mathsf{P}}(\overline{X}(0) \in (\cdot))$ for an arbitrary x as $t \to \infty$ was proved in semi-Markov case by Athreya, McDonald, and Ney (1978). There the coupling method was used in order to construct an artificial recurrent state. The distribution of the time between successive occurences of this state was assumed to be non-lattice.

The approach presented here to reliability analysis of complex systems with repair was developed by Nawrotzki (1975), Arndt and Franken (1977, 1979), Franken and Streller (1980). For comprehensive treatment cf. Beichelt and Franken (1983), Franken, Kirstein, and Streller (1984). For Markovian systems, i.e. if $\overline{X}(t)$ is a homogeneous Markov process, formula (4.3.14) was proved by Keilson (1974). Theorem 4.3.3 was established by Ross (1975) under the additional assumption that the vectors $[U_n^{(i)}, D_n^{(i)}]$ of the life and repair times of the i-th component consist of i.i.d. r.v.'s; Ross's proof is based on the assumption that the stochastic process describing the time evolution of the whole system possesses regeneration points. An interesting application of Theorem 4.3.3 is given in BRANDT and SULANKE (1986).

4.5. Equation (4.5.5) is a continuous time analogue of equation (4.4.1). Note, however, that the situation differs from the usual relation between difference and differential equations. The transition from the discrete to the continuous time equation is not carried out by means of a refinement of the discrete time scale. For this reason (4.5.5) cannot be transformed into a stochastic differential equation, in general. The influence of randomness is given by the MPP $([\overline{T}_n, \overline{Y}_n])$ controlling the system. That means that in the continuous time situation randomness acts at the discrete epochs $\overline{T}_n$ only; between them the behaviour of the system is deterministic.

A proof of Theorem 4.5.4 is given in LISEK (1982), where a more general model is treated: The time axis is no longer $\mathbb{R}$ or $\mathbb{Z}$, but an Abelian group $[\mathfrak{G}, +]$ endowed with an order relation $\leqq$. One has to claim that there is a countable subgroup $(g_j, j \in \mathbb{Z})$ of $\mathfrak{G}$ satisfying

(i) $g_j + g_k = g_{j+k}$, $j, k \in \mathbb{Z}$.
(ii) $g_j \leqq g_k$ if $j \leqq k$.
(iii) For every $g \in \mathfrak{G}$ there exists a $j \in \mathbb{Z}$ such that $g_j \leqq g$.

(In case of $\mathfrak{G} = \mathbb{Z}$ or $\mathfrak{G} = \mathbb{R}$ the conditions (i)–(iii) are fulfilled for $g_j = j$.) The statements of Corollary 1.4.6, Theorem 1.4.11, Lemma 1.5.1, and Theorem 1.5.4 can also be proved for continuous time equations.

5. Arrival-stationary queueing processes. Existence and uniqueness

5.1. Introduction and notation

In this chapter we investigate the time evolution of some standard and non-standard queueing systems. Here we focus our attention on the arrival epochs of customers. We discuss the existence and uniqueness of (embedded) stationary state processes and the limiting behaviour of these systems as the time tends to infinity.

Assume the arriving customers to be numbered and denote the arrival epoch of the n-th customer by T_n:

$$\cdots \leqq T_{-1} \leqq T_0 \leqq T_1 \leqq T_2 \leqq \cdots$$

If arrival epochs coincide we have batch arrivals. Consequently, the customers within an arbitrary batch are also numbered. We imagine $T_0 = 0$ as the arrival epoch of the customer with the number zero. The sequence (Φ, P), $\Phi = ([A_n, Y_n])$, is called the *input* (the traffic) of the queueing system, where $A_n = T_{n+1} - T_n$ is the interarrival time between the n-th and $(n+1)$-th arrival, and Y_n is a mark comprising the characteristics of the n-th customer which we need to keep track of, such as its required service time, priority number, and so forth.

The queueing systems considered in this chapter have the following property: Using an appropriately chosen state space $\mathbf{X}$ and denoting by X_n the (random) state of the system seen by the n-th arriving customer we can describe the evolution of the system at arrival instants by a recursive stochastic equation

$$X_{n+1} = f(X_n, [A_n, Y_n]), \qquad n \in \mathbf{Z}. \tag{5.1.1}$$

Moreover, throughout this chapter we assume that the input (Φ, P), $\Phi = [(A_n, Y_n])$, is *stationary*. For this reason we can apply the results of Chapter 1 to these queueing systems. The form of the function f depends on the type of the system and on the choice of the state space $\mathbf{X}$. It should be mentioned that in some cases the validity of an equation of type (5.1.1) can be forced by a suitable modification of the original input (more precisely, by introducing certain additional marks), cf. Sections 5.7, 5.8.

We are interested in stationary weak solutions of (5.1.1), i.e. in stationary sequences (Φ', P'), $\Phi' = \big([[A_n', Y_n'], X_n']\big)$, with the properties

$$\mathsf{P}'\big(([A_n', Y_n']) \in (\cdot)\big) = \mathsf{P}\big(([A_n, Y_n]) \in (\cdot)\big)$$

and

$$\mathsf{P}'(X'_{n+1} = f(X_n', [A_n', Y_n']), n \in \mathbb{Z}) = 1,$$

cf. Chapter 1. The marginal sequence (X_n') of a weak solution $\left(([A_n', Y_n'], X_n']), \mathsf{P}'\right)$ of (5.1.1) for a given input $\left(([A_n, Y_n]), \mathsf{P}\right)$ is called a state process of the system described by (5.1.1). We search for conditions on the input ensuring the existence of a (unique) stationary state process. Moreover, we investigate whether the state process of the system which starts working at $T_0 = 0$ with an arbitrary initial state evolves to the unique stationary state process as $n \to \infty$. For most of the stationary weak solutions which appear in this chapter the probability distribution of the corresponding state process can be calculated immediately by means of the probability distribution of the input (Φ, P), $\Phi = ([A_n, Y_n])$. This is true, if we have strong solutions (more precisely, X_n is a measurable function of $\left([A_k, Y_k], k < n\right)$, $n \in \mathbb{Z}$) or if we obtain the distribution of the state process from P by a suitable randomization, cf. Section 5.3. In these cases we use the simplified notation $\Phi' = \left([[A_n, Y_n], X_n]\right)$.

Extending the well-known Kendall notation, we use the symbol G/G/m/r to denote a queueing system with m servers in parallel ($m \leqq \infty$), r waiting places ($r \leqq \infty$) and a stationary input (Φ, P), $\Phi = ([A_n, S_n, K_n])$, where A_n are the interarrival times, S_n are the associated service times, and K_n the additional marks describing the quality of customers (such as e.g. priority class). If the marks K_n, $n \in \mathbb{Z}$, are not needed, we drop them. If $m = \infty$ we write G/G/∞. In the following we shall use these notation:

$\mathsf{A}(t) = \mathsf{P}(A_0 < t)$	–	the distribution function of the generic interarrival time,
$m_A = \int_0^\infty \left(1 - \mathsf{A}(t)\right) \mathrm{d}t$	–	the mean interarrival time,
$\lambda = 1/m_A$	–	the arrival intensity,
$\mathsf{S}(t) = \mathsf{P}(S_0 < t)$	–	the distribution function of the generic service time,
$m_S = \int_0^\infty \left(1 - \mathsf{S}(t)\right) \mathrm{d}t$	–	the mean service time,
$\mu = 1/m_S$	–	the service rate,
$\varrho = \lambda/\mu = m_S/m_A$	–	the traffic intensity.

No assumption on the independence in the sequence $([A_n, S_n, K_n])$ is made in general. However, in some special cases the sequence $([A_n, S_n])$ will also have some of the following properties:

(A 5.1.1) (A_n) and (S_n) are independent sequences;
(A 5.1.2) (A_n) is a sequence of i.i.d. r.v.'s;
(A 5.1.3) (S_n) is a sequence of i.i.d. r.v.'s.

If we have the properties (A 5.1.1) and (A 5.1.2), we call this GI/G/m/r; if the properties (A 5.1.1) and (A 5.1.3) are present, this is G/GI/m/r; in case of (A 5.1.1)–(A 5.1.3) this is GI/GI/m/r. (Notice that the last is GI/G/m/r in Kendall's notation.)

We use the symbols M or D instead of the first (second) GI if, in addition, the following asumption about the distribution function $\mathsf{A}(t)$ of A_0 (or $\mathsf{S}(t)$ of S_0) are fulfilled:

$$\text{M: } \mathsf{A}(t) = 1 - \mathrm{e}^{-\lambda t} \qquad \left(\mathsf{S}(t) = 1 - \mathrm{e}^{-\mu t}\right),$$

$$\text{D: } \mathsf{A}(t) = \begin{cases} 0 \text{ if } t \leqq m_A, \\ 1 \text{ if } t > m_A \end{cases} \qquad \left(\mathsf{S}(t) = \begin{cases} 0 \text{ if } t \leqq m_S, \\ 1 \text{ if } t > m_S \end{cases}\right).$$

Consider a queueing system with waiting room, i.e. a system G/G/m/r with $r > 0$, $m < \infty$. The so-called queueing discipline determines the rule according to which customers in the queue (if there is one) will be chosen for service. The following queueing disciplines are often used:

FCFS (first-come-first-served) — customers are served in the order of their arrival. (For the system G/G/1/r this discipline is also called FIFO: first-in-first-out.)

LCFS (last-come-first-served) — customers are served in the reversed order of their arrival.

SJF (shortest-job-first) — when a server becomes idle, the waiting customer with the shortest service time will be served next.

SIRO (service in random order).

Notice that in order to treat a SIRO queue by means of an equation of type (5.1.1), one needs the additional marks K_n to describe the randomness of the choice of the next customer.

A queueing discipline is called work conserving iff:

(i) The service speed is always equal to 1.
(ii) As long as there are customers in the queue, the servers are working.
(iii) No additional work (service time) can arise during the service process of a customer (e.g. by interruptions).

For instance, FCFS, LCFS, SJF, and SIRO are work conserving queueing disciplines.

In what follows we use unified notations for the main queueing quantities: the letters V, W, Q, and L are reserved for *work load*, *waiting time*, *number of customers in the queue* and *number in the system* (in the queue and in service), respectively.

We often need the following notation for the iterates of f, cf. Chapter 1: Let $\varphi = ([a_n, y_n])$ be a sample path of the input and $x \in \mathbf{X}$ an arbitrary initial state. Then

$$\begin{aligned} &x_0(x) = x, \\ &x_1(x, [a_0, y_0]) = f(x, [a_0, y_0]), \ldots, \\ &x_n(x, [a_0, y_0], \ldots, [a_{n-1}, y_{n-1}]) = f\Big(x_{n-1}\big(x, [a_0, y_0], \ldots, \\ &\qquad [a_{n-2}, y_{n-2}]\big), [a_{n-1}, y_{n-1}]\Big), \qquad n \geqq 2. \end{aligned} \tag{5.1.2}$$

5.2. The system G/G/∞

Consider a G/G/∞ queue. Let the state X_n seen by the n-th arriving customer be the sequence $X_n = (X_{n1}, X_{n2}, \dots)$, where the components are the residual service times at servers arranged in descending order (the residual service time of an idle server is equal to O). This means that the state space is

$$\mathbb{X} = \{x = (x_1, x_2, \dots) : x_k \in \mathbb{R}_+, x_k \geqq x_{k+1}, \ k = 1, 2, \dots\}.$$

Endowed with the metric

$$\varrho(x, x') = \sum_{i=1}^{\infty} \frac{|x_i - x_i'|}{1 + |x_i - x_i'|} 2^{-i}, \qquad x, x' \in \mathbb{X},$$

$\mathbb{X}$ becomes a complete separable metric space. Equation (5.1.1) has the form

$$X_{n+1} = \overline{R}(S_n - A_n, X_{n1} - A_n, X_{n2} - A_n, \dots)_+, \qquad n \in \mathbb{Z}, \tag{5.2.1}$$

where $(x_1, x_2, \dots)_+ = \big(\max(0, x_1), \max(0, x_2), \dots\big)$ and $\overline{R}(x)$ consists of the components of x rearranged in descending order.

5.2.1. Theorem. *Consider a* G/G/∞ *system with the stationary input* (Φ, P), $\Phi = ([A_n, S_n])$, *satisfying the conditions*

$$\lim_{n\to-\infty} \frac{1}{n} S_n = 0 \quad \mathsf{P}\text{-}a.s., \qquad \lim_{n\to-\infty} \frac{1}{-n} \sum_{i=n}^{-1} A_i > 0 \quad \mathsf{P}\text{-}a.s.$$

(These conditions are valid if $m_S < \infty$, $0 < m_A$ *and* (A_n) *is ergodic.)*
Then the sequence (X_n),

$$X_n = x_n^{\min}(\Phi) = \lim_{r\to\infty} x_r(0, [A_{n-r}, S_{n-r}], \dots, [A_{n-1}, S_{n-1}])$$

$$= \overline{R}(S_{-1} - A_{-1}, S_{-2} - A_{-1} - A_{-2}, S_{-3} - A_{-1} - A_{-2} - A_{-3}, \dots)_+,$$

(where $0 = (0, 0, \dots)$*) is the unique stationary state process of the system. Moreover,*

$$\mathsf{P}\left(\bigcup_{m=1}^{\infty} \{x_k(x, [A_0, S_0], \dots, [A_{k-1}, S_{k-1}]) = x_k^{\min}(\Phi), k \geqq m\}\right) = 1 \tag{5.2.2}$$

for every $x \in \mathbb{X}$.

Proof. Before we start proving the theorem, we remark that $m_S < \infty$ ensures $S_n/n \xrightarrow[n\to-\infty]{} 0$ a.s. Namely,

$$\lim_{n\to-\infty} \frac{1}{n} S_n = \lim_{n\to-\infty} \frac{1}{n} \sum_{i=n}^{-1} S_i - \lim_{n\to-\infty} \frac{n+1}{n} \frac{1}{n+1} \sum_{i=n+1}^{-1} S_i = 0,$$

in view of the individual ergodic theorem.

1. The assumptions imply

$$\lim_{r\to\infty} \frac{1}{r} \left(s_{n-r} - \sum_{k=1}^{r} a_{n-k}\right) < 0$$

for P-a.e. $\varphi = ([a_k, s_k])$ and all n. Thus, the sequence

$$\left(s_{n-r} - \sum_{k=1}^{r} a_{n-k}, r \geqq 1\right)$$

has at most finitely many positive elements. This means that the number

$$r(n, \varphi) \underset{\text{df}}{=} \inf \{r : r \geqq 1, s_{n-m} - \sum_{k=1}^{m} a_{n-k} < 0 \text{ for all } m \geqq r\}$$

is finite for almost all φ and all n. For all $r \geqq r(n, \varphi)$ it follows that

$$\begin{aligned} & x_r(0, [a_{n-r}, s_{n-r}], \ldots, [a_{n-1}, s_{n-1}]) \\ &= \overline{R}\left(s_{n-r} - \sum_{k=1}^{r} a_{n-k}, s_{n-r+1} - \sum_{k=1}^{r-1} a_{n-k}, \ldots, s_{n-1} - a_{n-1}, 0, 0, \ldots\right)_+ \\ &= \overline{R}\left(s_{n-r(n,\varphi)} - \sum_{k=1}^{r(n,\varphi)} a_{n-k}, \ldots, s_{n-1} - a_{n-1}, 0, 0, \ldots\right)_+ \\ &= x_{r(n,\varphi)}(0, [a_{n-r(n,\varphi)}, s_{n-r(n,\varphi)}], \ldots, [a_{n-1}, s_{n-1}]). \end{aligned} \tag{5.2.3}$$

Formula (5.2.3) shows that

$$x_n^{\min}(\varphi) = \lim_{r \to \infty} x_r(0, [a_{n-r}, s_{n-r}], \ldots, [a_{n-1}, s_{n-1}])$$

belongs to $\mathbb{X}$ for almost all φ and all n.

2. The conditions (A 1.3.1)–(A 1.3.5) are fulfilled, cf. Section 1.3. Thus, the monotonicity method works. By Theorem 1.3.1, $(X_n) = \left(x_n^{\min}(\Phi)\right)$ is a stationary strong solution of (5.2.1). Formula (5.2.3) shows that for almost every φ each n is a 0-renewing epoch of φ (cf. Definition 1.3.7 with $l(n, 0, \varphi) = r(n, \varphi)$), where $r(n, \varphi)$ is defined above.

Now let $x = (x_1, x_2, \ldots) \in \mathbb{X}$ be an arbitrary state, φ a fixed sample path of the input, and $n \in \mathbb{Z}$. For all $r \geqq r(n, \varphi)$ with the additional property

$$\sum_{k=1}^{r} a_{n-k} \geqq x_1$$

we obtain

$$\begin{aligned} & x_r(x, [a_{n-r}, s_{n-r}], \ldots, [a_{n-1}, s_{n-1}]) \\ &= x_{r(n,\varphi)}(0, [a_{n-r(n,\varphi)}, s_{n-r(n,\varphi)}], \ldots, [a_{n-1}, s_{n-1}]). \end{aligned}$$

Hence all $n \in \mathbb{Z}$ are renewing epochs in the sense of Definition 1.3.10. Theorem 1.3.11 yields (5.2.2).

3. It remains to show the uniqueness statement. Let (Φ', P'), $\Phi' = ([[A_n', S_n'], X_n'])$, $X_n' = (X'_{n1}, X'_{n2}, \ldots)$, be a weak solution of (5.2.1), where (X_n') is assumed to be stationary. Then

$$\mathsf{P}'\left(\bigcup_{k=1}^{\infty} \{\# \{n : n \leqq 0, X'_{n1} \leqq k\} = \infty\}\right) = 1, \tag{5.2.4}$$

cf. Corollary A 1.1.3. Thus, for P'-a.e. $\varphi' = ([[a_j, s_j], x_j])$ and each n, there exists an $r_0 \geqq r(n, ([a_j, s_j]))$ such that $x_{n-r_0,1} \leqq k$ and

$$a_{n-r_0} + \cdots + a_{n-1} \geqq k .$$

This implies

$$\begin{aligned} x_n &= x_{n-r_0}(x_{n-r_0}, [a_{n-r_0}, s_{n-r_0}], \ldots, [a_{n-1}, s_{n-1}]) \\ &= x_{n-r_0}(0, [a_{n-r_0}, s_{n-r_0}], \ldots, [a_{n-1}, s_{n-1}]) \\ &= x_n^{\min}(([a_j, s_j])) . \end{aligned}$$

That means, $X_n' = x_n^{\min}(([A_n', S_n']))$ P'-a.s. In particular, $(X_n') \stackrel{\mathcal{D}}{=} (x_n^{\min}(\Phi))$. This finishes the proof. ∎

5.2.2. Remark. We have in fact shown a little bit more: If $([[A_n', S_n'], X_n'])$ is a weak solution of (5.2.1) and (X_n') is stationary, then

$$X_n' = x_n^{\min}(([A_j', S_j'])) \quad \text{a.s.}$$

In particular, $([[A_n, S_n], x_n^{\min}(\Phi)])$ is the only stationary weak solution.

Later we shall need the following result.

5.2.3. Theorem. *Consider a G/GI/∞ system with the stationary input* (Φ, P), $\Phi = ([A_n, S_n])$. *Assume* (A_n) *to be ergodic,* $m_s < \infty$ *and*

$$\mathsf{P}(S_0 < A_0) > 0 . \tag{5.2.5}$$

Then the stationary state process $X_n = x_n^{\min}(\Phi)$ *possesses infinitely many empty points, i.e.*

$$\mathsf{P}(\# \{n: n \leqq 0, X_n = 0\} = \# \{n: n > 0, X_n = 0\} = \infty) = 1 .$$

Proof. The sequence Φ is ergodic, cf. Theorem A 1.2.6. By Lemma A 2.7 $([A_n, S_n, X_n])$ is ergodic too, and hence it remains to show that

$$\mathsf{P}(X_0 = 0) > 0 .$$

By (5.2.5), there is a positive real number c such that

$$\mathsf{P}(S_0 < c) > 0 , \tag{5.2.6}$$

$$\mathsf{P}(A_0 > c) > 0 . \tag{5.2.7}$$

Together with the ergodicity of (A_n), (5.2.7) implies

$$\mathsf{P}(\# \{n \geqq 0: c < A_n\} = \infty) = 1 .$$

This yields the existence of an n_0 such that

$$\mathsf{P}\left(\# \left\{n: 0 \leqq n \leqq n_0, \sum_{i=0}^{n} A_i > X_{01}, A_n > c\right\} \geqq 1\right) > 0 \tag{5.2.8}$$

(where X_{01} is the first component of X_0) because

$$\sum_{n=1}^{\infty} A_n = \infty \quad \mathsf{P}\text{-a.s.}$$

Remember, that $X_0 = x_0^{\min}(\Phi)$ depends on Φ only via $[A_{-1}, S_{-1}], [A_{-2}, S_{-2}], \ldots$ From this, (5.2.6), (5.2.8), and the independence assumptions we obtain

$$\mathsf{P}\left(\# \left\{n: 0 \leqq n \leqq n_0, \sum_{i=0}^{n} A_i > X_{01}, A_n > c\right\} \geqq 1, S_0 < c, \ldots, S_{n_0} < c\right) > 0.$$

If

$$\sum_{i=0}^{n} A_i > X_{01}, A_n > c, S_0 < c, \ldots, S_n < c,$$

then

$$X_{n+1} = 0.$$

Thus,

$$\mathsf{P}\left(\bigcup_{n=1}^{n_0+1} \{X_n = 0\}\right) > 0$$

which immediately implies $\mathsf{P}(X_0 = 0) > 0$. ∎

The number L_n of customers in the system is determined by

$$L_n = \sum_{i<n} \mathbf{1}\{S_i > A_i + \cdots + A_{n-1}\}. \tag{5.2.9}$$

Under the assumptions of Theorem 5.2.1 the quantity L_n is almost surely finite.

5.2.4. Remark. Assumption (5.2.5) is valid if $\mathsf{S}(t) = \mathsf{P}(S_0 < t) > 0$ for all $t > 0$. In this case the statement of Theorem 5.2.3 holds without the ergodicity of (A_n), cf. Franken (1969, 1970), [FKAS, Lemma 2.3.4]. In general, the ergodicity assumption can not be omitted, as the following example shows.

5.2.5. Example. Consider a G/GI/∞ system with the input $\Phi = ([A_n, S_n])$ satisfying $\mathsf{P}(S_n \equiv 2) = 1$, $\mathsf{P}(A_n \equiv 1, n \in \mathbb{Z}) = \mathsf{P}(A_n \equiv 3, n \in \mathbb{Z}) = 1/2$. Then (5.2.5) is fulfilled, but with probability 1/2 the sequence $(x_n^{\min}(\Phi))$ does not have infinitely many empty points.

5.3. The system G/G/m/0

Consider a G/G/m/0 queue. Let the state X_n seen by the n-th customer be the vector $X_n = [X_{n1}, \ldots, X_{nm}]$, where the components are the residual service times at the servers arranged in descending order (the residual service time of an idle server is equal to 0). That means, the state space is

$$\mathbf{X} = \{x = [x_1, \ldots, x_m]: x \in \mathbb{R}_+^m, x_1 \geqq x_2 \geqq \cdots \geqq x_m\}.$$

Equation (5.1.1) has the form

$$X_{n+1} = \overline{R}(X_n + S_n \mathbf{1}\{0 = X_{nm}\} e_m - A_n \mathsf{1})_+, \quad n \in \mathbb{Z}, \tag{5.3.1}$$

where $e_m = [0, \ldots, 0, 1]$, $\mathsf{1} = [1, \ldots, 1]$, and $\overline{R}(x)$ consists of the components of x rearranged in descending order.

5.3.1. Theorem. *Consider a G/G/m/0 system with the stationary input* (Φ, P), $\Phi = ([A_n, S_n])$, *satisfying the conditions*

$$\lim_{n\to-\infty} \frac{1}{n} S_n = 0 \quad \mathsf{P}\text{-a.s.}, \quad \lim_{n\to-\infty} \frac{1}{-n} \sum_{i=n}^{-1} A_i > 0 \quad \mathsf{P}\text{-a.s.}$$

(These conditions are valid if $m_S < \infty$, *and* $0 < m_A$ *and* (A_n) *is ergodic, cf. proof of Theorem* 5.3.1.)
Then there exists a stationary weak solution $(\hat{\Phi}, \mathsf{Q})$, $\hat{\Phi} = \left([[\hat{A}_n, \hat{S}_n], \hat{X}_n]\right)$, *of* (5.3.1).

If in addition (Φ, P) *is ergodic, then there is a finite number* $q \geqq 1$ *and measurable mappings* $x(j, \varphi)$, $j = 1, \ldots, q$, *such that*

$$\mathsf{Q}\left(\left([[\hat{A}_n, \hat{S}_n], \hat{X}_n], n \geqq 0\right) \in (\cdot)\right)$$

$$= \frac{1}{q} \sum_{j=1}^{q} \mathsf{P}\left(\left([[A_n, S_n], x_n\left(x(j, \Phi), [A_0, S_0], \ldots, [A_{n-1}, S_{n-1}]\right)], n \geqq 0\right) \in (\cdot)\right). \tag{5.3.2}$$

Proof. The proof will be carried out by using a solution generating system, cf. Section 1.4. For a sample path $\varphi = ([a_j, s_j])$ of the input and $n \in \mathbb{Z}$ define

$$I(n, \varphi) = \left\{j : j < n, s_j > \sum_{i=j}^{n-1} a_i\right\}$$

and

$$A(n, \varphi) = \left\{x = [x_1, \ldots, x_m] : x \in \mathbf{X}; x_k = s_j - \sum_{i=j}^{n-1} a_i, \right.$$
$$\left. j \in I(n, \varphi) \text{ or } x_k = 0; k = 1, \ldots, m\right\}.$$

The sets $I(n, \varphi)$ and $A(n, \varphi)$ are finite for P-a.e. φ. Namely,

$$\# I(n, \varphi) = \# \left\{j : j < n, \frac{s_j}{n-j} > \frac{1}{n-j} \sum_{i=j}^{n-1} a_i\right\} < \infty,$$
$$\# A(n, \varphi) \leqq m^{1+\# I(n,\varphi)}.$$

It is easy to see that the sets $A(n, \varphi)$ form an SGS, cf. Definition 1.4.2. Now the assertions of the theorem follow from Theorem 1.4.3 and Corollary 1.4.6. ∎

5.3.2. Remark. Let the assumptions of Theorem 5.3.1 be fulfilled. Consider a r.v. Λ uniformly distributed on $\{1, \ldots, q\}$ and independent of Φ. Then the sequence $\left([[A_n, S_n], X_n]\right)$ with

$$X_n = x_n\left(x(\Lambda, \Phi), [A_0, S_0], \ldots, [A_{n-1}, S_{n-1}]\right), \qquad n \geqq 0,$$

has the distribution (5.3.2).

5.3.3. Remark. From the construction of the SGS $\mathscr{S}$ given in the proof of Theorem 5.3.1 it is easy to see that every stationary weak solution of (5.3.1) is concentrated on

$$\mathcal{A}(\mathscr{S}) = \left\{\left([[a_n, s_n], x_n]\right) : x_n \in A\left(n, ([a_k, s_k])\right), n \in \mathbb{Z}\right\},$$

cf. Section 1.5.

The following example shows that (5.3.2) is not the only stationary weak solution of equation (5.3.1) and that $(\mathring{X}_n)$ is not the only stationary process of the loss system, in general.

5.3.4. Example. Consider the system D/D/2/0 with $A_n \equiv 1$ and $S_n \equiv 7/2$. It is easy to check that $q = 6$,

$$x(1, \Phi) \equiv [0, 1/2], \quad x(2, \Phi) \equiv [0, 3/2], \quad x(3, \Phi) \equiv [0, 5/2],$$
$$x(4, \Phi) \equiv [1/2, 3/2], \quad x(5, \Phi) \equiv [1/2, 5/2],$$
$$x(6, \Phi) \equiv [3/2, 5/2].$$

Obviously,

$$\mathsf{Q}_1(\cdot) = \frac{1}{4} \sum_{j \in \{1,3,4,6\}} \mathsf{P}\Big(\big([[A_n, S_n], x_n\big(x(j, \Phi), [A_0, S_0], \ldots, [A_{n-1}, S_{n-1}]\big)\big], n \geqq 0\big) \in (\cdot)\Big)$$

and

$$\mathsf{Q}_2(\cdot) = \frac{1}{2} \sum_{j \in \{2,5\}} \mathsf{P}\Big(\big([[A_n, S_n], x_n\big(x(j, \Phi), [A_0, S_0], \ldots, [A_{n-1}, S_{n-1}]\big)\big], n \geqq 0\big) \in (\cdot)\Big)$$

define two different stationary (ergodic) weak solutions.

Now we look for conditions ensuring the uniqueness of the stationary weak solution and discuss limiting properties of the loss system. This will be done by applying the method of renewing epochs, cf. Section 1.3.

5.3.5. Theorem. *Consider a G/G/m/0 system with the stationary, ergodic input* (Φ, P), $\Phi = ([A_n, S_n])$, *Assume:*

(A 5.3.1) *There is a* $k \in \{0, \ldots, m-1\}$ *such that*

$$\mathsf{P}\Bigg(\sum_{i=-\infty}^{-k-1} 1\{S_i > A_i + \cdots + A_{-k-1}\} \leqq k,$$
$$\sum_{i=-\infty}^{-k-1} 1\{S_i > A_i + \cdots + A_{-k}\} \leqq k - 1,$$
$$\ldots,$$
$$\sum_{i=-\infty}^{-k-1} 1\{S_i > A_i + \cdots + A_{-2}\} \leqq 1,$$
$$\sum_{i=-\infty}^{-k-1} 1\{S_i > A_i + \cdots + A_{-1}\} = 0\Bigg) > 0,$$

or the weaker condition

(A 5.3.2) $$\mathsf{P}\Bigg(\sum_{i=-\infty}^{n-1} 1\{S_i > A_i + \cdots + A_{n-1}\} \leqq m - 1 \textit{ for all } n < 0 \textit{ with } S_n > A_n + \cdots + A_{-1}\Bigg) > 0$$

is true.

Then there is exactly one stationary state process (X_n). This state process has the form $X_n = \hat{x}_n(\Phi) = \hat{x}_0(\theta^n\Phi)$, $n \in \mathbb{Z}$. Here $\hat{x}_0(\Phi)$ is a measurable function that depends on $\varphi = ([a_n, s_n])$ only via the past $([a_n, s_n], n < 0)$. Moreover,

$$\mathsf{P}\left(\bigcup_{m=1}^{\infty} \{x_r(x, [A_0, S_0], \ldots, [A_{r-1}, S_{r-1}]) = \hat{x}_r(\Phi), r \geqq m\}\right) = 1 \qquad (5.3.3)$$

for every $x \in \mathbb{X}$.

Proof. It is easy to see that (A 5.3.1) yields

$$\mathsf{P}(0 \text{ is a } (0,k)\text{-renewing epoch of } \Phi) > 0.$$

Now assume (A 5.3.2). The set $I(0, \varphi) = \{n : n < 0, s_n > a_n + \cdots + a_{-1}\}$ is finite for a.e. $\varphi = ([a_n, s_n])$, cf. the proof of Theorem 5.3.1 and Example 1.4.1. If φ satisfies

$$\sum_{i=-\infty}^{n-1} 1\{s_i > a_i + \cdots + a_{n-1}\} \leqq m - 1, \qquad n \in I(0, \varphi),$$

then 0 is a 0-renewing epoch of φ with $l(0, 0, \varphi) = -\min\left(I(0, \varphi) \cup \{0\}\right)$. *Thus*

$$\mathsf{P}(0 \text{ is a 0-renewing epoch of } \Phi) > 0.$$

Theorem 1.3.6 resp. Theorem 1.3.9 show that there is a stationary strong solution $\left([[A_n, S_n], \hat{x}_n(\Phi)]\right)$ of equation (5.3.1), satisfying $\hat{x}_n(\Phi) = \hat{x}_0(\theta^n\Phi)$ a.s. (cf. the proof of Theorem 1.3.6), and that (5.3.3) is true.

It remains to show the uniqueness statement. Consider the SGS given in the proof of Theorem 5.3.1 and the corresponding sets $B(n, \Phi)$ defined in the proof of Theorem 1.4.3. It is easy to see that (A 5.3.2) implies $\mathsf{P}\left(\# B(0, \Phi) = 1\right) > 0$ and hence

$$\# B(0, \Phi) = 1 \qquad \mathsf{P}\text{-a.s.}$$

in view of the ergodicity of Φ. Thus (cf. Theorem 1.4.3, Lemma A 1.2.7)

$$\mathsf{Q}(\cdot) = \mathsf{P}\left(\left([[A_n, S_n], \hat{x}_n(\Phi)]\right) \in (\cdot)\right),$$

is ergodic too. Now the uniqueness statement follows by means of Lemma 1.5.1 and Theorem 1.5.6. ∎

5.3.6. Theorem. *Consider a* GI/GI/m/0 *queue with the input $\Phi = ([A_n, S_n])$ satisfying $\mathsf{E}S_0 < \infty$ and*

$$\mathsf{P}(S_0 < mA_0) > 0. \qquad (5.3.4)$$

Then the assertions of Theorem 5.3.5 *are valid.*

Proof. By (5.3.4), there is a positive real number c such that

$$\mathsf{P}(S_0 < c) > 0, \qquad \mathsf{P}(c < mA_0) > 0. \qquad (5.3.5)$$

The quantity

$$X = \max\left\{(S_i - (A_i + \cdots + A_{-1}))_+ : i < 0\right\}$$

(the maximal residual service time at time 0 in GI/GI/∞ with the input Φ) is a.s. finite, cf. the proof of Theorem 5.2.1. Thus, there is a number n_0 such that

$$\mathsf{P}\left(X < \frac{c}{m}\, n_0\right) > 0.$$

In view of the independence assumptions and (5.3.5), we obtain

$$\mathsf{P}\left(X < \frac{c}{m}\, n_0 ; S_i < c < mA_i,\, i = 0, \ldots, n_0 + m - 1\right) > 0.$$

Hence

$$\begin{aligned}\mathsf{P}\Bigg(&S_i - (A_i + \cdots + A_{n_0-1}) < 0,\, i < 0;\\ &S_{n_0-1} - A_{n_0-1} < \frac{c}{m}\,(m-1),\\ &S_{n_0-2} - A_{n_0-2} - A_{n_0-1} < \frac{c}{m}\,(m-2),\\ &\ldots,\\ &S_{n_0-m+1} - A_{n_0-m+1} - \cdots - A_{n_0-1} < \frac{c}{m},\\ &S_{n_0-i} - A_{n_0-i} - \cdots - A_{n_0-1} < 0,\, m \leqq i \leqq n_0\Bigg) > 0.\end{aligned}$$

(This means that with positive probability, at n_0 there are at most the last $m - 1$ customers in the GI/GI/∞ queue with input Φ.) From this and the stationarity of Φ, (A 5.3.1) follows with $k = m - 1$. Now the assertion follows immediately from Theorem 5.3.5. ∎

As the following example shows, the independence of the interarrival times cannot be omitted in Theorem 5.3.6.

5.3.7. Example. Consider the G/D/2/0 queue with the input (Φ, P): $S_n \equiv 5$ and $\mathsf{P}(A_{3n+i} = A_{3n+i+1} = 1,\ A_{3n+i+2} = 3,\ n \in \mathbb{Z}) = 1/3,\ i = 0, 1, 2$. Then $\mathsf{P}(S_0 < 2A_0) = 1/3$ but there are three stationary ergodic strong solutions, and (5.3.3) does not hold.

In the case of G/GI/m/0 queues we need a stronger assumption than (5.3.4) for ensuring the validity of (5.3.3) and the uniqueness of the stationary state process:

5.3.8. Theorem. *Consider a* G/GI/m/0 *queue with the stationary input* (Φ, P), $\Phi = ([A_n, S_n])$. *Assume* (A_n) *to be ergodic,* $\mathsf{E}S_0 < \infty$ *and*

$$\mathsf{P}(S_0 < A_0) > 0. \tag{5.3.6}$$

Then the assertions of Theorem 5.3.5 *are valid.*

Proof. Consider the G/GI/∞ queue with the input (Φ, P). By Theorem 5.2.3, this system has infinitely many empty points. Thus, with positive probability,

0 is an empty point for the system G/GI/m/0. Now (A 5.3.2) follows immediately. Since Φ is ergodic (cf. Theorem A 1.2.6) Theorem 5.3.5 yields the assertions. ∎

For the GI/GI/1/0 queue we have the following result:

5.3.9. Theorem. *Consider a* GI/GI/1/0 *queue with* $\mathsf{E}S_0 < \infty$, $\mathsf{E}A_0 > 0$. *Then there is exactly one stationary weak solution of* (5.3.1). *Moreover, there is a unique stationary state process.*

Proof. The first assertion has already been proved in Example 2.2.9. The latter can be obtained as follows: Let $([[A_n', S_n'], X_n'])$ be a weak solution with (X_n') stationary. Since the state process (X_n') has infinitely many empty points, $([[A_n', S_n'], X_n'])$ is concentrated on $\mathcal{A}(\mathcal{S}) = \{([[a_n, s_n], x_n]) : x_n \in A(n, ([a_k, s_k])), n \in \mathbb{Z}\}$. Now the uniqueness of the stationary state process follows by means of Theorem 1.5.6. ∎

The following example shows that neither the independence assumptions of the interarrival times nor those of the service times can be omitted in Theorem 5.3.9.

5.3.10. Example. Consider the GI/G/1/0 loss system with $A_n = 1$, $n \in \mathbb{Z}$, and (S_n) being a stationary Markov chain with the states 4 and 6, and transition probabilities $p_{12} = p_{21} = 1$. Then we have infinitely many stationary weak solutions, cf. Example 1.5.2. The same result is true for the G/GI/1/0 queue if we assume $S_n = 1$, $n \in \mathbb{Z}$, and (A_n) to be a stationary Markov chain with the states 3/5 and 4/5 and $p_{12} = p_{21} = 1$.

In the case of M/GI/1/0 it is not difficult to determine explicitly the distribution of the unique stationary state process (X_n), cf. Theorem 5.3.6.

5.3.11. Theorem. *For the system* M/GI/1/0 *with* $0 < \lambda = 1/m_A < \infty$, $m_S < \infty$ *it holds that*

$$\mathsf{P}(X_0 = 0) = \frac{m_A}{m_A + m_S},$$

$$\mathsf{P}(X_0 \leqq x \mid X_0 > 0) = \frac{1}{m_S} \int_0^x \big(1 - \mathsf{S}(u)\big)\, du, \quad x > 0.$$

Proof. Let Y_0 be a non-negative random variable with the distribution

$$\mathsf{P}(Y_0 = 0) = \frac{m_A}{m_A + m_S},$$

$$\mathsf{P}(Y_0 \leqq y \mid Y_0 > 0) = \frac{1}{m_S} \int_0^y \big(1 - \mathsf{S}(u)\big)\, du, \quad y > 0.$$

Assume that Y_0 is independent of $([A_n, S_n], n \geqq 0)$, and define

$$Y_1 = f(Y_0, [A_0, S_0]) = (Y_0 + S_0 \mathbf{1}\{Y_0 = 0\} - A_0)_+ .$$

The distribution of Y_1 can easily be calculated:

$$\begin{aligned}
\mathsf{P}(Y_1 = 0) &= \mathsf{P}(0 < Y_0 \leqq A_0) + \mathsf{P}(Y_0 = 0)\,\mathsf{P}(S_0 \leqq A_0)\\
&= \frac{m_S}{m_A + m_S}\,\mathsf{P}(Y_0 \leqq A_0 \mid Y_0 > 0) + \frac{m_A}{m_A + m_S}\,\mathsf{P}(S_0 \leqq A_0)\\
&= \frac{m_S}{m_A + m_S}\int_0^\infty \frac{1}{m_S}\big(1 - \mathsf{S}(y)\big)\,e^{-\lambda y}\,dy\\
&\quad + \frac{m_A}{m_A + m_S}\int_0^\infty \lambda\, e^{-\lambda a}\mathsf{S}(a)\,da = \frac{m_A}{m_A + m_S}.
\end{aligned}$$

and

$$\begin{aligned}
\mathsf{P}(Y_1 \leqq y \mid Y_1 > 0) &= \frac{m_A + m_S}{m_S}\,\mathsf{P}(Y_1 \leqq y, Y_1 > 0)\\
&= \frac{m_A + m_S}{m_S}\bigg(\frac{m_A}{m_A + m_S}\,\mathsf{P}(0 < S_0 - A_0 \leqq y)\\
&\quad + \frac{m_S}{m_A + m_S}\,\mathsf{P}(0 < Y_0 - A_0 \leqq y \mid Y_0 > 0)\bigg)\\
&= \frac{m_A}{m_S}\,\mathsf{P}(0 < S_0 - A_0 \leqq y)\\
&\quad + \mathsf{P}\big((Y_0 - y)_+ \leqq A_0 < Y_0 \mid Y_0 > 0\big)\\
&= \frac{1}{m_S}\bigg(\int_0^\infty \Big(1 - \mathsf{S}(s) - \big(1 - \mathsf{S}(s + y)\big)\Big)_+ e^{-\lambda s}\,ds\\
&\quad + \int_0^\infty \big(1 - \mathsf{S}(s)\big)\big(e^{-\lambda(s-y)_+} - e^{-\lambda s}\big)\,ds\bigg)\\
&= \frac{1}{m_S}\int_0^y \big(1 - \mathsf{S}(s)\big)\,ds.
\end{aligned}$$

Thus, $Y_1 = f(Y_0, [A_0, S_0]) \overset{\mathcal{D}}{=} Y_0$. Hence $\big(x_n(Y_0, [A_0, S_0], \ldots, [A_{n-1}, S_{n-1}]),\ n \geqq 0\big)$ defines a stationary state process. Since the stationary state process (X_n') of the system M/GI/1/0 is unique, it follows that $X_0' \overset{\mathcal{D}}{=} Y_0$. ∎

5.4. The single server queue G/G/1/∞

Consider a single server queue G/G/1/∞ with the stationary input (Φ, P), $\Phi = ([A_n, S_n])$, and with an arbitrary work-conserving queueing discipline. First we describe the state of the system by the work load, i.e. by the total sum of the service times associated with the customers in the queue plus the residual service time of the customer on service. Let V_n be the work load found by the n-th arriving customer. In the case of the FCFS queueing discipline, V_n is the actual waiting time W_n of the n-th customer. Independently of the queueing discipline, the equation

$$V_{n+1} = (V_n + S_n - A_n)_+, \qquad n \in \mathbb{Z}, \tag{5.4.1}$$

holds. (This means that equation (5.1.1) has the form (5.4.1) if we choose the work load as the state of the system.)

We use the notation

$$U_n = [A_n, S_n], \quad u_n = [a_n, s_n], \quad C_n = S_n - A_n, \quad c_n = s_n - a_n,$$

$$v_0(v) = v, \quad v_1(v, u_0) = (v + c_0)_+, \quad v_2(v, u_0, u_1) = \big((v + c_0)_+ + c_1\big)_+, \ldots,$$

$$v_{n+1}(v, u_0, \ldots, u_n) = \big(v_n(v, u_0, \ldots, u_{n-1}) + c_n\big)_+, \qquad n \geqq 0.$$

Notice that $v_k(v, u_0, \ldots, u_{k-1})$ depends on $u_0, \ldots, u_{k-1}$ only via $c_0, \ldots, c_{n-1}$.

5.4.1. Theorem. *Consider a G/G/1/∞ queue with an arbitrary work-conserving queueing discipline. Assume that the input (Φ, P), $\Phi = (U_n)$, $U_n = [A_n, S_n]$, is stationary and satisfies the condition*

$$\lim_{n\to\infty} \frac{1}{n} \sum_{i=1}^{n} (S_i - A_i) < 0 \qquad \mathsf{P}\text{-a.s.} \tag{5.4.2}$$

Then

$$V_n = v_n^{\min}(\Phi) = \lim_{r\to\infty} v_r(0, U_{n-r}, \ldots, U_{n-1})$$

$$= \left(\sup_{r\geqq 1} \sum_{j=1}^{r} (S_{n-j} - A_{n-j})\right)_+, \qquad n \in \mathbb{Z},$$

is the only stationary work load process. Moreover,

$$\mathsf{P}\left(\bigcup_{m=1}^{\infty} \{v_k(v, U_0, \ldots, U_{k-1}) = V_k, k \geqq m\}\right) = 1 \tag{5.4.3}$$

for every $v \in \mathbb{R}_+$.

If (Φ, P) is ergodic, then $\varrho < 1$, $\varrho = m_S/m_A$, is sufficient for the validity of (5.4.2).

Proof. By induction one can show that

$$v_n(v, u_0, \ldots, u_{n-1})$$
$$= \max(v + c_0 + \cdots + c_{n-1}, c_1 + \cdots + c_{n-1}, c_2 + \cdots + c_{n-1}, \ldots, c_{n-1}, 0). \tag{5.4.4}$$

From (5.4.2),

$$\lim_{r\to\infty} \frac{1}{r} \sum_{i=n-r}^{n-1} C_i < 0 \quad \mathsf{P}\text{-a.s.}, \qquad n \in \mathbb{Z}. \tag{5.4.5}$$

Together with (5.4.4), this yields

$$V_n = \lim_{r\to\infty} v_r(0, U_{n-r}, \ldots, U_{n-1})$$
$$= \lim_{r\to\infty} \max (C_{n-r} + \cdots + C_{n-1}, C_{n-r+1} + \cdots + C_{n-1},$$
$$\ldots, C_{n-1}, 0) < \infty \quad \text{P-a.s.}$$

Since the function $(v + c)_+$ is continuous, $\big([A_n, S_n], v_n^{\min}(\Phi)]\big)$ is a stationary strong solution of (5.4.1). Let $([u_n, v_n])$ fulfil $v_{n+1} = (v_n + c_n)_+$, $n \in \mathbb{Z}$. Since $(v + c)_+$ is non-decreasing in v,

$$v_r(0, u_{n-r}, \ldots, u_{n-1}) \leqq v_r(v_{n-r}, u_{n-r}, \ldots, u_{n-1}) = v_n$$

for all $r \geqq 1$. Hence $v_n^{\min}\big((u_k)\big) \leqq v_n$, $n \in \mathbb{Z}$. That means, $V_n = v_n^{\min}(\Phi)$ is the minimal stationary work load process. (Notice that the previous conclusions are a particular case of Theorem 1.3.1.)

Let $v \in \mathbb{R}_+$ be fixed. From (5.4.4) we obtain

$$0 \leqq v_n(v, U_0, \ldots, U_{n-1}) - v_n(0, U_0, \ldots, U_{n-1})$$
$$\leqq (v + C_0 + \cdots + C_{n-1})_+ .$$

Thus in view of (5.4.2),

$$|v_n(v, U_0, \ldots, U_{n-1}) - v_n(V_0, U_0, \ldots, U_{n-1})|$$
$$\leqq (v + C_0 + \cdots + C_{n-1})_+ + (V_0 + C_0 + \cdots + C_{n-1})_+ = 0$$

for sufficiently large n P-a.s. Assertion (5.4.3) is proved.

It remains to show the uniqueness statement. Let $\big(([U_n', V_n']), \mathsf{P}'\big)$, $U_n' = [A_n', S_n']$, be a weak solution of equation (5.4.1), where (V_n') is stationary. We have

$$\mathsf{P}'\left(\bigcup_{d=1}^{\infty} \{\# \{n : n \leqq 0, V_n' \leqq d\}\}\right) = 1. \tag{5.4.6}$$

Let $([u_n, v_n])$ be a sample path of $([U_n', V_n'])$ and (v_{n_i}) the subsequence of (v_n) with $v_{n_i} \leqq d$, $n_i \to -\infty$. (The existence of such a number d follows from (5.4.6).) In view of the minimality of $\big(v_n^{\min}\big((u_k)\big)\big)$, the monotonicity of $(v + c)_+$, (5.4.4), and (5.4.5), we obtain

$$0 \leqq v_n - v_n^{\min}\big((u_k)\big)$$
$$\leqq v_{n-n_i}(d, u_{n_i}, \ldots, u_{n-1}) - v_{n-n_i}(0, u_{n_i}, \ldots, u_{n-1})$$
$$\leqq (d + c_{n_i} + \cdots + c_{n-1})_+ = 0$$

for sufficiently small n_i. Thus, $V_n' = v_n^{\min}\big((U_k')\big)$ P'-a.s. ∎

5.4.2. Remark. In fact we have shown: If $([U_n', V_n'])$ is a weak solution of (5.4.1) and (V_n') is stationary, then

$$V_n' = v_n^{\min}\big((U_k')\big) \quad \text{a.s.}$$

In particular, $\big([U_n, v_n^{\min}(\Phi)]\big)$ is the only stationary weak solution.

Under the assumption (5.4.2) it is easy to see that the stationary single server queue becomes empty infinitely often: Consider $E_n = (A_n - V_n - S_n)_+$ the time during which the system is empty between the arrival of the n-th and $(n+1)$-st customer. Assume $E_n = 0$, $n \in \mathbb{Z}$, a.s. Then $V_{n+1} = V_n + S_n - A_n$, $n \in \mathbb{Z}$, and thus

$$0 \leqq V_n = V_0 + \sum_{i=0}^{n-1} (S_i - A_i),$$

which contradicts (5.4.2). Hence, in view of the stationarity of (E_n),

$$\# \{n : n \leqq 0, E_n > 0\} = \# \{n : n > 0, E_n > 0\} = \infty \quad \text{a.s.} \tag{5.4.7}$$

and

$$\sum_{n=-\infty}^{0} E_n = \sum_{n=1}^{\infty} E_n = \infty \quad \text{a.s.} \tag{5.4.8}$$

Now we consider other descriptions of the state of a single server queue with a stationary sequence $([A_n, S_n])$ of interarrival and service times and with an arbitrary work-conserving queueing discipline. Assume there is a suitable description of the state (e.g. in case of FCFS or LCFS the vector formed by the required and residual service times of the customers in the system arranged in the order of their arrivals), a stationary sequence $([A_n, S_n, K_n])$ (which will be considered as the input), and a function f such that the evolution of the system at the arrival epochs is given by the equation

$$X_{n+1} = f(X_n, [A_n, S_n, K_n]), \qquad n \in \mathbb{Z}. \tag{5.4.9}$$

Moreover, we assume that $V_n = 0$ if and only if $X_n = \mathbf{0}$ (where $\mathbf{0}$ denotes the idle state). Then the state of the system between consecutive empty points is uniquely determined by the input between these empty points. In view of (5.4.7) and Theorem 5.4.1, (5.4.3) one obtains:

5.4.3. Theorem. *Consider a single server queue with an arbitrary work conserving queueing discipline and the input* $\Phi = ([A_n, S_n, K_n])$. *Assume* (5.4.2) *is satisfied. Then there exists a uniquely determined stationary state process* (X_n). *Moreover,*

$$\mathsf{P}\left(\bigcup_{m=1}^{\infty} \{x_k(x, [A_0, S_0, K_0], \ldots, [A_{k-1}, S_{k-1}, K_{k-1}]) = X_k, k \geqq m\}\right) = 1$$

for every state x.

The stationary processes (L_n) and (Q_n) of the number of customers in the system and in the queue, respectively, will be introduced in Section 5.5.

5.5. The system G/G/m/∞ with FCFS queueing discipline

Consider a G/G/m/∞ queue with FCFS queueing discipline and the stationary input (Φ, P), $\Phi = ([A_n, S_n])$. The state of the system seen by the n-th arriving customer is the so-called Kiefer-Wolfowitz work load vector $V_n = [V_{n1}, \ldots, V_{nm}]$, $V_{n1} \leqq \cdots \leqq V_{nm}$; here V_{nj} denotes the time between the arrival of the

n-th customer and the time instant at which j servers would become idle for the first time assuming that the input was stopped at the n-th arrival, $1 \leqq j \leqq m$. Thus, V_{n1} is the actual waiting time of the n-th customer: $W_n = V_{n1}$. Then we have

$$\mathbb{X} = \{v = [v_1, \ldots, v_m] : v \in \mathbb{R}_+^m, v_1 \leqq v_2 \leqq \cdots \leqq v_m\}$$

as the state space, and equation (5.1.1) has the form

$$V_{n+1} = f(V_n, [A_n, S_n]) = R(V_n + S_n e_1 - A_n \mathbf{1})_+, \quad n \in \mathbb{Z}, \tag{5.5.1}$$

where $e_1 = [1, 0, \ldots, 0]$, $\mathbf{1} = [1, \ldots, 1]$ and $R(v)$ is the vector formed by re-arranging the components of the given vector $v = [v_1, \ldots, v_m]$ in increasing order.

In accordance with previous notation (cf. (5.1.2)) let

$$v_0(v) = v,$$

$$v_{n+1}(v, [a_0, s_0], \ldots, [a_n, s_n])$$

$$= f\big(v_n(v, [a_0, s_0], \ldots, [a_{n-1}, s_{n-1}]), [a_n, s_n]\big), \qquad n \geqq 0,$$

where f is defined by (5.5.1).

5.5.1. Theorem. *Consider a* G/G/m/∞ *queue with FCFS queueing discipline and with the stationary ergodic input* (Φ, P), $\Phi = ([A_n, S_n])$. *If*

$$\varrho = \frac{m_S}{m_A} < m,$$

then the sequence (V_n),

$$V_n = v_n^{\min}(\Phi) = \lim_{r \to \infty} v_r(0, [A_{n-r}, S_{n-r}], \ldots, [A_{n-1}, S_{n-1}]),$$

is a stationary state process of the system $(0 = [0, \ldots, 0])$. *It has the following minimality property: For every* $\varphi = ([a_n, s_n])$ *and every sequence* (v_n) *satisfying* $v_{n+1} = f(v_n, [a_n, s_n])$, $n \in \mathbb{Z}$, *it holds that*

$$v_n^{\min}(\varphi) \leqq v_n.$$

Proof. Denote by V_n^r the workload vector found by the n-th customer if customer $-r$ finds an empty system, $n \geqq -r$, i.e.

$$V_n^r = [V_{n1}^r, \ldots, V_{nm}^r] = v_{n+r}(0, [A_{-r}, S_{-r}], \ldots, [A_{n-1}, S_{n-1}]).$$

Since f is non-decreasing in the first argument, the limit

$$v_n^{\min}(\Phi) = [v_{n1}^{\min}(\Phi), \ldots, v_{nm}^{\min}(\Phi)] = \lim_{r \to \infty} V_n^r$$

exists. (However, $v_n^{\min}(\Phi)$ may have infinite components in general.)

We have

$$\mathsf{E}(V_0^r - V_1^r) = \mathsf{E}(V_0^r - V_1^{r-1}) + \mathsf{E}(V_1^{r-1} - V_1^r)$$

$$= \mathsf{E}(V_1^{r-1} - V_1^r) \leqq 0, \qquad r \geqq 1. \tag{5.5.2}$$

Using $(a - b)_+ = a - a \wedge b$, $a \wedge b = \min(a, b)$, we obtain

$$\sum_{j=1}^{m} V_{1j}^r = (V_{01}^r + S_0 - A_0)_+ + \sum_{j=2}^{m} (V_{0j}^r - A_0)_+$$

$$= V_{01}^r - V_{01}^r \wedge (A_0 - S_0) + \sum_{j=2}^{m} V_{0j}^r - (V_{0j}^r \wedge A_0)$$

$$= \sum_{j=1}^{m} V_{0j}^r - \left(V_{01}^r \wedge (A_0 - S_0) + \sum_{j=2}^{m} V_{0j}^r \wedge A_0\right).$$

Together with (5.5.2) this implies

$$\mathsf{E}\left(V_{01}^r \wedge (A_0 - S_0) + \sum_{j=2}^{m} V_{0j}^r \wedge A_0\right) \leqq 0, \qquad r \geqq 1.$$

Taking the limit as $r \to \infty$, we get

$$\mathsf{E}\left(v_{01}^{\min}(\Phi) \wedge (A_0 - S_0) + \sum_{j=2}^{m} v_{0j}^{\min}(\Phi) \wedge A_0\right) \leqq 0. \tag{5.5.3}$$

Since $\left(v_n^{\min}(\Phi)\right)$ fulfills the equation (5.5.1), the events $\{\varphi: v_{01}^{\min}(\varphi) = \infty\}$ and $\{\varphi: v_{0m}^{\min}(\varphi) = \infty\}$ are invariant. Thus, $v_{0m}^{\min}(\Phi) = \infty$ a.s. or $v_{0m}^{\min}(\Phi) < \infty$ a.s., and $v_{01}^{\min}(\Phi) = \infty$ a.s. or $v_{01}^{\min}(\Phi) < \infty$ a.s., because Φ is ergodic. Assume $v_{01}^{\min}(\Phi) = \infty$ a.s. Then (5.5.3) yields $\mathsf{E}S_0 \geqq m\mathsf{E}A_0$, which contradicts our assumption. Hence we have shown that $v_{01}^{\min}(\Phi) < \infty$ a.s.

Let $Z_r = (V_{01}^r + S_0 - A_0)_+$, $r \geqq 1$; $Z_\infty = (v_{01}^{\min}(\Phi) + S_0 - A_0)_+$. From $V_{1m}^r = Z_r \vee (V_{0m}^r - A_0)_+ = V_{0m}^r - (V_{0m}^r - Z_r) \wedge A_0$ and $Z_r \uparrow Z_\infty$, where $a \vee b = \max(a, b)$, we obtain

$$V_{1m}^r \leqq V_{0m}^r - (V_{0m}^r - Z_\infty) \wedge A_0.$$

By means of (5.5.2) it follows that

$$\mathsf{E}\left((V_{0m}^r - Z_\infty) \wedge A_0\right) \leqq 0,$$

and thus

$$\mathsf{E}\left(\left(v_{0m}^{\min}(\Phi) - Z_\infty\right) \wedge A_0\right) \leqq 0. \tag{5.5.4}$$

Assume $v_{0m}^{\min}(\Phi) = \infty$ a.s. Then, from (5.5.4) and $Z_\infty < \infty$ we obtain $m_A \leqq 0$, which is impossible. Thus, $v_{0m}^{\min}(\Phi) < \infty$ a.s. The minimality property follows immediately from the monotonicity of f, see also Theorem 1.3.1. ∎

From Theorem 5.5.1 we know that there always exists a minimal stationary state process for G/G/m/∞ with $\varrho < m$. However, uniqueness does not hold in general. This will be demonstrated by the following example, due to LOYNES (1962).

5.5.2. Example. Define $\varphi = ([a_n, s_n])$ by

$$a_n \equiv 1, \quad n \in \mathbb{Z}, \quad \text{and} \quad s_n = \begin{cases} 2 \text{ if } n \text{ is even}, \\ 3/2 \text{ if } n \text{ is odd}. \end{cases}$$

Consider the D/G/2/∞ queue with the stationary input (Φ, P) satisfying $\mathsf{P}(\Phi = \varphi) = \mathsf{P}(\Phi = \theta\varphi) = 1/2$. Define the functions $h_n^{(a)}$, $n \in \mathbb{Z}$, for φ and $\theta\varphi$ by

$$h_n^{(a)}(\varphi) = \begin{cases} [a - 1, 1/2] & \text{if } n \text{ is even}, \\ [0, a] & \text{if } n \text{ is odd}, \end{cases}$$

$$h_n^{(a)}(\theta\varphi) = h_{n+1}^{(a)}(\varphi),$$

where $1 \leqq a \leqq 3/2$. Then, for fixed $a \in [1, 3/2]$, $\Phi^{(a)} = \big([[A_n, S_n], h_n^{(a)}(\Phi)]\big)$ is a stationary strong solution of (5.5.1). In particular, $\Phi^{(1)} = \big([[A_n, S_n], v_n^{\min}(\Phi)]\big)$.

Theorem 5.5.1 and Example 5.5.2 give rise to the question: Are there also other extremal stationary state processes? Using the ideas from Section 1.3, we shall show the existence of a maximal stationary weak solution. This maximal solution turns out to be a strong one. In order to construct it, we need the following lemma.

5.5.3. Lemma. *For arbitrary* $v_1, v_2 \in \mathbf{X}$, $a, s \in \mathbb{R}_+$,

$$\|f(v_1, [a, s]) - f(v_2, [a, s])\|_{\max} \leqq \|v_1 - v_2\|_{\max},$$

where $\|x\|_{\max} = \max(|x_1|, \ldots, |x_m|)$ *denotes the maximum norm of a vector* $x = [x_1, \ldots, x_m] \in \mathbb{R}^m$.

The proof of the lemma is referred to the reader. The conditions (A 1.3.1) to (A 1.3.8) are fulfilled with $\varrho(v_1, v_2) = \|v_1 - v_2\|_{\max}$ and $x_{\max}(v, \varepsilon) = v + \varepsilon\mathbf{1}$. For fixed $\delta > 0$, $n \in \mathbb{Z}$ and $\varphi = (u_k) = ([a_k, s_k])$ with $v_k^{\min}(\varphi) < \infty$, $k \in \mathbb{Z}$,

$$v_r(v_{n-r}^{\min}(\varphi) + \delta\mathbf{1}, u_{n-r}, \ldots, u_{n-1})$$

is non-increasing as $r \to \infty$, cf. Section 1.3. Thus, if $\varrho < m$, we have from Theorem 5.5.1 that the limit

$$v_n^{\delta}(\varphi) = \lim_{r \to \infty} v_r\big(v_{n-r}^{\min}(\varphi) + \delta\mathbf{1}, u_{n-r}, \ldots, u_{n-1}\big)$$

exists and that it is finite for a.e. φ. Furthermore, it follows immediately that $\big([[A_n, S_n], v_n^{\delta}(\Phi)]\big)$ is a stationary strong solution of (5.5.1). For fixed n, $v_n^{\delta}(\varphi)$ is non-decreasing as $\delta \to \infty$.

5.5.4. Theorem. *Under the assumptions of Theorem* 5.5.1, *the sequence* $(V_n^{\max})$,

$$V_n^{\max} = v_n^{\max}(\Phi) = \lim_{\delta \to \infty} v_n^{\delta}(\Phi),$$

is the maximal stationary state process of the G/G/m/∞ *queue.*

Proof. In view of Theorem 1.3.2 it remains to show the finiteness of $V_n^{\max}$. In order to prove this, we need the following

5.5.5. Lemma. (i) *For arbitrary* $u, v \in \mathbb{R}^m$, *and* $a, s \in \mathbb{R}_+$ *we have*

$$\|R(u) - R(v)\| \leqq \|u - v\| \tag{5.5.5}$$

and

$$\|f(u, [a, s]) - f(v, [a, s])\| \leqq \|u - v\|, \tag{5.5.6}$$

where $\|x\| = \sqrt{\sum_{i=1}^{m} x_i^2}$ *denotes the Euclidean norm of* $x = [x_1, \ldots, x_m] \in \mathbb{R}^m$.

(ii) *If*

$$\|R(u + se_1) - R(v + se_1)\| = \|u - v\| \tag{5.5.7}$$

for $u = [u_1, \ldots, u_m]$, $v = [v_1, \ldots, v_m] \in \mathbb{X}$, *and* $s \in \mathbb{R}_+$, *then there exists a permutation* π *of* $\{1, \ldots, m\}$ *such that*

$$R(u + se_1) = [u^*_{\pi(1)}, \ldots, u^*_{\pi(m)}], \quad R(v + se_1) = [v^*_{\pi(1)}, \ldots, v^*_{\pi(m)}], \tag{5.5.8}$$

where $u^* = u + se_1$ *and* $v^* = v + se_1$. *If, in addition,*

$$v_1 + s < v_m, \tag{5.5.9}$$

then π *can be chosen in such a way that* $\pi(m) = m$.

Proof of Lemma 5.5.5.

1. Obviousy, for $u, v \in \mathbb{R}^m$

$$\|u_+ - v_+\| < \|u - v\|. \tag{5.5.10}$$

For $u, v \in \mathbb{R}^m$, let π and σ be permutations of $\{1, \ldots, m\}$ such that $R(u) = [u_{\pi(1)}, \ldots, u_{\pi(m)}]$ and $R(v) = [v_{\sigma(1)}, \ldots, v_{\sigma(m)}]$, respectively. From Hardy et al. (1952), p. 261, we know that

$$\sum_{j=1}^{m} u_j v_j \leqq \sum_{j=1}^{m} u_{\pi(j)} v_{\sigma(j)}. \tag{5.5.11}$$

Using (5.5.11), we obtain (5.5.5) from the following calculation:

$$\begin{aligned}
\|R(u) - R(v)\|^2 &= \sum_{j=1}^{m} (u_{\pi(j)} - v_{\sigma(j)})^2 \\
&= \sum_{j=1}^{m} u_j^2 + \sum_{j=1}^{m} v_j^2 - \sum_{j=1}^{m} 2u_{\pi(j)} v_{\sigma(j)} \\
&\leqq \sum_{j=1}^{m} u_j^2 + \sum_{j=1}^{m} v_j^2 - \sum_{j=1}^{m} 2u_j v_j = \|u - v\|^2.
\end{aligned}$$

For given $u, v \in \mathbb{R}^m$, $a, s \in \mathbb{R}$, it follows by means of (5.5.10) and (5.5.5) that

$$\begin{aligned}
\|f(u, [a, s]) - f(v, [a, s])\| &= \|(R(u + se_1 - a\mathbf{1}))_+ - (R(v + se_1 - a\mathbf{1}))_+\| \\
&\leqq \|R(u + se_1 - a\mathbf{1}) - R(v + se_1 - a\mathbf{1})\| \\
&\leqq \|(u + se_1 - a\mathbf{1}) - (v + se_1 - a\mathbf{1})\| \\
&= \|u - v\|.
\end{aligned}$$

Thus (5.5.6) holds.

2. In case of $s = 0$, the second assertion of the lemma follows immediately. Let $s > 0$. Since $u = [u_1, \dots, u_m]$, $v = [v_1, \dots, v_m] \in \mathbf{X}$ there exist i, j such that

$$\begin{aligned} &u_1 \leqq \cdots \leqq u_i < s + u_1 \leqq u_{i+1} \leqq \cdots \leqq u_m, \\ &v_1 \leqq \cdots \leqq v_j < s + v_1 \leqq v_{j+1} \leqq \cdots \leqq v_m. \end{aligned} \tag{5.5.12}$$

Now we have to consider the following cases.

Case 1. $i = j$. Then the assertions of the second part of the lemma are obviously true.

Case 2. $i < j$. Then

$$\begin{aligned} &\|R(u + se_1) - R(v + se_1)\|^2 \\ &= \sum_{k=2}^{i} (u_k - v_k)^2 + \sum_{k=j+1}^{m} (u_k - v_k)^2 + (u_1 + s - v_{i+1})^2 \\ &\quad + \sum_{k=i+1}^{j-1} (u_k - v_{k+1})^2 + (u_j - v_1 - s)^2 \\ &= \sum_{k=2}^{i} (u_k - v_k)^2 + \sum_{k=j+1}^{m} (u_k - v_k)^2 + (u_1 + s)^2 - 2(u_1 + s)\, v_{i+1} + v_{i+1}^2 \\ &\quad + \sum_{k=i+1}^{j-1} (u_k^2 - 2u_k v_{k+1} + v_{k+1}^2) + u_j^2 - 2u_j(v_1 + s) + (v_1 + s)^2 \end{aligned} \tag{5.5.13}$$

and

$$\begin{aligned} \|u - v\|^2 &= \sum_{k=2}^{i} (u_k - v_k)^2 + \sum_{k=j+1}^{m} (u_k - v_k)^2 \\ &\quad + (u_1 + s)^2 - 2(u_1 + s)(v_1 + s) + (v_1 + s)^2 \\ &\quad + \sum_{k=i+1}^{j-1} (u_k^2 - 2u_k v_k + v_k^2) + u_j^2 - 2u_j v_j + v_j^2. \end{aligned} \tag{5.5.14}$$

From (5.5.7), (5.5.13), and (5.5.14) we get

$$(u_1 + s)\, v_{i+1} + \sum_{k=i+1}^{j-1} u_k v_{k+1} + u_j(v_1 + s) = (u_1 + s)(v_1 + s) + \sum_{k=i+1}^{j} u_k v_k. \tag{5.5.15}$$

In (5.5.11), equality holds iff both sequences of numbers are ordered in the same sense. Using this fact, together with (5.5.12) and (5.5.15), then

$$u_1 + s = u_{i+1} = \cdots = u_j.$$

Now we define

$$\pi(k) = \begin{cases} j & \text{if } k = 1, \\ k - 1 & \text{if } k \in \{2, \dots, j\}, \\ k & \text{if } k \in \{j + 1, \dots, m\}. \end{cases}$$

This permutation satisfies the required property (5.5.8). If (5.5.9) also holds, then $j+1 \leqq m$, i.e. $\pi(m) = m$.

Case 3. $j < i$. In this case the assertion follows analogously to case 2. ∎

Proof of the finiteness of $v_n^{\max}(\Phi)$.

1. From the monotonicity of $f(\cdot, [a, s])$ and the construction of $v_n^{\delta}(\varphi)$ it follows that $v_n^{\delta}(\varphi)$ is non-decreasing in δ. Thus we can assume that δ runs through the set $\mathbb{Z}_+$ of positive integers. From the contraction property (5.5.6), the stationarity of Φ and the construction of $v_n^{\min}(\Phi)$ and $v_n^{\delta}(\Phi)$, the sequence of distances $\|v_n^{\delta}(\Phi) - v_n^{\min}(\Phi)\|$ is stationary and non-increasing in n. Hence it is almost surely constant:

$$\|v_n^{\delta}(\Phi) - v_n^{\min}(\Phi)\| = \|v_1^{\delta}(\Phi) - v_1^{\min}(\Phi)\|, \quad n \in \mathbb{Z}, \quad \delta \in \mathbb{Z}_+ \quad \text{P-a.s.} \tag{5.5.16}$$

2. Set

$$n_1(\Phi) = \min\left\{i \geqq 1 : \left(R\left(v_{i-1}^{\min}(\Phi) + S_{i-1}e_1 - A_{i-1}\mathbf{1}\right)\right)_1 < 0\right\} \tag{5.5.17}$$

if the minimum exists, and $n_1(\Phi) = \infty$ otherwise. ($(x)_j$ denotes the j-th component of the vector $x = [x_1, \ldots, x_m]$.) From $\varrho < m$, stationarity and ergodicity of $\left(\left[[A_n, S_n], v_n^{\min}(\Phi)\right]\right)$ it follows that

$$n_1(\Phi) < \infty \qquad \text{P-a.s.} \tag{5.5.18}$$

3. Let $\varphi = ([a_n, s_n])$ be a sample path of the input with the following properties:

$$v_n^{\min}(\varphi) < \infty; \quad n_1 = n_1(\varphi) < \infty; \quad \lim_{n\to\infty} \frac{1}{n} \sum_{i=1}^{n} a_i = m_A; \tag{5.5.19}$$

$$\|v_n^{\delta}(\varphi) - v_n^{\min}(\varphi)\| = \|v_1^{\delta}(\varphi) - v_1^{\min}(\varphi)\| \quad \text{for } n \in \mathbb{Z}, \quad \delta \in \mathbb{Z}_+. \tag{5.5.20}$$

Assume there is a $\delta^* \in \mathbb{Z}_+$ such that

$$v_{n_1-1}^{\delta^*}(\varphi) + s_{n_1-1}e_1 - a_{n_1-1}\mathbf{1} \geqq 0.$$

Then in view of (5.5.17) and $n_1 < \infty$ we obtain

$$\|v_{n_1}^{\delta^*}(\varphi) - v_{n_1}^{\min}(\varphi)\| < \|v_{n_1-1}^{\delta}(\varphi) - v_{n_1-1}^{\min}(\varphi)\|$$

which contradicts (5.5.20). Thus

$$v_{n_1,1}^{\delta}(\varphi) = v_{n_1,1}^{\min}(\varphi) = 0 \quad \text{for } \delta \in \mathbb{Z}_+. \tag{5.5.21}$$

Now assume $v_1^{\max}(\varphi) = \infty$, i.e. $v_1^{\delta}(\varphi)$, $\delta \in \mathbb{Z}_+$, is not bounded. Then $\|v_1^{\delta}(\varphi) - v_1^{\min}(\varphi)\| \xrightarrow[\delta\to\infty]{} \infty$ which implies

$$v_{n_1,m}^{\delta}(\varphi) \xrightarrow[\delta\to\infty]{} \infty \tag{5.5.22}$$

in view of (5.5.20). Let $k > n_1$ be choosen such that

$$\sum_{i=n_1}^{k-1} a_i > v_{n_1,m}^{\min}(\varphi). \tag{5.5.23}$$

In view of (5.5.22), there is a number d such that

$$v^d_{n_1,m}(\varphi) > \sum_{i=n_1}^{k-1} s_i + \sum_{i=n_1}^{k-1} a_i . \tag{5.5.24}$$

From $v^d_{n_1,1} = 0$ (cf. (5.5.21)), (5.5.10), (5.5.24), (5.5.20), and Lemma 5.5.5 we obtain a sequence $\pi_{n_1}, \ldots, \pi_{k-1}$ of permutations of $\{1, \ldots, m\}$ such that

$$\pi_i(m) = m, \tag{5.5.25}$$

$$\left.\begin{aligned} v^{\min}_{i+1}(\varphi) &= [v^*_{i,\pi_i(1)} - a_i, \ldots, v^*_{i,\pi_i(m)} - a_i]_+ , \\ v^d_{i+1}(\varphi) &= [v^{d*}_{i,\pi_i(1)} - a_i, \ldots, v^{d*}_{i,\pi_i(m)} - a_i]_+ \end{aligned}\right\} \tag{5.5.26}$$

for $i = n_1, \ldots, k - 1$, where

$$v_i^* = v_i^{\min}(\varphi) + s_i e_1 \quad \text{and} \quad v_i^{d*} = v_i^d(\varphi) + s_i e_1 .$$

In view of (5.5.25), (5.5.24), and (5.5.26) we obtain $v^{d*}_{im} - a_i \geqq 0$ for $i = n_1, \ldots, k - 1$. Taking this into account, with (5.5.25), (5.5.26), and (5.5.20) we get $v^*_{im} - a_i \geqq 0$ for $i = n_1, \ldots, k - 1$. This, (5.5.26), and (5.5.23) lead to

$$v^{\min}_{k,m}(\varphi) = v^{\min}_{n_1,m}(\varphi) - \sum_{i=n_1}^{k-1} a_i < 0$$

which contradicts $v^{\min}_{k,m}(\varphi) \geqq 0$. Therefore $v_1^{\max}(\varphi) < \infty$.

Due to (5.5.16), (5.5.18), and the ergodicity of Φ, the assumptions (5.5.19) and (5.5.20) are satisfied for a.e. φ. Thus step 3 of the proof yields

$$V_n^{\max} = v_n^{\max}(\Phi) < \infty \quad \text{a.s.} \quad \blacksquare$$

There is a close relation between the uniqueness of the stationary solution of (5.5.1) and the limiting behaviour of the system considered. In view of Lemma 5.5.3 and Theorem 5.5.1 the assumptions of Theorem 1.3.3 are satisfied. Thus we have

5.5.6. Theorem. *Under the assumptions of Theorem* 5.5.1 *the following statements are equivalent:*

(i) $\big([[A_n, S_n], v_n^{\min}(\Phi)]\big)$ *is the unique stationary weak solution of* (5.5.1).

(ii) *For an arbitrary pair* $[V, (U_n')]$, $(U_n') = ([A_n', S_n'])$, *such that* $(U_n') \overset{\mathcal{D}}{=} ([A_n, S_n)]$ *and* V *is an* $\mathbf{X}$*-valued r.v.:*

$$\big\|v_n(V, U_0', \ldots, U'_{n-1}) - v_n^{\min}\big((U_j')\big)\big\|_{\max} \xrightarrow[n\to\infty]{} 0 \quad a.s.$$

(iii) *For an arbitrary pair* $[V, (U_n')]$, $U_n' = [A_n', S_n']$, *such that* $(U_n') \overset{\mathcal{D}}{=} ([A_n, S_n])$ *and* V *is an* $\mathbf{X}$*-valued r.v.:*

$$\big(v_{n+k}(V, U_0', \ldots, U'_{n+k-1}), n \geqq 0\big) \xrightarrow[k\to\infty]{\mathcal{D}} \big(v_n^{\min}(\Phi), n \geqq 0\big).$$

Recall that Example 5.5.2 shows that uniqueness of the stationary state process does not hold in general. However, under additional assumptions uniqueness can be shown:

5.5.7. Theorem. *Consider a G/GI/m/∞ queue with FCFS queueing discipline and with the stationary input (Φ, P), $\Phi = ([A_n, S_n])$, where (A_n) is ergodic. If $\varrho < m$ then the sequence $(V_n) = \big(v_n^{\min}(\Phi)\big)$ is the only stationary state process, and for every (random) initial state V:*

$$\|v_n(V, [A_0, S_0], \ldots, [A_{n-1}, S_{n-1}]) - V_n\|_{\max} \xrightarrow[n\to\infty]{} 0 \quad a.s., \tag{5.5.27}$$

$$v_n(V, [A_0, S_0], \ldots, [A_{n-1}, S_{n-1}]) \xrightarrow[n\to\infty]{\mathfrak{D}} V_0 .$$

In order to prove this theorem, we need the following

5.5.8. Lemma. *For every $\varphi = ([a_n, s_n])$, $s_n \equiv h$, satisfying the conditions*

$$\lim_{n\to\infty} \frac{1}{n} \sum_{i=1}^{n} a_i = a > 0, \qquad h < ma, \tag{5.5.28}$$

and all initial states $u, v \in \mathbb{X}$ there exists an $n_0 = n_0(u, v, \varphi)$ such that

$$v_n(u, [a_0, h], \ldots, [a_{n-1}, h]) = v_n(v, [a_0, h], \ldots, [a_{n-1}, h]) \quad \textit{for all} \ \ n \geqq n_0 .$$

Proof. Let $\varphi = ([a_n, h])$ satisfy the conditions of Lemma 5.5.8. Consider an initial state $\bar{v} = [b, \ldots, b]$, $b \in \mathbb{R}_+$. Then each server works like a single server queue with the sample path $\varphi_i = \Big(\big[a_n, h\mathbf{1}_{\{i\}}\big(1 + (n-1) \bmod m\big)\big]\Big)$ for some $i \in \{1, \ldots, m\}$. In view of

$$\lim_{k\to\infty} \frac{1}{k} \sum_{n=1}^{k} h\mathbf{1}_{\{i\}}\big(1 + (n-1) \bmod m\big) = \frac{h}{m} < a$$

and (5.5.28) each server becomes empty after a finite time, and thus, for sufficiently large n, it has the same workload as in case of the initial state 0. This yields the existence of an $n_0(b, \varphi)$ such that

$$v_n(\bar{v}, [a_0, h], \ldots, [a_{n-1}, h]) = v_n(0, [a_0, h], \ldots, [a_{n-1}, h]), \quad n \geqq n_0(b, \varphi). \tag{5.5.29}$$

Now consider arbitrary initial states $u = [u_1, \ldots, u_m]$, $v = [v_1, \ldots, v_m] \in \mathbb{X}$. Setting $b = \max\{u_m, v_m\}$ and taking into account the inequalities

$$\begin{aligned} v_n(0, [a_0, h], \ldots, [a_{n-1}, h]) &\leqq v_n(u, [a_0, h], \ldots, [a_{n-1}, h]) \\ &\leqq v_n(\bar{v}, [a_0, h], \ldots, [a_{n-1}, h]), \\ v_n(0, [a_0, h], \ldots, [a_{n-1}, h]) &\leqq v_n(v, [a_0, h], \ldots, [a_{n-1}, h]) \\ &\leqq v_n(\bar{v}, [a_0, h], \ldots, [a_{n-1}, h]), \end{aligned}$$

the assertion follows immediately from (5.5.29). ∎

Proof of Theorem 5.5.7.

Assume $\big(v_n^{\min}(\Phi)\big)$ is not the only stationary state process, i.e. $\mathsf{P}\big(v_0^{\max}(\Phi) > v_0^{\min}(\Phi)\big) > 0$. Then, in view of Lemma 5.5.3 and since Φ is stationary and ergodic, cf. Theorem A 1.2.6, we have

$$\|v_n^{\max}(\Phi) - v_n^{\min}(\Phi)\|_{\max} = \|v_0^{\max}(\Phi) - v_0^{\min}(\Phi)\|_{\max} = c, \quad n \in \mathbb{Z}, \quad \mathsf{P}\text{-a.s.}, \tag{5.5.30}$$

where $c > 0$ is a fixed number. From $\varrho < m$ it follows that there exists an $h^* \geqq 0$ such that for every $\delta > 0$

$$\mathsf{P}(h^* - \delta < S_0 < h^* + \delta) > 0 \tag{5.5.31}$$

and

$$0 < h^* < m m_A .$$

Lemma 5.5.8 with $h = h^*$, $a = m_A$, $u = v_n^{\max}(\varphi)$, $v = v_n^{\min}(\varphi)$, yields the existence of a k such that

$$\mathsf{P}\left(v_k(v_0^{\max}(\Phi), [A_0, h^*], \ldots, [A_{k-1}, h^*])\right.$$
$$\left. = v_k(v_0^{\min}(\Phi), [A_0, h^*], \ldots, [A_{k-1}, h^*])\right) > 0.$$

From this and since the mapping $v_k(v, [a_0, s_0], \ldots, [a_{k-1}, s_{k-1}])$ is continuous there is a $\delta^* > 0$ such that

$$p = \mathsf{P}\Big(\big\|v_k(v_0^{\max}(\Phi), [A_0, h_0], \ldots, [A_{k-1}, h_{k-1}])$$
$$- v_k(v_0^{\min}(\Phi), [A_0, h_0], \ldots, [A_{k-1}, h_{k-1}])\big\|_{\max} < \frac{c}{2}$$
$$\text{for } h^* - \delta^* < h_i < h^* + \delta^*, \quad i = 0, \ldots, k-1\Big) > 0. \tag{5.5.32}$$

Taking into account the independence assumptions of a G/GI/m/∞ queue and the construction of $v_k^{\min}(\Phi)$, $v_k^{\max}(\Phi)$, we obtain

$$\mathsf{P}\left(\|v_k^{\max}(\Phi) - v_k^{\min}(\Phi)\|_{\max} < \frac{c}{2}\right)$$
$$\geqq \mathsf{P}\Big(\big\|v_k(v_0^{\max}(\Phi), [A_0, S_0], \ldots, [A_{k-1}, S_{k-1}])$$
$$- v_k(v_0^{\min}(\Phi), [A_0, S_0], \ldots, [A_{k-1}, S_{k-1}])\big\|_{\max} < \frac{c}{2},$$
$$h^* - \delta^* < S_i < h^* + \delta^*, i = 0, \ldots, k-1\Big)$$
$$\geqq \mathsf{P}\Big(\big\|v_k(v_0^{\max}(\Phi), [A_0, h_0], \ldots, [A_{k-1}, h_{k-1}])$$
$$- v_k(v_0^{\min}(\Phi), [A_0, h_0], \ldots, [A_{k-1}, h_{k-1}])\big\|_{\max} < \frac{c}{2}$$
$$\text{for } h^* - \delta^* < h_i < h^* + \delta^*, i = 0, \ldots, k-1;$$
$$h^* - \delta^* < S_i < h^* + \delta^*, i = 0, \ldots, k-1\Big)$$
$$= p\big(\mathsf{P}(h^* - \delta^* < S_0 < h^* + \delta^*)\big)^k > 0,$$

in view of (5.5.32) and (5.5.31). But this contradicts (5.5.30). Hence $\left(v_n^{\min}(\Phi)\right)$ is the only stationary state process. The rest of the assertions follows from Theorem 5.5.6. ∎

In the GI/GI/m/∞ case we are able to prove an ergodic property somewhat stronger than (5.5.27).

5.5.9. Theorem. *Consider a* GI/GI/m/∞ *queue with FCFS queueing discipline and with the stationary input* (Φ, P), $\Phi = ([A_n, S_n])$. *Assume* $\varrho < m$, *and let* (V_n) *be the unique stationary state process, cf. Theorem* 5.5.7. *Then*

$$\mathsf{P}\left(\bigcup_{m=1}^{\infty}\{v_k(v, [A_0, S_0], \ldots, [A_{k-1}, S_{k-1}]) = V_k, k \geqq m\}\right) = 1 \tag{5.5.33}$$

for all initial states $v \in \mathbf{X}$.

Proof. Since A_0 and S_0 are independent and $\varrho < m$, there exist an $a > 0$ and a $\delta > 0$ such that $\delta < (m-1)\,a$ and

$$\mathsf{P}(A_0 > a) > 0, \qquad \mathsf{P}(S_0 < ma - \delta) > 0.$$

and there is a state $\bar{v} = [\bar{v}_1, \ldots, \bar{v}_m]$ such that $\mathsf{P}(V_0 + \delta 1 \leqq \bar{v}) > 0$. From this and the independence assumptions of a GI/GI/m/∞ queue we obtain

$$\mathsf{P}(V_0 + \delta 1 \leqq \bar{v}, A_j > a, S_j < ma - \delta, j = 0, \ldots, n) > 0 \tag{5.5.34}$$

for all n. (Notice that V_0 depends only on the past $[A_{-1}, S_{-1}], [A_{-2}, S_{-2}], \ldots$).

Let V_i^*, $i \geqq 0$, be the work load vector for the system with initial state $[\bar{v}_m, \ldots, \bar{v}_m]$ and the deterministic interarrival times a and service-times $ma - \delta$, respectively. From the arguments given at the beginning of the proof of Lemma 5.5.8 – with $a_n = a$, $h = ma - \delta$, $b = \bar{v}_1$ – we get an n_0 such that

$$v_n(\bar{v}, [a, h], \ldots, [a, h]) = v_n(0, [a, h], \ldots, [a, h]) \quad \text{for } n \geqq n_0. \tag{5.5.35}$$

On the other hand, we have $v_n(0, [a, h], \ldots, [a, h]) = \big[0, (a-\delta)_+, (2a-\delta)_+, \ldots, \big((m-1)\,a - \delta\big)_+\big]$ for $n \geqq m$. Taking into account (5.5.35) we obtain for $n^* = \max(m, n_0)$

$$V^*_{n^*+1,1} = V^*_{n^*+2,1} = \cdots = V^*_{n^*+m,1} = 0. \tag{5.5.36}$$

Observe that if $A_j > a$, $S_j < ma - \delta$, $j = 0, \ldots, n^* + m - 1$, then V_j^* majorizes V_j and $v_j(v_0, [A_0, S_0], \ldots, [A_{j-1}, S_{j-1}])$, $j = 0, \ldots, n^* + m$, if $V_0 \leqq \bar{v}$ and $v_0 \leqq \bar{v}$, respectively. Hence in view of (5.5.36),

$$v_{n^*+m}(v_0, [A_0, S_0], \ldots, [A_{n^*+m-1}, S_{n^*+m-1}]) = V_{n^*+m}$$

if $v_0 \leqq \bar{v}$ and $V_0 \leqq \bar{v}$. Together with (5.5.34) we have

$$\mathsf{P}(V_0 + \delta 1 \leqq \bar{v}, v_{n^*+m}(v_0, [A_0, S_0], \ldots, [A_{n^*+m-1}, S_{n^*+m-1}]) = V_{n^*+m} \text{ for } v_0 \leqq \bar{v}) > 0. \tag{5.5.37}$$

In view of (5.5.27), after a random time N we have

$$\|V_n - v_n(v, [A_0, S_0], \ldots, [A_{n-1}, S_{n-1}])\|_{\max} < \delta \quad \text{for } n \geqq N,$$

i.e.

$$v_n(v, [A_0, S_0], \ldots, [A_{n-1}, S_{n-1}]) \leqq V_n + \delta 1 \quad \text{for } n \geqq N. \tag{5.5.38}$$

Furthermore, the ergodicity of Φ and (5.5.37) imply that there is an infinite sequence of indices $N_k \xrightarrow[k\to\infty]{} \infty$ a.s. such that

$$V_{N_k} + \delta 1 \leqq \bar{v}, \quad v_{n^*+m}(v_0, [A_{N_k}, S_{N_k}], \ldots, [A_{N_k+n^*+m-1}, S_{N_k+n^*+m-1}])$$
$$= V_{N_k+n^*+m} \quad \text{for } v_0 \leqq \bar{v}.$$

This and (5.5.38) yield the assertion. ∎

The number of customers in the system L_n and in the queue Q_n is determined by the input Φ and the work load vector process $V_n = v_n^{\min}(\Phi)$:

$$L_n = l_n^{\min}(\Phi) = \sum_{i<n} 1\left\{V_{i1} + S_i > \sum_{j=i}^{n-1} A_j\right\},$$

$$Q_n = q_n^{\min}(\Phi) = \sum_{i<n} 1\left\{V_{i1} > \sum_{j=i}^{n-1} A_j\right\}.$$

5.5.10. Theorem. *Under the assumptions of Theorem* 5.5.1, *the quantities* L_n *and* Q_n *are almost surely finite.*

Proof. Notice that

$$\mathsf{P}(V_{01} = 0) > 0, \tag{5.5.39}$$

since, if (5.5.39) did not hold, then we would have from (5.5.1)

$$\sum_{i=1}^{m} V_{ni} = \sum_{i=1}^{m} V_{0i} + \sum_{i=0}^{n-1} (S_i - mA_i).$$

This, the ergodicity of Φ, and $\varrho < m$ yield $\sum_{i=1}^{m} V_{ni} \to -\infty$ as $n \to \infty$, which contradicts $V_{ni} \geqq 0$. Since $(V_n) = \left(v_n^{\min}(\Phi)\right)$ is stationary and ergodic, we get from (5.5.39) that there are infinitely many indices $n_i = n_i(\Phi)$, $i \in \mathbb{Z}$, left and right from zero such that $V_{n_i,1} = 0$ $(\cdots < n_0 < 0 \leqq n_1 < \cdots)$. At the arrival of the n_0-th customer there are at most $m - 1$ customers in the system. Thus,

$$L_0 \leqq (m-1) - n_0 < \infty.$$

Since $Q_0 \leqq L_0$ the proof is complete. ∎

The arguments in the proof of Theorem 5.5.10 remain valid for an arbitrary work-conserving queueing discipline.

5.6. The system G/G/m/∞ with cyclic queueing discipline

Consider a G/G/m/∞ queue with the following cyclic queueing discipline: Let the servers be numbered from 1 up to m. Assume that the customer with index zero is randomly assigned to one of the m servers with equal probabilities $1/m$, independently of the sequence $([A_n, S_n])$ of interarrival and service times. If the n-th customer is assigned to the i-th server, then the $(n+1)$-st customer is assigned to the server with number $(i \bmod m) + 1$.

Denoting by J_n the (random) number of the server to which the n-th customer is assigned, we have

$$P(J_0 = i) = 1/m, \qquad i = 1, \ldots, m,$$
$$J_{n+1} = (J_0 + n) \bmod m + 1, \qquad n \in \mathbb{Z},$$

and J_0 and $([A_n, S_n])$ are independent.

The input of the queue is described by the sequence (Φ, P), $\Phi = ([A_n, S_n, J_n])$. The state seen by the n-th arriving customer is defined as the work load vector $V_n{}^c = [V_{n1}^c, \ldots, V_{nm}^c]$, where the component V_{ni}^c is the work load of the i-th server at the arrival of the n-th customer. The state space is $\mathbb{X} = \mathbb{R}_+{}^m$. Equation (5.1.1) has the form

$$V_{n+1}^c = \left(V_n{}^c + S_n e(J_n) - A_n \mathbf{1}\right)_+ = f(V_n{}^c, [A_n, S_n, J_n]), \qquad n \in \mathbb{Z}, \tag{5.6.1}$$

where $e(j) = [\mathbf{1}_{\{1\}}(j), \ldots, \mathbf{1}_{\{m\}}(j)]$.

Obviously, the cyclic discipline decomposes the input into m components $\Phi^{(i)}$. The i-th component $\Phi^{(i)}$ is served by the i-th server. For our needs it is more convenient to write the i-th component of the input in the form

$$\Phi^{(i)} = ([A_n{}^{(i)}, S_n{}^{(i)}]) = \left([A_n, S_n \mathbf{1}_{\{i\}}(J_n)]\right). \tag{5.6.2}$$

(This means that we imagine that all customers are "served" by the i-th server. But, those who are in fact assigned to other servers require the "service time" zero at the i-th server.) It is clear that the m server queue with the cyclic queueing discipline works like m separate single server queues with the inputs $(\Phi^{(i)}, \mathsf{P}^{(i)})$, cf. Section 5.4;

$$V_{n+1,i}^c = (V_{ni}^c + S_n{}^{(i)} - A_n{}^{(i)})_+, \qquad i = 1, \ldots, m.$$

Before we investigate the existence and uniqueness of the stationary state process for the cyclic m server queue, we shall deal with the decomposition (5.6.2). Obviously, $\Phi^{(i)}$, $i = 1, \ldots, m$, are stationary. However, the ergodicity of $([A_n, S_n])$ does not imply the ergodicity of $\Phi^{(i)}$ in general:

5.6.1. Example. Consider the stationary and ergodic sequence $([A_n, S_n])$ given by $A_n \equiv 1$ and $\mathsf{P}(S_{2n} = 5/2, S_{2n+1} = 1/2, n \in \mathbb{Z}) = \mathsf{P}(S_{2n} = 1/2, S_{2n+1} = 5/2, n \in \mathbb{Z}) = 1/2$. Choosing $m = 2$, for $(\Phi^{(1)}, \mathsf{P}^{(1)})$, $\Phi^{(1)} = ([A_n{}^{(1)}, S_n{}^{(1)}])$, we obtain $A_n{}^{(1)} \equiv 1$ and

$$\mathsf{P}^{(1)}(S_{2n}^{(1)} = 5/2, S_{2n+1}^{(1)} = 0) = \mathsf{P}^{(1)}(S_{2n}^{(1)} = 0, S_{2n+1}^{(1)} = 5/2) = 1/4,$$
$$\mathsf{P}^{(1)}(S_{2n}^{(1)} = 1/2, S_{2n+1}^{(1)} = 0) = \mathsf{P}^{(1)}(S_{2n}^{(1)} = 0, S_{2n+1}^{(1)} = 1/2) = 1/4.$$

Evidently, $\Phi^{(1)}$ is not ergodic.

The ergodicity of $([A_n, S_n])$ and $\varrho < m$ do not ensure the existence of a weak solution of (5.6.1), in general. To show this, consider the D/G/2/∞ queue with cyclic queueing discipline and the stationary input $\Phi = ([A_n, S_n, J_n])$ with $([A_n, S_n])$ as in Example 5.6.1. Then $([A_n, S_n])$ is ergodic, $\varrho = 3/2 < 2$, but

$$\mathsf{P}^{(1)}\left(\lim_{n\to\infty} \frac{1}{n} \sum_{i=1}^{n} A_i{}^{(1)} < \lim_{n\to\infty} \frac{1}{n} \sum_{i=1}^{n} S_i{}^{(1)}\right) \geqq 1/4. \tag{5.6.3}$$

As the two-server queue with the cyclic queueing discipline works like two single server queues with the inputs $\Phi^{(1)}$ and $\Phi^{(2)}$, respectively, we obtain from (5.6.3) that there is no weak solution of (5.6.1). However, stronger assumptions on $([A_n, S_n])$ ensure the existence of weak solutions.

5.6.2. Theorem. *Consider a G/G/m/∞ queue with cyclic queueing discipline and the stationary input* (Φ, P), $\Phi = (U_n) = ([A_n, S_n, J_n])$ *with the property* $\varrho < m$. *Assume that one of the following two conditions is fulfilled:*

(i) $([A_n, S_n])$ *is mixing, i.e.*

$$\mathsf{P}\big(([A_n, S_n]) \in A \cap \theta^k B\big) \xrightarrow[k\to\infty]{} \mathsf{P}\big(([A_n, S_n]) \in A\big)\, \mathsf{P}\big(([A_n, S_n]) \in B\big)$$

for all measurable sets A and B.

(ii) (A 5.1.1) *and* (A 5.1.3) *are fulfilled* (*i.e. the queue is of the type* G/GI/m/∞) *and the sequence* (A_n) *is ergodic.*

Then the sequence (V_n),

$$V_n^c = v_n^{\min}(\Phi) = \lim_{r\to\infty} v_r(0, U_{n-r}, \ldots, U_{n-1}),$$

is the only stationary work load process. (Here $v_k(0, U_0, \ldots, U_{k-1})$ *denotes the work load seen by the k-th customer if the system starts empty at* 0.) *Moreover,*

$$\mathsf{P}\left(\bigcup_{m=1}^{\infty} \{v_k(v, U_0, \ldots, U_{k-1}) = V_k^c,\ k \geqq m\}\right) = 1$$

for every $v \in \mathbb{R}_+^m$.

Proof. 1. Assume that $([A_n, S_n])$ is mixing. Since the sequence (J_n) is ergodic and independent of $([A_n, S_n])$, the input $\Phi = ([A_n, S_n, J_n])$ is ergodic (cf. Theorem A 1.2.6). Hence the stationary sequences $\Phi^{(i)} = ([A_n^{(i)}, S_n^{(i)}])$, cf. (5.6.2), are ergodic. From this and $\mathsf{E}S_0^{(i)} = \mathsf{E}S_0/m < \mathsf{E}A_0^{(i)}$, the assertions of the theorem follow immediately from Theorem 5.4.1, since the cyclic queue works like m single server queues with the inputs $\Phi^{(i)}$.

2. Now assume that Φ satisfies (A 5.1.1) and (A 5.1.3), i.e. we consider a cyclic G/GI/m/∞ queue. Let

$$\eta_i = \min\{j : j \geqq 1, J_j = i\}, \qquad i = 1, \ldots, m.$$

Obviously, for $i = 1, \ldots, m$, the sequence $(S_{\eta_i + jm}, j \in \mathbb{Z})$ consists of i.i.d. r.v.'s, and

$$\lim_{n\to\infty} \frac{1}{n} \sum_{k=1}^{n} S_k \mathbf{1}_{\{i\}}(J_k) = \lim_{n\to\infty} \frac{1}{n} \sum_{k=0}^{[n/m]-1} S_{\eta_i + km} = \frac{\mathsf{E}S_0}{m} < \mathsf{E}A_0 \quad \text{a.s.}$$

Now the assertions follow again from Theorem 5.4.1. ∎

5.7. The system G/M/m/r

In this section we consider a FCFS queueing system with m servers in parallel and r waiting places ($m + r < \infty$). The sequence (A_n) of interarrival times is assumed to be stationary and ergodic. The service times are independent of

(A_n) and i.i.d. exponentially distributed with the parameter $\mu = 1/m_S$. The assumptions imply that the number of customers in the system behaves like a death process between consecutive arrivals; the death intensity is $\min(m, i)/m_S$ if there are i customers in the system. Thus the dynamic of the queue can be described in the following manner: Let

(A 5.7.1) M_{ni}, $-\infty < n < \infty$, $1 \leqq i \leqq m + r$, be exponentially distributed r.v.'s with parameters $\min(m, i)/m_S$, i.e.

$$\mathsf{P}(M_{ni} \leqq t) = 1 - e^{-(\min(m,i)/m_S)t}.$$

Furthermore we assume

(A 5.7.2) $(M_n) = ([M_{n,1}, \ldots, M_{n,m+r}])$ is independent of (A_n)

and

(A 5.7.3) M_{ni}, $1 \leqq i \leqq m + r$, are independent r.v.'s.

Then the number L_n of customers in the system seen by the n-th arriving customer satisfies the following recursive stochastic equation:

$$L_{n+1} = \begin{cases} L_n' & \text{if } A_n = 0, \\ \min\left\{l \geqq 1 : \sum_{i=l}^{L_n'} M_{n,i} < A_n\right\} - 1 & \text{otherwise} \end{cases}$$

$$= f(L_n, [A_n, M_n]), \tag{5.7.1}$$

where

$$L_n' = \min(m + r, L_n + 1) \quad \text{and} \quad \sum_{i=l}^{L_n'} M_{n,i} = 0 \quad \text{for } l > L_n'.$$

In accordance to our conventions equation (5.7.1) describes the temporal behaviour of a G/M/m/r queue with the "input" (Φ, P), $\Phi = (U_n) = ([A_n, M_n])$, and the state space $\mathbb{X} = \{0, \ldots, m + r\}$. The question of existence and uniqueness of the stationary process of the number of customers in a G/M/m/r queue can be transformed into the question of the existence and uniqueness of the stationary solution of equation (5.7.1) with the stationary and ergodic input $([A_n, M_n])$ satisfying (A 5.7.1)–(A 5.7.3). In accordance with (5.1.2), define

$$l_1(l, u_0) = f(l, u_0),$$

$$l_n(l, u_0, \ldots, u_{n-1}) = f\big(l_{n-1}(l, u_0, \ldots, u_{n-2}), u_{n-1}\big), \quad n \geqq 2.$$

5.7.1. Theorem. *Consider a* G/M/m/r *system,* $m < \infty$, $r < \infty$, *with the input* (Φ, P), $\Phi = (U_n) = ([A_n, M_n])$, *satisfying* $m_A > 0$, *i.e.* (A_n) *is stationary and ergodic and* (A 5.7.1)–(A 5.7.3) *are fulfilled. Then* (L_n),

$$L_n = l_n(\Phi) = \lim_{r\to\infty} l_r(0, U_{n-r}, \ldots, U_{n-1}),$$

is the only process of the number of customers in the system. Moreover,

$$\mathsf{P}\left(\bigcup_{m=1}^{\infty} \{l_k(l, U_0, \ldots, U_{k-1}) = L_k, k \geqq m\}\right) = 1 \tag{5.7.2}$$

for every $l \in \{0, \ldots, m + r\}$.

Proof. The proof is based on the method of solution generating system given in Section 1.4. First notice that (A 5.7.1)–(A 5.7.3) and $\mathsf{P}(A_0 > 0) > 0$ imply

$$\mathsf{P}\big(f(i, [A_0, M_0]) = 0, i = 0, \ldots, m + r\big) > 0. \tag{5.7.3}$$

Now let $A(n, \varphi) \equiv \mathbb{L} = \{0, \ldots, m + r\}$. Then it is easy to check that these sets form an SGS (cf. also Example 1.4.8). Consider the sets

$$B(n, \Phi) = \bigcap_{r \leq n} l_{n-r}(\mathbb{L}, U_r, \ldots, U_{n-1}) \tag{5.7.4}$$

defined in proof of Theorem 1.4.3. From (1.4.6) in the proof of Theorem 1.4.3 we get

$$1 \leq i(\Phi) = \# B(n, \Phi), \qquad n \in \mathbb{Z}, \quad \mathsf{P}\text{-a.s.}$$

From $f\big(B(n, \Phi), U_n\big) = B(n + 1, \Phi)$ and (5.7.3) we obtain $\mathsf{P}\big(\# B(1, \Phi) = 1\big) > 0$ and thus

$$\# B(n, \Phi) = 1, \qquad n \in \mathbb{Z}, \quad \mathsf{P}\text{-a.s.},$$

since $\{\varphi: \# B(n, \varphi) = i, n \in \mathbb{Z}\}$, $i = 1, \ldots, m + r + 1$, are invariant events and Φ is ergodic (cf. Theorem A 1.2.6). Hence $i(\Phi) \equiv 1$. Denote the single element of $B(n, \Phi)$ by $L_n = l_n(\Phi)$. In view of (5.7.4) we have

$$\{L_n\} = \lim_{r \to \infty} l_r(\mathbb{L}, U_{n-r}, \ldots, U_{n-1}). \tag{5.7.5}$$

From Remark 1.4.5 it follows that $([U_n, L_n])$ is a stationary strong solution.

Taking into account the stationarity of Φ, (5.7.4) and (5.7.5), we get

$$\begin{aligned}\mathsf{P}\big(l_k(l, U_0, \ldots, U_{k-1}) = L_k, k \geq m\big) &= \mathsf{P}\big(l_m(l, U_0, \ldots, U_{m-1}) = L_m\big) \\ &= \mathsf{P}\big(l_m(l, U_{-m}, \ldots, U_{-1}) = L_0\big) \uparrow 1 \text{ as } m \to \infty,\end{aligned}$$

which implies (5.7.2).

It remains to show the uniqueness statement. Assume $([\tilde{U}_n, \tilde{L}_n])$ is a weak solution of (5.7.1). Since $\tilde{L}_n \in \{0, \ldots, m + r\}$ and $(\tilde{L}_n)$ satisfies (5.7.1) with $\tilde{\Phi} = (\tilde{U}_n)$, we obtain from (5.7.4), (5.7.5)

$$\begin{aligned}&\mathsf{P}\big(\tilde{L}_n = l_n(\tilde{\Phi})\big) \\ &= \mathsf{P}\Big(l_k(\tilde{L}_{n-k}, \tilde{U}_{n-k}, \ldots, \tilde{U}_{n-1}) = l_k\big(l_{n-k}(\tilde{\Phi}), \tilde{U}_{n-k}, \ldots, \tilde{U}_{n-1}\big)\Big) \\ &\geq \mathsf{P}\big(\# l_k(\mathbb{L}, \tilde{U}_{n-k}, \ldots, \tilde{U}_{n-1}) = 1\big) \uparrow 1 \quad \text{as } k \to \infty.\end{aligned}$$

Thus $\tilde{L}_n = l_n(\tilde{\Phi})$ a.s., which finishes the proof. ∎

5.7.2. Remark. Theorem 5.7.1 and its proof remain valid for many other queueing disciplines, such as LCFS and SIRO.

5.8. The single server queue with warming-up

Consider a single server queue with infinitely many waiting places and FCFS queueing discipline in which the first customer of each busy period (i.e. who meets an empty system) receives an additional service time, cf. also Section

5.11. This additional service time can be interpreted as the warm up time of the service facility.

In the following we imagine that an arbitrary customer, say the n-th, receives a mark C_n that is equal to the additional service time which he would receive if he were to initiate a busy period. We call C_n the potential warm up time. Denote by W_n the actual waiting time of the n-th customer. If we consider $\Phi = ([A_n, S_n, C_n])$ as the input and W_n, $n \in \mathbb{Z}$, as the consecutive states then the time evolution of the single server queue with warming up is given by the following recursive equation

$$W_{n+1} = \begin{cases} W_n + S_n - A_n & \text{if } W_n + S_n - A_n > 0, \\ C_{n+1} & \text{otherwise} \end{cases}$$
$$= f(W_n, [A_n, S_n, C_n]). \tag{5.8.1}$$

(The waiting time of a customer initiating a busy period is defined as his own warm up time.)

Throughout this section we assume $\mathsf{E}C_0 < \infty$, $\mathsf{E}S_0 < \infty$ and $\varrho = m_S/m_A < 1$.

In the following we reduce the investigation of the G/G/1/∞ queue with warm up times to that of the usual G/G/1/∞ queue without warm up times and that of a suitably chosen G/G/1/0 loss system.

Consider the standard G/G/1/∞ queue with FCFS queueing discipline and with the input $([A_n, S_n])$. Denote by V_n the actual waiting time of the n-th customer in this system. According to Theorem 5.4.1, $\varrho < 1$ implies the existence of a unique stationary ergodic strong solution $([A_n, S_n, V_n])$ of the corresponding equation $V_{n+1} = (V_n + S_n - A_n)_+$, satisfying $\mathsf{P}(V_0 = 0) > 0$; cf. (5.4.7). Thus, the conditional distribution $\tilde{\mathsf{P}}(\cdot) = \mathsf{P}(\cdot \mid V_0 = 0)$ is well-defined. Let $(\tilde{Z}_n) = ([\tilde{A}_n, \tilde{S}_n, \tilde{C}_n, \tilde{V}_n])$ be a sequence distributed according to $\tilde{\mathsf{P}}$. Define

$$\tilde{\nu}_0 = 0; \quad \tilde{\nu}_k = \min\{i : i > \tilde{\nu}_{k-1}, \tilde{V}_i = 0\}, \qquad k \geqq 1;$$
$$\tilde{\nu}_k = \max\{i : i < \tilde{\nu}_{k+1}, \tilde{V}_i = 0\}, \qquad k \leqq -1. \tag{5.8.2}$$

Then

$$\tilde{\xi}_k = -\sum_{i=\tilde{\nu}_k}^{\tilde{\nu}_{k+1}-1} (\tilde{S}_i - \tilde{A}_i) \geqq 0 \tag{5.8.3}$$

is just the length of the k-th idle period conditioned by $V_0 = 0$. Define

$$\tilde{\eta}_k = \tilde{C}_{\tilde{\nu}_k}. \tag{5.8.4}$$

The sequence $\tilde{\Psi} = ([\tilde{\xi}_k, \tilde{\eta}_k])$ is stationary, ergodic and satisfies $0 < \mathsf{E}\tilde{\xi}_0$ and $\mathsf{E}\tilde{\eta}_0 < \infty$, cf. Theorem A 1.3.1. Now we consider the single server loss system G/G/1/0 with the input $\tilde{\Psi}$. Recall that the dynamic of this system is given by

$$X_{k+1} = (X_k + 1\{X_k = 0\}\, \tilde{\eta}_k - \tilde{\xi}_k)_+. \tag{5.8.5}$$

In this loss system the service times of the served customers are just the actual warm up times occuring in the single server queue with warming-up. According to Theorem 5.3.1 there are measurable mappings $x(j, \cdot)$, $j = 1, \ldots, q$, such that

$$\frac{1}{q}\sum_{j=1}^{q}\tilde{\mathsf{P}}\Big(\big([\tilde{\xi}_k,\tilde{\eta}_k,x_k(x(j,\tilde{\Psi}),[\tilde{\xi}_0,\tilde{\eta}_0],\dots,[\tilde{\xi}_{k-1},\tilde{\eta}_{k-1}])],\quad k\geqq 0\big)\in(\cdot)\Big) \tag{5.8.6}$$

defines the distribution $\mathbf{Q}$ of a stationary weak solution for this loss system. Let Λ be a r.v. uniformly distributed on $\{1,\dots,q\}$ and independent of $([\tilde{A}_n, \tilde{S}_n, \tilde{C}_n, \tilde{V}_n])$. Then the sequence $([\tilde{\xi}_k, \tilde{\eta}_k, X_k], k \geqq 0)$, where

$$X_k = x_k(x(\Lambda,\tilde{\Psi}),[\tilde{\xi}_0,\tilde{\eta}_0],\dots,[\tilde{\xi}_{k-1},\tilde{\eta}_{k-1}]),$$

has the distribution (5.8.6), cf. Remark 5.3.2. Moreover, the sequence $([\tilde{\xi}_k, \tilde{\eta}_k, X_k, \tilde{\varkappa}_k], k \geqq 0)$, where

$$\tilde{\varkappa}_k = [\tilde{\nu}_{k+1}-\tilde{\nu}_k, \tilde{Z}_{\tilde{\nu}_k}, \tilde{Z}_{\tilde{\nu}_k+1},\dots,\tilde{Z}_{\tilde{\nu}_{k+1}-1}], \tag{5.8.7}$$

is stationary. Extending this sequence to negative k we obtain a stationary sequence $([\tilde{\xi}_k, \tilde{\eta}_k, X_k, \tilde{\varkappa}_k], k \in \mathbb{Z})$. Starting from this sequence, a stationary weak solution of (5.8.1) can be constructed as follows: Define

$$\tilde{W}_n = \begin{cases}\tilde{V}_n + \tilde{\eta}_k & \text{if } X_k = 0,\\ \tilde{V}_n + X_k & \text{if } X_k > 0\end{cases} \quad \text{for } \tilde{\nu}_k \leqq n < \tilde{\nu}_{k+1}.$$
$$= \tilde{V}_n + \tilde{X}_k \tag{5.8.8}$$

where $\tilde{X}_k = X_k + \mathbf{1}\{X_k = 0\}\,\tilde{\eta}_k$ is the work load immediately after the k-th arrival in the G/G/1/0 loss system with the input $\tilde{\Phi}$ described above. It is easy to check that $\tilde{W}_n$ satisfies

$$\tilde{W}_{n+1} = f(\tilde{W}_n, [\tilde{A}_n, \tilde{S}_n, \tilde{C}_n]), \qquad n \geqq 0,$$

almost surely. Applying Theorem A 1.3.2 to the sequence $([\tilde{A}_n, \tilde{S}_n, \tilde{C}_n, \tilde{V}_n, \tilde{W}_n], n \in \mathbb{Z})$ we obtain a stationary sequence $([A_n', S_n', C_n', V_n', W_n'], n \in \mathbb{Z})$ which defines a stationary weak solution $\big([[A_n', S_n', C_n'], W_n']\big)$ of (5.8.1). Thus we have proved

5.8.1. Theorem. *Consider a single server queue with warming-up. Assume that the stationary ergodic input* $\Phi = ([A_n, S_n, C_n])$, *where* C_n *are the potential warm up times, fulfills* $\varrho = m_S/m_A < 1$ *and* $\mathsf{E}C_0 < \infty$. *Then there exists a stationary weak solution* $\Phi' = \big([[A_n', S_n', C_n'], W_n']\big)$ *of* (5.8.1).

The approach given above shows a little bit more:

5.8.2. Lemma. *Under the assumptions of Theorem* 5.8.1 *there is exactly one stationary weak solution of* (5.8.1) *iff there is a unique stationary weak solution for the single server loss system with the input* $\tilde{\Psi} = ([\tilde{\xi}_k, \tilde{\eta}_k])$. *The stationary weak solution* Φ' *of* (5.8.1) *given by Theorem* 5.8.1 *is a strong one iff the stationary weak solution of the corresponding loss system is a strong one.*

In the following we consider the single server queue with warming-up under the following independence assumption:

A 5.8.1) (C_n) is a sequence of i.i.d. r.v.'s, independent of $([A_n, S_n])$.

From Lemma 5.8.2 and Theorem 5.3.9 we immediately get the following result:

5.8.3. Theorem. *Consider a* GI/GI/1/∞ *queue with warm up times* (C_n) *satisfying* (A 5.8.1). *If* $\varrho < 1$ *and* $\mathsf{E}C_0 < \infty$, *then there is exactly one stationary weak solution of* (5.8.1).

However, under the assumptions of Theorem 5.8.3 the unique stationary weak solution given there is not a strong one, in general.

5.8.4. Theorem. *Let* $\Phi = (U_n) = ([A_n, S_n, C_n])$ *be the stationary ergodic input of a* G/G/1/∞ *queue with warm up times satisfying* (A 5.8.1). *Assume* $\varrho < 1$, $\mathsf{E}C_0 < \infty$, *and*

$$\mathsf{P}\left(C_0 < \frac{\mathsf{E}A_0 - \mathsf{E}S_0}{\mathsf{P}(V_0 = 0)}\right) > 0,$$

where V_0 *is the stationary work load of the standard* G/G/1/∞ *queue with the input* $([A_n, S_n])$. *Then there is a measurable mapping* $w_0(\cdot)$ *such that* $\big([[A_n, S_n, C_n], W_n]\big)$, *with* $W_n = w_n(\Phi) = w_0(\theta^n\Phi)$ *is a stationary strong solution of* (5.8.1). *It is the only stationary weak solution. Moreover,*

$$\mathsf{P}\left(\bigcup_{m=1}^{\infty} \{w_n(w, U_0, \ldots, U_{n-1}) = W_n, n \geqq m\}\right) = 1 \tag{5.8.9}$$

for every $w \in \mathbb{R}_+$, *where* $w_1(w, u_0) = f(w, u_0)$, *and* $w_{n+1}(w, u_0, \ldots, u_n) = f\big(w_n(w, u_0, \ldots, u_{n-1}), u_n\big)$, $n \geqq 0$, *and the function* f *is defined by* (5.8.1).

Proof. Since we have a queue with independent warm up times, the single server loss system with the input $\tilde{\Psi} = ([\tilde{\xi}_k, \tilde{\eta}_k])$ is of the type G/GI/1/0. Furthermore,

$$\tilde{\mathsf{P}}\left(\tilde{\eta}_0 < \frac{\mathsf{E}A_0 - \mathsf{E}S_0}{\mathsf{P}(V_0 = 0)}\right) > 0.$$

Applying formula (1.6.2a) to the sequence $([A_n - S_n, 1\{V_n = 0\}])$ we obtain

$$\mathsf{E}_{\tilde{\mathsf{P}}}\tilde{\xi}_0 = \mathsf{E}_{\tilde{\mathsf{P}}}\tilde{\nu}_1 \mathsf{E}_{\mathsf{P}}(A_0 - S_0) = \frac{\mathsf{E}_{\mathsf{P}}A_0 - \mathsf{E}_{\mathsf{P}}S_0}{\mathsf{P}(V_0 = 0)}$$

and thus

$$\tilde{\mathsf{P}}\left(\tilde{\xi}_0 \geqq \frac{\mathsf{E}A_0 - \mathsf{E}S_0}{\mathsf{P}(V_0 = 0)}\right) > 0.$$

The r.v.'s $\tilde{\eta}_0$ and $\tilde{\xi}_0$ are independent, hence $\tilde{\mathsf{P}}(\tilde{\eta}_0 < \tilde{\xi}_0) > 0$. With Φ also $([\tilde{\xi}_k, \tilde{\eta}_k, \tilde{\varkappa}_k])$ is an ergodic sequence. According to Theorem 5.3.8 the loss system G/GI/1/0 with the input $\tilde{\Psi}$ has a stationary strong solution $X_k = \hat{x}_k\big(([\tilde{\xi}_n, \tilde{\eta}_n])\big)$, which is the only stationary weak solution. Thus, the weak solution of the single server queue with warming-up constructed via (5.8.8) is a strong one. The uniqueness statement follows from Lemma 5.8.2. It remains to show the limiting property (5.8.9).

Let $V_n^r(w) = v_{n-r}(w, [A_r, S_r], \ldots, [A_{n-1}, S_{n-1}])$, $n > r$, be the waiting times in the G/G/1/∞ queue without warm up times and with initial state w at r. From Theorem 5.4.1 and from the stationarity of Φ we obtain

$$\lim_{r\to\infty} \mathsf{P}\big(Z_k^{-r}(w) = Z_k, k \geqq 0\big) = \lim_{r\to\infty} \mathsf{P}\big(Z_k^0(w) = Z_k, k \geqq r\big) = 1, \tag{5.8.10}$$

where $Z_k^{-r}(w) = [A_k, S_k, C_k, V_k^{-r}(w)]$, $Z_k = [A_k, S_k, C_k, V_k]$. Since $\mathsf{P}(V_0 = 0) > 0$, we obtain from (5.8.10)

$$\lim_{r\to\infty} \tilde{\mathsf{P}}\big(\tilde{Z}_k^{-r}(w) = \tilde{Z}_k, k \geqq 0\big) = 1,$$

where

$$\tilde{Z}_k^{-r}(w) = [\tilde{A}_k, \tilde{S}_k, \tilde{C}_k, \tilde{V}_k^{-r}(w)],$$

$$\tilde{V}_k^{-r}(w) = v_{k+r}(w, [\tilde{A}_{-r}, \tilde{S}_{-r}], \ldots, [\tilde{A}_{k-1}, \tilde{S}_{k-1}]).$$

Analogously to (5.8.2), (5.8.3), (5.8.4), (5.8.7), (5.8.5), and (5.8.8) we define the r.v.'s

$$\tilde{\nu}_k^{-r}(w), \tilde{\xi}_k^{-r}(w), \tilde{\eta}_k^{-r}(w), \tilde{\varkappa}_k^{-r}(w), \tilde{X}_k^{-r}(w), \tilde{W}_k^{-r}(w), \qquad k \geqq 0,$$

from the process $(\tilde{Z}_k^{-r}(w), k \geqq 0)$. Then

$$\lim_{r\to\infty} \tilde{\mathsf{P}}([\tilde{\xi}_k^{-r}(w), \tilde{\eta}_k^{-r}(w), \tilde{\varkappa}_k^{-r}(w)] = [\tilde{\xi}_k, \tilde{\eta}_k, \tilde{\varkappa}_k], k \geqq 0) = 1. \tag{5.8.11}$$

The sequence $\big([\tilde{\xi}_k^{-r}(w), \tilde{\eta}_k^{-r}(w), X_k^{-r}(w)], k \geqq 0\big)$ satisfies equation (5.8.5) of the single server loss system for $k \geqq 0$ with the state $X_0^{-r}(w)$ at 0, and we obtain by means of (5.8.11)

$$\lim_{r\to\infty} \tilde{\mathsf{P}}\big([\tilde{\xi}_k^{-r}, \tilde{\eta}_k^{-r}, X_k^{-r}(w)], k \geqq 0\big) \text{ satisfies (5.8.5)} = 1. \tag{5.8.12}$$

From (5.8.11), (5.8.12) it follows that

$$\lim_{r\to\infty} \tilde{\mathsf{P}}\Bigg(\bigcup_{n=0}^{\infty} \{\tilde{\xi}_i^{-r}(w) = \tilde{\xi}_i, \tilde{\eta}_i^{-r}(w) = \tilde{\eta}_i, \tilde{\varkappa}_i^{-r}(w) = \tilde{\varkappa}_i,$$
$$i \geqq 0, X_k^{-r}(w) \in A\big(k, ([\tilde{\xi}_j, \tilde{\eta}_j])\big), k \geqq n\}\Bigg) = 1,$$

cf. Section 5.3 for notation.

Since the stationary solution $([\tilde{\xi}_k, \tilde{\eta}_k, X_k])$ of (5.8.5) is unique and strong, we have $q = 1$. From this, one obtains

$$\lim_{r\to\infty} \tilde{\mathsf{P}}\Bigg(\bigcup_{n=0}^{\infty} \{\tilde{\xi}_i^{-r}(w) = \tilde{\xi}_i, \tilde{\eta}_i^{-r}(w) = \tilde{\eta}_i, \tilde{\varkappa}_i^{-r}(w) = \tilde{\varkappa}_i,$$
$$i \geqq 0, X_k^{-r}(w) \in B\big(k, ([\tilde{\xi}_j, \tilde{\eta}_j])\big) = \{X_k\}, k \geqq n\}\Bigg) = 1.$$

(Namely, in view of the stationarity of $([\tilde{\xi}_j, \tilde{\eta}_j])$,

$$\lim_{n\to\infty} \tilde{\mathsf{P}}\Big(x_n\Big(A\big(0, ([\tilde{\xi}_j, \tilde{\eta}_j])\big), [\tilde{\xi}_0, \tilde{\eta}_0], \ldots, [\tilde{\xi}_{n-1}, \tilde{\eta}_{n-1}]\Big) = B\big(n, ([\tilde{\xi}_j, \tilde{\eta}_j])\big)\Big)$$

$$= \lim_{n\to\infty} \tilde{\mathsf{P}}\Big(x_n\Big(A\big(-n, ([\tilde{\xi}_j, \tilde{\eta}_j])\big), [\tilde{\xi}_{-n}, \tilde{\eta}_{-n}], \ldots, [\tilde{\xi}_{-1}, \tilde{\eta}_{-1}]\Big)$$

$$= B\big(0, ([\tilde{\xi}_j, \tilde{\eta}_j])\big)\Big) = 1.)$$

Consequently

$$\lim_{r\to\infty} \tilde{\mathsf{P}}\left(\bigcup_{n=0}^{\infty} \{\widetilde{W}_k^{-r}(w) = \widetilde{W}_k, k \geqq n\}\right) = 1.$$

Together with

$$\mathsf{P}\big(([Z_n, W_n], n \in \mathbb{Z}) \in (\cdot)\big)$$

$$= \frac{1}{\mathsf{E}\tilde{\nu}_1} \sum_{j=0}^{\infty} \tilde{\mathsf{P}}\big(\tilde{\nu}_1 > j, ([\tilde{Z}_{n+j}, \widetilde{W}_{n+j}], n \in \mathbb{Z}) \in (\cdot)\big),$$

cf. Theorem A 1.3.1, and the stationarity of $([A_n, S_n, C_n, W_n])$ we obtain assertion (5.8.9) as follows:

$$\mathsf{P}\left(\bigcup_{m=0}^{\infty} \{w_n(w, U_0, U_1, \ldots, U_{n-1}) = W_n, n \geqq m\}\right)$$

$$= \lim_{r\to\infty} \mathsf{P}\left(\bigcup_{m=0}^{\infty} \{w_{n+r}(w, U_{-r}, \ldots, U_{n-1}) = W_n, n \geqq m\}\right)$$

$$= \lim_{r\to\infty} \frac{1}{\mathsf{E}\tilde{\nu}_1} \sum_{j=0}^{\infty} \tilde{\mathsf{P}}\left(\tilde{\nu}_1 > j, \bigcup_{m=0}^{\infty} \{\widetilde{W}_n^{-r+j}(w) = \widetilde{W}_n, n \geqq m + j\}\right)$$

$$\geqq \frac{1}{\mathsf{E}\tilde{\nu}_1} \lim_{r\to\infty} \sum_{j=0}^{\infty} \tilde{\mathsf{P}}\left(\tilde{\nu}_1 > j, \bigcup_{m=0}^{\infty} \{\widetilde{W}_n^{-r+j}(w) = \widetilde{W}_n, n \geqq m\}\right)$$

$$= \frac{1}{\mathsf{E}\tilde{\nu}_1} \sum_{j=0}^{\infty} \tilde{\mathsf{P}}(\tilde{\nu}_1 > j) = 1. \quad \blacksquare$$

Without any independence assumption on the input one can prove

5.8.5. Theorem. *Consider a single server queue with warming-up. Assume that the stationary and ergodic input* $\Phi = ([A_n, S_n, C_n])$ *fulfills*

$$\mathsf{E}S_0 + \mathsf{E}C_0 < \mathsf{E}A_0. \tag{5.8.13}$$

Then the assertions of Theorem 5.8.4 *are true.*

Proof. The proof will only be sketched. Consider the usual G/G/1/∞ queue with the input $\Psi = ([A_n, S_n + C_n])$. Then, by Theorem 5.4.1 and (5.4.7) the unique stationary state process of this queue has infinitely many empty points, in view of (5.8.13). Obviously, these empty points also have to be empty points in the single server queue with warming-up. Thus there is a stationary strong solution of (5.8.1) which is the only stationary weak solution. The limiting property (5.8.9) follows by (5.4.3) and the fact that the usual G/G/1/∞ queue with the input Ψ majorizes the queue with warming-up and input Φ. ∎

5.9. The many server loss system with repeated call attempts

Consider an m-server system with the stationary input

$$\Phi = (U_n) = ([A_n, Y_n]),$$

where as usual the A_n are the interarrival times of customers. The marks Y_n of the form

$$Y_n = \left[S_n^1, [R_n^2, S_n^2], \ldots, [R_n^{\nu_n}, S_n^{\nu_n}]\right], \qquad n \in \mathbb{Z},$$

characterize the behaviour of the customer: $\nu_n \geqq 1$ is a random integer, denoting the maximal possible number of call attempts of the n-th customer. The n-th arriving customer occupies one of the idle servers (if there are any) for a random time S_n^1 and leaves the system after service completion. If the n-th incoming customer finds all servers busy and $\nu_n = 1$, he leaves the system unserved, in case of $\nu_n \geqq 2$ he occupies a place in an unbounded waiting room and forms a so-called source of repeated call attempts (abbr. SRC). After the random time R_n^2 he starts the next call attempt. If there is a server free now, then the customer receives an amount of service S_n^2, after which he leaves the system; otherwise he continues to wait for a time R_n^2, after which he looks for service again. If there is a server free then the customer receives an amount of service S_n^3, after which he leaves the system, otherwise ... This procedure will be repeated ν_n times. (Notice that if $\nu_n = 1$ then the n-th customer makes no repeated call attempts. Thus, we have the standard m-server loss system if $\nu_n \equiv 1$ a.s.)

In teletraffic theory the sequences (A_n) and (Y_n) are assumed to be independent and the Y_n are i.i.d. r.v.'s. The probabilities $p_1, p_2, \ldots$ with

$$p_k = \mathsf{P}(\nu_n \geqq k + 1 \mid \nu_n \geqq k)$$

form the so-called perseverance function, cf. e.g. Deul (1982). Thus, in this case an arbitrary customer leaves the system unserved with the probability $1 - p_k$, if his k-th call attempt was not successful. The state

$$X_n = [X_{n1}, \ldots, X_{nm}, g(1), \ldots, g(Q_n)]$$

of the system at the arrival of the n-th customer is described by the residual service times X_{ni}, $1 \leqq i \leqq m$, with the servers arranged in decreasing order, the number Q_n of SRC's in the waiting room and the marks $g(i)$, $1 \leqq i \leqq Q_n$, describing the SRC's arranged in the order of their arrival. The fate of the i-th SRC is described by

$$g(i) = [R(1, i), S(1, i), R(2, i), S(2, i), \ldots, R(r, i), S(r, i)],$$

where $R(1, i)$ is the residual waiting time to the next call attempt, $S(1, i)$ is the corresponding service time; $R(2, i), R(3, i), \ldots$ are the time distances between the next consecutive repeated call attempts, $S(2, i), S(3, i), \ldots$ are the required service times. (If there are no SRC's, i.e. if $Q_n = 0$, then the state of the system is described by $[X_{n1}, \ldots, X_{nm}]$.)

In view of the dynamics of the system, there exists a measurable function f such that the states X_n of the system at the customer arrivals satisfy the equation:

$$X_{n+1} = f(X_n, U_n), \qquad n \in \mathbb{Z}. \tag{5.9.1}$$

The explicit form of f is not important here.

5.9.1. Theorem. *Consider an m-server loss system with repeated call attempts as described above. If the stationary and ergodic input* $\Phi = (U_n) = ([A_n, Y_n])$ *satisfies*

$$A_0 > 0 \quad a.s.,$$

$$\mathsf{E}\left(\max_{1 \le i \le \nu_0} S_0{}^i\right) < \infty; \tag{5.9.2}$$

$$\mathsf{E} \sum_{i=2}^{\nu_0} R_0{}^i < \infty, \tag{5.9.3}$$

then there exists a stationary weak solution of (5.9.1) *and hence there exists a stationary state process of the system considered.*

Proof. The proof will be given by means of the method of solution generating systems, described in Section 1.4. Let

$$R_j{}^k(n) = \left(S_j{}^k + \sum_{i=2}^{k} R_j{}^i - \sum_{i=j}^{n-1} A_i\right)_+, \qquad 1 \le k \le \nu_j,$$

with the convention $\sum_{i=2}^{1} R_j{}^i = 0$. Obviously, in the case of $\sum_{i=2}^{k} R_j{}^i < \sum_{i=j}^{n-1} A_i$ the quantity $R_j{}^k(n)$ is the residual service time of the j-th customer at the n-th arrival if his k-th call attempt was successful, $k = 1, \ldots, \nu_j$. (The arrival of the customer is his first call attempt.) Let

$$\varkappa_j(n) = \max\left\{1 \le l \le \nu_j : \sum_{i=2}^{l} R_j{}^i < \sum_{i=j}^{n-1} A_i\right\}$$

be the potential number of repeated call attempts of the j-th customer up to the arrival of the n-th customer. If $\varkappa_j(n) < \nu_j$, then

$$K_j(n) = \left[\sum_{i=2}^{\varkappa_j(n)+1} R_j{}^i - \sum_{i=j}^{n-1} A_i, S_j{}^{\varkappa_j(n)+1}, R_j{}^{\varkappa_j(n)+2}, S_j{}^{\varkappa_j(n)+2}, \ldots, R_j{}^{\nu_j}, S_j{}^{\nu_j}\right]$$

is the "state" of the j-th customer at the n-th arrival if he has not been served yet. The set

$$I(n, \Phi) = \left\{j < n : \varkappa_j(n) < \nu_j \text{ or } S_j{}^{\nu_j} + \sum_{i=2}^{\nu_j} R_j{}^i - \sum_{i=j}^{n-1} A_i > 0\right\}$$

contains the indices of all those customers who are possibly in service or are SRC's. Conditions (5.9.2) and (5.9.3) imply

$$\# I(n, \Phi) < \infty \quad \text{a.s.}$$

This follows analogously to considerations in Example 1.4.1. Define

$$A(n, \Phi) = \{[X_1, \ldots, X_m, g(1), \ldots, g(l)]: 0 \leqq l \leqq \# I(n, \Phi);$$
$$X_i \in R(n), i = 1, \ldots, m; g(i) \in K(n), i = 1, \ldots, m\},$$

where

$$R(n) = \{R_j^k(n): j \in I(n, \Phi), 1 \leqq k \leqq \varkappa_j(n)\} \cup \{0\},$$

$$K(n) = \{K_j(n): j \in I(n, \Phi), \varkappa_j(n) < \nu_j\}.$$

(If $K(n) = \emptyset$ then there are no SRC's, i.e. the state is $[X_1, \ldots, X_m]$.) It is easy to check that the sets $A(n, \Phi)$ form an SGS, cf. Definition 1.4.2. Now the assertion follows from Theorem 1.4.3. ∎

The stationary state process mentioned in Theorem 5.9.1 need not be unique. However, under additional assumptions we have a unique state process which appears as the limiting process as n tends to ∞ regardless of the initial condition.

5.9.2. Theorem. *Consider an m-server loss system with repeated calls. Assume that the stationary and ergodic input* $\Phi = ([A_n, Y_n])$ *satisfies the assumptions of Theorem* 5.9.1. *If additionally*

(A_n) *and* (Y_n) *are independent;*

(Y_n) *is a sequence of i.i.d. r.v.'s.;*

$$\mathsf{P}\left(A_0 > \max_{j=1,\ldots,\nu_0} \left(S_0^j + \sum_{i=2}^{j} R_0^i\right)\right) > 0,$$

then the sequence (X_n),

$$X_n = x_n(\Phi) = \lim_{r\to\infty} x_r(0, U_{n-r}, \ldots, U_{n-1}),$$

is the only stationary state process, where 0 *denotes the state of the empty system. Moreover,*

$$\mathsf{P}\left(\bigcup_{m=1}^{\infty} \{x_k(x, U_0, \ldots, U_{k-1}) = X_k, k \geqq m\}\right) = 1,$$

regardless of the initial state x.

The proof of this Theorem is in the spirit of Theorems 5.2.3, 5.3.5, and 5.3.8 and is not outlined here.

5.10. Open networks of loss systems

Consider a queueing network consisting of k nodes. The j-th node is an m_j-server loss system as described in Section 5.3, $j = 1, \ldots, k$. An arriving customer from outside — say the n-th — is characterized by the mark

$$Y_n = [R_n^1, S_n^1, R_n^2, S_n^2, \ldots, R_n^{\nu_n}, S_n^{\nu_n}]$$

describing the total service requirement of the customer. Here $R_n^1, \ldots, R_n^{\nu_n}$ denote the random route of the random length ν_n through the network, $1 \leqq R_n^i \leqq k$, $1 \leqq i \leqq \nu_n$. The value S_n^i is the desired service time at the node R_n^i. The n-th customer enters the network at the node R_n^1. If all servers at this node are busy, the customer leaves the network and is lost. Otherwise he obtains service of the length S_n^1, and after completing service he moves to the node R_n^2, etc. If the customer reaches the node $R_n^{\nu_n}$ and one of the servers at this node is free, he gets service of the length $S_n^{\nu_n}$, and after finishing service at $R_n^{\nu_n}$ he leaves the network. Thus, the input of the queueing network considered is given by the sequence

$$\Phi = (U_n) = ([A_n, Y_n]).$$

In order to avoid some technical difficulties, in the following we restrict ourselves to the case $m_1 = \cdots = m_k = 1$ (single server loss systems at all nodes). However, the results remain valid for arbitrary $m_1, \ldots, m_k$.

The state of the system is described by

$$\boldsymbol{X} = [X_1, Y(1), \ldots, X_k, Y(k)],$$

where X_i $(i = 1, \ldots, k)$ is the residual service time of the single server loss system at node i and

$$Y(i) = [R(1, i), S(1, i), R(2, i), S(2, i), \ldots, R(r, i), S(r, i)]$$

is the "residual" mark of the customer who is in service at node i; $R(1, i), \ldots, R(r, i)$ is the (required) remaining route through the network and $S(1, i), \ldots, S(r, i)$ are the corresponding service times. (If there is no customer at node i, then we set $[X_i, Y(i)] = [0, 0]$; if the customer at node i has no further requirement, we set $[X_i, Y(i)] = [X_i, 0]$). Denote by $\boldsymbol{X}_n$ the state of the system at the arrival of the n-th customer. In view of the dynamic of the network there is a measurable function f such that

$$\boldsymbol{X}_{n+1} = f(\boldsymbol{X}_n, U_n), \qquad n \in \mathbb{Z}. \tag{5.10.1}$$

The explicit form of f is complicated but we make no use of it here.

The existence of a steady state can be shown under weak and quite natural assumptions.

5.10.1. Theorem. *Consider an open network of k single server loss systems with the stationary and ergodic input (Φ, P), $\Phi = (U_n) = ([A_n, Y_n])$. If*

$$A_0 > 0 \quad a.s. \tag{5.10.2}$$

and

$$\mathsf{E} \sum_{i=1}^{\nu_0} S_0^i < \infty, \tag{5.10.3}$$

then there exists a stationary state process $(\boldsymbol{X}_n)$ of the network considered.

Proof. We construct a stationary weak solution of (5.10.1) by means of the method of solution generating system described in Section 1.4. Denote by $\varkappa_j(n)$ the number of nodes already passed by the j-th customer (if he is not lost) before the arrival of the n-th customer, $j < n$,

$$\varkappa_j(n) = \max\left\{0 \leqq l \leqq \nu_j : \sum_{i=1}^{l} S_j^i < \sum_{i=j}^{n-1} A_i\right\}.$$

The j-th customer is still in the system at the n-th arrival if and only if $\varkappa_j(n) < \nu_j$ and if he is not lost. In this case he is at node $R_j^{\varkappa_j(n)+1}$, his residual service time is

$$X_n^j = \sum_{i=1}^{\varkappa_j(n)+1} S_j^i - \sum_{i=j}^{n-1} A_i,$$

and his residual mark is

$$Y_n^j = [R_j^{\varkappa_j(n)+2}, S_j^{\varkappa_j(n)+2}, R_j^{\varkappa_j(n)+3}, S_j^{\varkappa_j(n)+3}, \ldots, R_j^{\nu_j}, S_j^{\nu_j}].$$

Denote by

$$I(n, \Phi) = \{j < n : \varkappa_j(n) < \nu_j\}$$

the set of indices of customers who are potentially in the system at the n-th arrival. Similarly the single loss system, cf. Example 1.4.1 and the proof of Theorem 5.3.1, conditions (5.10.2) and (5.10.3) imply $\# I(n, \Phi) < \infty$ a.s. Let

$$K_i(n) = \{[X_n^j, Y_n^j] : j \in I(n, \Phi), R_j^{\varkappa_j(n)+1} = i\} \cup \{[0, 0]\}.$$

Then it is easy to check that the sets

$$A(n, \Phi) = \{[X_1, Y(1), \ldots, X_k, Y(k)] : [X_i, Y(i)] \in K_i(n), i = 1, \ldots, k\}$$

form an SGS. Now the assertion follows from Theorem 1.4.3. ∎

To show the uniqueness of the stationary state process and the convergence to this under arbitrary initial states, we need some additional assumptions.

5.10.2. Theorem. *Let the assumptions of Theorem* 5.10.1 *be fulfilled. If additionally*

(A_n) *and* (Y_n) *are independent sequences;*

(Y_n) *is a sequence of i.i.d. r.v.'s,*

and

$$\mathsf{P}\left(\sum_{i=1}^{\nu_0} S_0^i < A_0\right) > 0,$$

then the sequence $(\boldsymbol{X}_n)$,

$$\boldsymbol{X}_n = \lim_{r\to\infty} x_r(\boldsymbol{0}, U_{n-r}, U_{n-r+1}, \ldots, U_{n-1}),$$

is the only stationary state process, where $\boldsymbol{0}$ *denotes the state of the empty system. Moreover,*

$$\mathsf{P}\left(\bigcup_{m=1}^{\infty} \{x_k(\boldsymbol{x}, U_0, \ldots, U_{k-1}) = \boldsymbol{X}_k, k \geqq m\}\right) = 1$$

for every initial state $\boldsymbol{x}$.

The proof of Theorem 5.10.2 is similar to those of Theorems 5.2.3, 5.3.5, and 5.3.8 and will be omitted.

5.11. Remarks and references

5.2. The existence of a (time-) stationary state process for G/G/∞ was proved in Franken (1969, 1970), where the input was assumed to be a stationary marked point process (without multiple points) and $\varrho = \mathsf{E}S_0/\mathsf{E}A_0 < \infty$. (The theorem was formulated for G/GI/∞, but the proof is valid for G/G/∞, too.) Using the point process approach, the arrival-stationary state process was derived.

Independently, Borovkov (1972a, b) investigated the arrival-stationary (in case of no batch arrivals) and the "batch-stationary" (in case of batch arrivals) G/G/∞ queue, cf. Section 7.1 for the explanation of terms related with the batch-stationary situation. In case of no batch arrivals, he proved the corresponding part of Theorem 5.2.1 under the assumptions that (A_n) is ergodic and S_n has finite expectation. In case of batch arrivals, the analogous result was shown under the assumptions that the stationary sequence of "inter-batch-times" is ergodic and the expectation of the sum of the service times belonging to a batch is finite.

Gutjahr (1977b) generalized Theorem 5.2.3 to non-ergodic (A_n). He showed that

$$\mathsf{P}\left(\bigcup_{n=0}^{\infty} \{S_n < A_n\}\right) = 1$$

is necessary and sufficient for the existence of infinitely many empty points, see also Kaplan (1975) for GI/GI/∞. The proof of Theorem 5.2.3 given here is new and seems to give more insight than that given in [FKAS].

5.3. Theorem 5.3.1 was first proved in Lisek (1979a, b) for the G/G/m/0 queue (in the time-stationary case, no batch arrivals).

For a discussion of other types of renewing epochs than those given in Theorem 5.3.5 we refer to Borovkov (1972a, b, 1978, 1980).

Theorem 5.3.6 is from Borovkov (1972a, b), Theorem 5.3.8 was proved in Franken (1969, 1970). Theorem 5.3.9 is due to Lisek (1985a, b). A different proof of this result is given in Wirth (1983).

For systems GI/GI/m/0 Joffe and Ney (1963) investigated the arrival- and time-stationary number of customers in the system by using a condition which corresponds to (5.3.6).

Concerning results for the rate of convergence to the stationary distribution we refer to Foss (1986).

5.4. The existence of a uniquely determined stationary work load process $(v_n^{\min}(\Phi),\ n \in \mathbb{Z})$ was shown for G/G/1/∞ queues with FCFS discipline and ergodic input with $\varrho < 1$ by Loynes (1962), see also Borovkov (1972a), Schassberger (1973). Under these conditions the existence of infinitely many empty points was also pointed out by Loynes (1962). Kalähne (1976) carried out similar considerations for arbitrary work conserving queueing disciplines, cf. also Miyazawa (1977, 1979). Similar arguments were used by Wirth (1986b) in order to investigate equation (5.1.1) for single server queues with varying service rate, i.e. if the service rate depends on the work load or on the number in the system.

5.5. Theorem 5.5.1 was first proved by Loynes (1962), see also Borovkov (1972a, 1980), Schassberger (1973). The proof presented here stems from Neveu (1984), see also Baccelli and Brémaud (1987). The results of Theorem 5.5.4, Theorem 5.5.6, and Theorem 5.5.7 are from Brandt (1983, 1985a, b). Independently, Foss (1983) investigated stationary solutions of (5.5.1) by using the technique of renewing epochs. In particular, he proved the equivalence of Theorem 5.5.6. (i) and Theorem 5.5.6 (iii). It is shown that (5.5.33) implies the existence of x-renewing epochs. Furthermore, Foss showed that if the $[A_n, S_n]$, $n \in \mathbb{Z}$, are i.i.d. and $\varrho < m$, then Theorem 5.5.6 (i) and (5.5.33) are true. Finally he announced that Theorem 5.5.6 (i) is equivalent to the strong ergodic property (5.5.33). Theorem 5.5.9 was proved by Borovkov (1980) using the method of renewing epochs, see also Akhmarov and Leontyeva (1976). For GI/GI/m/∞ queues with $\varrho < m$ Kiefer and Wolfowitz (1955) proved Theorem 5.5.6 (i) and Theorem 5.5.6 (iii). For the FCFS G/GI/m/∞ queue (without batch arrivals) Gutjahr (1977a) established a necessary and sufficient condition for the existence of infinitely many empty points. If (A_n) is ergodic, this criterion has the form

$$\mathsf{P}\left(\bigcap_{j=1}^{\infty}\left\{\sum_{i=-m(j-1)-1}^{-1} A_i + jh < 0\right\}\right) > 0,$$

where $h = \inf\{x \in \mathbb{R}_+ : \mathsf{S}(x) > 0)\}$. If $h = 0$, then the ergodicity of (A_n) is not needed. For the GI/GI/m/∞ queue this condition has the form

$$\mathsf{P}(A_0 > S_0) > 0,$$

cf. Whitt (1972), Gutjahr (1977a). Concerning the queue length process for G/G/m/∞ queues with FCFS and other queueing disciplines see Miyazawa (1977, 1979). Recently, Nakatsuka (1986) proved the so-called substability (cf. Loynes (1962)) for G/G/m/∞ queues with $\varrho < m$ and arbitrary work conserving queueing discipline without interruptions of service. In particular, in the case of M/G/m/∞ one gets the existence of a uniquely determined stationary state process and the ergodic property (5.5.33). Concerning estimates of the rate of convergence in ergodic theorems for many server queues we refer to Foss (1985, 1986). Systems of the type G/G/m/∞ with limited waiting time were investigated by Krupin (1975).

5.7. These considerations are based on Brandt (1987b). Nakatsuka (1986) investigated G/G/m/r queues with arbitrary work-conserving queueing disciplines where interruptions of service are not allowed. He showed the substability of such queues provided $\varrho < m$ or if the queueing discipline is FCFS. Besides other useful ergodic criteria, he proved that the condition

$$\mathsf{P}(S_0 \leqq x_0) = 1, \quad \mathsf{P}\big(A_0 > (m + r)\, x_0\big) > 0 \text{ for some } x_0 \in \mathbb{R}_+$$

ensures a uniquely determined stationary state process and ergodic properties of the kind (5.5.33).

Using the method of renewing epochs, Akhmarov and Leontyeva (1976) analyzed GI/GI/m/r queues with FCFS queueing discipline. They proved that the condition

$$\mathsf{P}(S_0 \geqq x + y \mid S_0 \geqq y) < c\,\mathrm{e}^{-\alpha x} \quad \text{for all } x \geqq x_0 \text{ and all } y$$

$$\text{with } \mathsf{P}(S_0 \geqq y) > 0 \quad \text{for some } c < \infty \text{ and } \alpha > 0$$

(i.e. S_0 decreases uniformly exponentially), together with $\mathsf{P}(S_0 < mA_0) > 0$, is sufficient for the existence of a stationary state process and ergodic properties of the kind (5.5.33). The same statements are valid if $\mathsf{P}(S_0 < A_0) > 0$ and $\varrho < m$. Systems of the type G/G/m/r were also investigated by Krupin (1974, 1975) using the idea of renewing epochs.

5.8. Single server queues in which the first customer of each busy period receives exceptional service have been investigated in several papers, e.g. Finch (1959a), Yeo (1962), Miller (1964), Welch (1964), Rossberg and Siegel (1974), Siegel (1974), Dewess (1975), Do Le Minh (1980), [FKAS, Sect. 5.4], Wirth (1983), Levy and Kleinrock (1986). A good survey and further references are in Doshi (1986).

The approach given here is based on Wirth (1984b) and uses a (sample path) decomposition of queues with warming-up. For the delay analysis by decomposition see also Levy and Kleinrock (1986). Decomposition ideas can also be used for analyzing the single server with vacation periods, cf. Doshi (1985), Levy and Kleinrock (1986), and Fuhrmann (1984). A good survey concerning decomposition of queues with vacations is in Doshi (1986). A general decomposition scheme for complicated queueing systems is outlined in Fuhrmann and Cooper (1985).

5.9. For more special systems with repeated calls see Deul (1982) and the references therein.

In the literature the existence of stationary solutions and ergodic properties of several other queueing models have been investigated. See e.g. Afanas'eva (1965), Afanas'eva and Martynov (1969), Krupin (1974, 1975), Akhmarov (1979a), Baccelli, Boyer and Hebuterne (1984) for queues with impatient customers. For some recent results concerning quite complicated queues and communication systems we refer to Falin (1987), Borovkov (1986, 1987, 1988, 1989), Berezner and Malyshev (1989), Kelbert and Sukhov (1983, 1985).

6. Relationships between arrival-, time-, and departure-stationary queueing processes

6.1. Introduction

In Chapter 5 we introduced arrival-stationary queueing models. Chapter 4 enables us to define the corresponding time-stationary models and to discuss the relations between arrival- and time-stationary queueing quantities. First let us define the time-stationary model of a given queueing system. Consider a stationary and ergodic sequence $\Phi = ([A_n, Y_n])$ describing the input of this queueing system; $A_n \geqq 0$ is the interarrival time between the n-th and $(n+1)$-st arrival, and Y_n is a mark expressing the quality of the n-th customer (e.g. its required service time). As already mentioned, Φ describes the stationary input viewed by an arbitrary arriving customer. Throughout this chapter we assume

$$0 < \mathsf{E}A_0 < \infty .$$

Assume that the evolution of the system under consideration is characterized by the recursive stochastic equation

$$X_{n+1} = f(X_n, [A_n, Y_n]), \qquad n \in \mathbb{Z}, \tag{6.1.1}$$

where X_n is interpreted as the state of the system seen by the n-th arriving customer, cf. Chapter 5. The function $f(x, [a, y])$ is assumed to be left continuous in a. Now consider an arbitrary but fixed stationary weak (or strong) solution of (6.1.1) (assuming there is one). In the following we omit the index "prime" used in the previous chapters for indicating weak solutions and denote this solution by $\big([[A_n, Y_n], X_n]\big)$ and its distribution by P. (In order to avoid confusion, the reader may imagine $\big([[A_n, Y_n], X_n]\big)$ as a strong solution of (6.1.1).) Define

$$T_0 = 0, \quad T_n = \sum_{k=0}^{n-1} A_k \text{ for } n > 0, \quad T_n = -\sum_{k=n}^{-1} A_k \text{ for } n < 0.$$

Suppose that, given the states X_n observed by the arriving customers, the state $X(t)$ of the system at an arbitrary time t is determined by

$$X(t) = f(X_n, [t - T_n, Y_n]) \quad \text{for } T_n < t \leqq T_{n+1}, \tag{6.1.2}$$

where the function f is the same as in (6.1.1). The process $X(t)$ describes the evolution of the system in continuous time provided the 0-th customer arrives at $T_0 = 0$. (Note that for all queueing systems introduced in Chapter 5 the state space $\mathbb{X}$ and the function f were chosen in such a way that the given interpretation of $X(t)$ is correct.)

The stochastic process $X(t)$ is not stationary, in general. The pair $\chi = \left[(X(t), t \in \mathbb{R}), \cdot \left([T_n, [X_n, Y_n]], n \in \mathbb{Z}\right)\right]$ is a synchronous process with an embedded point process (PEMP), cf. Section 4.2. We call χ the *arrival-stationary model* of the queueing system. From Theorem 4.4.1 we obtain the existence of a unique stationary PEMP $(\bar{\chi}, \bar{\mathsf{P}})$, $\bar{\chi} = \left[(\bar{X}(t)\, t \in \mathbb{R}), \left([\bar{T}_n, [\bar{X}_n, \bar{Y}_n]], n \in \mathbb{Z}\right)\right]$, satisfying

$$\bar{\mathsf{P}}\left(\left([\bar{T}_n, [\bar{X}_n, \bar{Y}_n]], n \in \mathbb{Z}\right) \in (\cdot)\right)$$
$$= (\mathsf{E}A_0)^{-1} \int_0^\infty \mathsf{P}\left(A_0 > t, \left([T_n - t, [X_n, Y_n]], n \in \mathbb{Z}\right) \in (\cdot)\right) \mathrm{d}t, \tag{6.1.3}$$

$$\mathsf{P}\left(\left([T_n, [X_n, Y_n]], n \in \mathbb{Z}\right) \in (\cdot)\right)$$
$$= \frac{\mathsf{E}_{\bar{\mathsf{P}}} \sum_{0<\bar{T}_n<1} \mathbf{1}_{(\cdot)}\left(\left([\bar{T}_{n+k}, [\bar{X}_{n+k}, \bar{Y}_{n+k}]], k \in \mathbb{Z}\right)\right)}{\mathsf{E}_{\bar{\mathsf{P}}} \# \{n : 0 < \bar{T}_n \leqq 1\}}, \tag{6.1.4}$$

$$\bar{\mathsf{P}}\left(\bar{X}_{n+1} = f(\bar{X}_n, [\bar{T}_{n+1} - \bar{T}_n, \bar{Y}_n]), n \in \mathbb{Z}\right) = 1,$$

$$\bar{\mathsf{P}}\left(\bar{X}(t) = f(\bar{X}_n, [t - \bar{T}_n, \bar{Y}_n]) \text{ for } \bar{T}_n < t \leqq \bar{T}_{n+1}, n \in \mathbb{Z}\right) = 1.$$

If χ is ergodic, then

$$t^{-1} \int_0^t h(X(u))\, \mathrm{d}u \xrightarrow[t\to\infty]{} \mathsf{E}_{\bar{\mathsf{P}}} h(\bar{X}(0)) \quad \mathsf{P}\text{-a.s.}, \quad h \in F_+(\mathbb{X}), \tag{6.1.5}$$

$$n^{-1} \sum_{k=1}^n h(\bar{X}_k) \xrightarrow[n\to\infty]{} \mathsf{E}_{\mathsf{P}} h(X_0) \quad \bar{\mathsf{P}}\text{-a.s.}, \quad h \in F_+(\mathbb{X}). \tag{6.1.6}$$

The PEMP $(\bar{\chi}, \bar{\mathsf{P}})$ is called the *time-stationary model* of the queueing system under consideration. In particular, $(\bar{X}(t))$ is the corresponding continuous time stationary state process of the queueing system. The marked point process $([\bar{T}_n, \bar{Y}_n])$ is called the time-stationary input associated with $\Phi = ([A_n, Y_n])$, cf. Section 4.5. All stationary strong solutions constructed in Chapter 5 have the form

$$X_n = \lim_{r\to\infty} x_r(x_0, [A_{n-r}, Y_{n-r}], \ldots, [A_{n-1}, Y_{n-1}]), \qquad n \in \mathbb{Z}, \tag{6.1.7}$$

where x_0 is an appropriately chosen initial state. (In most queueing systems we put $x_0 = \mathsf{O}$, where O denotes the state "the system is empty". That means the state X_n depends on the past $[A_{n-1}, Y_{n-1}], [A_{n-2}, Y_{n-2}], \ldots$ of the input only. Formula (6.1.7) can be interpreted as follows: The system started its work at time $-\infty$ with the initial state x_0. As one would expect, the same interpretation is true in the corresponding time-stationary model, too. More generally, we have

6.1.1. Lemma. *Let* $\left([[A_n, Y_n], X_n]\right)$ *be a stationary strong solution of* (6.1.1), $0 < \mathsf{E}A_0 < \infty$.

(i) *Assume that there is a measurable mapping g such that*

$$X_n = g([A_{n-1}, Y_{n-1}], [A_{n-2}, Y_{n-2}], \ldots), \quad n \in \mathbb{Z}, \quad \mathsf{P}\text{-a.s.}$$

Then for the corresponding time-stationary model $(\bar{\chi}, \bar{\mathsf{P}})$, $\bar{\chi} = \left[(\bar{X}(t)), ([\bar{T}_n, [\bar{X}_n, \bar{Y}_n]])\right]$ *it holds that* $\bar{X}_n = g([\bar{T}_n - \bar{T}_{n-1}, \bar{Y}_{n-1}], [\bar{T}_{n-1} - \bar{T}_{n-2}, \bar{Y}_{n-2}], \ldots)$, $n \in \mathbb{Z}$, $\bar{\mathsf{P}}$*-a.s., and thus*

$$\bar{X}(t) = f(g([\bar{T}_n - \bar{T}_{n-1}, \bar{Y}_{n-1}], [\bar{T}_{n-1} - \bar{T}_{n-2}, \bar{Y}_{n-2}], \ldots), [t - \bar{T}_n, \bar{Y}_n]) \text{ for } \bar{T}_n < t \leqq \bar{T}_{n+1}, \quad n \in \mathbb{Z}, \quad \bar{\mathsf{P}}\text{-a.s.}$$

(ii) *Let* $(\bar{\chi}, \bar{\mathsf{P}})$, $\bar{\chi} = \left[(\bar{X}(t)), ([\bar{T}_n, [\bar{X}_n, \bar{Y}_n]])\right]$ *be the time-stationary model corresponding to* $([[A_n, Y_n], X_n])$. *Assume that there is a measurable mapping g such that*

$$\bar{X}_n = g([\bar{T}_n - \bar{T}_{n-1}, \bar{Y}_{n-1}], [\bar{T}_{n-1} - \bar{T}_{n-2}, \bar{Y}_{n-2}], \ldots), \quad n \in \mathbb{Z}, \quad \bar{\mathsf{P}}\text{-a.s.}$$

Then

$$X_n = g([A_{n-1}, Y_{n-1}], [A_{n-2}, Y_{n-2}], \ldots), \quad n \in \mathbb{Z}, \quad \mathsf{P}\text{-a.s.}$$

Lemma 6.1.1 is an immediate consequence of (6.1.3) and (6.1.4) Later we shall use the following notation for time-stationary queueing characteristics. (Some of the notation will be introduced in a more formal way when it is needed.)

$\lambda_{\bar{\mathsf{P}}} = (\mathsf{E}A_0)^{-1}$	–	the arrival intensity,
$\bar{V}(t)$	–	the time-stationary workload,
$\bar{W}(t)$	–	the time-stationary virtual waiting time (in FCFS single server queues $\bar{W}(t) = \bar{V}(t)$),
$\bar{L}(t)$	–	the time-stationary number of customers in the system,
$\bar{Q}(t)$	–	the time-stationary number of customers in the queue (if there is one),

$$\bar{\mathsf{V}}(x) = \bar{\mathsf{P}}(\bar{V}(0) \leqq x), \quad \bar{\mathsf{W}}(x) = \bar{\mathsf{P}}(\bar{W}(0) \leqq x), \quad \bar{p}_k = \bar{\mathsf{P}}(\bar{L}(0) = k).$$

This chapter presents some relations between arrival- and time-stationary characteristics of queueing systems. First we prove that, under quite natural assumptions, "Poisson arrivals see time averages" (PASTA). In our setting this means that $\bar{X}(0) \overset{\mathcal{D}}{=} X_0$ (Section 6.2). In Section 6.3 we discuss relations centered around the famous Little's formula "$L = \lambda W$", more precisely $\mathsf{E}_{\bar{\mathsf{P}}}\bar{L}(0) = \lambda_{\bar{\mathsf{P}}} \mathsf{E}_{\mathsf{P}} W_0$. Formulae of this type are important e.g. for simulating stochastic models, since $\mathsf{E}_{\mathsf{P}} W_0$ can be determined by discrete event simulation, but $\mathsf{E}_{\bar{\mathsf{P}}}\bar{L}(0)$ by continuous time observations only. Relations among the numbers of customers in the system at arrival epochs, at departure epochs, and at an arbitrary time are investigated in Section 6.4. Section 6.5 contains some results on busy cycles. In Section 6.6 the well-known Takács and Pollaczek-Khinchin formulae are verified for general single server queues. Similar results are obtained in Section 6.7 for queues with warming-up.

6.2. Poisson arrivals see time averages (PASTA)

Queueing systems with Poisson arrivals, in particular M/GI/m/r queues, are often discussed in queueing literature. Consider a certain queueing system starting its work at time 0 with an arbitrary but fixed initial state x. The customers arrive at the epochs $0 < T_1 < T_2 < \cdots$ and carry the marks $Y_1, Y_2, \ldots$ The sequence $(T_n, n \geqq 1)$ is assumed to be a homogeneous Poisson process on $\mathbb{R}_+$. By $X(t, x)$ we denote the state of the system at time t; the state $X(T_n, x) = X(T_n - 0, x)$ seen by the n-th arriving customer will be denoted by $X_n(x)$. The most relevant queueing characteristics are the time fraction of being in a certain set B of states

$$\lim_{t\to\infty} t^{-1} \int_0^t \mathbf{1}_B\big(X(u, x)\big) \, du$$

and the fraction of arriving customers who see states from B:

$$\lim_{n\to\infty} n^{-1} \sum_{i=1}^{n} \mathbf{1}_B\big(X_i(x)\big).$$

From properties of the homogeneous Poisson process one expects that these averages are a.s. equal (perhaps, under some additional conditions). Results of this type are called "Poisson arrivals see time averages" (abbr. PASTA), cf. Wolff (1982).

We prove a relationship between the time- and arrival-stationary models of queueing systems with Poisson arrivals. This statement provides a PASTA result which is, as we believe, sufficiently general. (However, there are other approaches providing more general results, cf. Section 6.8).

First we give a lemma, which is an immediate consequence of Lemma 3.5.6.

6.2.1. Lemma. *Let $\Phi = ([A_n, Y_n])$ be a stationary sequence with the properties*
(A 6.2.1) *the A_n are i.i.d. exponentially distributed r.v.'s with parameter λ,*
(A 6.2.2) *(A_n) and (Y_n) are independent,*
and let $\overline{\Psi} = ([\overline{T}_n, \overline{Y}_n])$ be the corresponding time-stationary input. Then

$$h([A_{-1}, Y_{-1}], [A_{-2}, Y_{-2}], \ldots) \overset{\mathcal{D}}{=} h([-\overline{T}_0, \overline{Y}_0], [\overline{T}_0 - \overline{T}_{-1}, \overline{Y}_{-1}], [\overline{T}_{-1} - \overline{T}_{-2}, \overline{Y}_{-2}], \ldots)$$

for all measurable functions $h\colon (\mathbb{R}_+ \times \mathbb{Y})^{\mathbb{Z}_+} \to \mathbb{X}$.

6.2.2. Theorem. *Let $([A_n, Y_n], n \in \mathbb{Z})$ be the stationary input of a queueing system and let $\big([[A_n, Y_n], X_n]\big)$ be a stationary strong solution of equation* (6.1.1). *Assume* (A 6.2.1), (A 6.2.2) *and*

(A 6.2.3) *there is a measurable function g such that $X_n = g([A_{n-1}, Y_{n-1}], [A_{n-2}, Y_{n-2}], \ldots)$, i.e. X_n depends on the past $([A_i, Y_i], i < n)$ of the input.*

Let $\big(\overline{X}(t)\big)$ be the associated time-stationary state process, cf. Section 6.1. *Then $\overline{X}(0) \overset{\mathcal{D}}{=} X_0$.*

Proof. Define

$$h([a_1, y_1], [a_2, y_2], \ldots) = f\big(g([a_2, y_2], [a_3, y_3], \ldots), [a_1, y_1]\big),$$

$a_i \in \mathbb{R}_+$, $y_i \in \mathbf{Y}$. Then by Lemma 6.2.1, (6.1.2), and Lemma 6.1.1, (i) we obtain the assertion:

$$\begin{aligned} X_0 &= f(X_{-1}, [A_{-1}, Y_{-1}]) = h([A_{-1}, Y_{-1}], [A_{-2}, Y_{-2}], \ldots) \\ &\overset{\mathcal{D}}{=} h([-\overline{T}_0, \overline{Y}_0], [\overline{T}_0 - \overline{T}_{-1}, \overline{Y}_{-1}], [\overline{T}_{-1} - \overline{T}_{-2}, \overline{Y}_{-2}], \ldots) \\ &= f(\overline{X}_0, [-\overline{T}_0, \overline{Y}_0]) = \overline{X}(0). \quad \blacksquare \end{aligned}$$

A further consequence of Lemma 6.2.1 is

6.2.3. Theorem. *Consider a queueing system in the steady state. Assume that the (arrival-stationary) number L_0 of customers in the system, the number Q_0 of customers in the queue, the workload V_0, and the actual waiting time W_0, respectively, are measurable functions of the past $([A_n, Y_n], n < 0)$ of the input, i.e.*

$$L_0 = h_1([A_{n-1}, Y_{n-1}], \ldots), \qquad Q_0 = h_2([A_{n-1}, Y_{n-1}], \ldots),$$

$$V_0 = h_3([A_{n-1}, Y_{n-1}], \ldots), \qquad W_0 = h_4([A_{n-1}, Y_{n-1}], \ldots).$$

If the corresponding time-stationary quantities $\overline{L}(0)$, $\overline{Q}(0)$, $\overline{V}(0)$, and $\overline{W}(0)$ are given by the same functions, i.e.

$$\overline{L}(0) = h_1([-\overline{T}_0, \overline{Y}_0], [\overline{T}_{-1} - \overline{T}_0, \overline{Y}_{-1}], \ldots),$$

$$\overline{Q}(0) = h_2([-\overline{T}_0, \overline{Y}_0], [\overline{T}_{-1} - \overline{T}_0, \overline{Y}_{-1}], \ldots),$$

$$\overline{V}(0) = h_3([-\overline{T}_0, \overline{Y}_0], [\overline{T}_{-1} - \overline{T}_0, \overline{Y}_{-1}], \ldots),$$

$$\overline{W}(0) = h_4([-\overline{T}_0, \overline{Y}_0], [\overline{T}_{-1} - \overline{T}_0, \overline{Y}_{-1}], \ldots),$$

then

$$\overline{L}(0) \overset{\mathcal{D}}{=} L_0, \quad \overline{Q}(0) \overset{\mathcal{D}}{=} Q_0, \quad \overline{V}(0) \overset{\mathcal{D}}{=} V_0 \quad \textit{and} \quad \overline{W}(0) \overset{\mathcal{D}}{=} W_0,$$

respectively.

6.2.4. Corollary. *Let the assumptions of Theorem 6.2.2 be fulfilled. If, in addition, (Y_n) is ergodic, then*

$$\lim_{n\to\infty} n^{-1} \sum_{i=1}^{n} 1_{(\cdot)}(X_i) = \lim_{t\to\infty} t^{-1} \int_0^t 1_{(\cdot)}\big(\overline{X}(u)\big)\, du \quad a.s.$$

Since the ergodicity of (Y_n) yields the ergodicity of the arrival-stationary model, Corollary 6.2.4 follows immediately from (6.1.5), (6.1.6), and Theorem 6.2.2.

Corollary 6.2.4 can be interpreted as a "stationary version" of PASTA. The following considerations lead to PASTA results for a fixed initial state.

Consider the ergodic arrival-stationary model $\big[(X(t)), ([T_n, [X_n, Y_n]])\big]$ and the corresponding time-stationary model $\big[(\overline{X}(t)), ([\overline{T}_n, [\overline{X}_n, \overline{Y}_n]])\big]$, cf. Section 6.1. For the arrival-stationary ergodic input $\Phi = ([A_n, Y_n])$ we use the notations

$X_n(x)$ – the state of the system seen by the n-th arriving customer if the state at time 0 is x,

$\left(X_n(x) = x_n(x, [A_0, Y_0], \ldots, [A_{n-1}, Y_{n-1}])\right)$,

$X(t, x)$ – the state of the system at time t if the state at time 0 is x.

For the time-stationary ergodic input $([\overline{T}_n, \overline{Y}_n])$ associated with Φ we denote the corresponding r.v.'s by $\overline{X}_n(x)$ and $\overline{X}(t, x)$, respectively.

In Chapter 5 one can find theorems stating

$$\mathsf{P}\left(\bigcup_{m=1}^{\infty} \{X_n(x) = X_n, n \geqq m\}\right) = 1, \qquad x \in \mathbf{X},$$

for several queueing systems. (We point out that in all standard queueing systems with Poisson arrivals this is true.) In particular, for all measurable sets B

$$\begin{aligned} \lim_{n\to\infty} \frac{1}{n} \sum_{i=1}^{n} \mathbf{1}_B\left(X_i(x)\right) &= \lim_{n\to\infty} \frac{1}{n} \sum_{i=1}^{n} \mathbf{1}_B(X_i) \\ &= \mathsf{P}(X_0 \in B) \qquad \mathsf{P}\text{-a.s.}, \quad x \in \mathbf{X}. \end{aligned} \tag{6.2.1}$$

By the same arguments that yielded (6.2.1), in many cases one can prove

$$\lim_{t\to\infty} t^{-1} \int_0^t \mathbf{1}_B\left(X(u, x)\right) du = \lim_{t\to\infty} t^{-1} \int_0^t \mathbf{1}_B\left(X(u)\right) du \qquad \mathsf{P}\text{-a.s.}, \quad x \in \mathbf{X}, \tag{6.2.2}$$

$$\lim_{n\to\infty} n^{-1} \sum_{i=1}^{n} \mathbf{1}_B\left(\overline{X}_i(x)\right) = \lim_{n\to\infty} n^{-1} \sum_{i=1}^{n} \mathbf{1}_B(\overline{X}_i) \qquad \overline{\mathsf{P}}\text{-a.s.}, \quad x \in \mathbf{X}, \tag{6.2.3}$$

$$\begin{aligned} \lim_{t\to\infty} t^{-1} \int_0^t \mathbf{1}_B\left(\overline{X}(u, x)\right) du &= \lim_{t\to\infty} t^{-1} \int_0^t \mathbf{1}_B\left(\overline{X}(u)\right) du \quad \overline{\mathsf{P}}\text{-a.s.}, \qquad x \in \mathbf{X}. \\ &= \overline{\mathsf{P}}\left(\overline{X}(0) \in B\right) \end{aligned} \tag{6.2.4}$$

Suppose (6.2.1)–(6.2.4) are true. From (6.2.2), Theorem 4.2.6, and (6.2.3), Theorem 3.6.6 we obtain

$$\lim_{t\to\infty} t^{-1} \int_0^t \mathbf{1}_B\left(X(u, x)\right) du = \overline{\mathsf{P}}\left(\overline{X}(0) \in B\right) \quad \mathsf{P}\text{-a.s.}, \quad x \in \mathbf{X}, \tag{6.2.5}$$

and

$$\lim_{n\to\infty} n^{-1} \sum_{i=1}^{n} \mathbf{1}_B\left(\overline{X}_i(x)\right) = \mathsf{P}(X_0 \in B) \quad \overline{\mathsf{P}}\text{-a.s.}, \quad x \in \mathbf{X}, \tag{6.2.6}$$

respectively. From (6.2.1) and (6.2.5) or (6.2.4) and (6.2.6) one observes that in the case of an arrival-stationary or time-stationary input the PASTA problem reduces to the question of whether $\overline{X}(0)$ and X_0 have the same distribution, and this question is answered positively by Theorem 6.2.2.

6.3. Little type formulae

Little's formula "$L = \lambda W$" in its original form states that the time-average number of customers in the system is equal to arrival rate $\times$ average delay (sojourn time) per customer. There are numerous results closely related to "$L = \lambda W$" (conservation laws). These formulae are of considerable practical importance, since time averages are hard to measure in real queueing systems (e.g. in teletraffic systems) but customer averages are relatively easy to record.

A unified approach to Little type formulae will be provided by the formula

$$\text{“}H = \lambda G\text{”}.$$

Consider an arbitrary MPP $\Psi = ([T_n, K_n], n \in \mathbb{Z})$ and a non-negative measurable function $g(u, k)$, $u \in \mathbb{R}$, $k \in \mathbb{K}$. Define

$$h(t, \Psi) = \sum_{n=-\infty}^{\infty} g(t - T_n, K_n), \qquad t \in \mathbb{R}, \tag{6.3.1}$$

$$g_n(\Psi) = \int_{-\infty}^{\infty} g(t - T_n, K_n)\, dt, \qquad n \in \mathbb{Z}. \tag{6.3.2}$$

For example, if $\Psi = ([T_n, W_n])$ is the MPP of arrivals and associated actual waiting times of customers and $g(u, w) = 1\{0 < u < w\}$, then

$$\begin{aligned} h(t, \Psi) &= \sum_{n=-\infty}^{\infty} 1\{0 < t - T_n < W_n\} \\ &= \sum_{n=-\infty}^{\infty} 1\,\{\text{the } n\text{-th customer is waiting in the queue at time } t\} \\ &= Q(t) \end{aligned}$$

is just the number of customers in the queue at time t, and

$$g_n(\Psi) = \int_{-\infty}^{\infty} 1\{0 < t - T_n < W_n\}\, dt = W_n.$$

(Note, for systems with interrupts $Q(t)$ is the number of customers who have arrived and have not received any service before t.)

Returning to the general case, we pose the question: When do the averages

$$\lambda(\Psi) = \lim_{t\to\infty} \frac{\max\{n : n > 0, T_n \leq t\}}{t}, \tag{6.3.3}$$

$$\bar{H}(\Psi) = \lim_{t\to\infty} \frac{1}{t} \int_0^t h(u, \Psi)\, du, \tag{6.3.4}$$

$$G(\Psi) = \lim_{n\to\infty} \frac{1}{n} \sum_{k=1}^{n} g_k(\Psi) \tag{6.3.5}$$

exist and satisfy the relation

$$\bar{H}(\Psi) = \lambda(\Psi)\, G(\Psi) \quad \text{a.s.?} \tag{6.3.6}$$

(Here no stationarity assumption on Ψ has to be made.) A positive answer for stationary and ergodic MPP's and their associated synchronous versions gives the following theorem.

6.3.1. Theorem. *Let* (Φ, P), $\Phi = ([A_n, K_n])$, *be a stationary sequence,* (Ψ, P) *the corresponding synchronous MPP and* $(\bar{\Psi}, \bar{\mathsf{P}})$ *the associated stationary MPP with the intensity* $\lambda = \lambda_{\bar{\mathsf{P}}}$. *Then*

$$\mathsf{E}_{\bar{\mathsf{P}}} h(0, \bar{\Psi}) = \lambda \mathsf{E}_{\mathsf{P}} g_0(\Psi). \tag{6.3.7}$$

If, in addition, $(\bar{\Psi}, \bar{\mathsf{P}})$ *is ergodic, then*

$$\bar{H}(\bar{\Psi}) = \lambda(\bar{\Psi})\, G(\bar{\Psi}) \qquad \bar{\mathsf{P}}\text{-a.s.}, \tag{6.3.8}$$

$$\bar{H}(\Psi) = \lambda(\Psi)\, G(\Psi) \qquad \mathsf{P}\text{-a.s.} \tag{6.3.9}$$

Proof. From Theorem 3.2.7 and relation (3.5.6) we obtain the following representation of the intensity measure of $\bar{\mathsf{P}}$:

$$\nu_{\bar{\mathsf{P}}}(B \times C) = |B|\, \nu_{\bar{\mathsf{P}}}\big((0, 1] \times C\big) = |B|\, \lambda_{\bar{\mathsf{P}}} \mathsf{P}(K_0 \in C)$$

($|B|$ denotes the Lebesgue measure of the set B.) Using this and taking into account Campbell's Theorem 3.2.4 we get

$$\begin{aligned}
\mathsf{E}_{\bar{\mathsf{P}}} h(0, \bar{\Psi}) &= \mathsf{E}_{\bar{\mathsf{P}}} \sum_{n=-\infty}^{\infty} g(-\bar{T}_n, \bar{K}_n) = \int_{\mathbb{R}\times\mathbb{K}} g(-t, k)\, \nu_{\bar{\mathsf{P}}}(\mathrm{d}[t, k]) \\
&= \lambda_{\bar{\mathsf{P}}} \int_{\mathbb{K}} \int_{-\infty}^{\infty} g(-t, k)\, \mathrm{d}t \mathsf{P}(K_0 \in \mathrm{d}k) \\
&= \lambda_{\bar{\mathsf{P}}} \int_{\mathbb{K}} \int_{-\infty}^{\infty} g(t, k)\, \mathrm{d}t \mathsf{P}(K_0 \in \mathrm{d}k) \\
&= \lambda_{\bar{\mathsf{P}}} \mathsf{E}_{\mathsf{P}} \int_{-\infty}^{\infty} g(t, K_0)\, \mathrm{d}t \\
&= \lambda_{\bar{\mathsf{P}}} \mathsf{E}_{\mathsf{P}} g_0(\Psi).
\end{aligned}$$

Thus, (6.3.7) is proved.

From the ergodicity of the MPP $(\bar{\Psi}, \bar{\mathsf{P}})$ it follows that

$$\lambda(\bar{\Psi}) = \lambda_{\bar{\mathsf{P}}} \quad \text{a.s.}, \qquad \bar{H}(\bar{\Psi}) = \mathsf{E}_{\bar{\mathsf{P}}} h(0, \bar{\Psi}) \quad \text{a.s.}$$

The relation

$$G(\bar{\Psi}) = \mathsf{E}_{\mathsf{P}} g_0(\Psi) \quad \text{a.s.}$$

follows from Theorem 3.6.6, (3.6.11). Together with (6.3.7) this proves (6.3.8).

The equation

$$G(\Psi) = \mathsf{E}_{\mathsf{P}} g_0(\Psi) \quad \text{a.s.}$$

follows from the stationarity and ergodicity of $(g_n(\Psi))$ which follows from the ergodicity of (Φ, P).

$$\overline{H}(\Psi) = \mathsf{E}_{\overline{\mathsf{P}}} h(0, \overline{\Psi}) \quad \text{and} \quad \lambda(\Psi) = \lambda_{\overline{\mathsf{P}}} \quad \text{a.s.}$$

follow from Theorem 3.6.6, (3.6.9). Thus, (6.3.7) and (6.3.9) are equivalent. ∎

Theorem 6.3.1 leads to Little type formulae for queueing systems.

6.3.2. Theorem. *Consider a queueing system* G/G/m/∞ *in steady state and with a work conserving queueing discipline with no interruptions of service, i.e. the time- and arrival-stationary queueing characteristics introduced in Section* 6.1 *do exist. Then*

$$\mathsf{E}_{\overline{\mathsf{P}}} \overline{Q}(0) = \lambda_{\overline{\mathsf{P}}} \mathsf{E}_{\mathsf{P}} W_0 = \frac{\mathsf{E}_{\mathsf{P}} W_0}{m_A}, \tag{6.3.10}$$

$$\mathsf{E}_{\overline{\mathsf{P}}} \overline{L}(0) = \lambda_{\overline{\mathsf{P}}} (E_{\mathsf{P}} W_0 + m_S) = \frac{\mathsf{E}_{\mathsf{P}} W_0}{m_A} + \varrho, \tag{6.3.11}$$

$$\mathsf{E}_{\overline{\mathsf{P}}} \min\big(m, \overline{L}(0)\big) = \lambda_{\overline{\mathsf{P}}} m_S = \varrho, \tag{6.3.12}$$

$$\mathsf{E}_{\overline{\mathsf{P}}} \overline{V}(0) = \lambda_{\overline{\mathsf{P}}} \mathsf{E}_{\mathsf{P}} \left(W_0 S_0 + \frac{1}{2} S_0^2 \right). \tag{6.3.13}$$

Proof. Consider the stationary MPP $\overline{\Psi} = \big([\overline{T}_n, [\overline{S}_n, \overline{W}_n]]\big)$ and the functions

$$g_1(u, s, w) = 1\{0 < u < w\},$$

$$g_2(u, s, w) = 1\{0 < u < w + s\},$$

$$g_3(u, s, w) = 1\{0 < u < w + s, u \geqq w\},$$

$$g_4(u, s, w) = \begin{cases} s & \text{if } 0 < u < w, \\ w + s - u & \text{if } 0 \leqq w \leqq u < w + s, \\ 0 & \text{otherwise.} \end{cases}$$

Then

$$\overline{Q}(0) = \sum_{n=-\infty}^{\infty} g_1(-\overline{T}_n, \overline{S}_n, \overline{W}_n), \qquad W_0 = \int_{-\infty}^{\infty} g_1(t, S_0, W_0)\, dt,$$

$$\overline{L}(0) = \sum_{n=-\infty}^{\infty} g_2(-\overline{T}_n, \overline{S}_n, \overline{W}_n), \qquad W_0 + S_0 = \int_{-\infty}^{\infty} g_2(t, S_0, W_0)\, dt,$$

$$\min\big(m, \overline{L}(0)\big) = \sum_{n=-\infty}^{\infty} g_3(-\overline{T}_n, \overline{S}_n, \overline{W}_n), \qquad S_0 = \int_{-\infty}^{\infty} g_3(t, S_0, W_0)\, dt,$$

$$\overline{V}(0) = \sum_{n=-\infty}^{\infty} g_4(-\overline{T}_n, \overline{S}_n, \overline{W}_n), \qquad \int_{-\infty}^{\infty} g_4(t, S_0, W_0)\, dt = S_0 W_0 + \frac{1}{2} S_0^2.$$

Applying statement (6.3.7) to the stationary MPP $\overline{\Psi} = \big([\overline{T}_n, [\overline{S}_n, \overline{W}_n]]\big)$ and to the functions $g_1 - g_4$, we obtain (6.3.10)–(6.3.13). ∎

6.3.3. Corollary. *For a queueing system* G/G/1/∞ *in steady state with a work conserving queueing discipline and no interruptions of service we have*

$$\overline{p}_w = \overline{\mathsf{P}}\big(\overline{L}(0) \geqq 1\big) = \varrho,$$

where $\overline{p}_w$ *denotes the time-stationary waiting probability, i.e. the time-stationary probability that the server is busy. For an* M/G/1/∞ *queue with work conserving queueing discipline and without service interrupts*

$$\mathsf{P}(L_0 = 0) = \overline{\mathsf{P}}\big(\overline{L}(0) = 0\big) = 1 - \overline{p}_w = 1 - \varrho.$$

For G/G/m/∞ *with a work conserving queueing discipline and without service interrupts, we have*

$$\overline{p}_w = \overline{\mathsf{P}}\big(\overline{L}(0) \geqq m\big) \leqq \frac{\varrho}{m},$$

and for G/GI/m/∞ *with FCFS queueing discipline*

$$\mathsf{E}_{\overline{\mathsf{P}}}\overline{V}(0) = \varrho\left(\mathsf{E}_{\mathsf{P}}W_0 + \frac{\mathsf{E}_{\mathsf{P}}S_0^2}{2m_S}\right).$$

Proof. In case of M/G/1/∞ we obtain $\overline{L}(0) \overset{\mathfrak{D}}{=} L_0$ from Theorem 6.2.3. Thus, the first three assertions are immediate consequences of (6.3.12). For G/GI/m/∞ queues the actual waiting time W_0 is a measurable function of $([T_{n+1} - T_n, S_n],$ $n < 0)$, cf. Theorem 5.5.7. Thus, W_0 and S_0 are independent. Now the latter assertion of Corollary 6.3.3 follows from (6.3.13).

Results similar to those of Theorem 6.3.2 are valid in the case of finite waiting capacity:

6.3.4. Theorem. *For a queueing system* G/G/m/r *with a work-conserving queueing discipline and without service interrupts in steady state we have*

$$E_{\overline{\mathsf{P}}}\overline{Q}(0) = \lambda_{\overline{\mathsf{P}}}\,\mathsf{E}_{\mathsf{P}}W_0 \tag{6.3.14}$$

(the waiting time of a lost customer is equal to zero),

$$\mathsf{E}_{\overline{\mathsf{P}}}\overline{Q}(0) = \lambda_{\overline{\mathsf{P}}}\,\mathsf{P}(L_0 < m + r)\,\mathsf{E}_{\mathsf{P}}(W_0 \mid L_0 < m + r), \tag{6.3.15}$$

$$\mathsf{E}_{\overline{\mathsf{P}}}\overline{L}(0) = \lambda_{\overline{\mathsf{P}}}\,\mathsf{P}(L_0 < m + r)\,\big(\mathsf{E}_{\mathsf{P}}(W_0 \mid L_0 < m + r) + \mathsf{E}_{\mathsf{P}}(S_0 \mid L_0 < m + r)\big), \tag{6.3.16}$$

$$\mathsf{E}_{\overline{\mathsf{P}}}\min\big(m, \overline{L}(0)\big) = \lambda_{\overline{\mathsf{P}}}\,\mathsf{P}(L_0 < m + r)\,\mathsf{E}_{\mathsf{P}}(S_0 \mid L_0 < m + r), \tag{6.3.17}$$

$$\mathsf{E}_{\overline{\mathsf{P}}}\overline{V}(0) = \lambda_{\overline{\mathsf{P}}}\,\mathsf{P}(L_0 < m + r)\,\mathsf{E}_{\mathsf{P}}\left(W_0S_0 + \frac{1}{2}S_0^2 \mid L_0 < m + r\right). \tag{6.3.18}$$

Proof. (6.3.14) can be proved in a similar way to (6.3.10). To prove (6.3.15) to (6.3.18) it suffices to apply the same arguments as in the proof of Theorem 6.3.2 to the MPP that consists of all non-rejected customers. (It follows from Theorem 3.4.1 (formula (3.4.1)) and Theorem A 1.3.1 that the stationary MPP of non-rejected customers has the intensity $\lambda_{\overline{\mathsf{P}}}\,\mathsf{P}(L_0 < m + r)$ and the associated (according to Theorem 3.4.1) stationary sequence has the one-dimensional distribution

$$\mathsf{P}\left(\left[\sum_{i=0}^{\min\{j:j>0,L_j<m+r\}-1}(T_{i+1}-T_i),[S_0,W_0]\right]\in(\cdot)\mid L_0<m+r\right).) \quad \blacksquare$$

6.3.5. Corollary. *For a queueing system* G/G/m/r *with a work conserving queueing discipline and without service interrupts in steady state assume that* L_0 *and* S_0 *are independent. Then*

$$\mathsf{E}_{\bar{\mathsf{P}}}\min\left(m,\bar{L}(0)\right)=\lambda_{\mathsf{P}}\mathsf{P}(L_0<m+r)\,\mathsf{E}_{\mathsf{P}}S_0.$$

This is a consequence of (6.3.17).

6.3.6. Corollary. *For a queueing system* G/G/m/0 *in steady state assume that* L_0 *and* S_0 *are independent. (In the case of* G/GI/m/0 *this is true e.g. for the solution given by Theorem* 5.3.1.) *Then*

$$\mathsf{E}_{\bar{\mathsf{P}}}\bar{L}(0)=\lambda_{\bar{\mathsf{P}}}\mathsf{P}(L_0<m)\,\mathsf{E}_{\mathsf{P}}S_0,$$

$$\mathsf{E}_{\bar{\mathsf{P}}}\bar{V}(0)=\frac{1}{2}\lambda_{\bar{\mathsf{P}}}\mathsf{P}(L_0<m)\,\mathsf{E}_{\mathsf{P}}S_0^2.$$

(This follows from (6.3.16) and (6.3.18).)

Using Theorem 6.3.1, and statements (6.3.8) and (6.3.9), we can reformulate all the results of Theorem 6.3.2, Corollary 6.3.3, Theorem 6.3.4, Corollaries 6.3.5 and 6.3.6 in a sample path manner (for arrival- and time-stationary processes). Here we restrict ourselves to sample path versions of formula (6.3.10).

6.3.7. Corollary. *Consider a queueing system* G/G/m/∞ *with a work conserving queueing discipline and without service interrupts in steady state and assume that the time-stationary model (and thus, the arrival-stationary model) is ergodic. Then*

$$\lim_{t\to\infty}\frac{1}{t}\int_0^t\bar{Q}(u)\,du=\lim_{t\to\infty}\frac{1}{t}\max\{n:\bar{T}_n\leqq t\}\lim_{n\to\infty}\frac{1}{n}\sum_{k=1}^n\bar{W}_k\quad \bar{\mathsf{P}}\text{-a.s.},$$

$$\lim_{t\to\infty}\frac{1}{t}\int_0^t Q(u)\,du=\lim_{t\to\infty}\frac{1}{t}\max\{n:T_n\leqq t\}\lim_{n\to\infty}\frac{1}{n}\sum_{k=1}^n W_k\quad \mathsf{P}\text{-a.s.}$$

Now we consider a queueing system starting its work in the state x at time 0 and having a stationary ergodic input. Denote by $W_n(x)$ and $Q(t,x)$ the actual waiting times and the queue length process, respectively, if the input is arrival-stationary. For the associated time-stationary input we use the notations $\bar{W}_n(x)$ and $\bar{Q}(t,x)$. Assume

$$\lim_{n\to\infty}\frac{1}{n}\sum_{i=1}^n W_i(x)=\lim_{n\to\infty}\frac{1}{n}\sum_{i=1}^n W_i\quad \mathsf{P}\text{-a.s.},$$

$$\lim_{n\to\infty}\frac{1}{n}\sum_{i=1}^n\bar{W}_i(x)=\lim_{n\to\infty}\frac{1}{n}\sum_{i=1}^n\bar{W}_i\quad \mathsf{P}\text{-a.s.},$$

$$\lim_{t\to\infty} \frac{1}{t} \int_0^t Q(u, x)\, du = \lim_{t\to\infty} \frac{1}{t} \int_0^t Q(u)\, du \qquad \mathsf{P}\text{-a.s.},$$

$$\lim_{t\to\infty} \frac{1}{t} \int_0^t \bar{Q}(u, x)\, du = \lim_{t\to\infty} \frac{1}{t} \int_0^t \bar{Q}(u)\, du \qquad \bar{\mathsf{P}}\text{-a.s.}$$

Let the assumptions of Corollary 6.3.7 be fulfilled. Then we have the following sample path versions of (6.3.10) for non-stationary queues with a time- or arrival-stationary input:

$$\lim_{t\to\infty} \frac{1}{t} \int_0^t \bar{Q}(u, x)\, du = \lim_{n\to\infty} \frac{1}{n} \max\{n : \bar{T}_n \leqq t\} \lim_{n\to\infty} \frac{1}{n} \sum_{i=1}^{n} \bar{W}_i(x) \qquad \bar{\mathsf{P}}\text{-a.s.},$$

$$\lim_{t\to\infty} \frac{1}{t} \int_0^t Q(u, x)\, du = \lim_{n\to\infty} \frac{1}{n} \max\{n : T_n \leqq t\} \lim_{n\to\infty} \frac{1}{n} \sum_{i=1}^{n} W_i(x) \qquad \mathsf{P}\text{-a.s.}$$

Concerning the validity of the assumptions cf. Section 6.2.

6.4. The number of customers

In this section we investigate relationships between the distributions $(\bar{p}_j, j \geqq 0)$, $(p_j, j \geqq 0)$, and $(p_j^*, j \geqq 0)$ of the number of customers in a queueing system G/G/m/r at an arbitrary time instant, seen by an arbitrary arriving customer, and left behind by an arbitrary departing customer, respectively. First we have to introduce these probabilities. Consider a queueing system G/G/m/r with a work conserving queueing discipline without interrupts of service. Without loss of generality we assume the service times to be strictly positive, i.e. $\mathsf{P}(S_0 > 0) = 1$. Assume there is an arrival-stationary model from which the waiting times W_n and the quantities

$$C_n = \begin{cases} 1 & \text{if the } n\text{-th customer is served}, \\ 0 & \text{if the } n\text{-th customer is lost}, \end{cases}$$

$n \in \mathbb{Z}$, can be derived. (Existence theorems for some particular cases were given in Chapter 5.) The waiting time of a lost customer is equal to zero. Consider the stationary number

$$L_n = \sum_{m<n} \mathbf{1}\{C_m = 1, T_m + W_m + S_m > T_n\}$$

of customers in the system seen by the n-th arriving customer. We have $C_n = \mathbf{1}\{L_n < m + r\}$. Assume that the L_n, $n \in \mathbb{Z}$, are a.s. finite. The one-dimensional distribution

$$p_j = \mathsf{P}(L_0 = j), \qquad j = 0, 1, 2, \ldots,$$

is called the arrival-stationary distribution of the number of customers in the system.

Obviously, $([T_n, [S_n, W_n, C_n, L_n]])$ is a synchronous MPP, i.e. the sequence $([T_{n+1} - T_n, [S_n, W_n, C_n, L_n]])$ is stationary. The corresponding (according to Theorem 3.4.1) stationary MPP $(([\overline{T}_n, [\overline{S}_n, \overline{W}_n, \overline{C}_n, \overline{L}_n]]), \overline{\mathsf{P}})$ is a model for a queueing system in steady state, seen by an observer located at an arbitrary time instant, cf. Section 6.1. In particular, $\overline{L}_n$ is the number of customers in the system seen by the n-th arriving customer in the time-stationary model. The $\overline{C}_n$ have the same meaning as the C_n. We have

$$\overline{C}_n = 1\{\overline{L}_n < m + r\}$$

and

$$\overline{L}_n = \sum_{m<n} 1\{\overline{C}_m = 1, \overline{T}_m + \overline{W}_m + \overline{S}_m > \overline{T}_n\}.$$

Consider the number

$$\overline{L}(t) = \sum_{\overline{T}_n < t} 1\{\overline{C}_n = 1, \overline{T}_n + \overline{W}_n + \overline{S}_n > t\}$$

of customers who are in the system at time t and remain there at $t + 0$ (in the time-stationary model). The stochastic process $(\overline{L}(t))$ is stationary. The one-dimensional distribution

$$\overline{p}_j(t) = \overline{\mathsf{P}}(\overline{L}(0) = j), \qquad j = 0, 1, 2, \ldots,$$

is called the time-stationary distribution of the number of customers in the system.

Define

$$\overline{T}_n' = \begin{cases} \overline{T}_n + \overline{W}_n + \overline{S}_n & \text{if } \overline{L}_n < m + r \ \text{ (i.e. } \overline{C}_n = 1\text{)}, \\ \overline{T}_n & \text{otherwise} \end{cases}$$

and

$$\overline{L}_n' = \begin{cases} \overline{L}(\overline{T}_n') + \#\{j : \overline{T}_j' = \overline{T}_n', \overline{C}_j = 1, j > n\} & \text{if } \overline{C}_n = 1, \\ \overline{L}_n & \text{if } \overline{C}_n = 0. \end{cases}$$

The sequence $(\overline{T}_n')$ consists of all departure epochs (i.e. departures of served and lost customers) in the time-stationary model, ordered according to the arrivals of the corresponding customers. (The n-th leaving customer leaves $\overline{L}_n'$ customers in the system.) Now we renumber the pairs $[\overline{T}_n', \overline{L}_n']$ such that the $\overline{T}_n'$ are arranged in ascending order, and departures at the same time remain ordered according to the corresponding arrivals. With the convention that the first departure after time 0 gets the number 1, the renumbering rule is uniquely determined. In this way we obtain a stationary MPP $([\overline{T}_n^*, \overline{L}_n^*])$. By definition, the n-th departing customer leaves $\overline{L}_n^*$ customers in the system. Thus, the MPP $([\overline{T}_n^*, \overline{L}_n^*])$ defines the departure epochs and the numbers of customers remaining in the system immediately after the departures in the time-stationary model. This MPP has the intensity $\lambda_{\overline{\mathsf{P}}}$, which can easily be seen as follows. Denote the sojourn time of the n-th customer by $\bar{D}_n = \bar{T}_n' - \bar{T}_n$, $n \in \mathbb{Z}$. The MPP $([\bar{T}_n, \bar{D}_n])$ is stationary. Let ν be its intensity measure. Applying Campbell's Theorem 3.2.4 to the function $g(t, s) = 1\{0 < t + s \leq 1\}$, for the departure intensity λ^* we get

$$\lambda^* = \mathsf{E} \sum_n g(\bar{T}_n, \bar{D}_n) = \int_0^\infty \int_{-\infty}^\infty g(t, s)\, \nu(\mathrm{d}t \times \mathrm{d}s)$$

$$= \int_0^\infty \int_{-\infty}^\infty g(t, s)\, \mathrm{d}t\, \nu\big((0, 1] \times \mathrm{d}s\big) = \nu\big((0, 1] \times [0, \infty)\big) = \lambda_{\bar{\mathsf{P}}} . \qquad (6.4.1)$$

By Theorem 3.4.1 the stationary MPP $([\bar{T}_n^*, \bar{L}_n^*])$ uniquely (up to stochastic equivalence) determines a stationary sequence $([A_n^*, L_n^*])$ with the distribution P^* satisfying

$$\mathsf{P}^*\big(([A_n^*, L_n^*]) \in (\cdot)\big)$$

$$= \frac{\mathsf{E}_{\bar{\mathsf{P}}} \sum_{0 < \bar{T}_n^* \leq 1} 1\{([\bar{T}^*_{n+k+1} - \bar{T}^*_{n+k}, \bar{L}^*_{n+k}], k \in \mathbb{Z}) \in (\cdot)\}}{\mathsf{E}_{\bar{\mathsf{P}}} \# \{n : 0 < \bar{T}_n^* \leq 1\}} . \qquad (6.4.2)$$

According to the general interpretation given in Section 3.1, the sequence $([A_n^*, L_n^*])$ describes the steady state behaviour of the system at departures from the point of view of a "typical" departure (which is located at time zero); A_n^* is the distance between the n-th and the $(n + 1)$-th departure, and L_n^* is the number of customers in the system at (immediately after) the n-th departure. We call L_n^* the departure-stationary number of customers in the system, and the one-dimensional distribution

$$p_j^* = \mathsf{P}^*(L_0^* = j) = \frac{\mathsf{E} \# \{n : 0 < \bar{T}_n^* \leq 1, \bar{L}_n^* = j\}}{\lambda_{\bar{\mathsf{P}}}},$$

$$j = 0, 1, 2, \ldots,$$

is called the departure-stationary distribution of the number of customers in the system.

6.4.1. Remark. In an analogous way one can introduce

V_n^* – the departure-stationary workload,

W_n^* – the departure-stationary virtual waiting time (i.e. the waiting time of a fictitious customer whose arrival time coincides with the time of the n-th departure),

Q_n^* – the departure-stationary number of customers in the queue.

Now we are able to state the basic result that the arrival- and departure-stationary distribution of the number of customers in the system coincide.

6.4.2. Theorem. *For the queueing system* G/G/m/r

$$p_j = p_j^*, \qquad j = 0, 1, \ldots,$$

holds.

Proof. Consider the stationary MPP's $([\overline{T}_n, \overline{L}_n])$ and $([\overline{T}_n^*, \overline{L}_n^*])$. Define

$$\overline{N}_j(t) = \#\{n: 0 < \overline{T}_n \leqq t, \overline{L}_n = j\},$$
$$\overline{N}_j^* = \#\{n: 0 < \overline{T}_n^* \leqq t, \overline{L}_n^* = j\}, \qquad j = 0, 1, 2, \ldots$$

Obviously, $\overline{N}_j(t) = \overline{N}_j^*(t) = 0$ for $j > m + s$, and $\overline{N}_{m+r}(t) = \overline{N}^*_{m+r}(t)$. For $j < m + r$ the quantity $\overline{N}_j(t)$ $(\overline{N}_j^*(t))$ is the number of arriving (departing) customers in $(0, t]$ finding (leaving behind) j customers in the system. In view of the definition of p_j and p_j^*, cf. (3.4.2), (6.4.2), it suffices to show

$$\mathsf{E}\overline{N}_j(1) = \mathsf{E}\overline{N}_j^*(1), \qquad j = 0, 1, \ldots \tag{6.4.3}$$

It is easy to see that

$$|\overline{N}_j(t) - \overline{N}_j^*(t)| \leqq 1, \qquad j = 0, \ldots, m + r - 1, \qquad t > 0. \tag{6.4.4}$$

Since $([\overline{T}_n, \overline{L}_n])$ and $([\overline{T}_n^*, \overline{L}_n^*])$ are stationary MPP's, it holds that

$$\mathsf{E}\overline{N}_j(t) = t\mathsf{E}\overline{N}_j(1) \quad \text{and} \quad \mathsf{E}\overline{N}_j^*(t) = t\mathsf{E}\overline{N}_j^*(1),$$

$j = 0, \ldots, m + r - 1$, $t > 0$, and thus, in view of (6.4.4),

$$|\mathsf{E}\overline{N}_j(1) - \mathsf{E}\overline{N}_j^*(1)| = \frac{1}{t}\,|\mathsf{E}\overline{N}_j(t) - \mathsf{E}\overline{N}_j^*(t)| \leqq \frac{1}{t}.$$

Now (6.4.3) follows as $t \to \infty$. ∎

In some particular queueing systems, there is a simple relation between the time- and arrival-stationary distribution $(\overline{p}_j, j \geqq 0)$ and $(p_j, j \geqq 0)$.

6.4.3. Theorem. (i) *For a queueing system* G/GI/1/0 *assume that there is exactly one stationary state process. Then we have*

$$\overline{p}_1 = \varrho p_0 = \frac{m_S}{m_A}\,p_0.$$

(ii) *For the queueing system* G/M/m/r, $m \leqq \infty$, $r \leqq \infty$,

$$\min\{j, m\}\,\overline{p}_j = \varrho p_{j-1}, \qquad j = 1, \ldots, m + r,$$

holds.

Proof. (i) The construction of the stationary state process given in Section 5.3 shows that in the case of a G/GI/1/0 queue S_0 is independent of L_0. Now the first assertion is an immediate consequence of Corollary 6.3.5.

(ii) Consider a G/M/m/r queue. Remember,

$$p_j = \frac{\mathsf{E}\overline{N}_j(1)}{\lambda_{\overline{\mathsf{P}}}} = p_j^* = \frac{\mathsf{E}\overline{N}_j^*(1)}{\lambda_{\overline{\mathsf{P}}}}, \tag{6.4.5}$$

$$\overline{N}_j(t) = \#\{n: 0 < \overline{T}_n \leqq t, \overline{L}_n = j\},$$
$$\overline{N}_j^*(t) = \#\{n: 0 < \overline{T}_n^* \leqq t, \overline{L}_n^* = j\},$$

cf. (6.4.2) and Theorem 6.4.2. Applying Korolyuk's theorem (Theorem 3.5.7 (iii)) to the stationary point process consisting of all points $\overline{T}_n^*$ with

$\bar{L}_n{}^* = j$, we obtain

$$\bar{\mathsf{P}}(\bar{N}_j{}^*(t) = 1) = \bar{\mathsf{E}}\bar{N}_j{}^*(1)\, t + o(t). \tag{6.4.6}$$

Next we shall calculate $\bar{\mathsf{P}}(\bar{N}_j{}^*(t) = 1)$. It follows from the construction of the (unique) arrival-stationary state process (cf. Section 5.2 for G/M/∞, Section 5.5 for G/M/m/∞, and Section 5.7 for $m < \infty$, $r < \infty$) that, given L_0, the residual service times are independent and exponentially distributed. Thus, in view of (6.1.3), in the time-stationary model batch departures of served customers are excluded too, and

$$\{\bar{T}_n{}^*: n \in \mathbb{Z}, \bar{T}_n{}^* \text{ is a departure epoch of a served customer}\}$$
$$\cap \{\bar{T}_n : n \in \mathbb{Z}\} = \emptyset \quad \bar{\mathsf{P}}\text{-a.s.} \tag{6.4.7}$$

Let $i_1 = 1$, $i_n = \min\{k: \bar{T}_{i_{n-1}} < \bar{T}_k\}$ for $n \geqq 2$, $i_n = \min\{k: \bar{T}_k = \bar{T}_{j_n}\}$ for $n \leqq 0$, where $j_1 = 1$, $j_n = \max\{k: \bar{T}_k < \bar{T}_{j_{n+1}}\}$ for $n \leqq 0$, Then the i_n are the indices of the customers who arrive as the first customer of a batch. (A single arriving customer is to be considered as a batch of size one.) Notice that for the arrival instants $\bar{T}_{i_n}$ we have $\bar{T}_{i_n} < \bar{T}_{i_{n+1}}$. In view of (6.4.7)

$$\{\bar{T}_n{}^*: n \in \mathbb{Z}, \bar{T}_n{}^* \text{ is a departure epoch of a served customer}\}$$
$$\cap \{\bar{T}_{i_n} : n \in \mathbb{Z}\} = \emptyset \quad \bar{\mathsf{P}}\text{-a.s.}$$

Let $\hat{T}_n$ be the sequence of all points from

$$\{\bar{T}_n{}^*: n \in \mathbb{Z}, \bar{T}_n{}^* \text{ is a departure epoch of a served customer}\}$$
$$\cup \{\bar{T}_{i_n} : n \in \mathbb{Z}\},$$

renumbered according to $\cdots < \hat{T}_{-1} < \hat{T}_0 \leqq 0 < \hat{T}_1 < \cdots$. Define

$$\hat{L}_n = \begin{cases} \bar{L}_{i_k} & \text{if } \hat{T}_n = \bar{T}_{i_k}, \\ \bar{L}_k{}^* & \text{if } \hat{T}_n = \bar{T}_k{}^* \text{ and } \bar{T}_n{}^* \text{ is a departure epoch of a served customer.} \end{cases}$$

Then $([\hat{T}_n, \hat{L}_n])$ is a simple stationary MPP. Denote by

$$\hat{N}(t) = \#\{n: 0 < \hat{T}_n \leqq t\}$$

the number of points within the interval $(0, t]$. From Dobrushin's theorem (Theorem 3.5.7 (i)) we obtain

$$\bar{\mathsf{P}}(\hat{N}(t) > 1) = o(t). \tag{6.4.8}$$

Let $C(j, t)$, $t > 0$, be the following events:

$C(j, t) = \{\bar{L}(0) = j + 1$, and one of the $\min\{j + 1, m\}$ customers who is in service at time zero leaves the system within the interval $(0, t]\}$, $j = 0, \ldots, m + r - 1$.

Bearing in mind the construction of the state process, cf. Sections 5.2, 5.5, 5.7, and using formula (6.1.3), the reader can easily confirm that the residual service times at time zero are independent and exponentially distributed.

Moreover, $\bar{L}(0)$ and the residual service times are independent. Hence it follows that

$$\bar{\mathsf{P}}\big(C(j,t)\big) = \bar{p}_{j+1} \min\{j+1, m\}\, \mu t + o(t).$$

From this and (6.4.8) we get

$$\begin{aligned}\bar{\mathsf{P}}\big(\bar{N}_j^*(t) = 1\big) &= \bar{\mathsf{P}}\big(\bar{N}_j^*(t) = 1, \hat{N}(t) = 1\big) + o(t)\\ &= \bar{\mathsf{P}}\big(C(j,t), \hat{N}(t) = 1\big) + o(t)\\ &= \bar{\mathsf{P}}\big(C(j,t)\big) + o(t)\\ &= \bar{p}_{j+1} \min\{j+1, m\}\, \mu t + o(t), \; j = 0, \ldots, m+r-1.\end{aligned}$$

Now the second assertion of the Theorem 6.4.3 follows immediately from (6.4.5), (6.4.6):

$$\begin{aligned}p_j &= \frac{1}{\lambda_{\bar{\mathsf{P}}}} \lim_{t\to 0} \frac{\bar{\mathsf{P}}\big(\bar{N}_j^*(t) = 1\big)}{t} = \bar{p}_{j+1} \min\{j+1, m\} \frac{\mu}{\lambda_{\bar{\mathsf{P}}}}\\ &= \bar{p}_{j+1} \min\{j+1, m\}\, \varrho, \qquad j = 0, \ldots, m+r-1. \quad \blacksquare\end{aligned}$$

Now we shall prove some useful inequalities between $\bar{L}(0)$ and L_0. If Z_1 and Z_2 are two random variables with the distribution functions F_1 and F_2, respectively, and $\mathsf{F}_1(t) \geqq \mathsf{F}_2(t)$ for all t, then we say that Z_1 is stochastically smaller than Z_2, or $Z_1 \overset{\mathcal{D}}{\leqq} Z_2$. We say that a distribution function F with $\mathsf{F}(0-0) = 0$ is NBUE (NWUE) if

$$\big(1 - \mathsf{F}(t)\big) \geqq (\leqq) \frac{\int_t^\infty \big(1 - \mathsf{F}(u)\big)\, du}{\int_0^\infty \big(1 - \mathsf{F}(u)\big)\, du}, \quad \text{for all } t \text{ with } \mathsf{F}(t) < 1,$$

cf. Barlow and Proschan (1975), Chapter 6. (NBUE means "new is better than used in expectation".)

6.4.4. Theorem. *Consider a* GI/G/m/r *queue with a work conserving queueing discipline and assume*
(A 6.4.1) A_0 *and* L_0 *are independent.*
Then

$$\bar{L}(0) \overset{\mathcal{D}}{\leqq} \min\{L_0 + 1, m + r\}.$$

If, in addition, the distribution of the interarrival times is NBUE (NWUE) and the assumptions

(A 6.4.2) *there is a measurable function $h(t, x)$, non-increasing in the first argument, such that $\bar{L}(t) = h(t - \bar{T}_0, \bar{X}_0)$ for all t with $\bar{T}_0 < t \leqq 0$, where $\bar{X}_0$ is the state of the system seen by the customer with number 0 in the time-stationary model*

and

A 6.4.3) A_0 *and* X_0 *are independent*

are fulfilled, then we have

$$\bar{L}(0) \overset{\mathcal{D}}{\geqq} (\overset{\mathcal{D}}{\leqq})\, L_0 .$$

Proof. Obviously,

$$\begin{aligned}\bar{L}(0) &= \min\{\bar{L}_0 + 1, m + r\} - \text{(number of departures between } \bar{T}_0 \text{ and } 0) \\ &\overset{\mathcal{D}}{\leqq} \min\{\bar{L}_0 + 1, m + r\}.\end{aligned}$$

Under the assumption (A 6.4.1), Theorem 3.5.1 (i) yields $\bar{L}_0 = L_0$. Thus, the first assertion is true.

Now let the distribution of the interarrival times be NBUE. (The NWUE case is similar.) By means of (3.5.4) we obtain

$$-\bar{T}_0 \overset{\mathcal{D}}{\leqq} -T_{-1}. \tag{6.4.9}$$

From (A 6.4.3) and Theorem 3.5.1 (i) we conclude that

$$\bar{X}_0 \overset{\mathcal{D}}{=} X_0 \overset{\mathcal{D}}{=} X_{-1}. \tag{6.4.10}$$

From the independence of A_0 and X_0 we obtain by using (3.4.1) the independence of $\bar{X}_0$ and $-\bar{T}_0$.

In view of (A 6.4.2), (6.4.10), and (6.4.9)

$$\bar{L}(0) = h(-\bar{T}_0, \bar{X}_0) \overset{\mathcal{D}}{\geqq} h(-T_{-1}, X_{-1}) = L_0 .$$

(Note that in view of $\bar{\mathsf{P}}(\bar{T}_0 = 0) = 0$ and (6.4.9) we have $\mathsf{P}(T_{-1} = 0) = 0$.) ∎

6.4.5. Remark. For many queueing systems (e.g. FCFS GI/G/m/r queues) assumption (A 6.4.2) can be ensured by use of an appropriate (but complicated) state space. However, for some priority queues (A 6.4.2) fails.

Assumption (A 6.4.1) (assumption (A 6.4.3)) is fulfilled, e.g., if L_0 (X_0) is a measurable function of the past $[T_{-1}, S_{-1}], [T_{-2}, S_{-2}], \ldots$

6.5. The busy cycle

For a discussion of the generic busy cycle of a queueing system G/G/m/r with a work conserving queueing discipline in steady state, we start with the synchronous MPP $\left(([T_n, [S_n, W_n, C_n, L_n]]), \mathsf{P}\right)$ and its stationary equivalent $\left(([\bar{T}_n, [\bar{S}_n, \bar{W}_n, \bar{C}_n, \bar{L}_n]]), \bar{\mathsf{P}}\right)$ introduced in Section 6.4. Consider the arrival-stationary number

$$L(t) = \sum_{T_n < t} 1\{C_n = 1, T_n + W_n + S_n > t\}$$

of customers who are in the system at time t, and who will stay there at $t + 0$. Note that $\left[(L(t)), ([T_n, L_n])\right]$ is not a PEMP in sense of Definition 4.2.1; $L(t)$ is neither left nor right continuous.

In this section we assume

$$\mathsf{P}(S_0 > 0) = 1 \quad \text{and} \quad p_0 = \mathsf{P}(L_0 = 0) > 0\,.$$

(For G/G/1/∞ queues with $\varrho < 1$ we always have $p_0 > 0$, because there are infinitely many empty points, cf. (5.4.7). However, $p_0 > 0$ is not true in general for GI/GI/m/∞ queues with $\varrho < m$, cf. WHITT (1972).

Let $\left[\left(\tilde{L}(t)\right), \left(\left[\tilde{T}_n, [\tilde{S}_n, \tilde{L}_n]\right]\right)\right]$ be a pair consisting of a process $\left(\tilde{L}(t), t \in \mathbb{R}\right)$ and a sequence $\left(\left[\tilde{T}_n, [\tilde{S}_n, \tilde{L}_n]\right]\right)$ with the distribution $\tilde{\mathsf{P}}$ given by

$$\tilde{\mathsf{P}}\left(\left[\left(\tilde{L}(t)\right), \left(\left[\tilde{T}_n, [\tilde{S}_n, \tilde{L}_n]\right]\right)\right] \in (\cdot)\right)$$
$$= \mathsf{P}\left(\left[\left(L(t)\right), \left(\left[T_n, [S_n, L_n]\right]\right)\right] \in (\cdot) \mid L_0 = 0\right).$$

This object is an appropriate model of the queueing system in steady state viewed by an arbitrary customer who arrives when all servers are idle. Treating busy cycles, we shall use the following notation:

$\eta = \min\{t : t > 0, \tilde{L}(t) = 0\}$	— the length of the first busy period,
$\varkappa = \inf\{t : t \geqq \eta, \tilde{L}(t) > 0\} - \eta$	— the length of the first idle period,
$\eta + \varkappa$	— the length of the first busy cycle,
$\nu = \max\{n : n \geqq 0, \tilde{T}_n < \eta\} + 1$	— the number of customers arriving during the first busy period.

6.5.1. Theorem. *For a* G/G/m/r *queue with work conserving queueing discipline,* $m_A < \infty$, *and* $p_0 = \mathsf{P}(L_0 = 0) > 0$ *we have*

$$\mathsf{E}_{\tilde{\mathsf{P}}}\nu = (p_0)^{-1}, \tag{6.5.1}$$

$$\mathsf{E}_{\tilde{\mathsf{P}}}(\eta + \varkappa) = m_A \mathsf{E}_{\tilde{\mathsf{P}}}\nu = \lambda_{\bar{\mathsf{P}}}^{-1}\mathsf{E}_{\tilde{\mathsf{P}}}\nu, \tag{6.5.2}$$

$$\mathsf{E}_{\tilde{\mathsf{P}}}\varkappa = m_A \bar{p}_0 / p_0, \tag{6.5.3}$$

$$\bar{p}_0 = \frac{\mathsf{E}_{\tilde{\mathsf{P}}}\varkappa}{\mathsf{E}_{\tilde{\mathsf{P}}}(\eta + \varkappa)}. \tag{6.5.4}$$

For a G/G/1/∞ *we have, in addition,*

$$\mathsf{E}_{\tilde{\mathsf{P}}}\eta = m_S \mathsf{E}_{\tilde{\mathsf{P}}}\nu. \tag{6.5.5}$$

Proof. Observe that $\nu = \min\{n : n > 0, \tilde{L}_n = 0\}$. Now (6.5.1) follows from Theorem A 1.3.1. Since

$$\eta + \varkappa = \sum_{n=0}^{\nu-1} (\tilde{T}_{n+1} - \tilde{T}_n),$$

(6.5.2) is a consequence of the generalized Wald's identity (1.6.2a).

The process $\left(\tilde{L}(t - 0), t \in \mathbb{R}\right)$ is left continuous; $\tilde{L}(t)$ and $\tilde{L}(t - 0)$ differ only at departure epochs. Consider the subsequence $(\tilde{T}_{n_k})$ of (T_n) defined by the property $\tilde{L}_{n_k} = 0$, where we set $n_0 = 0$. That means, $\tilde{T}_{n_k}$ are just the beginnings of busy periods in the $\tilde{\mathsf{P}}$-model; in particular $\tilde{T}_{n_0} = T_0 = 0, \tilde{T}_{n_1} = \eta + \varkappa$.

The pair $\left[\left(\tilde{L}(t-0),\ t \in \mathbb{R}\right),\ ([\tilde{T}_{n_k}, \tilde{L}(\tilde{T}_{n_k}-0)],\ k \in \mathbb{Z})\right]$ is a synchronous PEMP. The associated stationary PEMP has the form $\left[\left(\overline{L}(t-0),\ t \in \mathbb{R}\right),\ ([\overline{T}_{n_k}, \overline{L}(\overline{T}_{n_k}-0)],\ k \in \mathbb{Z})\right]$, where $\left(\overline{L}(t)\right)$ is the time-stationary process of the number of customers in the system defined in Section 6.4, and $(\overline{T}_{n_k})$ is the subsequence of the sequence of arrival instants $(\overline{T}_n)$ in the time-stationary model defined by the property $\overline{L}_{n_k} = 0$. Applying the stochastic mean value theorem (4.2.10) to this PEMP and to the function $g(l) = \mathbf{1}\{l = 0\}$, we obtain

$$\overline{p}_0 = \overline{\mathsf{P}}\left(\overline{L}(0) = 0\right) = \overline{\mathsf{P}}\left(\overline{L}(0-0) = 0\right)$$
$$= \frac{1}{\mathsf{E}_{\tilde{\mathsf{P}}}(\tilde{T}_{n_1} - \tilde{T}_{n_0})} \mathsf{E}_{\tilde{\mathsf{P}}} \int_0^{\tilde{T}_{n_1}} \mathbf{1}\{\tilde{L}(t-0) = 0\}\, \mathrm{d}t = \frac{1}{\mathsf{E}_{\tilde{\mathsf{P}}}(\eta + \varkappa)} \mathsf{E}_{\tilde{\mathsf{P}}}\varkappa .$$

Thus, (6.5.4) is true. From (6.5.4), (6.5.2) and (6.5.1) we obtain (6.5.3).

Since in the queueing system G/G/1/∞

$$\eta = \sum_{n=0}^{\nu-1} \tilde{S}_n ,$$

(6.5.5) follows immediately from (1.6.2a). ∎

6.5.2. Corollary. *For* M/G/1/∞ *queues with work conserving queueing discipline we have*

$$\mathsf{E}_{\tilde{\mathsf{P}}}\eta = \frac{m_S}{1-\varrho} \quad \textit{and} \quad \tilde{\mathsf{P}}(\varkappa \leqq t) = 1 - \exp\left(-t/m_A\right).$$

Proof. By Corollary 6.3.3, $p_0 = \overline{p}_0 = 1 - \varrho$. Now the first assertion follows from (6.5.1), (6.5.5). From the fact that $\tilde{L}(t)$ is a measurable function of $(\tilde{S}_n)$ and $(\tilde{T}_n, \tilde{T}_n < t)$ we obtain $\tilde{\mathsf{P}}(\varkappa > x \mid \eta = u) = \mathsf{P}(T_{N(u)+1} - u > x \mid L_0 = 0,$ $\eta = u) = \mathrm{e}^{-x/m_A}$ for all u. Thus, $\varkappa \overset{\mathcal{D}}{=} A_0$. ∎

6.6. Single server queue: Takács and Pollaczek-Khinchin formulae

Consider a queueing system G/G/1/∞ with a work conserving queueing discipline in steady state. First we give some results concerning the time-stationary residual service time $\overline{S}(0)$ at the server. It follows from the construction of the stationary state process in Section 5.4 and from definition of the time-stationary model in Section 6.1 that the stationary process $\left(\overline{S}(t)\right)$ of the residual service time is well-defined and left continuous, and thus it can be put into the time-stationary model as an additional component. Moreover, $\overline{S}(0)$ is a function of $([\overline{T}_n, \overline{S}_n],\ n < 0)$.

6.6.1. Theorem. *For the residual service time* $\overline{S}(0)$ *of a queueing system* G/G/1/∞ *in steady state we have*

$$\overline{\mathsf{P}}\left(\overline{S}(0) \leqq y\right) = 1 - \varrho + \varrho \mathsf{S}_R(y) \qquad (6.6.1)$$

and

$$\bar{\mathsf{P}}\big(\bar{S}(0) \leqq y \mid \bar{S}(0) > 0\big) = \mathsf{S}_R(y), \tag{6.6.2}$$

where

$$\mathsf{S}_R(y) = \frac{1}{m_S} \int_0^y \big(1 - \mathsf{S}(u)\big)\,du.$$

Proof. We use the notation introduced in Section 6.5 in particular in the proof of Theorem 6.5.1. Consider the stationary PEMP $\big[\big(\bar{S}(t)\big), ([\bar{T}_{n_j}, \bar{S}_{n_j}])\big]$, where $\big(\bar{S}(t)\big)$ is the time-stationary process of the residual service time and $(\bar{T}_{n_j})$, $n_0 \leqq 0 < n_1$, is the sequence of empty points, and its synchronous version $\big(\big[\big(\tilde{S}(t)\big), ([\tilde{T}_{n_j}, \tilde{S}_{n_j}])\big], \tilde{\mathsf{P}}\big)$. Using the stochastic mean value theorem (4.2.10) and formula (6.5.2) in Theorem 6.5.1 we obtain

$$\begin{aligned}\bar{\mathsf{P}}\big(\bar{S}(0) > y\big) &= \frac{1}{\mathsf{E}_{\tilde{\mathsf{P}}}\tilde{T}_{n_1}}\,\mathsf{E}_{\tilde{\mathsf{P}}}\int_0^{\tilde{T}_{n_1}} 1\{\tilde{S}(t) > y\}\,dt\\ &= \frac{1}{\mathsf{E}_{\tilde{\mathsf{P}}}(\eta + \varkappa)}\,\mathsf{E}_{\tilde{\mathsf{P}}}\int_0^{\eta+\varkappa} 1\{\tilde{S}(t) > y\}\,dt\\ &= \frac{1}{m_A \mathsf{E}_{\tilde{\mathsf{P}}}\nu}\,\mathsf{E}_{\tilde{\mathsf{P}}}\sum_{n=0}^{\nu-1} (\tilde{S}_n - y)_+.\end{aligned}$$

Taking into account the generalized Wald's identity (1.6.2a) we can continue

$$\begin{aligned}\bar{\mathsf{P}}\big(\bar{S}(0) > y\big) &= \frac{\varrho}{m_S}\,\mathsf{E}_{\mathsf{P}}(S_0 - y)_+ = \frac{\varrho}{m_S}\int_0^\infty \mathsf{P}\big((S_0 - y)_+ > u\big)\,du\\ &= \frac{\varrho}{m_S}\int_y^\infty \mathsf{P}(S_0 > u)\,du = \varrho\big(1 - \mathsf{S}_R(y)\big).\end{aligned}$$

Formula (6.6.2) is an immediate consequence of (6.6.1). ∎

Now we want to prove the famous Takács' formula.

6.6.2. Theorem (Takács formula). *For the time- and customer-stationary workload distribution* $\bar{\mathsf{V}}(y) = \bar{\mathsf{P}}\big(\bar{V}(0) \leqq y\big)$ *and* $\mathsf{V}(y) = \mathsf{P}(V_0 \leqq y)$, *respectively, of a queueing system* G/GI/1/∞ *with a work conserving queueing discipline we have*

$$\bar{\mathsf{V}}(y) = 1 - \varrho + \varrho(\mathsf{V} * \mathsf{S}_R)(y), \qquad y \geqq 0,$$

where

$$\mathsf{S}_R(y) = \frac{1}{m_S}\int_0^y \big(1 - \mathsf{S}(u)\big)\,du.$$

Proof. Again we use the notation introduced in Section 6.5. Consider the stationary PEMP $\big[\big(\bar{V}(t)\big), ([\bar{T}_{n_j}, \bar{S}_{n_j}, \bar{V}_{n_j}])\big]$, where $(\bar{T}_{n_j})$ is the sequence of empty points, and its synchronous version $\left[\left(\tilde{V}(t)\right), \left([\tilde{T}_{n_j}, \tilde{S}_{n_j}, \tilde{V}_{n_j}]\right)\right]$. By means of the stochastic mean value theorem (4.2.10) we obtain

$$\overline{\mathsf{P}}\left(\overline{V}(0) \leqq y\right) = \frac{1}{\mathsf{E}_{\tilde{\mathsf{P}}}(\eta + \varkappa)} \mathsf{E}_{\tilde{\mathsf{P}}} \int_0^{\eta+\varkappa} 1\{\tilde{V}(t) \leqq y\} \, dt$$

$$= \frac{\mathsf{E}_{\tilde{\mathsf{P}}}\varkappa}{\mathsf{E}_{\tilde{\mathsf{P}}}(\eta + \varkappa)} + \frac{1}{\mathsf{E}_{\tilde{\mathsf{P}}}(\eta + \varkappa)} \mathsf{E}_{\tilde{\mathsf{P}}} \sum_{i=0}^{\nu-1} \min\{\tilde{S}_i, (y - \tilde{V}_i)_+\}$$

and in view of (6.5.2), (6.5.4), and $\overline{p}(0) = 1 - \overline{\mathsf{P}}\left(\overline{L}(0) > 0\right) = 1 - \varrho$

$$\overline{\mathsf{P}}\left(\overline{V}(0) \leqq y\right) = 1 - \varrho + \frac{\varrho}{m_S} \frac{1}{\mathsf{E}_{\tilde{\mathsf{P}}}\nu} \mathsf{E}_{\tilde{\mathsf{P}}} \sum_{i=0}^{\nu-1} \min\{\tilde{S}_i, (y - \tilde{V}_i)_+\} .$$

Formula (1.6.2a) yields

$$\overline{\mathsf{P}}\left(\overline{V}(0) \leqq y\right) = 1 - \varrho + \frac{\varrho}{m_S} \mathsf{E}_{\mathsf{P}} \min\{S_0, (y - V_0)_+\} . \tag{6.6.3}$$

From the construction of V_0 and from the independence of the service times we see that V_0 and S_0 are independent. Thus,

$$\overline{\mathsf{P}}\left(\overline{V}(0) \leqq y\right) = 1 - \varrho + \frac{\varrho}{m_S} \int_0^\infty \mathsf{P}(\min\{S_0, (y - V_0)_+\} > u) \, du$$

$$= 1 - \varrho + \frac{\varrho}{m_S} \int_0^\infty \left(1 - \mathsf{S}(u)\right) \mathsf{P}(V_0 < y - u) \, du$$

$$= 1 - \varrho + \frac{\varrho}{m_S} \int_0^y \left(1 - \mathsf{S}(u)\right) \mathsf{V}(y - u) \, du$$

$$= 1 - \varrho + \varrho(\mathsf{V} * \mathsf{S}_R)(y) . \quad \blacksquare$$

6.6.3. Remark. The proofs of Theorem 6.6.1 and Theorem 6.6.2 presented above are based, to a high degree, on the "classical" proofs for GI/GI/1/∞, which make use of the fact that the empty points are "regeneration points", cf. Cohen (1976a), Lemoine (1974). Now we show how the proof of formula (6.6.3), for example, can be altered by a more handy choice of the embedded points.

Consider the stationary PEMP $\left[\left(\overline{V}(t)\right), \left([\overline{T}_n, \overline{S}_n, \overline{V}_n]\right)\right]$ and its synchronous version $\left[\left(V(t)\right), \left([T_n, S_n, V_n]\right)\right]$. Then (4.2.10) leads to

$$\overline{\mathsf{P}}\left(\overline{V}(0) > y\right) = \lambda_{\overline{\mathsf{P}}} \mathsf{E}_{\mathsf{P}} \int_0^{T_1} 1\{V(t) > y\} \, dt$$

$$= \lambda_{\overline{\mathsf{P}}} \mathsf{E}_{\mathsf{P}}\left((V_0 + S_0 - y)_+ - (V_1 - y)_+\right)$$

$$= \frac{\varrho}{m_S} \mathsf{E}_{\mathsf{P}}\left((V_0 + S_0 - y)_+ - (V_0 - y)_+\right) \tag{6.6.4}$$

(note, $V_0 \stackrel{\mathcal{D}}{=} V_1$). Elementary calculations show that (6.6.4) is equivalent to (6.6.3). This proof did not use the generalized Wald's identity which has, in fact, nothing to do with Takács' formula.

Now we are in a position to give a simple proof of the famous Pollaczek-Khinchin formulae for M/GI/1/∞ queues. The Laplace-Stieltjes transforms of the distribution functions $\mathsf{V}, \overline{\mathsf{V}}, \mathsf{S}, \mathsf{S}_R, \ldots$ will be denoted by $\mathcal{V}(s)$, $\overline{\mathcal{V}}(s)$, $\mathcal{S}(s)$, $\mathcal{S}_R(s), \ldots$, respectively.

6.6.4. Theorem. *For* M/GI/1/∞ *queues with a work-conserving queueing discipline we have*

$$\mathcal{V}(s) = \overline{\mathcal{V}}(s) = \frac{1-\varrho}{1-\lambda_{\bar{\mathsf{P}}}\big(1-\mathcal{S}(s)\big)/s}, \qquad s > 0. \tag{6.6.5}$$

Moreover, for M/GI/1/∞ *queues with the queueing discipline FCFS it holds that*

$$\sum_{j=0}^{\infty} \overline{p}_j z^j = \sum_{j=0}^{\infty} p_j z^j = \sum_{j=0}^{\infty} p_j^* z^j = \frac{(1-\varrho)(1-z)}{1-z/\mathcal{S}\big(\lambda_{\bar{\mathsf{P}}}(1-z)\big)}. \tag{6.6.6}$$

Proof. From PASTA for the workload, cf. Theorem 6.2.3, we get $\overline{\mathsf{V}}(y) = \mathsf{V}(y)$. The Takács' formula, cf. Theorem 6.6.2, yields

$$\mathcal{V}(s) = \frac{1-\varrho}{1-\varrho\mathcal{S}_R(s)} = \frac{1-\varrho}{1-\dfrac{\varrho}{m_S s}\big(1-\mathcal{S}(s)\big)},$$

since

$$\mathcal{S}_R(s) = \frac{1}{m_S s}\big(1-\mathcal{S}(s)\big).$$

Thus, (6.6.5) is true.

In order to prove (6.6.6) we use some notation introduced in Section 6.4. Notice that the MPP $\big([\overline{T}_n, [\overline{D}_n, \overline{L}_n']]\big)$, where $\overline{D}_n = \overline{T}_n' - \overline{T}_n$, is stationary. In view of the FCFS queueing discipline, its synchronous version is just $\big([T_n, [D_n, L_n']]\big)$, $D_n = V_n + S_n$, $L_n' = \#\{j : j > n, T_j < T_n + D_n\}$. Using (6.4.1), (6.4.2) and applying Campbell's theorem to the function

$$g(t, d, l) = 1\{0 < t + d \leqq 1, l = j\},$$

we get

$$\begin{aligned} p_j^* &= \frac{1}{\lambda_{\bar{\mathsf{P}}}} \mathsf{E}_{\bar{\mathsf{P}}} \sum_{0<\overline{T}_n^*\leqq 1} 1\{\overline{L}_n^* = j\} = \frac{1}{\lambda_{\bar{\mathsf{P}}}} \mathsf{E}_{\bar{\mathsf{P}}} \sum_n g(\overline{T}_n, \overline{D}_n, \overline{L}_n') \\ &= \int_{\mathbb{R}\times\mathbb{R}_+\times\mathbb{Z}_+} g(t, d, l) \, \mathrm{d}t \mathsf{P}\big([D_0, L_0'] \in \mathrm{d}[d, l]\big) \\ &= \int_{\mathbb{R}_+\times\mathbb{Z}_+} 1\{l = j\} \, \mathsf{P}\big([D_0, L_0'] \in \mathrm{d}[d, l]\big) \\ &= \mathsf{P}(L_0' = j) = \mathsf{P}(\#\{n : n > 0, T_n < V_0 + S_0\} = j). \end{aligned} \tag{6.6.7}$$

It follows from the construction of V_0, cf. Section 5.4, and from the independence assumptions of M/GI/1/∞ that $(T_n, n > 0)$ is independent of $[V_0, S_0]$

and that V_0 is independent of S_0. Thus,

$$p_j^* = \int_0^\infty \frac{(\lambda_{\bar{\mathsf{P}}} t)^j}{j!} e^{-\lambda_{\bar{\mathsf{P}}} t} \, d(\mathsf{V} * \mathsf{S})(t). \tag{6.6.8}$$

This and (6.6.5) imply

$$\sum_{j=0}^\infty p_j^* z^j = \int_0^\infty e^{-\lambda_{\bar{\mathsf{P}}} t} \sum_{j=0}^\infty \frac{(\lambda_{\bar{\mathsf{P}}} t)^j}{j!} z^j \, d(\mathsf{V} * \mathsf{S})(t)$$

$$= \mathcal{V}\big(\lambda_{\bar{\mathsf{P}}}(1 - z)\big) \mathcal{S}\big(\lambda_{\bar{\mathsf{P}}}(1 - z)\big)$$

$$= \frac{(1 - \varrho)\, \mathcal{S}\big(\lambda_{\bar{\mathsf{P}}}(1 - z)\big)}{1 - \dfrac{\lambda_{\bar{\mathsf{P}}}}{\lambda_{\bar{\mathsf{P}}}(1 - z)} \Big(1 - \mathcal{S}\big(\lambda_{\bar{\mathsf{P}}}(1 - z)\big)\Big)}$$

$$= \frac{(1 - \varrho)(1 - z)}{1 - z/\mathcal{S}\big(\lambda_{\bar{\mathsf{P}}}(1 - z)\big)}.$$

The rest of the assertion follows from $p_j = p_j^*$, cf. Theorem 6.4.2, and PASTA, cf. Theorem 6.2.3. ∎

6.6.5. Remark. Consider an M/GI/1/∞ queue with a work conserving queueing discipline satisfying the following conditions:

(i) The service times are not interrupted.
(ii) When the server becomes idle, then the next customer is chosen from the queue independently of the potential service times of the waiting customers.

Then the process (L_n) of the number of customers has the same distribution as in case of FCFS. Thus, under the assumptions (i), (ii), the Pollaczek-Khinchin formula (6.6.6) is also true.

6.7. Queues with warming-up

The aim of this section is the derivation of a relationship for determining the waiting time distribution of the single server queue with warming-up. This can quite easily be done by using the approach given in Section 5.8. In the following we use the notation introduced there.

Consider a G/G/1/∞ queue with warm up times (i.e. a G/G/1/∞ queue with FCFS queueing discipline in which the first customer of each busy period (who meets an empty system) receives an additional service time, which is interpreted as the warm up time of the service facility). Assume that the input $\Phi = ([A_n, S_n, C_n])$ is stationary and ergodic; C_n is the potential warm up time of the n-th customer. The waiting times W_n of the n-th customer satisfy equation (5.8.1). If $\varrho = m_S/m_A < 1$ and $EC_0 < \infty$ then Theorem 5.8.1 yields a stationary sequence (W_n) of waiting times. The main idea was as follows:

We reduced the investigation of the queue with warming-up to that of the usual single server queue without warm up times (G/G/1/∞) and that of a suitable chosen single server loss system (G/G/1/0). The distribution of the waiting time process (W_n) is given by

$$\tilde{W}_n = \tilde{V}_n + \tilde{X}_k \quad \text{for} \quad \tilde{\nu}_k \leq n < \tilde{\nu}_{k+1},$$

cf. (5.8.2), (5.8.8), and

$$\mathsf{P}\big(([A_n, S_n, C_n, V_n, W_n]) \in (\cdot)\big)$$
$$= \frac{1}{\mathsf{E}\tilde{\nu}_1} \sum_{j=0}^{\infty} \tilde{\mathsf{P}}\big(\tilde{\nu}_1 > j, ([\tilde{A}_{n+j}, \tilde{S}_{n+j}, \tilde{C}_{n+j}, \tilde{V}_{n+j}, \tilde{W}_{n+j}]) \in (\cdot)\big), \tag{6.7.1}$$

where the V_n are the stationary waiting times in the associated G/G/1/∞ queue without warm up times and with input $([A_n, S_n])$, and $\tilde{X}_k = X_k + 1\{X_k = 0\}\, \tilde{\eta}_k$ is the work load immediately after the k-th arrival in the G/G/1/0 loss system with the input

$$\tilde{\Psi} = ([\xi_k, \tilde{\eta}_k]),$$

cf. (5.8.3)—(5.8.5), (5.8.8). Formula (6.7.1) implies

$$\mathsf{P}(W_0 \leq x) = \frac{1}{\mathsf{E}\tilde{\nu}_1} \sum_{j=0}^{\infty} \tilde{\mathsf{P}}(\tilde{\nu}_1 > j, \tilde{V}_j + \tilde{X}_0 \leq x), \quad x \geq 0. \tag{6.7.2}$$

We use the d.f.'s $\mathsf{S}(x) = \mathsf{P}(S_0 \leq x)$, $\mathsf{C}(x) = \mathsf{P}(C_0 \leq x)$, $\mathsf{V}(x) = \mathsf{P}(V_0 \leq x)$, $\mathsf{W}(x) = \mathsf{P}(W_0 \leq x)$, $\mathsf{G}(x) = \tilde{\mathsf{P}}(\tilde{X}_0 \leq x)$ and their Laplace-Stieltjes-transforms $\mathcal{S}(s)$, $\mathcal{C}(s)$, $\mathcal{V}(s)$, $\mathcal{W}(s)$, $\mathcal{G}(s)$.

Now the arguments leading to Theorem 5.8.1 enable us to prove

6.7.1. Theorem. *Consider a* GI/GI/1/∞ *queue with independent warm up times* (C_n), $\varrho = m_B/m_A < 1$, *and* $\mathsf{E}C_0 < \infty$ (*cf. Theorem* 5.8.3). *Then*

$$\mathsf{W} = \mathsf{V} * \mathsf{G}.$$

If in addition $A(x) = 1 - e^{-\mu x}$ (*i.e.* M/GI/1/∞ *with warming-up*), *then*

$$\mathcal{V}(s) = \frac{1 - \varrho}{1 - \dfrac{1 - \mathcal{S}(s)}{m_A s}} \tag{6.7.3}$$

and

$$\mathcal{G}(s) = \frac{m_A}{m_A + m_C} \mathcal{C}(s) + \frac{m_C}{m_A + m_C} \frac{1 - \mathcal{C}(s)}{m_C s}, \quad m_C = \mathsf{E}C_0. \tag{6.7.4}$$

Proof. Since in a GI/GI/1/∞ queue without warming-up the behaviour in different busy cycles is independent, the r.v.'s $\tilde{\nu}_1, \tilde{V}_j, j \geq 0$, depend only on $([\tilde{A}_n, \tilde{S}_n], n \geq 0)$, and $\tilde{X}_0$ depends only on $([\tilde{A}_n, \tilde{S}_n], n < 0)$. Thus (6.7.2) yields

$$\mathsf{W}(x) = \frac{1}{\mathsf{E}\tilde{\nu}_1} \sum_{j=0}^{\infty} \big(\tilde{\mathsf{P}}(\tilde{\nu}_1 > j, \tilde{V}_j \in (\cdot)\big) * \tilde{\mathsf{P}}\big(\tilde{X}_0 \in (\cdot)\big)(x) = (\mathsf{V} * \mathsf{G})(x).$$

Formula (6.7.3) is the well-known Pollaczek-Khintchin formula for M/GI/1/∞ queues, cf. Theorem 6.6.4.

Since $\mathsf{P}(A_0 > x) = \exp(-x/m_A)$, $x \geqq 0$, the idle periods of an M/GI/1/∞ queue without warm up times are exponentially distributed with mean m_A, and hence $\tilde{\mathsf{P}}(\xi_0 > x) = \exp(-x/m_A)$, $x \geqq 0$. Thus, $\tilde{\Psi} = ([\xi_k, \tilde{\eta}_k])$ is the input of an M/GI/1/0 system. Using Theorem 5.3.11 we obtain

$$\tilde{\mathsf{P}}(\tilde{X}_0 \leqq x) = \tilde{\mathsf{P}}(0 < X_0 \leqq x) + \tilde{\mathsf{P}}(X_0 = 0, \tilde{C}_0 \leqq x)$$

$$= \frac{m_C}{m_A + m_C} \frac{1}{m_C} \int_0^x \bigl(1 - \mathsf{C}(u)\bigr)\, du + \frac{m_A}{m_A + m_C} \mathsf{C}(x).$$

From this (6.7.4) follows immediately. ∎

6.8. Remarks and references

6.2. The proof of Theorem 6.2.2 is from [FKAS], where one can also find a different proof.

The most general version of PASTA is due to Wolff (1982). Let $(T_n, n \geqq 1)$ be a homogeneous Poisson process on $\mathbb{R}_+$ and $\bigl(Y(t), t \geqq 0\bigr)$ an arbitrary stochastic process defined on the same probability space. Assume that the sample paths of $\bigl(\mathbf{1}_B(Y(u)), u \geqq 0\bigr)$ are left continuous and have right-hand limits. Define

$$V(t) = t^{-1} \int_0^t \mathbf{1}_B\bigl(Y(u)\bigr)\, du,$$

$$Z(t) = \frac{\sum_{T_n < t} \mathbf{1}_B\bigl(Y(T_n)\bigr)}{\#\{n: n \geqq 1, T_n < t\}}.$$

With these definitions, $V(t)$ is the fraction of time during $[0, t]$ that $Y(\cdot)$ is in B, and $Z(t)$ is the fraction of arrivals on $[0, t]$ who find $Y(\cdot)$ in B. Wolff made the "lack of anticipation assumption" (LAA): For each $t \geqq 0$, $(\#\{n: t \leqq T_n < t + u, u \geqq 0)$ and $\bigl(\mathbf{1}_B(Y(v)), 0 \leqq v \leqq t\bigr)$ are independent.

Under (LAA) it holds that

$$V(t) \xrightarrow[t\to\infty]{} V(\infty) \text{ a.s. if and only if } Z(t) \xrightarrow[t\to\infty]{} V(\infty).$$

The same is true for non-stationary Poisson arrivals. Wolff's PASTA theorem shows that PASTA has nothing to do with stationarity or ergodicity. The only reason for making stronger assumptions such as stationarity or ergodicity is to prove the convergence of $V(t)$ or $Z(t)$ as $t \to \infty$.

There are examples, where non-Poisson arrivals see time averages. For the M/M/1/∞ queue with Bernoulli feedback, Burke (1976) has shown that the composite stream of exogenous Poisson arrivals and feedback customers is not Poisson, even though this stream sees time averages. For remarks on the development of this topic see [FKAS] and Wolff (1982). Recently, Melamed and Whitt (1990) investigated when arrivals see time averages (ASTA), i.e. when the stationary distribution of the embedded sequence coincides with the time stationary distribution of the process. They showed that a certain lack of bias assumption (LBA) (without the Poisson-assumption) is necessary and sufficient for ASTA in a stationary framework. LBA covers known examples of non-Poisson ASTA as well as the familiar PASTA. Concerning a martingale approach to this topic cf. MELAMED and WHITT (1989). Closely related work has been done by BRÉMAUD (1989, 1990), KÖNIG, SCHMIDT and VAN DOORN (1989), SERFOZO (1989a, b), STIDHAM and EL TAHA (1989) and VAN DOORN and REGTERSCHOT (1988).

6.3. A survey of the history of the Little type formulae is RAMALHOTO, AMARAL, and COCHITO (1983). The most important original references are LITTLE (1961), JEWELL (1967), ROSS (1970), BRUMELLE (1971b, 1972), STIDHAM (1972, 1974), FRANKEN (1976), HEYMAN and STIDHAM (1980), MIYAZAWA (1977, 1979). Theorem 6.3.1 in the present form is due to FRANKEN (1982); the idea goes back to FRANKEN (1976), where one can find Theorem 6.3.2. Formula (6.3.12) was stated by KRAKOWSKI (1973) by means of heuristic arguments; see also HARRIS (1974). Formula (6.3.13) is, in a somewhat different form, due to BRUMELLE (1971b).

In Section 6.3 we restricted ourselves to the most fundamental formulae of Little type. More complicated balance equations can be found, e.g. in [FKAS], MIYAZAWA (1985).

6.4. The proofs of Theorems 6.4.2 and 6.4.3(ii) presented here are from BRANDT (1988a); the ideas go back to TAKÁCS (1962) and FRANKEN (1975, 1976); see also FINCH (1959a, b), BOROVKOV (1972a), SHANBHAG and TAMBOURATZIS (1973), MIYAZAWA (1976), KÖNIG and SCHMIDT (1978, 1980a). For more general results on $G/G/m/r$ queues without batch arrivals we refer to [FKAS], sect. 4.3 and MIYAZAWA (1985).

For $r = \infty$ and i.i.d. service times the assertions of Theorem 6.4.4 were proved in MIYAZAWA (1976), see also KÖNIG, SCHMIDT, and STOYAN (1976). In KÖNIG and SCHMIDT (1980a) an analogous result for queues with state-dependent service rates was derived.

6.5. First results of the type stated in this section were obtained by RICE (1962) and MARSHALL (1968), see also IV'NIČKIJ (1971) and STOYAN (1977). There one can find proofs for $GI/GI/1/\infty$, cf. [FKAS] for further discussions.

6.6. This section is based on FRANKEN (1982) and [FKAS]. In [FKAS] the history of this material is sketched. Some fundamental papers are TAKÁCS (1963b), LEMOINE (1974), COHEN (1976b), KÖNIG et al. (1978).

6.7. This section continues the analysis of queues with warm up, started in Section 5. Our presentation is due to WIRTH (1984b). The fact that the distribution function W of the stationary waiting time in a $GI/GI/1/\infty$ queue with warm up times appears as the convolution of the d.f. V of the stationary waiting time in the associated usual $GI/GI/1/\infty$ queue and a certain d.f. G was shown in SIEGEL (1974) by means of analytic methods, cf. Theorem 6.7.1. This representation was used in DEWESS (1975) for discussing analytical properties of W. Formula (6.7.4) was obtained by FINCH (1959a).

7. Batch-arrival-stationary queueing processes

7.1. Introduction and preliminary results

Let $\Phi = ([A_n, Y_n])$ be a stationary sequence describing the input of a queueing system; $A_n \geqq 0$ is the interarrival time between the n-th and the $(n+1)$-th customer, and Y_n is a mark comprising the customer's characteristics which we want to keep track of (e.g. service time). As already discussed, Φ can be interpreted as a stationary input "seen by an arbitrary customer". (Notice that this interpretation is justified if Φ is ergodic.) For this reason, Φ is usually called an *arrival-stationary input*. Since we do not assume that A_n is strictly positive for every n, batch arrivals are allowed, cf. also Section 5.1. Since the marks (e.g. service times) are indexed, it follows that the customers within a batch are numbered, too.

Throughout this chapter we assume $\mathsf{P}(\# \{n: A_n > 0\} = \infty) = 1$. Then the conditional distribution $\mathsf{P}\big(\Phi \in (\cdot) \mid A_{-1} > 0\big)$ is well-defined. Let $([\tilde{A}_n, \tilde{Y}_n])$ be a sequence with this probability distribution. We may interpret $([\tilde{A}_n, \tilde{Y}_n])$ as the input of the queue under consideration viewed by the first customer of an arbitrary arriving batch. Now consider a stationary weak solution $\big([[A_n, Y_n], X_n]\big)$ of the equation

$$X_{n+1} = f(X_n, [A_n, Y_n]), \qquad n \in \mathbb{Z}. \tag{7.1.1}$$

Here the $X_n, n \in \mathbb{Z}$, are interpreted as the consecutive states of the queueing system seen by the arriving customers. (As in Chapter 6, we omit the notation $\big([[A_n', Y_n'], X_n']\big)$ for a weak solution here. The distribution of $\big([[A_n, Y_n], X_n]\big)$ will also be denoted by P. In order to avoid confusion, the reader may imagine $\big([[A_n, Y_n], X_n]\big)$ as a strong solution of (7.1.1).) Consider a sequence $\big([[\tilde{A}_n, \tilde{Y}_n], \tilde{X}_n]\big)$ distributed according to the conditional probability law

$$\tilde{\mathsf{P}}(\cdot) = \mathsf{P}\Big(\big([[A_n, Y_n], X_n]\big) \in (\cdot) \mid A_{-1} > 0\Big). \tag{7.1.2}$$

This sequence describes the time evolution of the system under consideration at the arrivals of customers, seen by the first customer of an arbitrary batch. Thus, for treating phenomena related to batches, it suffices to consider a stationary solution of (7.1.1) and to derive (7.1.2) from it.

However, if we study batch arrival queues, another model of the "input in steady state" seems to be natural. Consider a stationary sequence $\hat{\Phi} = ([\hat{A}_n, \hat{G}_n, \hat{\underline{Y}}_n])$, where $\hat{A}_n$ denotes the time between the arrivals of the n-th and $(n+1)$-th batch, $\hat{G}_n$ the size of the n-th batch, and $\hat{\underline{Y}}_n = [\hat{Y}_{n1},$

$\ldots, \hat{Y}_{n\hat{G}_n}]$ the vector of marks associated with the $\hat{G}_n$ customers of the n-th batch. Analogously to the notion "arrival-stationary input" for Φ, we call $\hat{\Phi}$ a *batch-arrival-stationary input*. Both models Φ and $\hat{\Phi}$ can be used for treating batch arrival queues. First we shall discuss relations between Φ and $\hat{\Phi}$. In order to do this, we transform $\hat{\Phi}$ into a sequence $\tilde{\Phi} = ([\tilde{A}_n, \tilde{Y}_n]) = \tilde{\varphi}(\hat{\Phi})$ by introducing artificial interarrival times between customers within a batch and by numbering them and their marks individually. For a sample path $\hat{\varphi} = ([a_k, \gamma_k, \underline{y}_k])$, $\underline{y}_k = [y_{k1}, \ldots, y_{k\gamma_k}]$, of $\hat{\Phi}$ we define

$$\tilde{a}_n(\hat{\varphi}) = \begin{cases} a_k & \text{if } n = \tilde{\nu}_k(\hat{\varphi}) - 1, \\ 0 & \text{if } n \neq \tilde{\nu}_k(\hat{\varphi}) - 1 \text{ for all } k, \end{cases}$$

$$\tilde{y}_n(\hat{\varphi}) = y_{kj} \text{ if } \tilde{\nu}_{k-1}(\hat{\varphi}) \leqq n < \tilde{\nu}_k(\hat{\varphi}), \quad j = n - \tilde{\nu}_{k-1}(\hat{\varphi}) + 1,$$

$$\tilde{\varphi}(\hat{\varphi}) = \big([\tilde{a}_n)\hat{\varphi}), \tilde{y}_n(\hat{\varphi})]\big),$$

where

$$\tilde{\nu}_{-1}(\hat{\varphi}) = 0, \quad \tilde{\nu}_k(\hat{\varphi}) = \sum_{j=0}^{k} \gamma_j \text{ for } k \geqq 0, \quad \tilde{\nu}_k(\hat{\varphi}) = -\sum_{j=-1}^{k+1} \gamma_j \text{ for } k \leqq -2.$$

The sequences $\hat{\Phi}$ and $\tilde{\Phi}$ are equivalent in the sense that there exists a one-to-one mapping between their sample paths. The inverse mapping of $\tilde{\varphi}(\hat{\varphi})$ may be expressed as follows: For $\tilde{\varphi} = ([\tilde{a}_i, \tilde{y}_i])$ with $\tilde{a}_{-1} > 0$ let $(\tilde{a}_{i_n})$, $i_n = i_n(\tilde{\varphi})$, be the subsequence of all strictly positive elements of $(\tilde{a}_i)$, where we claim $i_{-1} = i_{-1}(\tilde{\varphi}) = -1$. Define

$$\hat{a}_n(\tilde{\varphi}) = \tilde{a}_{i_n}, \qquad \hat{\gamma}_n(\tilde{\varphi}) = i_n(\tilde{\varphi}) - i_{n-1}(\tilde{\varphi}),$$

$$\hat{\underline{y}}_n(\tilde{\varphi}) = [\tilde{y}_{i_{n-1}(\tilde{\varphi})+1}, \ldots, \tilde{y}_{i_n(\tilde{\varphi})}], \quad \hat{\varphi}(\tilde{\varphi}) = \big([\hat{a}_n(\tilde{\varphi}), \hat{\gamma}_n(\tilde{\varphi}), \hat{\underline{y}}_n(\tilde{\varphi})]\big).$$

Then we have $\hat{\Phi} = \hat{\varphi}(\tilde{\Phi}) = \hat{\varphi}\big(\tilde{\varphi}(\hat{\Phi})\big)$, and in particular $\hat{\gamma}_0(\tilde{\Phi}) = \hat{G}_0$. Hence $\tilde{\Phi}$ describes the batch-arrival-stationary input $\hat{\Phi}$ as well, and for this reason we call $\tilde{\Phi}$ also "*batch-arrival-stationary input*". The stationarity of $\hat{\Phi}$ implies the following invariance property of $\tilde{\Phi} = ([\tilde{A}_n, \tilde{Y}_n])$:

$$([\tilde{A}_{n+\hat{\gamma}_0(\tilde{\Phi})}, \tilde{Y}_{n+\hat{\gamma}_0(\tilde{\Phi})}]) \overset{\mathcal{D}}{=} ([\tilde{A}_n, \tilde{Y}_n]). \tag{7.1.3}$$

By means of Theorem A 1.3.2 we can prove that there is a one-to-one correspondence between arrival-stationary input distributions and batch-arrival-stationary input distributions.

7.1.1. Theorem. *There exists a one-to-one mapping between the families*

$$\{(\Phi, \mathsf{P}): \Phi = ([A_n, Y_n]) \text{ is stationary and satisfies } \mathsf{P}(\#\{n: A_n > 0\} = \infty) = 1\}$$

and

$$\{(\hat{\Phi}, \hat{\mathsf{P}}): \hat{\Phi} = ([\hat{A}_n, \hat{G}_n, \hat{\underline{Y}}_n]) \text{ is stationary and satisfies } \hat{\mathsf{P}}(\hat{A}_0 > 0) = 1, \mathsf{E}\hat{G}_0 < \infty\}.$$

For a given arrival-stationary input (Φ, P) *the corresponding batch-arrival-stationary input distribution* $\hat{\mathsf{P}}$ *is given by*

$$\hat{\mathsf{P}}(\cdot) = \mathsf{P}\big(\hat{\varphi}(\Phi) \in (\cdot) \mid A_{-1} > 0\big). \tag{7.1.4}$$

For a given batch-arrival-stationary input $(\tilde{\Phi}, \tilde{\mathsf{P}})$ *the corresponding arrival-stationary input distribution* P *is determined by*

$$\mathsf{P}(\Phi \in (\cdot)) = \frac{1}{\mathsf{E}_{\tilde{\mathsf{P}}}\tilde{\nu}_0(\tilde{\Phi})} \sum_{j=0}^{\infty} \tilde{\mathsf{P}}\big(\tilde{\nu}_0(\tilde{\Phi}) > j, ([\tilde{A}_{n+j}, \tilde{Y}_{n+j}]) \in (\cdot)\big) \tag{7.1.5}$$

$$= \frac{1}{\mathsf{E}_{\hat{\mathsf{P}}}\hat{G}_0} \sum_{j=0}^{\infty} \hat{\mathsf{P}}\Big(\hat{G}_0 > j, \big([\tilde{a}_{n+j}(\hat{\Phi}), \tilde{y}_{n+j}(\hat{\Phi})]\big) \in (\cdot)\Big). \tag{7.1.6}$$

In particular,

$$\mathsf{P}(A_0 > 0) = \frac{1}{\mathsf{E}_{\tilde{\mathsf{P}}}\tilde{\nu}_0(\tilde{\Phi})} = \frac{1}{\mathsf{E}_{\hat{\mathsf{P}}}\hat{G}_0}. \tag{7.1.7}$$

Proof. It suffices to show the existence of a one-to-one mapping between P and $\tilde{\mathsf{P}}$, and the validity of formula (7.1.5).

Define $D_n = 1\{A_{n-1} > 0\}$, $\tilde{D}_n = 1\{\tilde{A}_{n-1} > 0\}$ where $\big(([A_n, Y_n]), \mathsf{P}\big)$ and $\big(([\tilde{A}_n, \tilde{Y}_n]), \tilde{\mathsf{P}}\big)$ are an arrival- and batch-arrival-stationary input, respectively. The distributions Q and $\tilde{\mathsf{Q}}$ of $([A_n, Y_n, D_n])$ and $([\tilde{A}_n, \tilde{Y}_n, \tilde{D}_n])$ are completely determined by P and by $\tilde{\mathsf{P}}$, respectively. The application of Theorem A 1.3.2 to Q, $\tilde{\mathsf{Q}}$ and $L = \mathbb{R}_+ \times \mathbb{Y} \times \{1\}$ immediately supplies the desired result. ∎

In addition to Theorem 7.1.1 we have the following ergodic statement, which follows immediately from Theorem A 1.3.4.

7.1.2. Theorem. *Let* (Φ, P), $\Phi = ([A_n, Y_n])$, *be an arrival-stationary, ergodic input. Then for every non-negative, measurable function h*

$$\frac{1}{n} \sum_{j=0}^{n-1} h([\tilde{A}_j, \tilde{Y}_j]) \xrightarrow[n\to\infty]{} \mathsf{E}f([A_0, Y_0]) \qquad \tilde{\mathsf{P}}\text{-a.s}$$

for, equivalently,

$$\frac{1}{n} \sum_{j=0}^{n-1} h\big([\tilde{a}_j(\hat{\Phi}), \tilde{y}_j(\hat{\Phi})]\big) \xrightarrow[n\to\infty]{} \mathsf{E}h([A_0, Y_0]) \qquad \hat{\mathsf{P}}\text{-a.s.}$$

Conversely,

$$\frac{1}{n} \sum_{j=0}^{n-1} h([A_{m+\nu_j}, Y_{m+\nu_j}]) \xrightarrow[n\to\infty]{} \mathsf{E}h([\tilde{A}_m, \tilde{Y}_m]), \quad m \in \mathbb{Z}, \qquad \mathsf{P}\text{-a.s.},$$

where $0 < \nu_0 < \nu_1 < \cdots$ *are the indices of the first customers (with positive index) of each batch in the arrival-stationary model, i.e.*

$$\nu_0 = \min\{n : n > 0, A_{n-1} > 0\},$$

$$\nu_j = \min\{n : n > \nu_{j-1}, A_{n-1} > 0\}, \qquad j \geq 1.$$

Theorem 7.1.2 shows that P arises as a limiting distribution in the model $(\hat{\Phi}, \hat{\mathsf{P}})$ (or $(\tilde{\Phi}, \tilde{\mathsf{P}})$) and vice versa. Intuitively, the relationship between an arrival-stationary ergodic input (Φ, P) and the associated batch-arrival-stationary input $(\tilde{\Phi}, \tilde{\mathsf{P}})$ can be described as follows. Consider a source generating arrival instants of customers and their associated marks (e.g. service times), and assume that the source is in steady state. Then (Φ, P) describes the output of this source viewed by an arbitrarily chosen customer. In other words, the origin of the time axis coincides with the arrival instant of an arbitrarily chosen customer. The sequence $(\tilde{\Phi}, \tilde{\mathsf{P}})$ describes the output of the same source, but viewed by an arbitrarily chosen batch — more precisely, by the first customer of an arbitrarily chosen batch. In other words, the origin of the time axis coincides with the arrival instant of an arbitrarily chosen batch.

Now we express some characteristics of the customer with the number 0 within the arrival-stationary input (Φ, P) by means of $\hat{\mathsf{P}}$-probabilities. Let

$$\nu_{-1} = \max\{n : n \leqq 0, A_{n-1} > 0\}, \quad \nu_0 = \min\{n : n > 0, A_{n-1} > 0\},$$

$N_0 = 1 - \nu_{-1}$ — the position of the customer with number 0 in his own batch (in the arrival-stationary input), (7.1.8)

$G_0 = \nu_0 - \nu_{-1}$ — the size of the batch that contains the customer with number 0 (in the arrival-stationary input). (7.1.9)

7.1.3. Lemma. *The arrival-stationary distribution of the pair* $[N_0, G_0]$ *is determined by the corresponding batch-arrival-stationary distribution* $\hat{g}_n = \hat{\mathsf{P}}(\hat{G}_0 = n)$, $n \geqq 1$, *of the batch size* $\hat{G}_0$:

$$\mathsf{P}(G_0 = i, N_0 = j) = \frac{1}{\mathsf{E}\hat{G}_0} \hat{g}_i \mathbf{1}\{j \leqq i\}, \qquad i, j \geqq 1. \tag{7.1.10}$$

In particular,

$$r_n = \mathsf{P}(N_0 = n) = \frac{1}{\mathsf{E}\hat{G}_0} \sum_{k=n}^{\infty} \hat{g}_k, \qquad n \geqq 1, \tag{7.1.11}$$

$$g_n = \mathsf{P}(G_0 = n) = \frac{n}{\mathsf{E}\hat{G}_0} \hat{g}_n, \qquad n \geqq 1. \tag{7.1.12}$$

Proof. Using (7.1.6) we obtain (7.1.10) as follows:

$$\begin{aligned} \mathsf{P}(G_0 = i, N_0 = j) &= \mathsf{P}(\nu_0 - \nu_{-1} = i, 1 - \nu_{-1} = j) \\ &= \frac{1}{\mathsf{E}\hat{G}_0} \sum_{k=0}^{\infty} \hat{\mathsf{P}}(\hat{G}_0 = i > k, 1 + k = j) \\ &= \frac{1}{\mathsf{E}\hat{G}_0} \hat{g}_i \mathbf{1}\{j \leqq i\}. \quad \blacksquare \end{aligned}$$

7.1.4. Remark. The probability generating function (PGF) of the distribution $(r_n, n \geqq 1)$ has the form

$$R(s) = \sum_{n=1}^{\infty} r_n s^n = \frac{1}{\mathsf{E}\hat{G}_0} \sum_{n=1}^{\infty} \sum_{k=n}^{\infty} \hat{g}_k s^n = \frac{1}{\mathsf{E}\hat{G}_0} \frac{s(1 - \hat{\Gamma}(s))}{(1 - s)}, \tag{7.1.13}$$

where $\hat{\Gamma}(s) = \sum_{n=1}^{\infty} \hat{g}_n s^n$ is the PGF of the batch size distribution.

Consider the batch-arrival-stationary input $(\hat{\Phi}, \hat{\mathsf{P}})$, $\hat{\Phi} = ([\hat{A}_n, \hat{G}_n, \hat{\underline{S}}_n])$, where $\hat{\underline{S}}_n = [\hat{S}_{n1}, \ldots, \hat{S}_{n\hat{G}_n}]$ is the vector of service times of the $\hat{G}_n$ customers in the n-th batch. Later we shall need some of the following independence assumptions on the input $(\hat{\Phi}, \hat{\mathsf{P}})$:

(A 7.1.1) $(\hat{A}_n)$ is independent of $([\hat{G}_n, \hat{\underline{S}}_n])$.
(A 7.1.2) $(\hat{A}_n)$ is a sequence of i.i.d. r.v.'s.
(A 7.1.3) $([\hat{G}_n, \hat{\underline{S}}_n])$ is a sequence of i.i.d. r.v.'s.
(A 7.1.4) Let $(S^*_{nj},\ n \in \mathbb{Z},\ j = 1, 2, \ldots)$ be a field of i.i.d. r.v.'s, independent of $(\hat{A}_n)$ and $(\hat{G}_n)$. It holds that $[S^*_{n1}, \ldots, S^*_{n\hat{G}_n}] \overset{\mathcal{D}}{=} \hat{\underline{S}}_n,\ n \in \mathbb{Z}$.

Extending the notation introduced in Section 5.1, we use the notation $\mathrm{G}^{\mathrm{G}}/\mathrm{G}/m/r$ for queueing systems with m servers in parallel ($m \leqq \infty$), r waiting places ($r \leqq \infty$) and a batch-arrival-stationary input $\hat{\Phi} = ([\hat{A}_n, \hat{G}_n, \hat{\underline{S}}_n])$. $\mathrm{G}^{\mathrm{G}}/\mathrm{G}/m/r$ queues with (A 7.1.1), (A 7.1.3), and (A 7.1.4) are denoted by $\mathrm{G}^{\mathrm{GI}}/\mathrm{GI}/m/r$. In case of (A 7.1.1)–(A 7.1.4) we write $\mathrm{GI}^{\mathrm{GI}}/\mathrm{GI}/m/r$. The other conventions (e.g. the use of the symbol M) are similar to those given in 5.1. Furthermore we shall use the notation:

$m_A = \mathsf{E}A_0$	–	mean interarrival time between customers within the arrival-stationary input,
$m_S = \mathsf{E}S_0$	–	mean service time of customers within the arrival-stationary input,
$\varrho = m_S/m_A$	–	traffic intensity of the arrival-stationary input,
$m_{\hat{A}} = \mathsf{E}\hat{A}_0$	–	mean interarrival time between batches within the batch-arrival-stationary input,
$\hat{\mathcal{S}}_n = \sum_{j=1}^{\hat{G}_n} \hat{S}_{nj}$	–	total service time of all customers of the n-th batch (in the batch-arrival-stationary input),
$m_{\hat{\mathcal{S}}} = \mathsf{E}\hat{\mathcal{S}}_0$	–	mean total service time of all customers of a batch (in the batch-arrival-stationary input),
$m_{\hat{G}} = \mathsf{E}\hat{G}_0$	–	mean batch size,
$\hat{\varrho} = m_{\hat{\mathcal{S}}}/m_{\hat{A}}$	–	traffic intensity of the batch-arrival-stationary input.

By means of (A 1.13) we obtain

$$\mathsf{E}\hat{A}_0 = \mathsf{E}\hat{G}_0 \mathsf{E}A_0, \qquad \mathsf{E}\sum_{j=1}^{\hat{G}_0} \hat{S}_{0j} = \mathsf{E}\hat{G}_0 \mathsf{E}S_0, \tag{7.1.14}$$

and thus

$$\varrho = \frac{m_S}{m_A} = \frac{\mathsf{E}\sum_{j=1}^{\hat{G}_0} \hat{S}_{0j}}{\mathsf{E}\hat{A}_0} = \frac{m_{\hat{\mathcal{S}}}}{m_{\hat{A}}} = \hat{\varrho}. \tag{7.1.15}$$

There are two possible ways of describing the evolution of a batch arrival queue. The first possibility was already sketched at the beginning of this section: It starts with the arrival-stationary input Φ, solves the equation (7.1.1) and calculates the conditional distribution $\hat{\mathsf{P}}(\cdot) = \mathsf{P}(\cdot \mid A_{-1} > 0)$. The second way is the direct use of the batch-arrival-stationary input $\hat{\Phi}$. If the function f in (7.1.1), describing the dynamics of the system "customer by customer", is given, then the dynamics "batch by batch" can be described as follows: Define

$$\hat{f}(x, [a, \gamma, [y_1, \ldots, y_\gamma]]) = x_\gamma(x, [0, y_1], \ldots, [0, y_{\gamma-1}], [a, y_\gamma]).$$

(Here, just as in previous chapters, the right-hand side is the γ-fold iteration of f: $x_1(x, [a, y_1]) = f(x, [a, y_1])$, $x_\gamma(x, [0, y_1], \ldots, [0, y_{\gamma-1}], [a, y_\gamma]) = f(x_{\gamma-1}(x, [0, y_1], \ldots, [0, y_{\gamma-1}]), [a, y_\gamma])$.)

Let $\hat{X}_n$ be the state of the system viewed by the first customer of the n-th batch. Then the first customer of the $(n+1)$-th batch will find the state

$$\hat{X}_{n+1} = \hat{f}_n(\hat{X}_n, [\hat{A}_n, \hat{G}_n, \hat{\underline{Y}}_n]). \tag{7.1.16}$$

Given the batch-arrival-stationary input $(\hat{\Phi}, \hat{\mathsf{P}})$, one can try to solve equation (7.1.16). Analogously to Theorem 7.1.1 we obtain a one-to-one mapping between stationary weak solutions of (7.1.1) and (7.1.16):

7.1.5. Theorem. *Consider an arrival-stationary input* (Φ, P) *satisfying* $\mathsf{P}(\#\{n: A_n > 0\} = \infty) = 1$ *and the corresponding batch-arrival-stationary input* $(\hat{\Phi}, \hat{\mathsf{P}})$. *Then there is a one-to-one mapping between the distribution of stationary weak solutions* $([[A_n, Y_n], X_n])$ *of the equation*

$$X_{n+1} = f(X_n, [A_n, Y_n]), \qquad n \in \mathbb{Z}, \tag{7.1.17}$$

and the distributions of stationary weak solutions $([[\hat{A}_n, \hat{G}_n, \hat{\underline{Y}}_n], \hat{X}_n])$ *of the equation*

$$\hat{X}_{n+1} = \hat{f}(\hat{X}_n, [\hat{A}_n, \hat{G}_n, \hat{\underline{Y}}_n]), \qquad n \in \mathbb{Z}. \tag{7.1.18}$$

This mapping is given by

$$\hat{\mathsf{P}}\big(([[\hat{A}_n, \hat{G}_n, \hat{\underline{Y}}_n], \hat{X}_n]) \in (\cdot)\big) = \mathsf{P}\big(\phi(([[A_n, Y_n], X_n])) \in (\cdot) \mid A_{-1} > 0\big), \tag{7.1.19}$$

where

$$\phi(([[A_n, Y_n], X_n])) = \big([[\hat{a}_n(\Phi), \hat{g}_n(\Phi), \hat{\underline{y}}_n(\Phi)], \hat{x}_n(([[A_j, Y_j], X_j]))]\big),$$

$$\hat{x}_n(([[A_j, Y_j], X_j])) = X_{i_{n-1}(\Phi)+1},$$

or, conversely,

$$\mathsf{P}\big(([[A_n, Y_n], X_n], n \geq 0) \in (\cdot)\big)$$

$$= \frac{1}{\mathsf{E}\hat{G}_0} \sum_{j=0}^{\infty} \hat{\mathsf{P}}\Big(\hat{G}_0 > j, \big([[\bar{a}_{n+j}(\hat{\Phi}), \bar{y}_{n+j}(\hat{\Phi})], \bar{x}_{n+j}(([[\hat{A}_k, \hat{G}_k, \hat{\underline{Y}}_k], \hat{X}_k]))], n \geq 0\big) \in (\cdot)\Big), \tag{7.1.20}$$

where

$$\bar{x}_{n+j}(([[\hat{A}_k, \hat{G}_k, \hat{\underline{Y}}_k], \hat{X}_k]))$$
$$= x_{n+j}(\hat{X}_0, [\bar{a}_0(\hat{\Phi}), \bar{y}_0(\hat{\Phi})], \ldots, [\bar{a}_{n+j-1}(\hat{\Phi}), \bar{y}_{n+j-1}(\hat{\Phi})]).$$

If, in addition, $([[A_n, Y_n], X_n])$ *is ergodic (e.g. if* Φ *is ergodic and* $([[A_n, Y_n], X_n])$ *is a strong solution of* (7.1.17)), *then for every nonnegative, measurable function* h

$$\frac{1}{n}\sum_{j=0}^{n-1} h\Big([\bar{a}_j(\hat{\Phi}), \bar{y}_j(\hat{\Phi})], \bar{x}_j\big(([[\hat{A}_k, \hat{G}_k, \hat{\underline{Y}}_k], \hat{X}_k])\big)\Big)$$

$$\xrightarrow[n\to\infty]{} \mathsf{E}h([[A_0, Y_0], X_0]) \quad \hat{\mathsf{P}}\text{-a.s.}$$

and

$$\frac{1}{n}\sum_{j=0}^{n-1} h([[A_{m+\nu_j}, Y_{m+\nu_j}], X_{m+\nu_j}])$$

$$\xrightarrow[n\to\infty]{} \mathsf{E}h\Big([\bar{a}_m(\hat{\Phi}), \bar{y}_m(\hat{\Phi})], \bar{x}_m\big(([[\hat{A}_k, \hat{G}_k, \hat{\underline{Y}}_k], \hat{X}_k])\big)\Big), \quad m \in \mathbb{Z}, \quad \mathsf{P}\text{-a.s.},$$

where

$$\nu_0 = \min\{n : n > 0, A_{n-1} > 0\}, \qquad \nu_j = \min\{n : n > \nu_{j-1}, A_{n-1} > 0\},$$
$$j \geqq 1,$$

are the indices of the first customers of each batch in the arrival-stationary model.

Theorem 7.1.5 shows that both ways for constructing batch-arrival-stationary queueing processes lead to the same result. We call a stationary weak solution $([[\hat{A}_n, \hat{G}_n, \hat{\underline{Y}}_n], \hat{X}_n])$ of (7.1.18) a *batch-arrival-stationary queueing model* with the batch-arrival-stationary input $\hat{\Phi} = ([\hat{A}_n, \hat{G}_n, \hat{\underline{Y}}_n])$.

Combining Theorem 7.1.5, (6.1.3), and (6.1.4), we obtain a useful relationship between batch-arrival-stationary and time-stationary queueing models.

7.1.6. Theorem. *Consider a batch-arrival-stationary queueing model* $([[\hat{A}_n, \hat{G}_n, \hat{\underline{Y}}_n], \hat{X}_n])$ *satisfying* $\hat{\mathsf{P}}(\hat{A}_0 > 0) = 1$, $\mathsf{E}\hat{G}_0 < \infty$, *and the corresponding arrival-stationary queueing model* (χ, P), $\chi = [(X(t)), ([T_n, [X_n, Y_n]])]$, *where*

$$X(t) = f(X_n, [t - T_n, Y_n]) \quad \text{if} \quad T_n < t \leqq T_{n+1}.$$

Let $(\bar{\chi}, \bar{\mathsf{P}})$, $\bar{\chi} = [(\bar{X}(t)), ([\bar{T}_n, [\bar{X}_n, \bar{Y}_n]])]$, *be the associated time-stationary queueing model. Define*

$$\hat{\chi} = [(\hat{X}(t)), ([\tilde{T}_n, [\tilde{X}_n, \tilde{Y}_n]])],$$

where

$$\tilde{T}_0 = 0; \quad \tilde{T}_n = \sum_{k=0}^{n-1} \tilde{A}_k, \quad n > 0; \qquad \tilde{T}_n = \sum_{k=n}^{-1} \tilde{A}_k, \quad n < 0;$$

$$\tilde{A}_n = \bar{a}_n\big(([\hat{A}_j, \hat{\underline{Y}}_j])\big), \quad \tilde{Y}_n = \bar{y}_n\big(([\hat{A}_j, \hat{\underline{Y}}_j])\big), \quad n \in \mathbb{Z};$$

$$\tilde{X}_n = x_l(\hat{X}_k, [0, \hat{Y}_{k1}], \ldots, [0, \hat{Y}_{kl}]) \qquad \text{if}$$
$$\bar{\nu}_{k-1}\big(([\hat{A}_j, \hat{\underline{Y}}_j])\big) \leqq n < \bar{\nu}_k\big(([\hat{A}_j, \hat{\underline{Y}}_j])\big),$$
$$l = n - \bar{\nu}_{k-1}\big(([\hat{A}_j, \hat{\underline{Y}}_j])\big);$$

$$\hat{X}(t) = f(\tilde{X}_n, [t - \tilde{T}_n, \tilde{Y}_n]) \quad \text{if} \quad \tilde{T}_n < t \leqq \tilde{T}_{n+1}.$$

Then

$$\overline{\mathsf{P}}(\overline{\chi} \in (\cdot)) = \frac{1}{m_{\hat{A}}} \int_0^\infty \hat{\mathsf{P}}\left(\hat{A}_0 > t, \theta_t\hat{\chi} \in (\cdot)\right) dt. \tag{7.1.21}$$

If the batch-arrival-stationary model is ergodic, then for every nonnegative, measurable function h

$$\frac{1}{k} \sum_{j=1}^{k} h(\overline{X}_{n_j}) \xrightarrow[k\to\infty]{} \mathsf{E}h(\hat{X}_0) \qquad \overline{\mathsf{P}}\text{-}a.s.,$$

where n_j, $j \in \mathbb{Z}$, are the positive indices for which $\overline{T}_{n_j} - \overline{T}_{n_j-1} > 0$ holds, and, conversely,

$$\frac{1}{t} \int_0^t h\left(\hat{X}(u)\right) du \xrightarrow[k\to\infty]{} \mathsf{E}h\left(\overline{X}(0)\right) \qquad \hat{\mathsf{P}}\text{-}a.s.$$

The next sections contain some results for particular batch arrival queues. In accordance with Chapter 5, the letters V, W, L, and Q are reserved for work load quantities, waiting times, the number of customers in the system and in the queue, respectively.

7.2. The system $G^{GI}/GI/\infty$

Consider a $G^{GI}/GI/\infty$ system with the batch-arrival-stationary input $\hat{\Phi} = ([\hat{A}_n, \hat{G}_n, \hat{\underline{S}}_n])$, $\hat{\underline{S}}_n = [\hat{S}_{n1}, \ldots, \hat{S}_{n\hat{G}_n}]$, satisfying $\hat{\mathsf{P}}(\hat{A}_0 > 0) = 1$, $\mathsf{E}\hat{G}_0 < \infty$, $\hat{S}_{ni} > 0$. Then there is a unique stationary process $(\hat{X}_n)$, $\hat{X}_n = (\hat{X}_{n1}, X_{n2}, \ldots)$, $\hat{X}_{n1} \geqq \hat{X}_{n2} \geqq \cdots$, of residual service times at the batch arrivals. (This follows by means of Theorem 7.1.1, Theorem 5.2.1, and Theorem 7.1.5.) Here we are mainly interested in the number $\hat{L}_n = \#\{i: \hat{X}_{ni} > 0, i \geqq 1\}$ of customers in the system (i.e. the number of busy servers) seen by the first customer of the n-th arriving batch. Let $\Phi = ([A_n, S_n])$ be the arrival-stationary input associated with $\hat{\Phi}$ according to Theorem 7.1.1. Define the variables

$$\hat{Z}_n^{\,i} = \sum_{j=1}^{\hat{G}_{n-i}} 1\left\{\sum_{k=n-i}^{n-1} \hat{A}_k < \hat{S}_{n-i,j}\right\}, \qquad n \in \mathbb{Z}, \qquad i \geqq 1,$$

and

$$Z_n^{\,i} = 1\left\{\sum_{k=n-i}^{n-1} A_k < S_{n-i}\right\}, \qquad n \in \mathbb{Z}, \qquad i \geqq 1.$$

Then $\hat{L}_n$ and the number of customers L_n of the n-th customer arrival, respectively, are given by

$$\hat{L}_n = \sum_{i=1}^{\infty} \hat{Z}_n^{\,i}, \tag{7.2.1}$$

$$L_n = \sum_{i=1}^{\infty} Z_n^{\,i}. \tag{7.2.2}$$

From (7.1.6), (7.2.1), (7.2.2), (A 7.1.1), (A 7.1.3) and using the fact that $\hat{L}_0$ is a function of $([\hat{A}_n, \hat{G}_n, \hat{S}_n], n < 0)$, we obtain the following relationship for the probabilities $p_i = \mathsf{P}(L_0 = i)$, $\hat{p}_i = \hat{\mathsf{P}}(\hat{L}_0 = i)$, and $\hat{g}_i = \hat{\mathsf{P}}(\hat{G}_0 = i)$:

$$p_i = \frac{1}{m_{\hat{G}}} \sum_{j=0}^{\infty} \hat{\mathsf{P}}(\hat{G}_0 > j, \hat{L}_0 = i - j) = \frac{1}{m_{\hat{G}}} \sum_{j=0}^{i} \hat{\mathsf{P}}(\hat{G}_0 > j)\, \hat{p}_{i-j}$$
$$= \frac{1}{m_{\hat{G}}} \sum_{j=0}^{i} \hat{p}_{i-j} \sum_{k=j+1}^{\infty} \hat{g}_k. \tag{7.2.3}$$

From this and Remark 7.1.4, for the PFG $\Pi(s)$ of $(p_i, i \geqq 0)$ we get:

$$\Pi(s) = \sum_{i=0}^{\infty} p_i s^i = \frac{1}{m_{\hat{G}}} \sum_{i=0}^{\infty} \sum_{j=0}^{i} \hat{p}_{i-j} \left(\sum_{k=j+1}^{\infty} \hat{g}_k \right) s^i$$
$$= \frac{1}{m_{\hat{G}}} \sum_{j=0}^{\infty} \left(\sum_{k=j+1}^{\infty} \hat{g}_k \right) \sum_{i=0}^{\infty} \hat{p}_i s^{i+j}$$
$$= \frac{1}{m_{\hat{G}}} \sum_{j=0}^{\infty} \left(\sum_{k=j+1}^{\infty} \hat{g}_k \right) \hat{\Pi}(s)\, s^j = \frac{\hat{\Pi}(s)}{m_{\hat{G}}} \frac{1 - \hat{\Gamma}(s)}{1 - s} = \frac{\hat{\Pi}(s)\, R(s)}{s}, \tag{7.2.4}$$

where $\hat{\Pi}(s) = \sum_{i=0}^{\infty} \hat{p}_i s^i$, $\hat{\Gamma}(s) = \sum_{i=1}^{\infty} \hat{g}_i s^i$, and $R(s) = \sum_{i=1}^{\infty} r_i s^i$, cf. (7.1.11).

7.2.1. Remark. Obviously, the formulae (7.2.3) and (7.2.4) remain true for $G^{GI}/GI/m/\infty$ queues.

By Theorem 7.1.6 the time stationary process $(\bar{L}(t), t \in \mathbb{R})$ of the number of customers in the system is well-defined.

7.2.1. The system $G^{GI}/M/\infty$

Now we consider $G^{GI}/M/\infty$ queues, i.e. $G^{GI}/GI/\infty$ queues in which the service times of individual customers are exponentially distributed. In this case relations between the batch-stationary probabilities $\hat{p}_i$ and the time-stationary probabilities $\bar{p}_i = \bar{\mathsf{P}}(\bar{L}(t) = i)$ can easily be derived.

7.2.2. Theorem. *For $G^{GI}/M/\infty$ queues with finite mean batch size $m_{\hat{G}} < \infty$ we have*

$$\bar{p}_i = \frac{\varrho}{i m_{\hat{G}}} \sum_{j=0}^{i-1} \hat{p}_{i-j-1} \sum_{k=j+1}^{\infty} \hat{g}_k, \qquad i \geqq 1, \tag{7.2.5}$$

and for the PGF $\bar{\Pi}(s) = \sum_{i=0}^{\infty} \bar{p}_i s^i$

$$\frac{\mathrm{d}}{\mathrm{d}s} \bar{\Pi}(s) = \frac{\varrho}{m_{\hat{G}}} \hat{\Pi}(s) \frac{1 - \hat{\Gamma}(s)}{1 - s}. \tag{7.2.6}$$

Proof. Formula (7.2.5) follows immediately from Theorem 6.4.3 and (7.2.3). The second assertion follows from Theorem 6.4.3 and (7.2.4). ∎

Denote by $\hat{B}_r$, $\overline{B}_r$, and $\hat{C}_r$, $r \geqq 0$, the binomial moments of $(\hat{p}_i, i \geqq 0)$, $(\overline{p}_i, i \geqq 0)$, and $(\hat{g}_i, i \geqq 0)$, respectively, i.e.

$$\hat{B}_r = \mathsf{E}\binom{L_0}{r}, \quad \overline{B}_r = \mathsf{E}\binom{\overline{L}(0)}{r}, \quad \hat{C}_r = \mathsf{E}\binom{\hat{G}_0}{r}, \qquad r \geqq 0$$

$(\hat{B}_0 = \overline{B}_0 = \hat{C}_0 = 1)$.

By means of Theorem 7.2.2 we can derive a relation among these binomial moments. (Throughout this section we assume that all moments occuring in the theorems exist.)

7.2.3. Theorem. *For* $G^{GI}/M/\infty$ *queues it holds that*

$$\overline{B}_r = \frac{\varrho}{r m_{\hat{G}}} \sum_{k=1}^{r} \hat{C}_k \hat{B}_{r-k}, \qquad r \geqq 1.$$

In particular,

$$\overline{B}_1 = \mathsf{E}\overline{L}(0) = \varrho.$$

Proof. We denote the n-th derivative of an arbitrary function f by $f^{(n)}$, $n \geqq 0$, $f^{(0)} = f$: From (7.2.6) we get

$$\left(\overline{\Pi}(s)\right)^{(r)} = \frac{\varrho}{m_{\hat{G}}} \sum_{k=0}^{r-1} \binom{r-1}{k} \left(\hat{\Pi}(s)\right)^{(r-1-k)} \left(\frac{1-\hat{\Gamma}(s)}{1-s}\right)^{(k)}, \qquad r \geqq 1. \tag{7.2.7}$$

Since $\overline{B}_r = \frac{1}{r!} \lim_{s\to 1-0} \left(\overline{\Pi}(s)\right)^{(r)}$, $\hat{B}_r = \frac{1}{r!} \lim_{s\to 1-0} \left(\hat{\Pi}(s)\right)^{(r)}$, (7.2.7) yields

$$\overline{B}_r = \frac{\varrho}{m_{\hat{G}}} \sum_{k=0}^{r-1} \frac{1}{rk!} \lim_{s\to 1-0} \left(\frac{1-\hat{\Gamma}(s)}{1-s}\right)^{(k)} \hat{B}_{r-1-k}. \tag{7.2.8}$$

Now we have to find out the k-th derivative of $\left(1-\hat{\Gamma}(s)\right)/(1-s)$ at $s = 1-0$. In case of $k = 0$ we obtain

$$\lim_{s\to 1-0} \frac{1-\hat{\Gamma}(s)}{1-s} = \hat{\Gamma}^{(1)}(1)$$

by means of Bernoulli-L'Hospital's rule. Now let $k > 0$. Then

$$\left(\frac{1-\hat{\Gamma}(s)}{1-s}\right)^{(k)} = \frac{\sum_{j=0}^{k} \binom{k}{j} j!(1-s)^{k-j} \left(1-\hat{\Gamma}(s)\right)^{(k-j)}}{(1-s)^{k+1}} = \frac{g(s)}{h(s)},$$

where $g(s) = \sum_{j=0}^{k} \binom{k}{j} j!(1-s)^{k-j} \left(1-\hat{\Gamma}(s)\right)^{(k-j)}$

and $h(s) = (1-s)^{k+1}$.

Since $\hat{\Gamma}(1) = 1$, we get $g(1) = h(1) = 0$. Thus, in order to obtain the limit $\lim_{s\to 1-0} \frac{g(s)}{h(s)}$, we have to apply Bernoulli-L'Hospital's rule. The derivative of

$g(s)$ can be obtained in the following way:

$$\begin{aligned} g^{(1)}(s) &= \sum_{j=0}^{k} \binom{k}{j} j!\,(1-s)^{k-j} \left(1-\hat{\Gamma}(s)\right)^{(k-j+1)} \\ &\quad - \sum_{j=0}^{k-1} \binom{k}{j} j!\,(k-j)\,(1-s)^{k-j-1} \left(1-\hat{\Gamma}(s)\right)^{(k-j)} \\ &= \sum_{j=0}^{k} \binom{k}{j} j!\,(1-s)^{k-j} \left(1-\hat{\Gamma}(s)\right)^{(k-j+1)} \\ &\quad - \sum_{j=1}^{k} \binom{k}{j-1} (j-1)!\,(k-j+1)\left(1-s\right)^{k-j} \left(1-\hat{\Gamma}(s)\right)^{(k-j+1)} \\ &= (1-s)^{k} \left(1-\hat{\Gamma}(s)\right)^{(k+1)} \\ &\quad + \sum_{j=1}^{k} \left(\binom{k}{j} j! - \binom{k}{j-1} (j-1)!\,(k-j+1)\right) \\ &\quad \times (1-s)^{k-j} \left(1-\hat{\Gamma}(s)\right)^{(k-j+1)} \\ &= (1-s)^{k} \left(1-\hat{\Gamma}(s)\right)^{(k+1)}, \end{aligned}$$

since

$$\binom{k}{j} j! = \binom{k}{j-1} (j-1)!\,(k-j+1).$$

Thus, from

$$\begin{aligned} \lim_{s\to 1-0} \left(\frac{1-\hat{\Gamma}(s)}{1-s}\right)^{(k)} &= \lim_{s\to 1-0} \frac{g(s)}{h(s)} = \lim_{s\to 1-0} \frac{(1-s)^{k}\left(1-\hat{\Gamma}(s)\right)^{(k+1)}}{-(k+1)\,(1-s)^{k}} \\ &= \frac{\hat{\Gamma}^{(k+1)}(1)}{k+1} = k!\,\hat{C}_{k+1} \end{aligned} \tag{7.2.9}$$

and (7.2.8) the assertion follows immediately. ∎

7.2.2. The system $GI^{GI}/M/\infty$

Let us consider a $GI^{GI}/M/\infty$ queue, i.e. a $G^{GI}/M/\infty$ queue in which, in addition, the $\hat{A}_n$ are assumed to be independent r.v.'s. For such queues the binomial moments $\hat{B}_r$, $r \geqq 0$, of the number $\hat{L}_0$ of customers can be computed recursively.

7.2.4. Theorem. *For* $GI^{GI}/M/\infty$ *queues the binomial moments* $\hat{B}_r = \mathsf{E}\binom{\hat{L}_0}{r}$ *are determined by*

$$\hat{B}_r = \frac{\hat{\mathcal{A}}(\mu r)}{1-\hat{\mathcal{A}}(\mu r)} \sum_{k=1}^{r} \hat{C}_k \hat{B}_{r-k}, \qquad r \geqq 1, \quad \hat{B}_0 = 1,$$

where

$$\hat{\mathcal{A}}(s) = \int_0^\infty e^{-st}\, d\hat{A}(t)$$

denotes the Laplace-Stieltjes transform of the d.f. $\hat{\mathsf{A}}(t)$ *of* $\hat{A}_0$, $\mu = 1/\mathsf{E}\hat{S}_{01}$ *is the service intensity, and the* $\hat{C}_k$ *are the binomial moments of the batch size* $\hat{G}_0$.
In particular,

$$\hat{B}_1 = \mathsf{E}\hat{L}_0 = \frac{\hat{\mathcal{A}}(\mu)}{1 - \hat{\mathcal{A}}(\mu)}\, \mathsf{E}\hat{G}_0 .$$

Proof. The variables $\hat{Z}_0{}^i = \sum_{j=1}^{\hat{G}_{-i}} 1\left\{\sum_{k=-i}^{-1} \hat{A}_k < \hat{S}_{-i,j}\right\}$, $i \geqq 1$, arising on the right-hand side of (7.2.1) for $n = 0$, are the contributions of the previous batches to the number $\hat{L}_0$ of busy servers seen by the 0-th arriving batch. For a fixed sample path $(a_n, n < 0)$ of the past $(\hat{A}_n, n < 0)$ of the inter-batch times, the variables $\hat{Z}_0{}^i$ are independent by assumption. Furthermore, for fixed i the r.v.'s $\xi_{ij} = 1\{t_{-i} < \hat{S}_{-i,j}\}$, $t_{-i} = a_{-i} + \cdots + a_{-1}$, $j \geqq 1$, are independent of $\hat{G}_{-i}$, and

$$\mathsf{P}(\xi_{ij} = 1) = \mathrm{e}^{-\mu t_{-i}} .$$

From these facts we get for the PGF $\mathcal{F}_i(z)$ of $\xi_i = \sum_{j=1}^{\hat{G}_{-i}} \xi_{ij}$

$$\mathcal{F}_i(z) = \hat{\Gamma}\big(\mathrm{e}^{-\mu t_{-i}} z + (1 - \mathrm{e}^{-\mu t_{-i}})\big). \tag{7.2.10}$$

By means of (7.2.10) we obtain

$$\mathsf{E}\binom{\xi_i}{r} = \frac{1}{r!} \lim_{z \to 1-0} \frac{\mathrm{d}^r \mathcal{F}_i(z)}{\mathrm{d}z^r} = \hat{C}_r\, \mathrm{e}^{-r\mu t_{-i}} . \tag{7.2.11}$$

Applying Lemma A 3.1.3 and using (7.2.11) we get

$$\mathsf{E}\binom{\hat{L}_0}{r} = \sum_{l=1}^{r} \sum_{\substack{k_1+\cdots+k_l=r \\ k_i \geq 1}} \sum_{i_1=1}^{\infty} \cdots \sum_{i_l=1}^{\infty} \int_0^{\infty} \cdots \int_0^{\infty} \hat{C}_{k_1} \cdots \hat{C}_{k_l}$$
$$\times\, \mathrm{e}^{-\mu k_1 u_1}\, \mathrm{e}^{-\mu k_2 (u_1+u_2)} \cdots \mathrm{e}^{-\mu k_l (u_1+\cdots+u_l)}\, \mathrm{d}\hat{\mathsf{A}}^{*i_l}(u_l) \cdots \mathrm{d}\hat{\mathsf{A}}^{*i_1}(u_1), \tag{7.2.12}$$

where $\hat{\mathsf{A}}^{*i}(\cdot)$ denotes the i-fold convolution of $\hat{\mathsf{A}}(\cdot)$. The integrals in (7.2.12) can be expressed in terms of the Laplace-Stieltjes transform $\hat{\mathcal{A}}(t)$ of $\hat{\mathsf{A}}(t)$:

$$\int_0^{\infty} \cdots \int_0^{\infty} \mathrm{e}^{-\mu k_1 u_1}\, \mathrm{e}^{-\mu k_2 (u_1+u_2)} \cdots \mathrm{e}^{-\mu k_l (u_1+\cdots+u_l)}\, \mathrm{d}\hat{\mathsf{A}}^{*i_l}(u_l) \cdots \mathrm{d}\hat{\mathsf{A}}^{*i}(u_1)$$
$$= \big(\hat{\mathcal{A}}(\mu(k_1 + \cdots + k_l))\big)^{i_1} \big(\hat{\mathcal{A}}(\mu(k_2 + \cdots + k_l))\big)^{i_2} \cdots \big(\hat{\mathcal{A}}(\mu k_l)\big)^{i_l}.$$

In view of this, from (7.2.12) we obtain

$$\hat{B}_r = \mathsf{E}\binom{\hat{L}_0}{r} = \sum_{l=1}^{r} \sum_{\substack{k_1+\cdots+k_l=r \\ k_i \geq 1}} \hat{C}_{k_1} \cdots \hat{C}_{k_l} D\big(\mu(k_1 + \cdots + k_l)\big)$$
$$\times\, D\big(\mu(k_2 + \cdots + k_l)\big) \cdots D(\mu k_l), \tag{7.2.13}$$

where

$$D(t) = \frac{\hat{\mathcal{A}}(t)}{1 - \hat{\mathcal{A}}(t)} .$$

For $r = 1$ (7.2.13) reduces to

$$\hat{B}_1 = \frac{\mathcal{A}(\mu)}{1 - \hat{\mathcal{A}}(\mu)} \hat{C}_1 \hat{B}_0 ,$$

i.e. the assertion of the theorem is true for $r = 1$. Thus, we can assume $r \geqq 2$ in the following. Now, by splitting the summand $l = 1$, the right-hand side of (7.2.13) can be transformed in the following way:

$$\hat{B}_r = D(\mu r) \Bigg(\hat{C}_r + \sum_{l=2}^{r} \sum_{\substack{k_1+\cdots+k_l=r \\ k_i \geqq 1}} \hat{C}_{k_1} \cdots \hat{C}_{k_l} \times D\big(\mu(k_2 + \cdots + k_l)\big) \cdots D(\mu k_l) \Bigg)$$

$$= D(\mu r) \Bigg(\hat{C}_r + \sum_{k_1=1}^{r-1} \hat{C}_{k_1} \Bigg(\sum_{l'=1}^{r-k_1} \sum_{\substack{k_1'+\cdots+k_{l'}'=r-k_1 \\ k_i' \geqq 1}} \hat{C}_{k_1'} \cdots \hat{C}_{k_{l'}'} \times D\big(\mu(k_1' + \cdots + k_{l'}')\big) D\big(\mu(k_2' + \cdots + k_{l'}')\big) \cdots D(\mu k_{l'}') \Bigg) \Bigg). \tag{7.2.14}$$

Taking into account that the expression within the inner parentheses in (7.2.14) is just $\hat{B}_{r-k_1}$, cf. (7.2.13), we get

$$\hat{B}_r = \frac{\mathcal{A}(\mu r)}{1 - \hat{\mathcal{A}}(\mu r)} \left(\hat{C}_r + \sum_{k=1}^{r-1} \hat{C}_k \hat{B}_{r-k} \right) = \frac{\mathcal{A}(\mu r)}{1 - \hat{\mathcal{A}}(\mu r)} \sum_{k=1}^{r} \hat{C}_k \hat{B}_{r-k}$$

$(\hat{B}_0 = 1)$. ∎

7.2.5. Remark. Since $\hat{C}_r < \infty$ implies $\hat{C}_i < \infty$, $i = 1, \ldots, r - 1$, Theorem 7.2.4 yields that $\hat{B}_r < \infty$ if $\hat{C}_r < \infty$.

If

$$\sum_{r=i}^{\infty} \binom{r}{i} \hat{B}_r < \infty , \qquad i \geqq 0 , \tag{7.2.15}$$

then, by Theorem A 3.1.1, the probabilities $\hat{p}_i$ can be computed via the binomial moments:

$$\hat{p}_i = \sum_{r=i}^{\infty} (-1)^{r-i} \binom{r}{i} \hat{B}_r , \qquad i \geqq 0. \tag{7.2.16}$$

However, assumption (7.2.15) is not always satisfied, as we shall show in Example 7.2.8.

Recall that the time-stationary probabilities $\bar{p}_i$ and the batch-stationary probabilities $\hat{p}_i$ satisfy (7.2.5). For the binomial moments $\bar{B}_r$ and $\hat{B}_r$, we obtain the following result immediately from Theorem 7.2.3 and Theorem 7.2.4.

7.2.6. Theorem. *For* $\mathrm{GI}^{\mathrm{GI}}/\mathrm{M}/\infty$ *queues the binomial moments* $\hat{B}_r = \mathsf{E}\binom{\hat{L}_0}{r}$ *and* $\bar{B}_r = \mathsf{E}\binom{\bar{L}(0)}{r}$, $r \geqq 0$, *satisfy*

$$\bar{B}_r = \frac{\varrho}{r m_{\hat{G}}} \frac{1 - \hat{\mathcal{A}}(\mu r)}{\hat{\mathcal{A}}(\mu r)} \hat{B}_r, \qquad r \geqq 1. \tag{7.2.17}$$

In particular, $\bar{B}_1 = \varrho$.

In the following we shall treat two important special cases of $\mathrm{GI}^{\mathrm{GI}}/\mathrm{M}/\infty$ queues.

7.2.3. The system $\mathrm{GI}^{\mathrm{GI}}/\mathrm{M}/\infty$ with geometrically distributed batch size

Consider an $\mathrm{GI}^{\mathrm{GI}}/\mathrm{M}/\infty$ queue, and assume that the batch sizes are geometrically distributed with parameter p, $0 < p < 1$, i.e.

$$\hat{\mathsf{P}}(\hat{G}_0 = i) = (1 - p)\, p^{i-1}, \qquad i \geqq 1.$$

Such a queue will be denoted by the symbol $\mathrm{GI}^{\delta_1 * \mathrm{Geom}(p)}/\mathrm{M}/\infty$. For this queue the binomial moments $\hat{B}_r$ can be computed explicitely.

7.2.7. Theorem. *For* $\mathrm{GI}^{\delta_1 * \mathrm{Geom}(p)}/\mathrm{M}/\infty$ *queues it holds that*

$$\hat{B}_r = \left(\frac{p}{1-p}\right)^r \frac{\hat{\mathcal{A}}(\mu r)}{p\big(1 - \hat{\mathcal{A}}(\mu r)\big)} \prod_{j=1}^{r-1} \left(1 + \frac{\hat{\mathcal{A}}(\mu j)}{p\big(1 - \hat{\mathcal{A}}(\mu j)\big)}\right), \qquad r \geqq 1. \tag{7.2.18}$$

In particular,

$$\hat{B}_1 = \mathsf{E}\hat{L}_0 = \frac{\hat{\mathcal{A}}(\mu)}{(1-p)\big(1 - \hat{\mathcal{A}}(\mu)\big)}.$$

Proof. Since

$$\hat{C}_k = \sum_{i=k}^{\infty} \binom{i}{k} (1-p)\, p^{i-1} = \frac{1}{1-p} \left(\frac{p}{1-p}\right)^{k-1},$$

we get from Theorem 7.2.4

$$\begin{aligned} \hat{B}_r &= \frac{\hat{\mathcal{A}}(\mu r)}{1 - \hat{\mathcal{A}}(\mu r)} \sum_{k=1}^{r} \frac{1}{1-p} \left(\frac{p}{1-p}\right)^{k-1} \hat{B}_{r-k} \\ &= a_r \sum_{k=1}^{r} \left(\frac{p}{1-p}\right)^k \hat{B}_{r-k}, \qquad r \geqq 1, \end{aligned} \tag{7.2.19}$$

where

$$a_j = \frac{\hat{\mathcal{A}}(\mu j)}{p\big(1 - \hat{\mathcal{A}}(\mu j)\big)}, \qquad j \geqq 1.$$

Now (7.2.18) can be obtained by induction as follows. In the case $r = 1$, (7.2.19) is equivalent to (7.2.18), since $\hat{B}_0 = 1$. Assume (7.2.18) is valid for r.

Then, for $r + 1$, (7.2.19) provides

$$\hat{B}_{r+1} = a_{r+1} \sum_{k=1}^{r+1} \left(\frac{p}{1-p}\right)^k \left(\frac{p}{1-p}\right)^{r+1-k} a_{r+1-k} \prod_{j=1}^{r-k} (1 + a_j),$$

where we set $a_0 = 1$. By elementary algebra we can continue:

$$\hat{B}_{r+1} = \left(\frac{p}{1-p}\right)^{r+1} a_{r+1} \sum_{k=1}^{r+1} a_{r+1-k} \prod_{j=1}^{r-k} (1 + a_j)$$

$$= \left(\frac{p}{1-p}\right)^{r+1} a_{r+1} \prod_{j=1}^{r} (1 + a_j). \quad \blacksquare$$

If $p < 1/2$, then it is easy to check that

$$\sum_{r=i}^{\infty} \binom{r}{i} \hat{B}_r < \infty, \qquad i \geqq 0,$$

is satisfied and thus the probabilities $\hat{p}_i$ can be computed via the binomial moments:

$$\hat{p}_i = \sum_{r=i}^{\infty} (-1)^{r-i} \binom{r}{i} \hat{B}_r, \qquad i \geqq 0, \tag{7.2.20}$$

cf. Theorem A 3.1.1. In the case of $p > 1/2$ there are examples for which

$$\binom{r}{i} \hat{B}_r \xrightarrow[r\to\infty]{} \infty$$

and thus (7.2.20) is not satisfied.

7.2.8. Example. Consider an $\mathrm{M}^{\delta_1 * \mathrm{Geom}(2/3)}/\mathrm{M}/\infty$ queue with $\hat{\mathsf{A}}(t) = 1 - \mathrm{e}^{-\lambda t}$. Then

$$\hat{\mathcal{A}}(s) = \frac{\lambda}{\lambda + s}, \qquad a_j = \frac{\hat{\mathcal{A}}(\mu j)}{p(1 - \hat{\mathcal{A}}(\mu j))} = \frac{3\lambda}{2\mu j},$$

and in view of Theorem 7.2.7

$$\hat{B}_r = \left(\frac{p}{1-p}\right)^r a_r \prod_{j=1}^{r-1} (1 + a_j) = 2^r \frac{3\lambda}{2\mu r} \prod_{j=1}^{r-1} \left(1 + \frac{3\lambda}{2\mu j}\right) \geqq \frac{2^r}{r} \frac{3\lambda}{2\mu}.$$

Hence we obtain

$$\binom{r}{i} \hat{B}_r \geqq \frac{(r - i + 1)^i}{i!} \frac{2^r}{r} \frac{3\lambda}{2\mu} \xrightarrow[r\to\infty]{} \infty$$

for fixed i.

Taking into account that $\mathsf{E}\hat{G}_0 = 1/(1-p)$ and $\sum_{k=j+1}^{\infty} \hat{g}_k = p^j$, from Theorem 7.2.2 (formula (7.2.5)) we obtain the following relationship between the time-stationary probabilities $\overline{p}_i$ and the batch-stationary probabilities $\hat{p}_i$.

7.2.9. Theorem. *For* $\mathrm{GI}^{\delta_1 * \mathrm{Geom}(p)}/\mathrm{M}/\infty$ *queues we have*

$$\overline{p}_i = \frac{\varrho(1-p)}{i} \sum_{j=0}^{i-1} \hat{p}_{i-1-j} p^j, \qquad i \geqq 1,$$

and, conversely,

$$\hat{p}_0 = \frac{\bar{p}_1}{\varrho(1-p)},$$

$$\hat{p}_i = \frac{1}{\varrho(1-p)} \left((i+1)\,\bar{p}_{i+1} - pi\bar{p}_i\right), \qquad i \geqq 1.$$

7.2.4. The system $\mathrm{GI^{GI}/M/\infty}$ with constant batch size

Consider a $\mathrm{GI^{GI}/M/\infty}$ queue, and assume that the batches have constant size k, i.e. $\hat{\mathsf{P}}(\hat{G}_0 = k) = 1$. Such a queue will be denoted by the symbol $\mathrm{GI}^k/\mathrm{M}/\infty$. For this queue we have $\hat{C}_l = \binom{k}{l}$, and thus Theorem 7.2.4 yields

7.2.10. Theorem. *For* $\mathrm{GI}^k/\mathrm{M}/\infty$ *queues it holds that*

$$\hat{B}_r = \frac{\hat{\mathcal{A}}(\mu r)}{1-\hat{\mathcal{A}}(\mu r)} \sum_{l=1}^{\min(k,r)} \binom{k}{l} \hat{B}_{r-l} = \frac{\hat{\mathcal{A}}(\mu r)}{1-\hat{\mathcal{A}}(\mu r)} \sum_{l=(r-k)_+}^{r-1} \binom{k}{r-l} \hat{B}_l,$$

$$r \geqq 1; \qquad \hat{B}_0 = 1. \tag{7.2.21}$$

Since $\hat{\mathcal{A}}(\mu r) < 1$ and $\hat{\mathcal{A}}(\mu r) \downarrow 0$ as $r \to \infty$, from (7.2.21) and Lemma A 3.1.4 it follows that (7.2.15) is satisfied. Thus, the probabilities $\hat{p}_i$ can be computed via the binomial moments by means of (7.2.16).

Using Theorem 7.2.2 (formula (7.2.5)), we obtain the following relationship between the time-stationary probabilities $\bar{p}_i$ and the batch-stationary probabilities $\hat{p}_i$.

7.2.11. Theorem. *For* $\mathrm{GI}^k/\mathrm{M}/\infty$ *queues we have*

$$\bar{p}_i = \frac{\varrho}{ik} \sum_{j=0}^{\min(i,k)-1} \hat{p}_{i-j-1}, \qquad i \geqq 1, \qquad \bar{p}_0 = 1 - \sum_{i=1}^{\infty} \bar{p}_i, \tag{7.2.22}$$

and, conversely,

$$\hat{p}_{mk} = \frac{k}{\varrho}\,\bar{p}_1 + \frac{k}{\varrho} \sum_{l=1}^{m} \left((lk+1)\,\bar{p}_{lk+1} - lk\bar{p}_{lk}\right), \qquad m \geqq 0,$$

$$\hat{p}_{i+mk} = \frac{k}{\varrho} \sum_{l=0}^{m} \left((i+lk+1)\,\bar{p}_{i+lk+1} - (i+lk)\,\bar{p}_{i+lk}\right),$$

$$1 \leqq i < k, \qquad m \geqq 0.$$

The latter assertion follows from (7.2.22) by induction.

7.3. The single server queue with batch arrivals $\mathrm{G^G/G/1/\infty}$

Consider a single server queue with a work conserving queueing discipline and with a batch-arrival-stationary input

$$\hat{\Phi} = ([\hat{A}_n, \hat{G}_n, \hat{\underline{S}}_n]), \quad \hat{\underline{S}}_n = [\hat{S}_{n1}, \ldots, \hat{S}_{n\hat{G}_n}] \quad (\text{abbr. } \mathrm{G^G/G/1/\infty}).$$

(Recall that $\hat{A}_n$ is the time between the arrivals of the n-th and the $(n+1)$-th batch, $\hat{G}_n$ is the size of the n-th batch, and the $\hat{S}_{ni}$ are the required service times of the customers within the n-th batch.) Let $\Phi = ([A_n, S_n])$ be the corresponding arrival-stationary input of the queue considered, cf. Theorem 7.1.1. Remember (cf. (7.1.15)),

$$\varrho = \frac{m_S}{m_A} = \frac{m_{\hat{S}}}{m_{\hat{A}}} = \hat{\varrho}.$$

If $\varrho < 1$, which is always assumed to be true in the following, then there is a unique arrival and batch-arrival-stationary work load process (V_n) and $(\hat{V}_n)$ of the queue considered:

$$\begin{aligned} V_{n+1} &= (V_n + S_n - A_n)_+, \\ \hat{V}_{n+1} &= \left(\hat{V}_n + \sum_{j=1}^{\hat{G}_n} \hat{S}_{nj} - \hat{A}_n\right)_+ \end{aligned} \tag{7.3.1}$$

(cf. Theorem 5.4.1, Theorem 7.1.5). Analogously, we have uniquely determined processes $(\hat{L}_n)$ and (L_n) of the number of customers seen by the customer in front of an arbitrary arriving batch and seen by an arbitrary arriving customer, respectively.

As a first result one can provide a relation between the d.f. $\hat{\mathsf{V}}(t)$ of $\hat{V}_n$ and the d.f. $\mathsf{V}(t)$ of V_n.

7.3.1. Theorem. *If the batch-arrival-stationary input $\hat{\Phi}$ satisfies* (A 7.1.1) *and* (A 7.1.3) *(e.g. in case of* $G^{GI}/GI/1/\infty$*), then*

$$\mathsf{V} = \hat{\mathsf{V}} * \hat{\mathsf{R}}, \tag{7.3.2}$$

where $*$ denotes the convolution, and $\hat{\mathsf{R}}$ is given by

$$\hat{\mathsf{R}}(x) = \frac{1}{\mathsf{E}\hat{G}_0} \sum_{n=1}^{\infty} \hat{g}_n \sum_{j=0}^{n-1} \hat{\mathsf{P}}(\hat{S}_{01} + \cdots + \hat{S}_{0j} < x \mid \hat{G}_0 = n).$$

In case of $G^{GI}/GI/1/\infty$ *the LST $\hat{\mathfrak{R}}(x)$ of $\hat{\mathsf{R}}(x)$ has the form*

$$\int_0^\infty e^{-sx}\, d\hat{\mathsf{R}}(x) = R(\mathcal{S}(s))/\mathcal{S}(s),$$

with

$$\mathcal{S}(s) = \int_0^\infty e^{-sx}\, d\mathsf{S}(x), \quad R(s) = \frac{s(1 - \hat{\Gamma}(s))}{\mathsf{E}\hat{G}_0(1 - s)}, \tag{7.3.3}$$

cf. (7.1.13).

Proof. Theorem 7.1.5 yields

$$\begin{aligned} \mathsf{V}(x) = \mathsf{P}(V_0 < x) &= \frac{1}{m_{\hat{G}}} \sum_{j=0}^{\infty} \hat{\mathsf{P}}(\hat{G}_0 > j, \hat{V}_0 + \hat{S}_{01} + \cdots + \hat{S}_{0j} < x) \\ &= \frac{1}{m_{\hat{G}}} \sum_{j=0}^{\infty} \sum_{i=j+1}^{\infty} \hat{\mathsf{P}}(\hat{V}_0 + \hat{S}_{01} + \cdots + \hat{S}_{0j} < x \mid \hat{G}_0 = i)\, \hat{\mathsf{P}}(\hat{G}_0 = i) \end{aligned}$$

Since $\hat{V}_0$ depends on $([\hat{A}_n, \hat{G}_n, \hat{S}_n], n < 0)$ only, and since the independence assumptions (A 7.1.1) and (A 7.1.3) hold, we can continue as follows:

$$\begin{aligned} \mathsf{V}(x) &= \frac{1}{m_{\hat{G}}} \sum_{j=0}^{\infty} \sum_{i=j+1}^{\infty} \hat{\mathsf{P}}(\hat{G}_0 = i) \left[\hat{\mathsf{P}}(\hat{V}_0 \in (\cdot)) \right. \\ &\quad \left. * \hat{\mathsf{P}}(\hat{S}_{01} + \cdots + \hat{S}_{0j} < (\cdot) \mid \hat{G}_0 = i)\right] (x) \\ &= \left\{\hat{\mathsf{V}}(\cdot) * \left[\frac{1}{m_{\hat{G}}} \sum_{i=1}^{\infty} \hat{g}_i \sum_{j=0}^{i-1} \hat{\mathsf{P}}(\hat{S}_{01} + \cdots + \hat{S}_{0j} < (\cdot) \mid \hat{G}_0 = i)\right]\right\} (x). \end{aligned}$$

Thus, (7.3.2) is valid.

In case of $\mathrm{G}^{\mathrm{GI}}/\mathrm{GI}/1/\infty$, we can split up the LST $\hat{\mathscr{R}}(x)$ of $\hat{\mathsf{R}}(x)$ into factors:

$$\begin{aligned} \int_0^\infty \mathrm{e}^{-sx} \, \mathrm{d}\hat{\mathsf{R}}(x) &= \frac{1}{m_{\hat{G}}} \sum_{n=1}^{\infty} \hat{g}_n \sum_{j=0}^{n-1} \mathscr{S}^j(s) = \frac{1}{m_{\hat{G}}} \sum_{n=1}^{\infty} \hat{g}_n \frac{1 - \mathscr{S}^n(s)}{1 - \mathscr{S}(s)} \\ &= \frac{1 - \hat{\Gamma}(\mathscr{S}(s))}{m_{\hat{G}}(1 - \mathscr{S}(s))} = \frac{R(\mathscr{S}(s))}{\mathscr{S}(s)}. \quad \blacksquare \end{aligned}$$

7.3.2. Remark. $\mathsf{V}(x)$ appears as the convolution of the two batch stationary distributions $\hat{\mathsf{V}}(x)$ and $\hat{\mathsf{R}}(x)$. In view of (A 1.6a) we have

$$\begin{aligned} \hat{\mathsf{R}}(\hat{x}) &= \frac{1}{m_{\hat{G}}} \sum_{n=1}^{\infty} \hat{g}_n \sum_{j=0}^{n-1} \hat{\mathsf{P}}(\hat{S}_{01} + \cdots + \hat{S}_{0j} < x \mid \hat{G}_0 = n) \\ &= \frac{1}{m_{\hat{G}}} \sum_{j=0}^{\infty} \sum_{i=j+1}^{\infty} \hat{\mathsf{P}}(\hat{S}_{01} + \cdots + \hat{S}_{0j} < x \mid \hat{G}_0 = i) \, \hat{\mathsf{P}}(\hat{G}_0 = i) \\ &= \frac{1}{m_{\hat{G}}} \sum_{j=0}^{\infty} \hat{\mathsf{P}}(\hat{G}_0 > j, \hat{S}_{01} + \cdots + \hat{S}_{0j} < x) \\ &= \mathsf{P}(S_{\nu_{-1}} + S_{\nu_{-1}+1} + \cdots + S_{-1} < x), \end{aligned}$$

$$\nu_{-1} = \max\{n \leqq 0, A_{n-1} > 0\}. \tag{7.3.4}$$

Thus, in the case of FCFS queueing discipline, we can interpret $\hat{R}(x)$ as the distribution of the waiting time of an arbitrary customer from the beginning of service for his own batch until his own service.

The relationship between the queue size distributions $p_k = \mathsf{P}(L_0 = k)$ and $\hat{p}_k = \hat{\mathsf{P}}(\hat{L}_0 = k)$ is given by the following statement.

7.3.3. Theorem. *If the batch-arrival-stationary input $\hat{\Phi}$ satisfies the independence assumptions* (A 7.1.1) *and* (A 7.1.3) (*e.g. in the case of* $\mathrm{G}^{\mathrm{GI}}/\mathrm{GI}/1/\infty$), *then the PGF's* $\Pi(s) = \sum_{k=0}^{\infty} p_k s^k$, $\hat{\Pi}(s) = \sum_{k=0}^{\infty} \hat{p}_k s^k$ *satisfy*

$$\Pi(s) = \hat{\Pi}(s) \frac{R(s)}{s}, \tag{7.3.5}$$

where

$$R(s) = \sum_{n=1}^{\infty} r_n s^n = \frac{s(1 - \hat{\Gamma}(s))}{m_{\hat{G}}(1 - s)}.$$

Proof. The proof is the same as those of (7.2.4), cf. Remark 7.2.1. ∎

7.3.4. Remark. The interpretation of (7.3.5) is similar to that of (7.3.2) in Remark 7.3.2. The assumptions lead to

$$\hat{P}(\hat{L}_0 = k) = P(L_{\nu_{-1}} = k)$$

and

$$P(L_0 = k) = P(L_{\nu_{-1}} - \nu_{-1} = k) = \left[\hat{P}(\hat{L}_0 = (\cdot)) * (r_n, n \geqq 1)\right](k).$$

(The distribution of ν_{-1} is given by (7.3.4).)

Denote the steady state probability that an arbitrary departing customer leaves n customers in the system by p_n^* and the corresponding PGF by $\Pi^*(s) = \sum_{n=0}^{\infty} p_n^* s^n$.

7.3.5. Corollary. *If* (A 7.1.1) *and* (A 7.1.3) *are satisfied, then*

$$\Pi^*(s) = \hat{\Pi}(s) \frac{R(s)}{s}. \tag{7.3.6}$$

Proof. Relation (7.3.6) is an immediate consequence of (7.3.5) as well as of the fact that $p_n^* = p_n$, $n = 0, 1, \ldots$, cf. Theorem 6.4.2. ∎

The formulae of Takács and Pollaczek-Khinchin

So far we have derived relations among batch-arrival-, arrival- and departure-stationary quantities. In the following we are interested in time-stationary characteristics, too. The aim is to give the famous Takács formula and to derive an analogue of the well-known Pollaczek-Khinchin formula for the batch arrival queue.

By Theorem 7.1.6 the time-stationary model $(\overline{\Psi}, \overline{P})$, $\overline{\Psi} = ([\overline{T}_n, \overline{S}_n, \overline{V}_n, \overline{L}_n])$, is well-defined, where $\overline{V}_n$ is the work load and L_n the number of customers found by the n-th arriving customer. Furthermore, the time-stationary processes $(\overline{V}(t), t \in \mathbb{R})$ of the work load and $(\overline{L}(t), t \in \mathbb{R})$ of the number of customers in the system are both well-defined. Let

$$\overline{V}(x) = \overline{P}(\overline{V}(0) \leq x), \qquad \overline{v}(s) = \int_0^\infty e^{-sx} \, d\overline{V}(x),$$

$$\overline{p}_n = \overline{P}(\overline{L}(0) = n), \qquad \overline{\Pi}(s) = \sum_{n=0}^{\infty} \overline{p}_n s^n.$$

Considering the batch-arrival-stationary model $([\hat{A}_n, \hat{G}_n, \hat{\underline{S}}_n, \hat{V}_n])$ it is obvious that the sequence $(\hat{V}_n)$ of work loads is the same as in the "equivalent" system with the interarrival times $\hat{A}_n$ and the service times $\hat{S}_n = \hat{S}_{n1} + \cdots + \hat{S}_{n\hat{G}_n}$, cf. (7.3.1). Using this equivalent system from Takács' formula (Theorem 6.6.2) and Theorem 6.6.4 we obtain immediately:

7.3.6. Theorem. (Takács formula) *In a* $G^G/G/1/\infty$ *queue with work conserving queueing discipline and with the independence assumptions* (A 7.1.1) *and* (A 7.1.3) *(e.g. in the case of* $G^{GI}/GI/1/\infty$*) we have*

$$\overline{V}(y) = 1 - \varrho + \varrho(\hat{V} * \hat{S}_R)(y), \tag{7.3.7}$$

where

$$\hat{\mathsf{S}}_R(y) = \frac{1}{m_{\hat{\mathsf{S}}}} \int_0^y \left(1 - \hat{\mathsf{S}}(u)\right) du, \quad \hat{\mathsf{S}}(u) = \hat{\mathsf{P}}\left(\hat{S}_0 = \sum_{j=1}^{\hat{G}_0} \hat{S}_{0j} < u\right),$$

$$m_{\hat{\mathsf{S}}} = \mathsf{E}\hat{S}_0, \qquad \varrho = \mathsf{E}\hat{S}_0/\mathsf{E}\hat{A}_0 < 1.$$

If, in addition, (A 7.1.2) *is satisfied and* $\hat{\mathsf{P}}(\hat{A}_0 > x) = e^{-\hat{\lambda}x}$, *then it holds that*

$$\overline{\mathcal{V}}(s) = \hat{\mathcal{V}}(s) = \frac{1 - \varrho}{1 - \hat{\lambda}\left(1 - \hat{\mathcal{S}}(s)\right)/s}, \tag{7.3.8}$$

where $\hat{\mathcal{V}}(s)$, $\hat{\mathcal{S}}(s)$ *are the LST's of* $\hat{\mathsf{V}}$ *and* $\hat{\mathsf{S}}$, *respectively.*
In the case of an $\mathrm{M^{GI}/GI/1/\infty}$, *i.e. if* (A 7.1.1)–(A 7.1.4) *hold, we have*

$$\overline{\mathcal{V}}(s) = \hat{\mathcal{V}}(s) = \frac{1 - \varrho}{1 - \hat{\lambda}\left(1 - \hat{\Gamma}\left(\mathcal{S}(s)\right)\right)/s}. \tag{7.3.9}$$

Now we shall derive an analogue to the well-known Pollaczek-Khinchin formula.

7.3.7. Theorem. *In an* $\mathrm{M^{GI}/GI/1/\infty}$ *queue with FCFS queueing discipline we have the following relation for the PGF's of the distributions* $(\hat{p}_n, n \geqq 0)$ *and* $(\overline{p}_n, n \geqq 0)$:

$$\hat{\Pi}(s) = \overline{\Pi}(s) = \frac{(1 - \varrho)(1 - s)}{1 - \dfrac{s}{\mathcal{S}\left(\hat{\lambda} - \hat{\lambda}\hat{\Gamma}(s)\right)}}, \tag{7.3.10}$$

where $\hat{\lambda} = 1/m_{\hat{A}}$ *is the intensity of the batch arrivals,* $\hat{\Gamma}(s)$ *the PGF of the batch size* $\hat{G}_0$, $\mathcal{S}(s)$ *the LST of the d.f. of the service time* S_0 *(it holds that* $S_0 \overset{\mathcal{D}}{=} \hat{S}_{01}$), *and* $\varrho = \mathsf{E}\hat{G}_0\mathsf{E}S_0\hat{\lambda}$.

Proof. Since the $\hat{A}_n$ are exponentially distributed and since the dynamics of the queue can be considered from batch to batch, $\hat{p}_j = \overline{p}_j$, $j = 0, 1, \ldots$, cf. Theorem 6.2.2.

Consider the arrival-stationary MPP (Ψ, P), $\Psi = \left([T_n, [S_n, V_n, L_n]]\right)$, and the departure-stationary probability $p_n^* = \mathsf{P}^*(L_0^* = n)$ that an arbitrary departing customer leaves behind n customers in the system (for details cf. Section 6.4). Analogously to (6.6.7), we obtain

$$p_n^* = \mathsf{P}(\#\{k > 0 : T_k < V_0 + S_0\} = n).$$

Thus

$$p_n^* = \sum_{i=1}^{\infty} \sum_{j=1}^{i} \mathsf{P}(\#\{k > 0 : T_k < V_0 + S_0\} = n, G_0 = i, N_0 = j), \tag{7.3.11}$$

where G_0 is the size of the batch which contains the customer with the number 0 and N_0 is the position of this customer in his own batch in the arrival-stationary model, cf. (7.1.8), (7.1.9). Applying Theorem 7.1.1, (7.1.6) to the events on the right-hand side of (7.3.11), we get

$$\mathsf{P}(\#\{k > 0: T_k < V_0 + S_0\} = n, G_0 = i, N_0 = j)$$
$$= \frac{1}{m_{\hat G}} \sum_{l=0}^{\infty} \hat{\mathsf{P}}\left(\hat G_0 = i > l, l = j-1, \sum_{r=1}^{\infty} \hat G_r 1\{\hat T_r < \hat V_0 + \hat S_{01} + \cdots + \hat S_{0j}\} = n - (i-j)\right)$$
$$= \frac{\hat g_i}{m_{\hat G}} \hat{\mathsf{P}}\left(\sum_{r=1}^{\infty} \hat G_r 1\{\hat T_r < \hat V_0 + \hat S_{01} + \cdots + \hat S_{0j}\} = n - (i-j) \mid \hat G_0 = i\right)$$
$$\text{for } j \leqq i. \quad (7.3.12)$$

It follows from the construction of $\hat V_0$ and from the independence assumptions of $M^{GI}/GI/1/\infty$ that $([\hat T_n, \hat G_n], n > 0)$ is independent of $[\hat V_0, \hat S_{01}, \ldots, \hat S_{0j}]$ and that $\hat V_0, \hat S_{01}, \ldots, \hat S_{0j}$ are independent r.v.'s. Applying the formula of total probability, the right-hand side of (7.3.12) can be continued:

$$= \frac{\hat g_i}{m_{\hat G}} \int_0^{\infty} \sum_{k=0}^{\infty} (\hat g^{*k})_{n-i+j} \frac{(\hat\lambda t)^k}{k!} e^{-\hat\lambda t} \, d(\hat{V} * S^{*j})(t), \qquad (7.3.13)$$

where $((\hat g^{*k})_n, n \in \mathbb{Z}_+)$ denotes the k-fold convolution of the distribution $(\hat g_n, n \in \mathbb{Z}_+)$ of the batch size $\hat G_0$. From (7.3.11) and (7.3.13) we get

$$\Pi^*(s) = \sum_{n=0}^{\infty} p_n^* s^n = \sum_{n=0}^{\infty} \sum_{i=1}^{\infty} \sum_{j=1}^{i} \int_0^{\infty} \sum_{k=0}^{\infty} (\hat g^{*k})_{n-i+j} \frac{(\hat\lambda t)^k}{k!} e^{-\hat\lambda t} \, d(\hat{V} * S^{*j})(t) \frac{\hat g_i}{m_{\hat G}} s^n$$
$$= \sum_{r=0}^{\infty} \sum_{j=1}^{\infty} \int_0^{\infty} \sum_{k=0}^{\infty} \sum_{n=r+1}^{\infty} (\hat g^{*k})_{n-r} s^{n-r} \frac{(\hat\lambda t)^k}{k!} e^{-\hat\lambda t} \, d(\hat{V} * S^{*j})(t) \frac{\hat g_{r+j}}{m_{\hat G}} s^r$$
$$= \sum_{r=0}^{\infty} \sum_{j=1}^{\infty} \int_0^{\infty} \sum_{k=0}^{\infty} \frac{(\hat\Gamma(s))^k (\hat\lambda t)^k}{k!} e^{-\hat\lambda t} \, d(\hat{V} * S^{*j})(t) \, \hat g_{r+j} \frac{s^r}{m_{\hat G}}$$
$$= \sum_{r=0}^{\infty} \sum_{j=1}^{\infty} \hat{\mathcal{V}}\left(\hat\lambda(1 - \hat\Gamma(s))\right) \left(\mathcal{S}\left(\hat\lambda(1 - \hat\Gamma(s))\right)\right)^j \hat g_{r+j} \frac{s^r}{m_{\hat G}}$$
$$= \sum_{k=1}^{\infty} \hat{\mathcal{V}}\left(\hat\lambda(1 - \hat\Gamma(s))\right) \frac{1}{s} \mathcal{S}\left(\hat\lambda(1 - \hat\Gamma(s))\right) \frac{1 - \left(\dfrac{\mathcal{S}\left(\hat\lambda(1 - \hat\Gamma(s))\right)}{s}\right)^k}{1 - \dfrac{\mathcal{S}\left(\hat\lambda(1 - \hat\Gamma(s))\right)}{s}} \hat g_k \frac{s^k}{m_{\hat G}}$$
$$= \frac{\hat{\mathcal{V}}\left(\hat\lambda(1 - \hat\Gamma(s))\right) \mathcal{S}\left(\hat\lambda(1 - \hat\Gamma(s))\right)}{s - \mathcal{S}\left(\hat\lambda(1 - \hat\Gamma(s))\right)} \frac{1}{m_{\hat G}} \left(\sum_{k=1}^{\infty} \hat g_k s^k - \sum_{k=1}^{\infty} \hat g_k \left(\mathcal{S}\left(\hat\lambda(1 - \hat\Gamma(s))\right)\right)^k\right)$$
$$= \frac{1}{m_{\hat G}} \hat{\mathcal{V}}\left(\hat\lambda(1 - \hat\Gamma(s))\right) \mathcal{S}\left(\hat\lambda(1 - \hat\Gamma(s))\right) \frac{\hat\Gamma(s) - \hat\Gamma\left(\mathcal{S}\left(\hat\lambda(1 - \hat\Gamma(s))\right)\right)}{s - \mathcal{S}\left(\hat\lambda(1 - \hat\Gamma(s))\right)}.$$
$$(7.3.14)$$

Now the required formula (7.3.10) follows from (7.3.14), (7.3.9), (7.3.6), and (7.3.3). ∎

7.3.8. Remark. The assertion of Theorem 7.3.7 remains valid for queueing disciplines with the following properties:

(i) The service of a customer is not interrupted.
(ii) When the server becomes idle, then the next customer is chosen from the queue independently of the (required) service times of the waiting customers.

This follows directly from Theorem 7.3.7, since the process of the number of customers in a queue satisfying (i) and (ii) has the same distribution as in the FCFS queue.

Finally we prove a relation between the binomial moments of $\hat{L}_0$ and $\bar{L}(0)$ in the case of exponentially distributed service times.

7.3.9. Theorem. *For* $G^{GI}/M/1/\infty$ *we have*

$$\bar{B}_r = \frac{\varrho}{m_{\hat{G}}} \left(\sum_{j=1}^{r} (\hat{C}_{j+1} + \hat{C}_j)\, \hat{B}_{r-j} + \hat{C}_1 \hat{B}_r \right),$$

where

$$\hat{B}_r = \sum_{i=r}^{\infty} \binom{r}{i} \hat{p}_i, \quad \bar{B}_r = \sum_{i=r}^{\infty} \binom{r}{i} \bar{p}_i \quad \text{and} \quad \hat{C}_r = \sum_{i=r}^{\infty} \binom{r}{i} \hat{g}_i, \quad r \geqq 0,$$

are the binomial moments of the batch-arrival- and time-stationary numbers $\hat{L}_0$ *and* $\bar{L}(0)$ *of customers in the system, respectively, and* $\hat{C}_r$, $r \geqq 0$, *are the binomial moments of the batch size distribution.*

Proof. The proof is similar to that of Theorem 7.2.3. From Theorem 6.4.3 we have the following relation between the time- and arrival-stationary probabilities:

$$\bar{p}_i = \varrho p_{i-1}, \qquad i \geqq 1, \tag{7.3.15}$$

and

$$\bar{p}_0 = 1 - \sum_{i=1}^{\infty} \bar{p}_i = 1 - \varrho. \tag{7.3.16}$$

Combining (7.3.5), (7.3.15), and (7.3.16) we get

$$\bar{\Pi}(s) - 1 + \varrho = \bar{\Pi}(s) - \bar{p}_0 = \varrho s \Pi(s) = \frac{\varrho s(1 - \hat{\Gamma}(s))}{m_{\hat{G}}(1 - s)} \hat{\Pi}(s). \tag{7.3.17}$$

Since $\bar{B}_r = \frac{1}{r!} \lim_{s \to 1-0} \bar{\Pi}^{(r)}(s)$, $r \geqq 1$ ($f^{(r)}$ denotes the r-th derivative of a function f; $f^{(0)} = f$), we have to calculate the r-th derivative of the right-hand side of (7.3.17) for $s \to 1 - 0$. From (7.3.17) it follows that ($r \geqq 1$)

$$\bar{\Pi}^{(r)}(s) = \frac{\varrho}{m_{\hat{G}}} \sum_{j=0}^{r} \binom{r}{j} \left(\frac{1 - \hat{\Gamma}(s)}{1 - s} \right)^{(j)} \left(s\hat{\Pi}(s) \right)^{(r-j)}. \tag{7.3.18}$$

Since

$$\left(s\hat{\Pi}(s) \right)^{(i)} = s\hat{\Pi}^{(i)}(s) + i\hat{\Pi}^{(i-1)}(s), \qquad i \geqq 1,$$

and taking into account relation (7.2.9) from (7.3.18) we get

$$\bar{B}_r = \frac{\varrho}{r!\, m_{\hat{G}}} \left(\sum_{j=0}^{r-1} \binom{r}{j} j!\, \hat{C}_{j+1}\big((r-j)!\, \hat{B}_{r-j} + (r-j)!\, \hat{B}_{r-j-1}\big) + \hat{C}_{r+1} r! \right)$$

$$= \frac{\varrho}{m_{\hat{G}}} \left(\sum_{j=0}^{r-1} \hat{C}_{j+1}(\hat{B}_{r-j} + \hat{B}_{r-j-1}) + \hat{C}_{r+1} \right)$$

$$= \frac{\varrho}{m_{\hat{G}}} \left(\sum_{j=1}^{r} (\hat{C}_{j+1} + \hat{C}_j)\, \hat{B}_{r-j} + \hat{C}_1 \hat{B}_r \right). \quad \blacksquare$$

7.4. An application: the single server queue with instantaneous feedback

In this section we deal with feedback queues of the following kind (cf. Fig. 3): Customers arrive at the instants $\hat{T}_n$ $(\cdots < \hat{T}_{-1} < \hat{T}_0 = 0 < \hat{T}_1 < \cdots)$ at a single server queue with infinitely many waiting places and FCFS queueing discipline. Each arriving customer possesses a mark $\hat{K} = ([\hat{S}^i, \hat{D}^i],\ i \geqq 1)$, consisting of the potential service times $\hat{S}^i > 0$ (for more distinction we call these $\hat{S}^i$ the service quanta of the customer) and the following feedback decisions $\hat{D}^i$:

$$\hat{D}^i = \begin{cases} 1 & \text{if the customer departs after having received the service quantum } \hat{S}^i, \\ 0 & \text{if the customer feeds back after having received the service quantum } \hat{S}^i. \end{cases}$$

When the customer enters the service facility for the first time, he receives service of the length $\hat{S}^1$. After this the customer leaves the system if $\hat{D}^1 = 1$, otherwise he immediately joins the end of the queue again. The possible second service has the length $\hat{S}^2$ and the following feedback decision is $\hat{D}^2$, and so on.

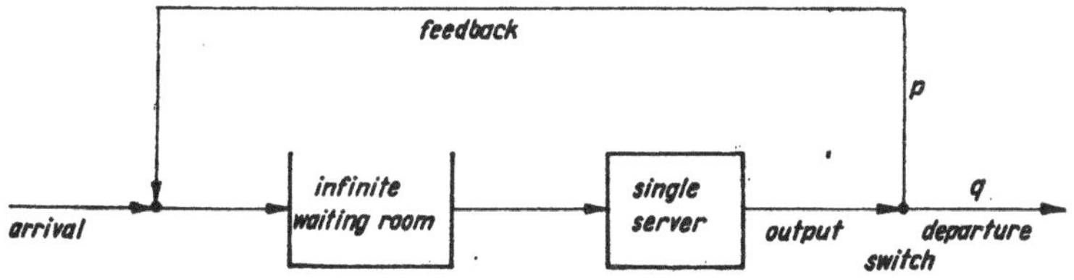

Fig. 3. Feedback queue

Queues of this kind are called *feedback queues with instantaneous feedback.* Obviously the service quanta $\hat{S}^i$ and the feedback decisions $\hat{D}^i$ with $i > \hat{G}$,

$$\hat{G} = \min \{j \geqq 1 : \hat{D}^j = 1\}, \tag{7.4.1}$$

have no influence on the behaviour of the system. The relevant part of the mark $\hat{K}$ is $\hat{Y} = [\hat{G}, \hat{S}^1, \ldots, \hat{S}^{\hat{G}}]$. Let $\hat{Y}_n = [\hat{G}_n, \hat{S}_n^{\,1}, \ldots, \hat{S}_n^{\,\hat{G}_n}]$ be the (reduced) mark of the n-th customer. The $\hat{Y}_n$ can be interpreted in the following manner:

A batch of $\hat{G}_n$ customers ($\hat{G}_n$ service quanta) arrives, and the service times (quanta) are $\hat{S}_n^i$, $i = 1, \ldots, \hat{G}_n$. Thus, a feedback queue with instantaneous feedback as described above can be considered as a batch arrival queue with an unusual queueing discipline. In the following we call this the ***batch-arrival-interpretation of the feedback queue.***

The input of the feedback queue described above is given by the sequence $(\hat{\Phi}, \hat{\mathsf{P}})$, $\hat{\Phi} = ([\hat{A}_n, \hat{K}_n])$, where $\hat{A}_n$ is the time between the arrival of the n-th and the $(n+1)$-th customer, $\hat{K}_n = ([\hat{S}_n^i, \hat{D}_n^i], i \geqq 1)$, and $\hat{S}_n^i > 0$. In order to avoid still more notation, we exclude batch arrivals of customers by assuming $\hat{A}_n > 0$. Assume that the input $\hat{\Phi}$ is a stationary sequence. Sometimes we need the following independence assumptions concerning the input $\hat{\Phi}$:

(A 7.4.1) $(\hat{A}_n)$ is independent of $(\hat{K}_n)$,
(A 7.4.2) $(\hat{A}_n)$ is a sequence of i.i.d. r.v.'s,
(A 7.4.3) $(\hat{K}_n)$ is a sequence of i.i.d. r.v.'s,
(A 7.4.4) In addition to (A 7.4.3), $(\hat{S}_n^i, i \geqq 1)$ and $(\hat{D}_n^i, i \geqq 1)$ are independent sequences of i.i.d. r.v.'s.

For the general feedback queue described above we use the notation $G^F/G/1/\infty$. Feedback queues with (A 7.4.1)–(A 7.4.4) and exponentially distributed $\hat{A}_n$ are denoted by $M^{BF}/GI/1/\infty$, where BF stands for Bernoulli feedback. (Notice that in the literature $M^{BF}/GI/1/\infty$ is called an $M/GI/1/\infty$ queue with *instantaneous Bernoulli feedback*, i.e. the usual single server queueing system with Poisson arrivals and independent service times, which additionally has a switch deciding independently each time whether a served customer departs (with probability $p = \hat{\mathsf{P}}(\hat{D}_0^1 = 1)$) or joins the queue again (feeds back with probability $q = \hat{\mathsf{P}}(\hat{D}_0^1 = 0)$). It is the simplest and most thoroughly investigated feedback queue.

In view of the batch arrival interpretation of the feedback queue, in accordance with Section 7.1 we use the following notation:

$\hat{G}_n = \min\{i \geqq 1 : \hat{D}_n^i = 1\}$,

$\hat{S}_n = \sum_{i=1}^{\hat{G}_n} \hat{S}_n^i$ – the total service time received by the n-th customer,

$m_{\hat{S}} = \mathsf{E}\hat{S}_0$ – the mean total service time received by the n-th customer,

$m_{\hat{A}} = \mathsf{E}\hat{A}_0$ – the mean interarrival time,

$\hat{\varrho} = m_{\hat{S}}/m_{\hat{A}}$ – traffic intensity,

$\hat{g}_n = \hat{\mathsf{P}}(\hat{G}_0 = n)$, $n = 1, 2, \ldots$ – distribution of the number of service quanta received by an arbitrary customer,

$m_{\hat{G}} = \mathsf{E}\hat{G}_0$.

The state $\hat{X}_n = [\hat{L}_n, \hat{X}_{n1}, \ldots, \hat{X}_{n\hat{L}_n}]$ of the feedback queue seen by the n-th customer consists of the number $\hat{L}_n$ of customers in the system and the sequences $\hat{X}_{ni} = ([\hat{S}_n(i, j), \hat{D}_n(i, j)], j \geqq 1)$ describing these customers, $i = 1, \ldots, \hat{L}_n$. Here the $\hat{S}_n(i, j)$ are the remaining (still unserved) service quanta, and the $\hat{D}_n(i, j)$ are the assigned feedback decisions of the i-th customer; $\hat{S}_n(1, 1)$

denotes the residual service time of the quantum currently in service. Then there is a function $\hat{f}$ such that the consecutive states of the system satisfy the equation

$$\hat{X}_{n+1} = \hat{f}(\hat{X}_n, [\hat{A}_n, \hat{K}_n]), \qquad n \in \mathbb{Z}. \tag{7.4.2}$$

The exact form of $\hat{f}$ is not important here. As usual, let $\hat{x}_r(x, [a_0, k_0], \ldots, [a_{r-1}, k_{r-1}])$ be the r-th iterative of $\hat{f}$.

7.4.1. Theorem. *Consider a single server feedback queue* ($\mathrm{G^F/G/1/\infty}$) *with the stationary input* $(\hat{\Phi}, \hat{\mathsf{P}})$, $\hat{\Phi} = ([\hat{A}_n, \hat{K}_n])$, $\hat{K}_n = ([\hat{S}_n{}^i, \hat{D}_n{}^i], \; i \geqq 1)$, *satisfying* $\hat{\varrho} < 1$ and

$$\lim_{n\to\infty} \frac{1}{n} \sum_{i=1}^{n} \hat{A}_n = m_{\hat{A}}, \qquad \lim_{n\to\infty} \frac{1}{n} \sum_{i=1}^{n} \hat{S}_i = m_{\hat{S}}.$$

Then the sequence $(\hat{X}_n)$,

$$\hat{X}_n = \hat{x}_n(\hat{\Phi}) = \lim_{r\to\infty} \hat{x}_r(\mathsf{O}, [\hat{A}_{n-r}, \hat{K}_{n-r}], \ldots, [\hat{A}_{n-1}, \hat{K}_{n-1}]), \tag{7.4.3}$$

satisfies (7.4.2), *and it is the unique stationary state process of the system* (O *denotes the state of the empty system*).

Proof. The feedback queue has the same empty points as the standard FCFS single server queue with service times $\hat{S}_n$ and interarrival times $\hat{A}_n$, in view of the batch arrival interpretation of the feedback queue. Consequently, cf. Section 5.4, (5.4.7), the system $\mathrm{G^F/G/1/\infty}$ has infinitely many empty points i_k: $\cdots < i_0 \leqq 0 < i_1 < \cdots$. Since the state of the feedback queue between consecutive empty points is uniquely determined by the input between these empty points, it is easy to see that

$$\hat{X}_n = \hat{x}_{n-i_k}(\mathsf{O}, [\hat{A}_{i_k}, \hat{K}_{i_k}], \ldots, [\hat{A}_{n-1}, \hat{K}_{n-1}]) \quad \text{for } i_k < n \leqq i_{k+1},$$

defines a stationary state process. The other assertions can easily be proved by means of Theorem 5.4.1, (5.4.3) and the batch arrival interpretation of a feedback queue, cf. also Theorem 5.4.3. ∎

Now we deal with the stationary distribution of the state of the system at (immediately after) an arbitrary *output instant* (i.e. at an instant at which the service of a quantum is finished) and an arbitrary *departure instant* (a departure instant is defined as an output instant at which a customer finally leaves the system). For this purpose we start with the time-stationary model: Let $(\bar{\Phi}, \bar{\mathsf{P}})$, $\bar{\Phi} = ([\bar{T}_n, \bar{K}_n])$, $\bar{K}_n = ([\bar{S}_n{}^i, \bar{D}_n{}^i], \; i \geqq 1)$, be the time-stationary input corresponding to $\hat{\Phi} = ([\hat{A}_n, \hat{K}_n])$ according to Theorem 3.4.1. Notice that $\mathsf{E}\hat{N}(1) = 1/\mathsf{E}\hat{A}_0$, where

$$\hat{N}(u) = \max\{n : \bar{T}_n \leqq u\}. \tag{7.4.4}$$

The time-stationary model $\left[(\bar{X}(t)), \; ([\bar{T}_n, [\bar{X}_n, \bar{K}_n]])\right]$, $\bar{X}(t) = [\bar{L}(t), \bar{X}_1(t), \ldots, \bar{X}_{\bar{L}(t)}(t)]$, $\bar{X}_i(t) = \left([\bar{S}(i, j)(t), \; \bar{D}(i, j)(t)], \; j \geqq 1\right)$, is given by Lemma 6.1.1. (Here $\bar{L}(t)$ is the number of customers in the system at time t.) Let

$$\bar{L}_q(t) = \sum_{i=1}^{\bar{L}(t)} \min\{j \geqq 1 : \bar{D}(i, j)(t) = 1\}$$

be the number of quanta in the system at time t in the time-stationary model. Then a point $\bar{t}^*$ is an output instant iff

$$\bar{L}_q(\bar{t}^* - 0) - \bar{L}_q(\bar{t}^* + 0) + \sum_{n=-\infty}^{\infty} \bar{G}_n 1\{\bar{t}^* = \bar{T}_n\} > 0;$$

$$(\bar{G}_n = \min\{i \geqq 1, \bar{D}_n^i = 1\});$$

$\bar{\bar{t}}^*$ is a departure instant iff

$$\bar{L}(\bar{\bar{t}}^* - 0) - \bar{L}(\bar{\bar{t}}^* + 0) + \sum_{n=-\infty}^{\infty} 1\{\bar{\bar{t}}^* = \bar{T}_n\} > 0.$$

In view of $\bar{S}_n^i > 0$, $n \in \mathbb{Z}$, $i \geqq 1$, there are no multiple output and departure instants, and hence we can number them as follows: $\cdots < \bar{T}_0^* \leqq 0 < \bar{T}_1^* < \cdots; \cdots < \bar{\hat{T}}_0^* \leqq 0 < \bar{\hat{T}}_1^* < \cdots$. Let $\bar{X}_n^* = \bar{X}(\bar{T}_n^* + 0)$ and $\bar{\hat{X}}_n^* = \bar{X}(\bar{\hat{T}}_n^* + 0)$ be the states of the system immediately after the output and departure epochs, respectively. Then $([\bar{T}_n^*, \bar{X}_n^*])$ and $([\bar{\hat{T}}_n^*, \bar{\hat{X}}_n^*])$ are stationary marked point processes (MPP's). Applying Theorem 3.4.1 to these point processes, we get the existence of stationary sequences (Φ^*, P^*), $\Phi^* = ([A_n^*, X_n^*])$, and $(\hat{\Phi}^*, \hat{\mathsf{P}}^*)$, $\hat{\Phi}^* = ([\hat{A}_n^*, \hat{X}_n^*])$, respectively. The r.v.'s A_n^* and $\hat{A}_n^*$ are the interoutput times and the interdeparture times, respectively. In particular, for the intensities we have

$$\lambda^* = \mathsf{E}N^*(1) = (\mathsf{E}A_0^*)^{-1}, \qquad \hat{\lambda}^* = \mathsf{E}\hat{N}^*(1) = (\mathsf{E}\hat{A}_0^*)^{-1}, \tag{7.4.5}$$

where

$$N^*(u) = \max\{n : \bar{T}_n^* \leqq u\}, \hat{N}^*(u) = \max\{n : \bar{\hat{T}}_n^* \leqq u\}. \tag{7.4.6}$$

The sequence (Φ^*, P^*) will be called the *output stationary model* and $(\hat{\Phi}^*, \hat{\mathsf{P}}^*)$ will be called the *departure stationary model.* In accordance with the interpretations given in Chapter 3, Φ^* describes the system viewed by an arbitrary service quantum leaving the server, and $\hat{\Phi}^*$ describes the system viewed by an arbitrary departing customer. Let

$$T_0^* = 0; \quad T_n^* = \sum_{i=0}^{n-1} A_i^*, \quad n > 0; \quad T_n^* = -\sum_{i=n}^{-1} A_i^*, \quad n < 0;$$

$$\hat{T}_0^* = 0; \quad \hat{T}_n^* = \sum_{i=0}^{n-1} \hat{A}_i^*, \quad n > 0; \quad \hat{T}_n^* = -\sum_{i=n}^{-1} \hat{A}_i^*, \quad n > 0.$$

Denote the service quantum finished at the instant T_n^* by S_n^* and the corresponding feedback decision at T_n^* by D_n^*. Thus, $([S_n^*, D_n^*])$ is the stationary sequence of service quanta with the assigned feedback decision which are served in this succession by the server. We denote the total output-stationary number of quanta demanded by the customer leaving the server at T_0^* by the additional variable G_0^*. The "output-stationary" variable N_0^* takes the value i ($i \geqq 1$) if the service quantum finished at T_0^* was the i-th quantum of the corresponding customer. G_0^* and N_0^* correspond to the r.v.'s G_0 and N_0 introduced in 7.1 in view of the batch arrival interpretation of the feedback queue. Let (Φ, P) be the corresponding "quanta stationary" input. In the following we need

7.4.2. Lemma. *For the feedback queue* $G^F/G/1/\infty$ *it holds that*

$$P(G_0 = i, N_0 = j) = P^*(G_0^* = i, N_0^* = j), \qquad i, j \geqq 0.$$

Proof. We start with the batch-arrival interpretation of the feedback queue. Denote by $([\overline{T}_n{}^q, \overline{S}_n{}^q])$ the time-stationary input of the corresponding batch arrival single server queue, $\overline{T}_n{}^q \leqq \overline{T}^q_{n+1}$. The $\overline{T}_n{}^q$ are the arrival instants of the service quanta, and the $\overline{S}_n{}^q$ are their required amounts of work. Let $\overline{R}_n$ be the sojourn time of the n-th service quantum in the system (i.e. $\overline{T}_n{}^q + \overline{R}_n$ is the output instant), $\overline{G}_n$ the total time-stationary number of quanta demanded by the customer (batch of quanta) to which the n-th quantum belongs, and $\overline{N}_n$ the time-stationary position of the n-th quantum in his own batch of service quanta. Then, $\overline{\Psi} = ([\overline{T}_n{}^q, [\overline{R}_n, \overline{G}_n, \overline{N}_n]])$ is a stationary MPP with mark space $\mathbb{R}_+ \times \mathbb{Z}_+ \times \mathbb{Z}_+$. Let ν be its intensity measure and

$$\lambda_{ij}^* = \mathsf{E} \# \{n: 0 \leqq \overline{T}_n{}^q + \overline{R}_n \leqq 1, \overline{G}_n = i, \overline{N}_n = j\},$$

$$\lambda_{ij} = \mathsf{E} \# \{n: 0 \leqq \overline{T}_n \leqq 1, \overline{G}_n = i, \overline{N}_n = j\}.$$

Applying Campbell's Theorem 3.2.4 to the function

$$h(t, r, g, k) = 1\{0 < t + r \leqq 1, g = i, k = j\}$$

we get

$$\lambda_{ij}^* = \mathsf{E} \sum_n h(\overline{T}_n{}^q, \overline{R}_n, \overline{G}_n, \overline{N}_n)$$

$$= \int_0^\infty \int_0^\infty \int_0^\infty \int_{-\infty}^\infty h(t, r, g, k)\, \nu(\mathrm{d}t \times \mathrm{d}r \times \mathrm{d}g \times \mathrm{d}k)$$

$$= \int_0^\infty \int_0^\infty \int_0^\infty \int_{-\infty}^\infty h(t, r, g, k)\, \mathrm{d}t\, \nu\big((01] \times \mathrm{d}r \times \mathrm{d}g \times \mathrm{d}k\big)$$

$$= \int_0^\infty \int_0^\infty \int_0^\infty 1\{g = i, k = j\}\, \nu\big((01] \times \mathrm{d}r \times \mathrm{d}g \times \mathrm{d}k\big)$$

$$= \nu\big((01] \times [0, \infty) \times \{i\} \times \{j\}\big) = \lambda_{ij}.$$

In particular,

$$\lambda = \mathsf{E} \# \{n: 0 < \overline{T}_n{}^q \leqq 1\} = \mathsf{E} \# \{n: 0 < \overline{T}_n{}^q + \overline{R}_n \leqq 1\} = \lambda^*.$$

By means of formula (3.4.2), the assertion follows immediately. ∎

7.4.3. Lemma. *For every feedback queue* $G^F/G/1/\infty$ *satisfying the conditions of Theorem* 7.4.1, *the probability* q_n *that an arbitrary customer departs after having finished the service of his n-th service quantum is determined by*

$$q_n = P^*(D_0^* = 1 \mid N_0^* = n) = \frac{\hat{g}_n}{\sum_{k=n}^{\infty} \hat{g}_k}. \tag{7.4.7}$$

The "unconditional" departure probability is

$$P^*(D_0^* = 1) = \frac{1}{m_{\hat{G}}}. \tag{7.4.8}$$

Proof. It holds that

$$\mathsf{P}^*(D_0^* = 1 \mid N_0^* = n) = \mathsf{P}^*(G_0^* = N_0^* \mid N_0^* = n)$$
$$= \frac{\mathsf{P}^*(G_0^* = N_0^* = n)}{\mathsf{P}^*(N_0^* = n)}. \tag{7.4.9}$$

Taking into account the batch arrival interpretation of the feedback queue, Lemma 7.4.2 and (7.1.10) yield

$$\mathsf{P}^*(G_0^* = i, N_0^* = j) = \frac{1}{m_{\hat{G}}} \hat{g}_i \mathbf{1}\{j \leqq i\}. \tag{7.4.10}$$

Thus we obtain

$$\mathsf{P}^*(G_0^* = N_0^* = n) = \frac{\hat{g}_n}{m_{\hat{G}}}, \qquad \mathsf{P}^*(N_0^* = n) = \frac{1}{m_{\hat{G}}} \sum_{k=n}^{\infty} \hat{g}_k, \tag{7.4.11}$$

which implies (7.4.7) in view of (7.4.9). Analogously we get

$$\mathsf{P}^*(D_0^* = 1) = \mathsf{P}^*(G_0^* = N_0^*) = \sum_{n=1}^{\infty} \mathsf{P}^*(G_0^* = N_0^* = n) = \frac{1}{m_{\hat{G}}}. \quad \blacksquare$$

7.4.4. Remark. The distribution $(\hat{g}_n, n \in \mathbb{Z}_+)$ of the number of service quanta demanded by an arbitrary customer is determined by

$$\hat{g}_n = p_1 p_2 \cdots p_{n-1} q_n, \qquad n = 1, 2, \ldots, \tag{7.4.12}$$

where $p_i = 1 - q_i$, $i = 1, 2, \ldots$, are the conditional feedback probabilities (under the condition $N_0^* = i$).

Namely, from (7.4.10) and (7.4.11) we get

$$p_i = \mathsf{P}^*(D_0^* = 0 \mid N_0^* = i)$$
$$= \frac{\mathsf{P}^*(G_0^* > i = N_0^*)}{\mathsf{P}^*(N_0^* = i)} = \sum_{j=i+1}^{\infty} \hat{g}_j \Big/ \sum_{j=i}^{\infty} \hat{g}_j, \qquad i \geqq 1,$$

and taking into account (7.4.7), relation (7.4.12) follows immediately.

In the following theorems we provide some characteristics for the general $\mathrm{G^F/G/1/\infty}$ feedback queue.

7.4.5. Theorem. *In a feedback queue* $\mathrm{G^F/G/1/\infty}$ *it holds that*

$$m_{\hat{A}} = \mathsf{E}\hat{A}_0 = \mathsf{E}\hat{A}_0^* = m_{\hat{G}} \mathsf{E} A_0^*, \tag{7.4.13}$$

$$m_{\hat{S}} = \mathsf{E}\hat{S}_0 = \mathsf{E} \sum_{i=1}^{\hat{G}_0} \hat{S}_0^i = m_{\hat{G}} \mathsf{E} S_0^*. \tag{7.4.14}$$

Thus,

$$\hat{\varrho} = \frac{m_{\hat{S}}}{m_{\hat{A}}} = m_{\hat{G}} \frac{\mathsf{E} S_0^*}{\mathsf{E}\hat{A}_0} = \frac{\mathsf{E} S_0^*}{\mathsf{E} A_0^*}.$$

Proof. Consider the input $\hat{\Phi} = ([\hat{A}_n, \hat{K}_n])$, $\hat{K}_n = ([\hat{S}_n^i, \hat{D}_n^i], i \geqq 1)$. As mentioned above, we can interpret $\hat{\Phi}$ as the input of a suitably chosen batch arrival queue (with an unusual queueing discipline). Let (Φ, P), $\Phi = ([A_n, S_n])$, be the corresponding arrival-stationary input in the sense of Theorem 7.1.1, where A_n is the interarrival time between arrivals of quanta (i.e. $A_n = 0$ if S_n and S_{n+1} belong to the same customer), and S_n is the length of the n-th service quantum. Analogously to the proof of (6.4.1), one can show that

$$\hat{\lambda} = \hat{\lambda}^*, \qquad \lambda = \lambda^*, \tag{7.4.15}$$

where

$$\hat{\lambda} = \mathsf{E}\hat{N}(1), \quad \hat{\lambda}^* = \mathsf{E}\hat{N}^*(1), \quad \lambda^* = \mathsf{E}N^*(1), \quad \lambda = \mathsf{E}\sum_{i=1}^{\hat{N}(1)} \bar{G}_i$$

are the intensity of the customer arrivals, the intensity of departures, the output intensity, and the intensity of quanta arrivals, respectively, cf. (7.4.4), (7.4.6). From (7.4.5), (7.4.4), and $\lambda = 1/\mathsf{E}A_0$ and using (7.4.15) we get $\mathsf{E}\hat{A}_0 = \mathsf{E}\hat{A}_0^*$ and $\mathsf{E}A_0 = \mathsf{E}A_0^*$. Thus, (7.4.13) follows from (7.1.14).

The proof of (7.4.14) runs in a similar way: For $x > 0$ we can consider the arrival instants of quanta S_n with $S_n > x$ and also the output instants of quanta S_n^* with $S_n^* > x$. Similarly to (7.4.15) we get the equality of the intensities of these point processes and thus $\mathsf{P}(S_0 > x) = \mathsf{P}^*(S_0^* > x)$ for all x. Hence $\mathsf{E}S_0 = \mathsf{E}S_0^*$, and (7.1.14) yields (7.4.14). ∎

We introduce the following stationary distributions of the queue size (with respect to different embedded instants):

$\hat{p}_k = \hat{\mathsf{P}}(\hat{L}_0 = k)$, $k = 0, 1, \ldots$ – the probability that there are k customers in the system immediately before a customer arrives;

$\hat{p}_k^* = \hat{\mathsf{P}}^*(\hat{L}_0^* = k)$, $k = 0, 1, \ldots$ – the probability that there are k customers in the system immediately after a departure;

$p_k^* = \hat{\mathsf{P}}^*(L_0^* = k)$, $k = 0, 1, \ldots$ – the probability that there are k customers in the system immediately after an output;

$\bar{p}_k = \bar{\mathsf{P}}(\bar{L}(0) = k)$, $k = 0, 1, \ldots$ – the probability that there are k customers in the system at an arbitrary instant.

7.4.6. Theorem. *In a* $\mathrm{G^F/G/1/\infty}$ *queue satisfying the conditions of Theorem* 7.4.1, *it holds that*

$$\hat{p}_k^* = \hat{p}_k, \qquad k = 0, 1, \ldots, \tag{7.4.16}$$

$$p_0^* = \frac{1}{m_{\hat{G}}}\hat{p}_0. \tag{7.4.17}$$

If the independence assumptions (A 7.4.1), (A 7.4.3), *and* (A 7.4.4) *hold, then*

$$p_k^* = \frac{1}{m_{\hat{G}}}\hat{p}_k + \left(1 - \frac{1}{m_{\hat{G}}}\right)\hat{p}_{k-1}, \qquad k = 1, 2, \ldots. \tag{7.4.18}$$

Proof. The proof of (7.4.16) is similar to that of Theorem 6.4.2. Since the departure instants are just the output instants T_n^* with $D_n^* = 1$, it holds that $\hat{\mathsf{P}}^*(\hat{L}_0^* = k) = \mathsf{P}^*(L_0^* = k \mid D_0^* = 1)$ according to Theorem A 1.3.2. Since the system cannot be empty after an output instant which is not a departure instant, $\mathsf{P}^*(L_0^* = 0, D_0^* = 0) = 0$. Thus

$$p_0^* = \mathsf{P}^*(L_0^* = 0 \mid D_0^* = 1)\,\mathsf{P}^*(D_0^* = 1) = \frac{1}{m_{\hat{G}}}\hat{p}_0^*, \tag{7.4.19}$$

in view of (7.4.8).

The number of customers immediately before the 0-th output instant is $L_0^* + D_0^*$. By the construction of $\hat{\Phi}$ and Φ^* and taking into account the independence assumptions, it follows that $L_0^* + D_0^*$ is independent of D_0^*. Thus we get

$$\begin{aligned}\mathsf{P}^*(L_0^* = k \mid D_0^* = 0) &= \mathsf{P}^*(L_0^* + D_0^* = k \mid D_0^* = 0)\\ &= \mathsf{P}^*(L_0^* + D_0^* = k \mid D_0^* = 1)\\ &= \mathsf{P}^*(L_0^* = k - 1 \mid D_0^* = 1),\end{aligned}$$

and taking into account (7.4.8), finally we obtain

$$\begin{aligned}p_k^* &= \frac{1}{m_{\hat{G}}} \mathsf{P}^*(L_0^* = k \mid D_0^* = 1) + \left(1 - \frac{1}{m_{\hat{G}}}\right) \mathsf{P}^*(L_0^* = k \mid D_0^* = 0)\\ &= \frac{1}{m_{\hat{G}}} \hat{\mathsf{P}}^*(\hat{L}_0^* = k) + \left(1 - \frac{1}{m_{\hat{G}}}\right) \hat{\mathsf{P}}^*(\hat{L}_0^* = k - 1), \quad k = 1, 2, \ldots\end{aligned} \tag{7.4.20}$$

Now we get the assertions (7.4.17) and (7.4.18) by applying (7.4.16) to (7.4.19) and (7.4.20). ∎

Now we deal with the work load in the feedback queue. Taking into account the batch arrival interpretation of the $G^F/G/1/\infty$ queue, it is obvious that the process of work load in the $G^F/G/1/\infty$ queue is just the same as the work load process in the $G^G/G/1/\infty$ batch arrival queue (without feedback) with the input $([\hat{A}_n, \hat{G}_n, \hat{\underline{S}}_n])$, $\hat{\underline{S}}_n = [\hat{S}_n^1, \ldots, \hat{S}_n^{\hat{G}_n}]$, and FCFS queueing discipline. Denote the distribution functions of the work load in the time-stationary model of the $G^F/G/1/\infty$ queue by $\bar{\mathsf{V}}(x)$ and in the batch-arrival-stationary model by $\hat{\mathsf{V}}(x)$. Their Laplace-Stieltjes transforms are denoted by $\bar{\mathcal{V}}(s)$ and $\hat{\mathcal{V}}(s)$, respectively. From Theorem 7.3.6 we obtain

7.4.7. Theorem. *In a feedback queue with independence assumptions* (A 7.4.1) *and* (A 7.4.3) *we have*

$$\bar{\mathsf{V}}(x) = 1 - \hat{\varrho} + \hat{\varrho}(\hat{\mathsf{V}} * \hat{\mathsf{S}}_R)(x),$$

where

$$\hat{\mathsf{S}}_R(x) = \frac{1}{m_{\hat{S}}} \int_0^x (1 - \hat{\mathsf{S}}(u))\, du, \quad \hat{\mathsf{S}}(u) = \hat{\mathsf{P}}(\hat{S}_0 < u) = \hat{\mathsf{P}}\left(\sum_{i=1}^{\hat{G}_0} \hat{S}_0^i < u\right).$$

If, in addition, (A 7.4.2) *and* $\hat{\mathsf{P}}(\hat{A}_0 < x) = 1 - e^{-\hat{\lambda} x}$, *then*

$$\bar{\mathcal{V}}(s) = \hat{\mathcal{V}}(s) = \frac{(1 - \hat{\varrho})\, s}{s - \hat{\lambda} + \hat{\lambda} \hat{\mathcal{S}}(s)},$$

where $\hat{\mathcal{S}}(s)$ *is the LST of* $\hat{\mathsf{S}}(x)$.

By means of Theorem 7.3.7 it is not very difficult to prove the following analogue of the well-known Pollaczek-Khinchin formula for $M/GI/1/\infty$ queues with instantaneous Bernoulli feedback.

7.4.8. Theorem. *In an* $M^{BF}/GI/1/\infty$ *queue with (instantaneous) Bernoulli feedback probability* p *it holds that*

$$\bar{\Pi}(s) = \hat{\Pi}(s) = \frac{(1-\hat{\varrho})(1-s)}{1 - \dfrac{s}{\mathscr{S}(\hat{\lambda} - \hat{\lambda}s)}},$$

where $\bar{\Pi}(s) = \sum_{k=0}^{\infty} \bar{p}_k s^k$, $\hat{\Pi}(s) = \sum_{k=0}^{\infty} \hat{p}_k s^k$ *are the PGF of the queue size at arbitrary and at arrival instants, respectively. The LST* $\hat{\mathscr{S}}(s)$ *of* $\hat{S}(x)$ *is given by* $\hat{\mathscr{S}}(s) = q\mathscr{S}(s)/(1 - p\mathscr{S}(s))$, *where* $\mathscr{S}(s)$ *is the LST of* $S(x) = \hat{P}(\hat{S}_0^1 < x)$, *and* $q = 1 - p$.

Proof. In an $M^{BF}/GI/1/\infty$ queue the number of service quanta is geometrically distributed, i.e. $\hat{g}_n = qp^{n-1}$, $n \geqq 1$. Because of the memoryless property of the geometric distribution, the number of unserved quanta of a customer in the system is also geometrically distributed. Thus the following relationship between the number of unserved quanta $\bar{L}_q(0)$ and the number $\bar{L}(0)$ of numbers in the system holds

$$\bar{P}(\bar{L}_q(0) = n) = \sum_{k=0}^{n} \bar{P}(\bar{L}(0) = k)\, \hat{g}_n^{*k}, \qquad n = 0, 1, \ldots,$$

where $(\hat{g}_n^{*k}, n \geqq k)$ is the k-fold convolution of the geometric distribution $(\hat{g}_n, n \geqq 1)$. Thus the PGF's obey

$$\bar{\Pi}_q(s) = \bar{\Pi}\left(\frac{qs}{1 - ps}\right) \tag{7.4.21}$$

(note that the PGF of $(\hat{g}_n, n \geqq 1)$ is $qs/(1 - ps)$). Since the length of the service quanta as well as the feedback decisions are i.i.d. r.v.'s, the number of service quanta in the $M^{BF}/GI/1/\infty$ queue is equal to the number of quanta in the corresponding FCFS batch arrival queue. From Theorem 7.3.7 we obtain

$$\bar{\Pi}_q(s) = \frac{(1-\hat{\varrho})(1-s)}{1 - \dfrac{s}{\mathscr{S}(\hat{\lambda} - \hat{\lambda} qs(1-ps)^{-1})}}.$$

Using (7.4.21), after some simple algebra we get

$$\bar{\Pi}(s) = \bar{\Pi}_q\left(\frac{s}{q + ps}\right) = \frac{(1-\hat{\varrho})(1-s)}{1 - \dfrac{s}{\mathscr{S}(\hat{\lambda} - \hat{\lambda}s)}}. \qquad \blacksquare$$

7.5. Comparing batch delays and customer delays

In an arbitrary queueing system the time between the arrival of a customer and the beginning of its service is called *delay*. When studying the performance of complicated communication systems in which messages are divided into packets for transmission, the packets can be interpreted as customers, the message as a batch, and the delay of a message is the delay of the last customer in this batch to enter service. Often the delay of a customer is easier to analyze than the batch delay. Hence it is useful to have relations between batch delays and customer delays. Our approach to this problem is based on results concerning stationary sequences with finite groups given in A 1.4.

First we consider an arbitrary queueing system with batch arrivals. There may be one or many servers, and the queueing discipline is arbitrary (not necessarily work conserving). In particular, dead periods are allowed, i.e. periods in which servers do not work though customers are waiting in the queue. We assume that the system is in steady state. Consider the stationary MPP $([\overline{T}_n', \overline{W}_n'])$, where $\overline{T}_n'$ are the beginnings of service of customers numbered according to their entering service ($\overline{T}_n' \leqq \overline{T}_{n+1}'$ and $\overline{T}_1' < 0 < \overline{T}_0'$) and W_n' denotes the delay of the n-th customer to enter service. (If many customers enter service simultaneously, they will be numbered according to an appropriate rule (e.g. randomly)). Define

$$\overline{B}_n' = \begin{cases} 0 & \text{if the } n\text{-th customer is the first one in his batch,} \\ n - k & \text{if } k\ (< n) \text{ is the number of the preceeding customer} \\ & \text{belonging to the same batch as the } n\text{-th one,} \end{cases}$$

$$\overline{C}_n' = \begin{cases} 0 & \text{if the } n\text{-th customer is the last one in his batch,} \\ s - n & \text{if } s\ (> n) \text{ is the number of the next customer belong-} \\ & \text{ing to the same batch as the } n\text{-th one.} \end{cases}$$

The MPP $\left([\overline{T}_n', [\overline{W}_n', \overline{B}_n', \overline{C}_n']]\right)$ is stationary, too. From Theorem 3.4.1 we obtain a stationary sequence $\left((K_n), \mathsf{P}\right)$, $K_n = [W_n, B_n, C_n]$, which describes the customer-stationary delays and the batch memberships of the customers. The quantity W_0 can be interpreted as the delay of an arbitrary customer entering service; the d.f. $\mathsf{W}(x) = \mathsf{P}(W_0 \leqq x)$ is called the *customer delay distribution*. From Theorem A 1.3.1 we obtain a sequence $\left((\tilde{K}_n), \tilde{\mathsf{P}}\right)$, $\tilde{K}_n = [\tilde{W}_n, \tilde{B}_n, \tilde{C}_n]$, with

$$\tilde{\mathsf{P}}(\cdot) = \mathsf{P}\left((K_n) \in (\cdot) \mid B_0 = 0\right).$$

The r.v. $\tilde{W}_n$ is the batch-stationary delay of the n-th customer; $\tilde{B}_n$ and $\tilde{C}_n$ are the distances between the n-th customer and his predecessor and successor in his own batch, respectively. It holds that (cf. Theorem A 1.3.1):

$$\tilde{\mathsf{P}}\left((\tilde{K}_n) \in (\cdot)\right) = \tilde{\mathsf{P}}\left((\tilde{K}_{n+\tilde{\nu}_b}) \in (\cdot)\right),$$

where

$$\tilde{\nu}_b = \min\{j : j \geqq 1, \tilde{B}_j = 0\},$$

and

$$\mathsf{P}\left((K_n) \in (\cdot)\right) = \frac{1}{\mathsf{E}\tilde{\nu}_b} \sum_{j=0}^{\infty} \tilde{\mathsf{P}}\left(\tilde{\nu}_b > j, (\tilde{K}_{n+j}) \in (\cdot)\right). \tag{7.5.1}$$

In order to define the batch delay distribution, we consider the conditiona probabilities

$$\check{\mathsf{P}}(\cdot) = \mathsf{P}\left((K_n) \in (\cdot) \mid C_0 = 0\right). \tag{7.5.2}$$

Let $(\check{K}_n) = ([\check{W}_n, \check{B}_n, \check{C}_n])$ be a sequence with the distribution $\check{\mathsf{P}}$. The sequence $(\check{K}_n)$ is "batch-stationary" in the following sense:

$$\check{\mathsf{P}}\big((\check{K}_n) \in (\cdot)\big) = \check{\mathsf{P}}\big((\check{K}_{n+\check{\nu}_c}) \in (\cdot)\big),$$

where

$$\check{\nu}_c = \min\{j : j \geqq 1, \check{C}_j = 0\}.$$

Thus $\check{\mathsf{W}}_n$ are the customer delays from the point of view of the last customer of an arbitrary batch. We call $\check{\mathsf{W}}(x) = \check{\mathsf{P}}(\check{W}_0 \leqq x)$ the *batch delay distribution.*

Remember that in general the customers of a batch are not served in succession. For this reason formulae (7.5.1), (7.5.2) are not convenient for investigating relations between batch and customer delay distributions. However, from Section A 1.4, Theorem A 1.4.1 we see that P can be obtained from $\tilde{\mathsf{P}}$ or $\check{\mathsf{P}}$ by averaging:

$$\begin{aligned}\mathsf{P}\big((K_n) \in (\cdot)\big) &= \frac{1}{\mathsf{E}\tilde{G}_0} \sum_{j=0}^{\infty} \tilde{\mathsf{P}}\big(\tilde{G}_0 > j, \tilde{\theta}^j(\tilde{K}_n) \in (\cdot)\big) \\ &= \frac{1}{\mathsf{E}\check{G}_0} \sum_{j=0}^{\infty} \check{\mathsf{P}}\big(\check{G}_0 > j, \check{\theta}^j(\check{K}_n) \in (\cdot)\big),\end{aligned} \tag{7.5.3}$$

where $\tilde{\theta}$ and $\check{\theta}$ are the random shifts defined by

$$\tilde{\theta}(\tilde{K}_n) = (\tilde{K}_{n+\tilde{C}_0}), \qquad \check{\theta}(\check{K}_n) = (\check{K}_{n-\check{B}_0}),$$

and $\tilde{G}_0$ and $\check{G}_0$ are the numbers of customers of the batch containing the customer with number zero in $(\tilde{K}_n)$ and $(\check{K}_n)$, respectively. In particular,

$$\mathsf{P}(B_0 = 0) = \mathsf{P}(C_0 = 0) = 1/\mathsf{E}\tilde{G}_0 = 1/\mathsf{E}\check{G}_0. \tag{7.5.4}$$

Let $\tilde{n}_j$, $j = 1, \ldots, \tilde{G}_0$, be the (random) indices of customers belonging to the batch containing the customer with number zero in $(\tilde{K}_n)$. Then (7.5.3) yields

$$\begin{aligned}\mathsf{W}(x) &= \frac{1}{\mathsf{E}\tilde{G}_0} \sum_{j=1}^{\infty} \tilde{\mathsf{P}}(\tilde{G}_0 \geqq j, \tilde{W}_{\tilde{n}_j} \leqq x) \\ &= \sum_{j=1}^{\infty} \sum_{i=j}^{\infty} \frac{1}{m_{\tilde{G}}} \tilde{g}_i \tilde{\mathsf{P}}(\tilde{W}_{\tilde{n}_j} \leqq x \mid \tilde{G}_0 = i),\end{aligned} \tag{7.5.5}$$

where $m_{\tilde{G}} = \mathsf{E}\tilde{G}_0$ and $\tilde{g}_i = \tilde{\mathsf{P}}(\tilde{G}_0 = i)$. From (7.5.2), (7.5.3), and (7.5.4), for the batch delay d.f. $\check{\mathsf{W}}(x)$ we get

$$\begin{aligned}\check{\mathsf{W}}(x) &= \frac{\mathsf{P}(W_0 \leqq x, C_0 = 0)}{\mathsf{P}(C_0 = 0)} = \sum_{j=1}^{\infty} \tilde{\mathsf{P}}(\tilde{G}_0 \geqq j, \tilde{W}_{\tilde{n}_j} \leqq x, \tilde{C}_{\tilde{n}_j} = 0) \\ &= \sum_{j=1}^{\infty} \tilde{\mathsf{P}}(\tilde{G}_0 = j, \tilde{W}_{\tilde{n}_j} \leqq x) = \sum_{j=1}^{\infty} \tilde{g}_j \tilde{\mathsf{P}}(\tilde{W}_{\tilde{n}_j} \leqq x \mid \tilde{G}_0 = j).\end{aligned} \tag{7.5.6}$$

So far the considerations hold for arbitrary queueing disciplines. Now we introduce a class of queueing disciplines for which some useful relations between customer and batch delays can be derived.

7.5.1. Definition. Consider a batch arrival queueing system with an arbitrary queueing discipline for which the batch-stationary sequence $\left(([\tilde{W}_n, \tilde{B}_n, \tilde{C}_n]), \tilde{\mathsf{P}}\right)$ is uniquely determined. The queueing discipline is called *impartial* if for all $j \geqq 1$ and all Borel subsets $D \subseteq \mathbb{R}_+$ the event $\{\tilde{W}_{\tilde{n}_j} \in D, \tilde{G}_0 \geqq j\}$ is independent of the events $\{\tilde{G}_0 = j\}, \{\tilde{G}_0 = j+1\}, \ldots$

The definition of an impartial queueing discipline means that the delay of the j-th customer of a batch does not depend on how many customers belonging to the same batch follow him. The word "impartial queueing discipline" is due to the idea that such a discipline should have the following properties:

(i) it selects batches independently of their size,
(ii) it selects customers independently of their service times,
(iii) the assignment of customers to servers is independent of the present batch sizes,
(iv) the lengths of the dead periods are independent of the present batch sizes.

Examples of impartial disciplines are (cf. Halfin (1983))

1. FCFS for batches and for customers within batches,
2. LCFS for batches and for customers within batches,
3. random choice of a batch, and then a random choice of a customer from this batch (but not a random choice among all waiting customers because this would favour large batches),
4. random choice of a batch, and then serving all the customers of this batch in FCFS, LCFS, or random order.

We recall that a discrete probability distribution $(p_n, n \geqq 1)$ with finite mean m_P has the property NBUE (NWUE) iff

$$(q_n, n \geqq 1) \overset{\mathcal{D}}{\leqq} (\overset{\mathcal{D}}{\geqq}) (p_n, n \geqq 1),$$

where

$$q_n = m_P^{-1} \sum_{k=n}^{\infty} p_k,$$

i.e.

$$\sum_{n=1}^{j} q_n \geqq (\leqq) \sum_{n=1}^{j} p_n, \qquad j = 1, 2, \ldots,$$

or, equivalently, iff

$$\sum_{n=1}^{\infty} f_n q_n \leqq (\geqq) \sum_{n=1}^{\infty} f_n p_n \tag{7.5.7}$$

holds for all real-valued, non-decreasing sequences $(f_n, n \geqq 1)$ for which the sums in (7.5.7) converge, cf. e.g. Stoyan (1983).

7.5.2. Theorem. *Consider a batch arrival queueing system with the following properties:*

– *The batch sizes form a sequence of i.i.d. r.v.'s which is independent of the interarrival times and of the service times.*

— *The service times are independent of the interarrival times.*
— *The queueing discipline is impartial.*

Then we have the following statements for the customer and batch delay distributions $\mathsf{W}(x)$ *and* $\check{\mathsf{W}}(x)$*:*

(i) *If the batch size distribution* $(\bar{g}_i, i \geqq 1)$ *has the property NBUE (NWUE), then* $\mathsf{W} \overset{\mathfrak{D}}{\leqq} (\overset{\mathfrak{D}}{\geqq}) \check{\mathsf{W}}$, *i.e.*

$$\mathsf{W}(x) \geqq (\leqq) \check{\mathsf{W}}(x) \quad \textit{for all} \quad x \geqq 0. \tag{7.5.8}$$

(ii) *If the batch size is geometrically distributed, then*

$$\mathsf{W} = \check{\mathsf{W}}. \tag{7.5.9}$$

Proof. Since the queueing discipline is impartial, (7.5.5) and (7.5.6) reduce to

$$\mathsf{W}(x) = \sum_{j=1}^{\infty} r_j \tilde{\mathsf{P}}(\tilde{W}_{\tilde{n}_j} \leqq x), \qquad \check{\mathsf{W}}(x) = \sum_{j=1}^{\infty} \bar{g}_j \tilde{\mathsf{P}}(\tilde{W}_{\tilde{n}_j} \leqq x), \tag{7.5.10}$$

where

$$r_j = \frac{1}{m_{\bar{G}}} \sum_{i=j}^{\infty} \bar{g}_i .$$

Since the customers are arranged in the order of entering service, the delay $\tilde{W}_{\tilde{n}_j}$ is non-decreasing with $j = 1, 2, \ldots$, i.e. $\tilde{\mathsf{P}}(\tilde{W}_{\tilde{n}_j} > x)$ is non-decreasing in j. Now let $(\bar{g}_j, j \geqq 1)$ be NBUE (the NWUE case is analogous). Then assertion (7.5.8) follows immediately from (7.5.7) and (7.5.10):

$$1 - \mathsf{W}(x) = \sum_{j=1}^{\infty} r_j \tilde{\mathsf{P}}(\tilde{W}_{\tilde{n}_j} > x) \leqq \sum_{j=1}^{\infty} \bar{g}_j \tilde{\mathsf{P}}(\tilde{W}_{\tilde{n}_j} > x) = 1 - \check{\mathsf{W}}(x).$$

The second assertion of the theorem is a consequence of the first one, because a geometric distribution is simultaneously NBUE and NWUE. ∎

7.6. Remarks and references

7.1. The idea of using elementary properties of stationary sequences for a rigorous discussion (no independence assumption on interarrival and service times) of batch-arrival-stationary queueing processes and their relations to arrival-stationary queueing processes is due to Wirth (1982). Relations (7.1.11) and (7.1.12) appear obvious in the framework of this book. For a direct approach under strong independence assumptions we refer to Burke (1975).

7.2. The $\mathrm{GI}^{\mathrm{GI}}/\mathrm{M}/\infty$ queue with Poisson arrivals and/or geometrically distributed batch sizes has been treated by several authors, e.g. Gallither et al. (1959), Mills (1980), Murari (1969), Reynolds (1968), and Rathmann (cf. Gnedenko and König (1984), p. 85). Theorem 7.2.4 (cf. also Brandt (1987a), the proof of which is based on ideas from Takács (1956) and Franken and Kerstan (1968), generalizes earlier results of Takács (1956) for $\mathrm{GI}/\mathrm{M}/\infty$ queues without batches and the results of Rathmann for geometrically distributed batch sizes (Theorem 7.2.7, (7.2.20)) and Brandt and Sulanke (1987) for constant batch sizes (Theorem 7.2.10, Lemma A 3.1.4, (7.2.16)). For further general results concerning binomial moments in G/M/∞ queues we refer to FRANKEN and KERSTAN (1968). Different proofs of Theorem 7.2.4 are given in VAN DOORN (1981) and HOLMAN et al. (1982), where also the particular cases $GI^{\delta_1 * Geom(p)}/M/\infty$, $M^{GI}/M/\infty$, and $M^{\delta_1 * Geom(p)}/M/\infty$ are discussed. A generalization of Theorem 7.2.4 to the situation that each arriving customer (of a batch) requires one of s different exponential service times (abbr. $GI^{GI}/M^s/\infty$) is given in BRANDT (1988). An application concerning the results for the $GI^k/M/\infty$ queue is given in BRANDT and SULANKE (1988).

7.3. The approach goes back to WIRTH (1982). It allows us to generalize some known results under weaker independence assumptions and provides simpler proofs. Using the representation (7.3.2) for the distribution $\mathsf{V}(x)$, the formula

$$\mathcal{V}(s) = \hat{\mathcal{V}}(s)\,\frac{1 - \hat{\mathcal{T}}(\mathcal{S}(s))}{\mathsf{E}\hat{G}_0(1 - \mathcal{S}(s))}$$

was obtained in BURKE (1975) for $G^{GI}/GI/1/\infty$ queues; for $GI^{GI}/GI/1/\infty$ queues it was also shown in COHEN (1976a). In [FKAS, pp. 150–151] the validity of (7.3.2) was shown for $G^{GI}/GI/1/\infty$, see also SCHMIDT (1978). Relation (7.3.6) between the queue sizes immediately before batch arrivals and immediately after departures of customers was shown in the special case of $M^{GI}/GI/1/\infty$ in CHAUDHRY (1979). For $G^{GI}/GI/1/\infty$ formula (7.3.7) was proved in [FKAS, p. 151, (5.2.3)]. Theorem 7.3.7, which is an analogue to the well-known Pollaczek-Khinchin formula, was given in CHAUDHRY (1979). Concerning further results and references for single server queues with batch arrivals we refer e.g. to CHAUDHRY and TEMPLETON (1983).

7.4. The approach to feedback queues given here is due to WIRTH (1984a). It allows one to illuminate the relations between feedback queues and batch arrival queues and to generalize some known results under weaker independence assumptions.

Formulae (7.4.17) and (7.4.18) were shown in DISNEY et al. (1980) for $M^{BF}/GI/1/\infty$ queues. Theorem 7.4.8, which is an analogue to the well-known Pollaczek-Khinchin formula, was shown by TAKÁCS (1963a). Concerning further results and references for feedback queues (also for feedback with delays) we refer to BURKE (1976), DISNEY et al. (1980), FOLEY and DISNEY (1983), and DISNEY and KÖNIG (1985).

7.5. The approach presented here is due to WIRTH (1985) and WIRTH (1986a). Assertion (ii) of Theorem 7.5.2 was first proved in HALFIN (1983) by using a discrete-time analogue of PASTA. In HALFIN's paper one can find further examples of impartial disciplines, in particular some contention schemes. HALFIN remarks that the impartial contention schemes which are important in modelling computer and communication systems, include e.g. round robin (token) and Carrier Sense Multiple Access (CSMA) schemes. Assertion (i) of Theorem 7.5.2 was proved by WHITT (1983). His approach corresponds in some sense to that given here. However, the difference is as follows: WHITT starts with a stationary sequence of random vectors $([B_k, X_{k1}, \ldots, X_{kB_k}], k \geqq 1)$, where B_k represents the size of the k-th batch indexed in the order of the arrivals, and X_{kj} represents the random quantity of interest (e.g. delay) associated with the j-th customer to enter service in the k-th batch. In the approach given here the customers are a-priori ordered according to their beginnings of service. However, under the independence assumptions of Theorem 7.5.2 the results coincide.

8. Continuity of queueing models

8.1. Introduction

In Section 1.7 we gave an introduction to the problem of model continuity. Now we consider arrival-stationary queueing processes, i.e. stationary weak or strong solutions $([[A_n, Y_n], X_n])$ of the equation

$$X_{n+1} = f(X_n, [A_n, Y_n]), \qquad n \in \mathbb{Z}, \tag{8.1.1}$$

of a queueing system, cf. Chapter 5. (As in Chapters 6 and 7, we omit the primes occuring in the notation of a weak solution in Chapters 1–5.) Let $([A_{kn}, Y_{kn}])$, $k = 1, 2, \ldots$, be a sequence of inputs such that

$$([A_{kn}, Y_{kn}]) \xrightarrow[k\to\infty]{\mathcal{D}} ([A_n, Y_n]),$$

and let $([[A_{kn}, Y_{kn}], X_{kn}])$, $k = 1, 2, \ldots$, be the corresponding stationary weak solutions of (8.1.1). (For convenience assume that there is an exactly one stationary weak solution for each input.) The main problem is: Under what additional assumptions does

$$([[A_{kn}, Y_{kn}], X_{kn}]) \xrightarrow[k\to\infty]{\mathcal{D}} ([[A_n, Y_n], X_n]) \tag{8.1.2}$$

hold.

A theorem that states (8.1.2) is called a continuity theorem for the arrival-stationary queueing process of a certain queueing system. Some methods for proving continuity theorems were discussed in Section 1.7. Using these methods, in the present chapter we prove continuity theorems for some standard queues. In case of G/G/∞ queues we also provide an estimate of the rate of convergence in (8.1.2) (in the sense of the Prokhorov distance of probability measures). By means of the method of metric modification this would be possible for other queueing systems, too, but we omit the investigation of rates of convergence for more complicated queues, since we want to avoid fussy notation.

If one is interested in continuity theorems for time-stationary queueing models, one can choose two different approaches. First, one remembers that in many cases time-stationary queueing models can be interpreted as solutions of a continuous time analogue of equation (8.1.1) and that these solutions have a structure that is analogous to those of solutions of (8.1.1), cf. Section 4.5. Hence it is not astonishing that the methods for proving continuity theorems given in Section 1.7 work for continuous time models too (of course, in an appropriately modified form). Thus, one is able to prove continuity theorems for time-stationary models in a direct way. However, a second way seems to be more elegant. From Theorem 3.7.1 and the continuous mapping Theorem A 2.4 we immediately obtain

8.1.1. Theorem. *Let* (χ, P), $\chi = \left[(X(t)), ([T_n, [X_n, Y_n]])\right]$, *be an arrival-stationary queueing model with* $a_{\mathsf{P}} = \mathsf{E}_{\mathsf{P}} T_1 < \infty$, *and* $(\bar{\chi}, \bar{\mathsf{P}})$, $\bar{\chi} = \left[(\bar{X}(t)), ([\bar{T}_n, [\bar{X}_n, \bar{Y}_n]])\right]$, *the corresponding time-stationary model. (Here*

$$X(t) = f(X_n, [t - T_n, Y_n]), T_n < t \leqq T_{n+1}, \quad \mathsf{P}\text{-a.s.} \tag{8.1.3}$$

and

$$\bar{X}(t) = f(\bar{X}_n, [t - \bar{T}_n, \bar{Y}_n]), \quad \bar{T}_n < t \leqq \bar{T}_{n+1}, \quad \bar{\mathsf{P}}\text{-a.s.}, \tag{8.1.4}$$

and $f(x, [a, y])$ *has to be left continuous in a, cf. Section 4.4.) Consider a sequence of arrival-stationary models* (χ_k, P_k),

$$\chi_k = \left[(X_k(t)), ([T_{kn}, [X_{kn}, Y_{kn}]])\right], \qquad a_{\mathsf{P}_k} = \mathsf{E}_{\mathsf{P}_k} T_{k1} < \infty,$$

$k = 1, 2, \ldots$, *and the sequence of the corresponding time-stationary models* $(\bar{\chi}_k, \bar{\mathsf{P}}_k)$, $\bar{\chi}_k = \left[(\bar{X}_k(t)), ([\bar{T}_{kn}, [\bar{X}_{kn}, \bar{Y}_{kn}]])\right]$.

Assume (8.1.2) *and* $a_{\mathsf{P}_k} \to a_{\mathsf{P}}$. *Then*

(i) $([\bar{T}_{kn}, [\bar{X}_{kn}, \bar{Y}_{kn}]]) \xrightarrow[k\to\infty]{\mathcal{D}} ([\bar{T}_n, [\bar{X}_n, \bar{Y}_n]])$,

$$\lambda_{\bar{\mathsf{P}}_k} \xrightarrow[k\to\infty]{} \lambda_{\bar{\mathsf{P}}}.$$

(ii) *If, in addition, for arbitrary* $u_1, \ldots, u_r \in \mathbb{R}$, $r \geqq 1$, *the mapping*

$$\eta \to [\bar{x}(u_1, \eta), \ldots, \bar{x}(u_r, \eta)],$$

where

$$\bar{x}(u, \eta) = f(x_n, [u - t_n, y_n]) \text{ if } t_n < u \leqq t_{n+1},$$
$$\eta = ([t_j, [x_j, y_j]]),$$

is continuous for $\bar{\mathsf{P}}$*-a.e.* η, *then*

$$[\bar{X}_k(u_1), \ldots, \bar{X}_k(u_r)] \xrightarrow[k\to\infty]{\mathcal{D}} [\bar{X}(u_1), \ldots, \bar{X}(u_r)],$$

$$u_1, \ldots, u_r \in \mathbb{R}, \qquad r \geqq 1.$$

8.2. The system G/G/∞

Consider a queueing system G/G/∞ with the stationary input (Φ, P), $\Phi = ([A_n, S_n])$, $m_S < \infty$, $\Phi \in S_R$ P-a.s. For a sample path $\varphi \in S_R{}^*$ we use the notation $\varphi = ([a_n(\varphi), s_n(\varphi)])$, cf. Section 3.2, and

$$t_0(\varphi) = 0, \qquad t_n(\varphi) = \sum_{j=0}^{n-1} a_j(\varphi) \quad \text{for } n > 0,$$

$$t_n(\varphi) = -\sum_{j=n}^{-1} a_j(\varphi) \quad \text{for } n < 0.$$

Consider the sets

$$I(n, \varphi) = \{j : j < n, t_j(\varphi) + s_j(\varphi) > t_n(\varphi)\}, \qquad n \in \mathbb{Z},$$

that are finite for a.e. φ, cf. Section 5.2. The quantity

$$l_n(\varphi) = \# I(n, \varphi)$$

can be interpreted as the number of customers in the system seen by the n-th arriving customer, provided that the sample path of the input is φ. Let the elements of $I(n, \varphi)$ be arranged in ascending order:

$$I(n, \varphi) = \{j(1, n, \varphi), \ldots, j(l_n(\varphi), n, \varphi)\}.$$

Then

$$x_n(\varphi) = x_n^{\min}(\varphi) = \overline{R}\big(t_{j(1,n,\varphi)}(\varphi) + s_{j(1,n,\varphi)}(\varphi) - t_n(\varphi), \ldots,$$
$$t_{j(l_n(\varphi),n,\varphi)}(\varphi) + s_{j(l_n(\varphi),n,\varphi)}(\varphi) - t_n(\varphi), 0, 0, \ldots\big)$$

is the state of the system seen by the n-th arriving customer for the given sample path φ. Here we use the state space

$$\mathbb{X} = \{x = (x_1, x_2, \ldots) : x_i \in \mathbb{R}_+; i = 1, 2, \ldots; x_i \downarrow 0\}$$

with the metric

$$d_x(x, x') = \max_i |x_i - x_i'|. \tag{8.2.1}$$

With respect to d_x, $\mathbb{X}$ is complete and separable. Notice that the metric d_x differs from that used in Section 5.2.

Before we start with the verification of continuity theorems, we remark that $l_n(\varphi)$ is discontinuous with respect to the natural metric d on S_R^* (which is defined by (3.2.1)), and thus we cannot directly use the continuous mapping theorem: Let $\varphi \in S_R^*$ be an arbitrary sample path, and define φ_k, $k \geqq 1$, by

$$[a_n(\varphi_k), s_n(\varphi_k)] = \begin{cases} [a_n(\varphi), s_n(\varphi)] & \text{if } n > -k, \\ [0, t_{-k+1}(\varphi) + 1] & \text{if } n = -k, \\ [a_{n+1}(\varphi), s_{n+1}(\varphi)] & \text{if } n < -k. \end{cases}$$

This means that if one adds a customer with service time $s_{-k}(\varphi) = t_{-k+1}(\varphi) + 1$ to φ at time $t_{-k+1}(\varphi)$, then one obtains φ_k. Obviously,

$$d(\varphi_k, \varphi) \xrightarrow[k \to \infty]{} 0,$$

but $l_0(\varphi_k) = l_0(\varphi) + 1$. Thus, $l_0(\varphi)$ is discontinuous.

In order to prove continuity theorems for the processes $(x_n(\Phi))$ and $(l_n(\Phi))$, we shall use the method of metric modification. For this reason we introduce the set

$$\mathbb{U}_1 = \{\varphi : \varphi \in S_R^*, l_n(\varphi) < \infty, n \in \mathbb{Z}\}$$

and the metric

$$\bar{d}_1(\varphi, \varphi') = \sum_{n=-\infty}^{+\infty} \frac{1}{2^{|n|}} d_n(\varphi, \varphi'),$$

$$\begin{aligned} d_n(\varphi, \varphi') = \min \Big\{1, &|t_n(\varphi) - t_n(\varphi')| + |s_n(\varphi) - s_n(\varphi')| \\ &+ \max \{|t_j(\varphi) - t_j(\varphi')| + |s_j(\varphi) - s_j(\varphi')| : \\ &j \in I(n, \varphi) \cup I(n, \varphi')\}\Big\}, \qquad \varphi, \varphi' \in \mathbb{U}_1. \end{aligned}$$

(It is easy to see that $\bar{d}_1$ is indeed a metric on $\mathbb{U}_1$.) Now we prove the properties (A 1.7.7)–(A 1.7.9).

8.2.1. Lemma. *With respect to the metric $\bar{d}_1$, $\mathbb{U}_1$ is complete and separable.*

Proof. (i) First we prove that $\mathbb{U}_1$ is separable. Consider the set $\tilde{\mathbb{U}}_1$ of all sequences $\varphi_k \in S_R{}^*$, where $a_n(\varphi_k)$ and $s_n(\varphi_k)$ are rational numbers, and

$$[a_n(\varphi_k), s_n(\varphi_k)] = [1, 0] \quad \text{for} \quad |n| > k,$$

$k = 1, 2, \ldots$ The denumerable set $\tilde{\mathbb{U}}_1$ is dense in $\mathbb{U}_1$.

(ii) Now we have to prove the completeness of $\mathbb{U}_1$. Consider a Cauchy sequence (φ_k). This sequence is convergent if and only if it has a convergent subsequence. Thus, without loss of generality, we can assume that

$$\bar{d}_1(\varphi_k, \varphi_{k+1}) < \frac{1}{2^k},$$

i.e.

$$d_n(\varphi_k, \varphi_{k+1}) < \frac{1}{2^{k-|n|}}, \qquad k = |n| + 1, |n| + 2, \ldots$$

Since d_n fulfils the triangular inequality, we obtain

$$\begin{aligned} d_n(\varphi_k, \varphi_{k+s}) &\leqq d_n(\varphi_k, \varphi_{k+1}) + d_n(\varphi_{k+1}, \varphi_{k+2}) + \cdots + d_n(\varphi_{k+s-1}, \varphi_{k+s}) \\ &< \frac{1}{2^{k-|n|}} + \cdots + \frac{1}{2^{k+s-1-|n|}} \\ &< \frac{1}{2^{k-|n|}} + \frac{1}{2^{k+1-|n|}} + \cdots = \frac{1}{2^{k-1-|n|}} \end{aligned} \tag{8.2.2}$$

for $s \geqq 1$, $k > |n|$. Since $\bar{d}_1(\varphi, \varphi') \geqq d(\varphi, \varphi')$, (φ_k) is a Cauchy sequence with respect to d, too. Hence there is a $\varphi \in S_R{}^*$ such that $d(\varphi_k, \varphi) \xrightarrow[k\to\infty]{} 0$, namely

$$t_n(\varphi) = \lim_{k\to\infty} t_n(\varphi_k), \qquad s_n(\varphi) = \lim_{k\to\infty} s_n(\varphi_k).$$

We have to show that $\varphi \in \mathbb{U}_1$ and $\bar{d}_1(\varphi_k, \varphi) \xrightarrow[k\to\infty]{} 0$. First consider $d_0(\varphi_k, \varphi)$. Let $1 > \varepsilon > 0$. In view of (8.2.1) we can choose a k such that

$$d_0(\varphi_k, \varphi_{k+s}) < \varepsilon, \qquad s \geqq 0. \tag{8.2.3}$$

We distinguish two cases:

(i) For all k satisfying (8.2.3): $|t_l(\varphi_k)| \leqq 2\varepsilon$, $l < 0$.
(ii) There is a k satisfying (8.2.3) and an $l < 0$ with $|t_l(\varphi_k)| > 2\varepsilon$.

Assume (i) is true. Let k be choosen in such a way that (8.2.3) is fulfilled. Then there is an n_0 such that $s_n(\varphi_k) \leqq 2\varepsilon$ for all $n \leqq n_0$, since $I(0, \varphi_k)$ is finite. Now let $n \leqq n_0$ and $s \geqq 0$. Condition (i) yields $|t_n(\varphi_{k+s})| \leqq 2\varepsilon$. If $|t_n(\varphi)| > 2\varepsilon$ then there is an s' satisfying $|t_n(\varphi_{k+s'})| > 2\varepsilon$, which contradicts (i). Thus, $|t_n(\varphi)| \leqq 2\varepsilon$, and hence

$$|t_n(\varphi_{k+s}) - t_n(\varphi)| \leqq 2\varepsilon, \qquad n \leqq n_0. \tag{8.2.4}$$

From (8.2.3) we obtain

$$s_n(\varphi_{k+s}) < \begin{cases} 3\varepsilon & \text{if } n \in I(0, \varphi_{k+s}), \\ |t_n(\varphi_{k+s})| \leqq 2\varepsilon & \text{if } n \notin I(0, \varphi_{k+s}). \end{cases}$$

Since this is true for all s, it follows that $s_n(\varphi) \leqq 3\varepsilon$, and thus

$$|s_n(\varphi_{k+s}) - s_n(\varphi)| \leqq 3\varepsilon, \qquad n \leqq n_0. \tag{8.2.5}$$

For $n = n_0 + 1, \ldots, -1$ we find an s_0 such that

$$\begin{aligned} &|t_0(\varphi_{k+s}) - t_0(\varphi)| + |s_0(\varphi_{k+s}) - s_0(\varphi)| \\ &+ |t_n(\varphi_{k+s}) - t_n(\varphi)| + |s_n(\varphi_{k+s}) - s_n(\varphi)| \leqq \varepsilon \quad \text{for } s \geqq s_0. \end{aligned}$$

This and (8.2.4), (8.2.5) yield $d_0(\varphi_{k+s}, \varphi) \leqq 6\varepsilon$ for sufficiently large s. This means that if for all $\varepsilon > 0$ (i) holds, then

$$d_0(\varphi_k, \varphi) \xrightarrow[k\to\infty]{} 0, \quad \text{and} \quad t_n(\varphi) = s_n(\varphi) = 0, \qquad n < 0.$$

Now assume that for an arbitrary, small ε (ii) is true. Let k satisfy (8.2.3), and let $l < 0$ be a fixed index with $|t_l(\varphi_k)| > 2\varepsilon$. From (8.2.3) we obtain

$$I(0, \varphi_{k+s}) \subseteqq I(l, \varphi_k) \cup \{l, \ldots, -1\}, \qquad s \geqq 0.$$

Obviously, $I(0, \varphi) \subseteqq I(l, \varphi_k) \cup \{l, \ldots, -1\}$. Thus,

$$\begin{aligned} &\max\{|t_n(\varphi_{k+s}) - t_n(\varphi)| + |s_n(\varphi_{k+s}) - s_n(\varphi)| : n \in I(0, \varphi_{k+s}) \cup I(0, \varphi)\} \\ &\leqq \max\left\{|t_n(\varphi_{k+s}) - t_n(\varphi)| + |s_n(\varphi_{k+s}) - s_n(\varphi)| : n \in I(l, \varphi_k) \cup \{l, \ldots, -1\}\right\} < \varepsilon \end{aligned}$$

for sufficiently large s. This yields $d_0(\varphi_k, \varphi) \xrightarrow[k\to\infty]{} 0$. Analogously one shows that $d_n(\varphi, \varphi_k) \to 0$ and $\# I(n, \varphi) < \infty$ for all n. ∎

8.2.2. Lemma. *Let $\mathcal{U}_1$ be the Borel σ-field on $\mathbb{U}_1$ with respect to $\bar{d}_1$, and let $\mathcal{U}^*$ be the Borel σ-field on $S_R{}^*$ with respect to d. Then*

$$\mathcal{U}_1 = \{B : B \in \mathcal{U}^*, B \subseteqq \mathbb{U}_1\}.$$

Proof. The σ-field $\{B: B \in \mathcal{U}^*, B \subseteq \mathbb{U}_1\}$ is generated by the sets

$$\{\varphi: \varphi \in \mathbb{U}_1, [t_n(\varphi), s_n(\varphi)] \in (\cdot)\}, \qquad n \in \mathbb{Z}.$$

Denote the family of all these sets by $\mathbf{B}^*$. The σ-field $\mathcal{U}_1$ is generated by the sets

$$\Big\{\varphi: \varphi \in \mathbb{U}_1, I(n, \varphi) = \{j_1, \ldots, j_k\}, \big[[t_{j_1}(\varphi), s_{j_1}(\varphi)], \ldots,$$
$$[t_{j_k}(\varphi), s_{j_k}(\varphi)], [t_n(\varphi), s_n(\varphi)]\big] \in (\cdot)\Big\}, \quad j_1, \ldots, j_k \in \mathbb{Z}, \quad k \in \mathbb{Z}_+, \quad n \in \mathbb{Z}.$$

(This can be proved in the same way as the analogous result for the natural topology.) Let $\mathbf{B}_1$ be the family of all these sets. Obviously, $\mathbf{B}^* \subseteq \mathbf{B}_1$, and it is easy to see that $\mathbf{B}_1 \subseteq \sigma(\mathbf{B}^*)$. ∎

For $\varphi \in \mathbb{U}_1$ define

$$\inf \varphi = \min \{|t_j(\varphi) + s_j(\varphi)| : j \in \mathbb{Z}\}.$$

Now we can formulate continuity properties of the mappings $l_0(\varphi)$ and $x_0(\varphi)$.

8.2.3. Lemma. (i) *If* $\bar{d}_1(\varphi, \varphi') < 1$, *then*

$$d_x\big(x_0(\varphi), x_0(\varphi')\big) \leqq \bar{d}_1(\varphi, \varphi').$$

(ii) *If* $\bar{d}_1(\varphi, \varphi') < \min\{1, \inf \varphi\}$, *then*

$$l_0(\varphi) = l_0(\varphi').$$

Proof. (i) If $\bar{d}_1(\varphi, \varphi') < 1$, then $d_0(\varphi, \varphi') < 1$, and thus

$$d_x\big(x_0(\varphi), x_0(\varphi')\big)$$
$$\leqq \max\Big\{\big|t_j(\varphi) + s_j(\varphi) - \big(t_j(\varphi') + s_j(\varphi')\big)\big| : j \in I(0, \varphi) \cup I(0, \varphi')\Big\}$$
$$\leqq \max\{|t_j(\varphi) - t_j(\varphi')| + |s_j(\varphi) - s_j(\varphi')| : j \in I(0, \varphi) \cup I(0, \varphi')\}$$
$$\leqq \bar{d}_1(\varphi, \varphi').$$

(ii) Assume $\bar{d}_1(\varphi, \varphi') < \min\{1, \inf \varphi\}$. Then

$$\big|t_j(\varphi) + s_j(\varphi) - \big(t_j(\varphi') + s_j(\varphi')\big)\big| < \inf \varphi$$

for all $j \in I(0, \varphi) \cup I(0, \varphi')$. It follows that $I(0, \varphi) = I(0, \varphi')$, and thus, $l_0(\varphi) = l_0(\varphi')$. ∎

Consider the inputs (Φ, P), $\Phi = ([A_n, S_n])$, and (Φ_k, P_k), $\Phi_k = ([A_{kn}, S_{kn}])$, $k = 1, 2, \ldots$, with $\mathsf{P}_k(\Phi_k \in \mathbb{U}_1) = \mathsf{P}(\Phi \in \mathbb{U}_1) = 1$. Let π_1 be the Prokhorov metric for distributions on $\mathbb{U}_1$ with respect to $\bar{d}_1$. Remember, $\pi_1(\mathsf{P}_k, \mathsf{P}) \xrightarrow[k\to\infty]{} 0$

$$\text{iff } \inf\left\{\varepsilon: \varepsilon > 0, \mathsf{P}_k(\Phi_k \in B) < \mathsf{P}\left(\Phi \in \bigcup_{\varphi \in B} \{\varphi': \bar{d}_1(\varphi, \varphi') < \varepsilon\}\right) + \varepsilon\right\} \xrightarrow[k\to\infty]{} 0$$

for all closed sets $B \in \mathcal{U}_1$

and

$$\inf\left\{\varepsilon:\varepsilon>0,\ \mathsf{P}(\Phi\in B)<\mathsf{P}_k\left(\Phi_k\in\bigcup_{\varphi\in B}\{\varphi':\bar{d}_1(\varphi,\varphi')<\varepsilon\}\right)+\varepsilon\right\}\xrightarrow[k\to\infty]{}0$$

for all closed sets $B\in\mathcal{U}_1$

iff $\int h(\varphi)\ \mathsf{P}_k(\mathrm{d}\varphi)\xrightarrow[k\to\infty]{}\int h(\varphi)\ \mathsf{P}(\mathrm{d}\varphi)$

for all continuous (with respect to $\bar{d}_1$), bounded functions h.

The family of continuous functions with respect to $\bar{d}_1$ is greater than the family of continuous functions with respect to d. Thus, the π_1-convergence is stronger than the usual convergence in distribution (i.e. the convergence with respect to Prokhorov distance induced by d).

Later we shall need

8.2.4. Lemma. (i) *If*

$$\pi_1\left(\mathsf{P}_k\big(\eta_n(\Phi_k)\in(\cdot)\big),\ \mathsf{P}\big(\eta_n(\Phi)\in(\cdot)\big)\right)\xrightarrow[k\to\infty]{}0,\qquad n\geqq 0,\tag{8.2.6}$$

where

$$\eta_n(\varphi)=\left(\big[a_i\big(\eta_n(\varphi)\big),\ s_i\big(\eta_n(\varphi)\big)\big],\ i\in\mathbb{Z}\right),$$

$$a_i\big(\eta_n(\varphi)\big)=\begin{cases}a_i(\varphi) & \text{if } \min\{-n,j(1,-n,\varphi)\}\leqq i\leqq n,\\ 1 & \text{otherwise},\end{cases}$$

$$s_i\big(\eta_n(\varphi)\big)=\begin{cases}s_i(\varphi) & \text{if } \min\{-n,j(1,-n,\varphi)\}\leqq i\leqq n,\\ 0 & \text{otherwise},\end{cases}$$

$\varphi\in\mathbb{U}_1$, $n\in\mathbb{Z}$, *then* $\pi_1(\mathsf{P}_k,\mathsf{P})\xrightarrow[k\to\infty]{}0$.

(ii) *If* $\pi_1(\mathsf{P}_k,\mathsf{P})\xrightarrow[k\to\infty]{}0$ *and*

$$\mathsf{P}\left(\bigcup_{n\in\mathbb{Z}}\bigcup_{j<n}\{T_j+S_j=T_n\}\right)=0,\tag{8.2.7}$$

then (8.2.6) *holds.*

Proof. (i) Let $\varepsilon>0$. Since

$$\bar{d}_1\big(\Phi,\eta_n(\Phi)\big)\leqq\sum_{|k|>n}\frac{1}{2^{|k|}}=2^{1-n},$$

$$\bar{d}_1\big(\Phi_k,\eta_n(\Phi_k)\big)\leqq 2^{1-n},$$

we have $\bar{d}_1\big(\Phi,\eta_n(\Phi)\big)\xrightarrow[n\to\infty]{}0$ a.s. and $\bar{d}_1\big(\Phi_k,\eta_n(\Phi_k)\big)\xrightarrow[n\to\infty]{}0$ a.s., and thus

$$\pi_1\left(\mathsf{P}(\Phi\in(\cdot)),\ \mathsf{P}\big(\eta_n(\Phi)\in(\cdot)\big)\right)+\pi_1\left(\mathsf{P}_k\big(\eta_n(\Phi_k)\in(\cdot)\big),\ \mathsf{P}_k\big(\Phi_k\in(\cdot)\big)\right)<\frac{\varepsilon}{2}$$

for sufficiently large n. From (8.2.6) and

$$\begin{aligned}\pi_1(\mathsf{P}_k, \mathsf{P}) &\leqq \pi_1\left(\mathsf{P}(\Phi \in (\cdot)), \mathsf{P}(\eta_n(\Phi) \in (\cdot))\right)\\ &\quad + \pi_1\left(\mathsf{P}(\eta_n(\Phi) \in (\cdot)), \mathsf{P}_k(\eta_n(\Phi_k) \in (\cdot))\right)\\ &\quad + \pi_1\left(\mathsf{P}_k(\eta_n(\Phi_k) \in (\cdot)), \mathsf{P}_k(\Phi_k \in (\cdot))\right)\\ &< \frac{\varepsilon}{2} + \pi_1\left(\mathsf{P}(\eta_n(\Phi) \in (\cdot)), \mathsf{P}_k(\eta_n(\Phi_k) \in (\cdot))\right)\end{aligned}$$

we obtain $\pi_1(\mathsf{P}_k, \mathsf{P}) < \varepsilon$ for sufficiently large k.

(ii) Under the assumption (8.2.7) the mappings $\varphi \to \eta_n(\varphi)$ are continuous with respect to $\bar{d}_1$. Thus (8.2.6) follows from the continuous mapping theorem, cf. Theorem A 2.4. ∎

The assertion of Lemma 8.2.4 is analogous to the well-known fact that the usual weak convergence of random sequences is equivalent to the convergence of the finite-dimensional distributions. Now we can formulate a relation between the convergence

$$\pi_1(\mathsf{P}_k, \mathsf{P}) \to 0$$

and the usual weak convergence (based on d).

8.2.5. Lemma. *Let (Φ, P), (Φ_k, P_k), $k = 1, 2, \ldots$, be stationary ergodic inputs satisfying*

$$m_{Sk} = \mathsf{E}_{\mathsf{P}_k} S_{k0} \xrightarrow[k\to\infty]{} \mathsf{E}_{\mathsf{P}} S_0 = m_S < \infty \tag{8.2.8}$$

and

$$m_{Ak} = \mathsf{E}_{\mathsf{P}_k} A_{k0} > 0, \qquad \mathsf{E}_{\mathsf{P}} A_0 = m_A > 0. \tag{8.2.9}$$

Assume $\Phi_k \xrightarrow[k\to\infty]{\mathfrak{D}} \Phi$ *and* $\mathsf{P}\left(\bigcup_{j<0} \{T_j + S_j = 0\}\right) = 0$. *Then*

$$\pi_1(\mathsf{P}_k, \mathsf{P}) \xrightarrow[k\to\infty]{} 0.$$

Proof. In view of Lemma 8.2.4 it suffices to show that

$$\pi_1\left(\mathsf{P}_k(\eta_n(\Phi_k) \in (\cdot)), \mathsf{P}(\eta_n(\Phi) \in (\cdot))\right) \xrightarrow[k\to\infty]{} 0, \qquad n \geqq 0.$$

The triangular inequality yields

$$\begin{aligned}&\pi_1\left(\mathsf{P}_k(\eta_n(\Phi_k) \in (\cdot)), \mathsf{P}(\eta_n(\Phi) \in (\cdot))\right)\\ &\leqq \pi_1\left(\mathsf{P}_k(\eta_n(\Phi_k) \in (\cdot)), \mathsf{P}_k\left(\eta_n(\alpha_m(\Phi_k)) \in (\cdot)\right)\right)\\ &\quad + \pi_1\left(\mathsf{P}_k\left(\eta_n(\alpha_m(\Phi_k))\right) \in (\cdot)\right), \mathsf{P}\left(\eta_n(\alpha_m(\Phi)) \in (\cdot)\right)\Big)\\ &\quad + \pi_1\left(\mathsf{P}\left(\eta_n(\alpha_m(\Phi))\right) \in (\cdot)\right), \mathsf{P}(\eta_n(\Phi) \in (\cdot))\Big),\end{aligned} \tag{8.2.10}$$

where

$$\alpha_m(\varphi) = \left(\left[a_i(\alpha_m(\varphi)), s_i(\alpha_m(\varphi))\right], i \in \mathbb{Z}\right),$$

$$a_i(\alpha_m(\varphi)) = \begin{cases} a_i(\varphi) & \text{if } -m \leqq i \leqq m, \\ 1 & \text{otherwise}, \end{cases}$$

$$s_i(\alpha_m(\varphi)) = \begin{cases} s_i(\varphi) & \text{if } -m \leqq i \leqq m, \\ 0 & \text{otherwise}. \end{cases}$$

Let $m > n$. From the definition of the Prokhorov metric we obtain

$$\begin{aligned} &\pi_1\left(\mathsf{P}\left(\eta_n(\alpha_m(\Phi)) \in (\cdot)\right), \mathsf{P}(\eta_n(\Phi) \in (\cdot))\right) \\ &\leqq \mathsf{P}\left(\eta_n(\alpha_m(\Phi)) \neq \eta_n(\Phi)\right) \\ &= \mathsf{P}\left(\sum_{j<-m} 1\{T_j + S_j > T_{-n}\} > 0\right) \\ &= \mathsf{P}\left(\bigcup_{j<n-m} \{T_j + S_j > 0\}\right) \\ &\leqq \mathsf{P}\left(\bigcup_{j<n-m} (\{T_j > jb\} \cup \{S_j > -jb\})\right) \\ &\leqq \mathsf{P}\left(\bigcup_{j<n-m} \{T_j > jb\}\right) + \mathsf{P}\left(\bigcup_{j<n-m} \{S_j > -jb\}\right) \end{aligned} \tag{8.2.11}$$

for each $b > 0$.

Analogously,

$$\begin{aligned} &\pi_1\left(\mathsf{P}_k\left(\eta_n(\alpha_m(\Phi_k)) \in (\cdot)\right), \mathsf{P}_k(\eta_n(\Phi_k) \in (\cdot))\right) \\ &\leqq \mathsf{P}_k\left(\bigcup_{j<n-m} \{T_{kj} > jb\}\right) + \mathsf{P}_k\left(\bigcup_{j<n-m} \{S_{kj} > -jb\}\right), \qquad b > 0. \end{aligned} \tag{8.2.12}$$

Let $\varepsilon > 0$. If we choose $b < m_A$, then

$$\mathsf{P}\left(\bigcup_{j<n-m} \{T_j > jb\}\right) < \frac{\varepsilon}{4} \tag{8.2.13}$$

for sufficiently large m in view of the law of large numbers. Since $\mathsf{E}_{\mathsf{P}} S_0 < \infty$,

$$\begin{aligned} \mathsf{P}\left(\bigcup_{j<n-m} \{S_j > -jb\}\right) &\leqq \sum_{j=-\infty}^{n-m} \mathsf{P}(S_j > -jb) \\ &= \sum_{j=m-n}^{\infty} \mathsf{P}\left(\frac{S_0}{b} > j\right) < \frac{\varepsilon}{4} \end{aligned} \tag{8.2.14}$$

for sufficiently large m.

Now we want to estimate the terms in (8.2.12). For the last term we obtain

$$\mathsf{P}_k\left(\bigcup_{j<n-m}\{S_{kj}>-jb\}\right)\leqq\sum_{j=m-n}^{\infty}\mathsf{P}_k(S_{k0}>jb)$$

$$<\int_{m-n-1}^{\infty}\mathsf{P}_k(S_{k0}>xb)\,\mathrm{d}x$$

$$=\frac{1}{b}\int_{b(m-n-1)}^{\infty}\mathsf{P}_k(S_{k0}>x)\,\mathrm{d}x<\frac{\varepsilon}{4} \tag{8.2.15}$$

for sufficiently large m, because $m_{Sk}\to m_S$, cf. Lemma A 2.1. Consider the first term on the right hand side of (8.2.12). We have

$$\mathsf{P}_k\left(\bigcup_{j<n-m}\{T_{kj}>jb\}\right)=\mathsf{P}_k\left(\max_{j<n-m}\sum_{l=j}^{-1}(b-A_{kl})>0\right).$$

This, Lemma A 2.8, and Theorem A 2.9 yield

$$\mathsf{P}_k\left(\bigcup_{j<n-m}\{T_{kj}>jb\}\right)<\frac{\varepsilon}{4} \tag{8.2.16}$$

for sufficiently large m and $b<m_A$. Thus we have shown that

$$\pi_1\left(\mathsf{P}_k(\eta_n(\Phi_k)\in(\cdot)),\mathsf{P}_k\left(\eta_n(\alpha_m(\Phi_k))\in(\cdot)\right)\right)$$

$$+\pi_1\left(\mathsf{P}\left(\eta_n(\alpha_m(\Phi))\in(\cdot)\right),\mathsf{P}(\eta_n(\Phi)\in(\cdot))\right)<\varepsilon,\quad k\geqq1, \tag{8.2.17}$$

for sufficiently large m, cf. (8.2.11)–(8.2.16).

In view of Skorokhod's Representation Theorem A 2.6 we can choose Φ^*, Φ_k^*, $k=1,2,\ldots$, such that $\Phi^*\overset{\mathcal{D}}{=}\Phi$, $\Phi_k^*\overset{\mathcal{D}}{=}\Phi_k$, and

$$d(\Phi_k^*,\Phi^*)\xrightarrow[k\to\infty]{}0\quad\text{a.s.}$$

From $\mathsf{P}\left(\bigcup_{j<l}\{T_j+S_j=T_l\}\right)=0$, $l\in\mathbb{Z}$, we get

$$I\left(j,\alpha_m(\Phi_k^*)\right)=I\left(j,\alpha_m(\Phi^*)\right),\qquad j=-n,\ldots,n,$$

if $d(\Phi_k^*,\Phi^*)$ is sufficiently small. Thus,

$$\bar{d}_1\left(\eta_n(\alpha_m(\Phi_k^*)),\eta_n(\alpha_m(\Phi^*))\right)\xrightarrow[k\to\infty]{}0,$$

and hence

$$\pi_1\left(\mathsf{P}_k\left(\eta_n(\alpha_m(\Phi_k))\in(\cdot)\right),\mathsf{P}\left(\eta_n(\alpha_m(\Phi))\in(\cdot)\right)\right)\xrightarrow[k\to\infty]{}0. \tag{8.2.18}$$

From (8.2.10), (8.2.17), and (8.2.18),

$$\pi_1\left(\mathsf{P}_k(\eta_n(\Phi_k)\in(\cdot)),\mathsf{P}(\eta_n(\Phi)\in(\cdot))\right)\xrightarrow[k\to\infty]{}0,\qquad n\geqq0.\ \blacksquare$$

The method of metric modification, described in Section 1.7, yields immediately

8.2.6. Theorem. (i) *Let* (Φ, P), (Φ_k, P_k), $k = 1, 2, \ldots$, *be arrival-stationary, ergodic inputs of a queueing system* G/G/∞ *satisfying*

$$m_{Sk} = \mathsf{E}_{\mathsf{P}_k} S_{k0} \xrightarrow[k\to\infty]{} \mathsf{E}_{\mathsf{P}} S_0 = m_S < \infty, \tag{8.2.19}$$

$$m_{Ak} = \mathsf{E}_{\mathsf{P}_k} A_{k0} > 0, \qquad \mathsf{E}_{\mathsf{P}} A_0 = m_A > 0 \tag{8.2.20}$$

and

$$\Phi_k \xrightarrow[k\to\infty]{\mathcal{D}} \Phi.$$

Then

$$\big([[A_{kn}, S_{kn}], x_n(\Phi_k)]\big) \xrightarrow[k\to\infty]{\mathcal{D}} \big([[A_n, S_n], x_n(\Phi)]\big). \tag{8.2.21}$$

If, in addition,

$$\mathsf{P}\Big(\bigcup_{j<0} \{T_j + S_j = 0\}\Big) = 0, \tag{8.2.22}$$

then

$$\big([[A_{kn}, S_{kn}], l_n(\Phi_k)]\big) \xrightarrow[k\to\infty]{\mathcal{D}} \big([[A_n, S_n], l_n(\Phi)]\big). \tag{8.2.23}$$

(ii) *Let* $(\overline{\Psi}, \overline{\mathsf{P}})$, $(\overline{\Psi}_k, \overline{\mathsf{P}}_k)$, $k = 1, 2, \ldots$, *be time-stationary, ergodic inputs of a queueing system* G/G/∞ *satisfying* (8.2.19), (8.2.20), $m_{A_k} \to m_A$ *(notice that these are conditions on the corresponding arrival-stationary inputs), and* $\overline{\Psi}_k \xrightarrow[k\to\infty]{\mathcal{D}} \overline{\Psi}$. *Consider the time-stationary models* $\overline{\chi}_k$, $\overline{\chi}$,

$$\overline{\chi}_k = \Big[\big(\overline{X}_k(t)\big), \big([\overline{T}_{kn}, [\overline{X}_{kn}, \overline{S}_{kn}]]\big)\Big], \qquad k = 1, 2, \ldots,$$

$$\overline{\chi} = \Big[\big(\overline{X}(t)\big), \big([\overline{T}_n, [\overline{X}_n, \overline{S}_n]]\big)\Big],$$

corresponding to $\big([[A_{kn}, S_{kn}], x_n(\Phi_k)]\big)$ *and* $\big([[A_n, S_n], x_n(\Phi)]\big)$, *respectively. Then*

$$\big([\overline{T}_{kn}, [\overline{X}_{kn}, \overline{S}_{kn}]]\big) \xrightarrow[k\to\infty]{\mathcal{D}} \big([\overline{T}_n, [\overline{X}_n, \overline{S}_n]]\big), \tag{8.2.24}$$

$$[\overline{X}_k(u_1), \ldots, \overline{X}_k(u_r)] \xrightarrow[k\to\infty]{\mathcal{D}} [\overline{X}(u_1), \ldots, \overline{X}(u_r)], \tag{8.2.25}$$

$$[l(u_1, \overline{\Psi}_k), \ldots, l(u_r, \overline{\Psi}_k)] \xrightarrow[k\to\infty]{\mathcal{D}} [l(u_1, \overline{\Psi}), \ldots, l(u_r, \overline{\Psi})], \tag{8.2.26}$$

$$u_1, \ldots, u_r \in \mathbb{R}, \quad r \geqq 1,$$

where $l(u, \psi) = \#\{j : t_j < u, t_j + s_j > u\}$, $u \in \mathbb{R}$, $\psi = ([t_j, s_j])$. *($l(u, \overline{\Psi}_k)$ and $l(u, \overline{\Psi})$, respectively, are the time-stationary numbers of customers in the system at time u.) If, in addition,* (8.2.22) *or, equivalently,* $\overline{\mathsf{P}}\Big(\bigcup_{j\leqq 0} \{\overline{T}_j + \overline{S}_j = \overline{T}_0\}\Big) = 0$, *then*

$$\big([\overline{T}_{kn}, [\overline{S}_{kn}, \overline{L}_{kn}]]\big) \xrightarrow[k\to\infty]{\mathcal{D}} \big([\overline{T}_n, [\overline{S}_n, \overline{L}_n]]\big), \tag{8.2.27}$$

where

$$\overline{L}_{kn} = \#\{j : j < n, \overline{T}_{kj} + \overline{S}_{kj} > \overline{T}_{kn}\},$$

$$\overline{L}_n = \#\{j : j < n, \overline{T}_j + \overline{S}_j > \overline{T}_n\}, \qquad n \in \mathbb{Z},$$

are the (time-stationary) numbers of customers in the system at $\overline{T}_{kn}$ *and* $\overline{T}_n$, *respectively.*

Proof. (i) Let π be the Prokhorov metric for distributions on $\mathbb{X}$. Then

$$\begin{aligned}
&\pi\Big(\mathsf{P}_k\big(x_0(\Phi_k) \in (\cdot)\big), \mathsf{P}\big(x_0(\Phi) \in (\cdot)\big)\Big)\\
&\leqq \pi\Big(\mathsf{P}_k\big(x_0(\Phi_k) \in (\cdot)\big), \mathsf{P}\big(x_0(\alpha_m(\Phi_k)) \in (\cdot)\big)\Big)\\
&\quad + \pi\Big(\mathsf{P}\big(x_0(\Phi) \in (\cdot)\big), \mathsf{P}\big(x_0(\alpha_m(\Phi)) \in (\cdot)\big)\Big)\\
&\quad + \pi\Big(\mathsf{P}_k\big(x_0(\alpha_m(\Phi_k)) \in (\cdot)\big), \mathsf{P}\big(x_0(\alpha_m(\Phi)) \in (\cdot)\big)\Big)\\
&\leqq \mathsf{P}_k\Big(\sum_{j<-m} 1\{T_{kj} + S_{kj} > 0\}\Big) + \mathsf{P}\Big(\sum_{j<-m} 1\{T_j + S_j > 0\}\Big)\\
&\quad + \pi\Big(\mathsf{P}_k\big(x_0(\alpha_m(\Phi_k)) \in (\cdot)\big), \mathsf{P}\big(x_0(\alpha_m(\Phi)) \in (\cdot)\big)\Big),
\end{aligned} \tag{8.2.28}$$

where $\alpha_m(\cdot)$ is defined in the proof of Lemma 8.2.5. In the proof of Lemma 8.2.5 it was shown that the first two summands of (8.2.28) become small uniformly in k for large m. Since $\alpha_m(\Phi_k) \xrightarrow[k\to\infty]{\mathfrak{D}} \alpha_m(\Phi)$ and since the mapping $\varphi \to x_0(\varphi)$ is continuous (with respect to the metric d) on $\alpha_m(S_R{}^*) = \{\varphi : \varphi \in S_R{}^*, \alpha_m(\varphi) = \varphi\}$, the continuous mapping theorem yields

$$\pi\Big(\mathsf{P}_k\big(x_0(\alpha_m(\Phi_k)) \in (\cdot)\big), \mathsf{P}\big(x_0(\alpha_m(\Phi)) \in (\cdot)\big)\Big) \xrightarrow[k\to\infty]{} 0.$$

Thus,

$$x_0(\Phi_k) \xrightarrow[k\to\infty]{\mathfrak{D}} x_0(\Phi).$$

In the same way one shows the convergence of all finite-dimensional distributions of the sequences $\big([[A_{kn}, S_{kn}], x_n(\Phi_k)]\big)$ to those of $\big([[A_n, S_n], x_n(\Phi)]\big)$, and hence (8.2.21) is true.

Now assume (8.2.22). Then by Lemma 8.2.5, $\pi_1(\mathsf{P}_k, \mathsf{P}) \to 0$. In view of Lemma 8.2.3 (ii), the continuous mapping theorem yields (8.2.23).

(ii) Let (Φ, P), (Φ_k, P_k), $k = 1, 2, \ldots$, be the arrival-stationary inputs corresponding to $(\overline{\Psi}, \overline{\mathsf{P}})$ and $(\overline{\Psi}_k, \overline{\mathsf{P}}_k)$, $k = 1, 2, \ldots$, respectively. From Theorem 3.7.1 we conclude that

$$\Phi_k \xrightarrow[k\to\infty]{\mathfrak{D}} \Phi.$$

Thus, in view of part (i) of Theorem 8.2.6, (8.2.21) is true. Let

$$\begin{aligned}
&X_{kn} = x_n(\Phi_k), && n \in \mathbb{Z},\\
&X_n = x_n(\Phi), && n \in \mathbb{Z},\\
&X_k(t) = f(X_{kn}, [t - T_{kn}, S_{kn}]), && T_{kn} < t \leqq T_{k,n+1},\\
&X(t) = f(X_n, [t - T_n, S_n]), && T_n < t \leqq T_{n+1}.
\end{aligned}$$

The time-stationary models $\bar{\chi}$ and $\bar{\chi}_k$, $k = 1, 2, \ldots$, are the stationary PEMP's corresponding to the arrival-stationary models (i.e. synchronous PEMP's) $\left[(X(t)), \left([T_n, [X_n, S_n]]\right)\right]$ and $\left[(X_k(t)), \left([T_{kn}, [X_{kn}, S_{kn}]]\right)\right]$, $k = 1, 2, \ldots$, respectively, cf. Section 4.4. Thus, Theorem 8.1.1 (i) yields (8.2.24). The function $f(x, [a, s])$ of the queueing system G/G/∞ is continuous in all arguments. Hence (8.2.25) follows from Theorem 8.1.1 (ii). The mapping

$$\eta = \left([t_n, [x_n, s_n]]\right) \to [l(u_1, \eta), \ldots, l(u_r, \eta)]$$

is continuous on the set

$$\{\eta = \left([t_n, [x_n, s_n]]\right) \in M_k, \mathbf{K} = \mathbf{X} \times \mathbf{R}_+, l(u_1, \eta) < \infty, \ldots, l(u_r, \eta) < \infty,$$
$$t_n \neq u_i, t_n + s_n \neq u_i, i = 1, \ldots, r; n \in \mathbf{Z}\}.$$

(Notice that with probability 1 arrivals and departures do not occur at $u_1, \ldots, u_r$.) The continuous mapping Theorem A 2.4 yields (8.2.26).

Now assume (8.2.22). The corresponding to $\left([[A_n, S_n], l_n(\Phi)]\right)$ and $\left([[A_{kn}, S_{kn}], l_n(\Phi_k)]\right)$, $k = 1, 2, \ldots$, stationary MPP's are just $\left([\bar{T}_n, [\bar{S}_n, \bar{L}_n]]\right)$ and $\left([\bar{T}_{kn}, [\bar{S}_{kn}, \bar{L}_{kn}]]\right)$, $k = 1, 2, \ldots$, respectively. Together with (8.2.23), Theorem 3.7.1 yields (8.2.27). ∎

Next we shall show how Lemma 8.2.3 leads to an estimate of the rate of convergence in $x_0(\Phi_k) \xrightarrow{\mathcal{D}} x_0(\Phi)$ and $l_0(\Phi_k) \xrightarrow{\mathcal{D}} l_0(\Phi)$, respectively. Let π_X and π_Z be the Prokhorov distances for probability laws on $\mathbf{X}$ and $\mathbf{Z}_+$ with respect to the metrics d_x, cf. (8.2.1) on $\mathbf{X}$ and $|k - l|$, $k, l \in \mathbf{Z}_+$, respectively.

8.2.7. Theorem. *Let* (Φ_1, P_1) *and* (Φ_2, P_2) *be arrival-stationary inputs of a queueing system* G/G/∞ *satisfying* $m_{S1} < \infty$, $m_{S2} < \infty$, $\sum_{n=-\infty}^{-1} A_{1n} = \infty$ P_1*-a.s.*, $\sum_{n=-\infty}^{-1} A_{2n} = \infty$ P_2*-a.s. If* $\pi_1(\mathsf{P}_1, \mathsf{P}_2) < 1$, *then*

$$\pi_X\left(\mathsf{P}_1(x_0(\Phi_1) \in (\cdot)), \mathsf{P}_2(x_0(\Phi_2) \in (\cdot))\right) \leqq \pi_1(\mathsf{P}_1, \mathsf{P}_2) \tag{8.2.29}$$

and

$$\pi_Z\left(\mathsf{P}_1(l_0(\Phi_1) \in (\cdot)), \mathsf{P}_2(l_0(\Phi_2) \in (\cdot))\right) \leqq \pi_1(\mathsf{P}_1, \mathsf{P}_2)$$
$$+ \max\left\{\mathsf{P}_1(\inf \Phi_1 \leqq 2\pi_1(\mathsf{P}_1, \mathsf{P}_2)), \mathsf{P}_2(\inf \Phi_2 \leqq 2\pi_1(\mathsf{P}_1, \mathsf{P}_2))\right\}. \tag{8.2.30}$$

Proof. (i) Let $\pi_1(\mathsf{P}_1, \mathsf{P}_2) < \varepsilon < 1$, i.e.

$$\mathsf{P}_1(Y) < \mathsf{P}_2\left(\bigcup_{\varphi \in Y} \{\varphi' : \bar{d}_1(\varphi, \varphi') < \varepsilon\}\right) + \varepsilon,$$
$$\mathsf{P}_2(Y) < \mathsf{P}_1\left(\bigcup_{\varphi \in Y} \{\varphi' : \bar{d}_1(\varphi, \varphi') < \varepsilon\}\right) + \varepsilon,$$

for all closed subsets $Y \subseteqq \mathbf{U}_1$. For an arbitrary closed subset $B \subseteqq \mathbf{X}$ we have

$$\begin{aligned}\mathsf{P}_1(x_0(\Phi_1) \in B) &\leqq \mathsf{P}_1(\Phi_1 \in \mathrm{cl}\,\{\varphi : x_0(\varphi) \in B\}) \\ &< \mathsf{P}_2\left(\bigcup_{\varphi \in \mathrm{cl}\{\varphi'' : x_0(\varphi'') \in B\}} \{\varphi' : \bar{d}_1(\varphi, \varphi') < \varepsilon\}\right) + \varepsilon \\ &= \mathsf{P}_2\left(\bigcup_{\varphi \in \{\varphi'' : x_0(\varphi'') \in B\}} \{\varphi' : \bar{d}_1(\varphi, \varphi') < \varepsilon\}\right) + \varepsilon \\ &\leqq \mathsf{P}_2\left(\bigcup_{\varphi \in \{\varphi'' : x_0(\varphi'') \in B\}} \{\varphi' : d_x(x_0(\varphi), x_0(\varphi')) < \varepsilon\}\right) + \varepsilon,\end{aligned} \tag{8.2.31}$$

cf. Lemma 8.2.3 (i), and analogously

$$\mathsf{P}_2\big(x_0(\Phi_2) \in B\big) < \mathsf{P}_1\Big(\bigcup_{\varphi \in \{\varphi'' : x_0(\varphi'') \in B\}} \big\{\varphi' : d_x\big(x_0(\varphi), x_0(\varphi')\big) < \varepsilon\big\}\Big) + \varepsilon .$$

Thus,

$$\pi_X\Big(\mathsf{P}_1\big(x_0(\Phi_1) \in (\cdot)\big), \mathsf{P}_2\big(x_0(\Phi_2) \in (\cdot)\big)\Big) \leqq \varepsilon ,$$

and (8.2.29) is true.

(ii) Let $\pi_1(\mathsf{P}_1, \mathsf{P}_2) < \varepsilon < 1$, and let B an arbitrary subset of $\mathbb{Z}_+$. Analogously to (8.2.31), from Lemma 8.2.3 (ii) we get

$$\begin{aligned} \mathsf{P}_1\big(l_0(\Phi_1) \in B\big) &< \mathsf{P}_2\Big(\bigcup_{\varphi \in \{\varphi'' : l_0(\varphi'') \in B\}} \{\varphi' : \bar{d}_1(\varphi, \varphi') < \varepsilon\}\Big) + \varepsilon \\ &\leqq \mathsf{P}_2\Big(\bigcup_{\varphi \in \{\varphi'' : l_0(\varphi'') \in B\}} \{\varphi' : l_0(\varphi') = l_0(\varphi)\} \cup \{\varphi' : l_0(\varphi') \neq l_0(\varphi), \\ &\qquad \inf \varphi \leqq \bar{d}_1(\varphi, \varphi') < \varepsilon\}\Big) + \varepsilon . \end{aligned} \tag{8.2.32}$$

Consider two sample paths $\varphi, \varphi' \in \mathbb{U}_1$ with

$$\inf \varphi \leqq \bar{d}_1(\varphi, \varphi') < \varepsilon \text{ and } l_0(\varphi') \neq l_0(\varphi). \text{ Then}$$

$$\big|t_j(\varphi) + s_j(\varphi) - \big(t_j(\varphi') + s_j(\varphi')\big)\big| < \varepsilon$$

for all $j \in I(0, \varphi) \cup I(0, \varphi')$. Let j_0 be an index such that $|t_{j_0}(\varphi) + s_{j_0}(\varphi)| = \inf \varphi$. We want to estimate $\inf \varphi'$. In order to do this we have to distinguish several cases.

If $j_0 \in I(0, \varphi) \cup I(0, \varphi')$, then $|t_{j_0}(\varphi') + s_{j_0}(\varphi')| < \inf \varphi + \varepsilon < 2\varepsilon$, and thus $\inf \varphi' < 2\varepsilon$. Now assume $j_0 \notin I(0, \varphi) \cup I(0, \varphi')$. First consider the case that $\# I(0, \varphi) < \# I(0, \varphi')$. Let $j_1 \in I(0, \varphi')$, but $j_1 \notin I(0, \varphi)$. Then $t_{j_1}(\varphi) + s_{j_1}(\varphi) \leqq 0$, and thus $0 < t_{j_1}(\varphi') + s_{j_1}(\varphi') < \varepsilon$. Hence $\inf \varphi' < \varepsilon$.

Now we consider the case that $\# I(0, \varphi) > \# I(0, \varphi')$. Let $j_2 \in I(0, \varphi)$, but $j_2 \notin I(0, \varphi')$. Then $t_{j_2}(\varphi) + s_{j_2}(\varphi) > 0$, and thus $|t_{j_2}(\varphi') + s_{j_2}(\varphi')| < \varepsilon$. Hence $\inf \varphi' < \varepsilon$. Altogether we have obtained

$$\{\varphi' : l_0(\varphi') \neq l_0(\varphi), \inf \varphi \leqq \bar{d}_1(\varphi, \varphi') < \varepsilon\} \subseteqq \{\varphi' : \inf \varphi' < 2\varepsilon\} .$$

Now (8.2.32) yields

$$\begin{aligned} &\mathsf{P}_1\big(l_0(\Phi_1) \in B\big) \\ &< \mathsf{P}_2\Big(\bigcup_{\varphi \in \{\varphi'' : l_0(\varphi'') \in B\}} \{\varphi' : l_0(\varphi') = l_0(\varphi)\} \cup \{\varphi' : \inf \varphi' < 2\varepsilon\}\Big) + \varepsilon \\ &= \mathsf{P}_2\big(l_0(\Phi_2) \in B\big) + \mathsf{P}_2(\inf \Phi_2 < 2\varepsilon) + \varepsilon . \end{aligned}$$

Analogously we get

$$\mathsf{P}_2\big(l_0(\Phi_2) \in B\big) < \mathsf{P}_1\big(l_0(\Phi_1) \in B\big) + \mathsf{P}_1(\inf \Phi_1 < 2\varepsilon) + \varepsilon .$$

Thus,

$$\begin{aligned} &\pi_Z\Big(\mathsf{P}_1\big(l_0(\Phi_1) \in (\cdot)\big), \mathsf{P}_2\big(l_0(\Phi_2) \in (\cdot)\big)\Big) \\ &\leqq \pi_1(\mathsf{P}_1, \mathsf{P}_2) \\ &\quad + \max\Big\{\mathsf{P}_1\big(\inf \Phi_1 \leqq 2\pi_1(\mathsf{P}_1, \mathsf{P}_2)\big), \mathsf{P}_2\big(\inf \Phi_2 \leqq 2\pi_1(\mathsf{P}_1, \mathsf{P}_2)\big)\Big\} . \quad \blacksquare \end{aligned}$$

8.3. The system G/G/m/0

Consider a queueing system G/G/m/0 with the stationary ergodic input (Φ, P), $\Phi = ([A_n, S_n])$, $m_S < \infty$, $m_A > 0$. In this section we shall use the notation introduced in Section 8.2; in particular, we shall exploit the properties of the metric $\bar{d}_1$ on $\mathbb{U}_1$. In order to be compatible with Section 8.2, we use the state space $\mathbb{X} = \{(x_i, i \geqq 1): x_i \in \mathbb{R}_+, i = 1, 2, \ldots, x_i \downarrow 0\}$ with the metric d_x. (Of course, in the loss system G/G/m/0 only states (x_i) with $x_{m+1} = x_{m+2} = \cdots = 0$ are possible. All considerations from Section 5.3, in particular Theorem 5.3.1, remain true.) Consider the sets

$$\begin{aligned} A(n, \varphi) & \\ &= \Big\{(x_i, i \geqq 1): (x_i, i \geqq 1) \in \mathbb{X}, x_1 = t_{j_1} + s_{j_1} - t_n, \ldots, \\ &\quad x_l = t_{j_l} + s_{j_l} - t_n, x_{l+1} = 0; \\ &\quad l \in \big\{0, \ldots, \min\{m, l_n(\varphi)\}\big\}, \{j_1, \ldots, j_l\} \subseteqq I(n, \varphi)\Big\}, \qquad n \in \mathbb{Z}, \quad \varphi \in \mathbb{U}_1. \end{aligned}$$

Denote the elements of $A(n, \varphi)$, arranged in lexicographical order, by

$$x(1, n, \varphi), \ldots, x\big(\# A(n, \varphi), n, \varphi\big).$$

If we assume $\mathsf{P}(\inf \Phi = 0) = 0$, then the mappings

$$\varphi \to \# A(n, \varphi), \qquad n \in \mathbb{Z},$$

and

$$\begin{aligned} &\varphi \to x_n\big(x(j, 0, \varphi), [a_0(\varphi), s_0(\varphi)], \ldots, [a_{n-1}(\varphi), s_{n-1}(\varphi)]\big), \\ &j = 1, \ldots, \# A(n, \varphi); \quad n \geqq 0, \end{aligned}$$

are almost surely continuous with respect to $\bar{d}_1$, cf. Lemma 8.2.3. Together with Lemma 8.2.5, Theorem 1.7.2 (with $d_M = \bar{d}_1$, $M = \mathbb{U}_1$) yields

8.3.1. Theorem. *Let (Φ, P), (Φ_k, P_k), $k = 1, 2, \ldots$, be arrival-stationary, ergodic inputs of a queueing system* G/G/m/0 *satisfying*

$$m_{Sk} \xrightarrow[k\to\infty]{} m_S < \infty, \tag{8.3.1}$$

$$m_{Ak} > 0, \qquad m_A > 0, \tag{8.3.2}$$

$$\mathsf{P}\Big(\bigcup_{j\leqq 0}\{T_j + S_j = 0\}\Big) = 0, \tag{8.3.3}$$

and

$$\Phi_k \xrightarrow[k\to\infty]{\mathcal{D}} \Phi.$$

Furthermore, let q and q_k, $k = 1, 2, \ldots$, be the numbers given by Theorem 5.3.1 *for the inputs Φ and Φ_k, $k = 1, 2, \ldots$, respectively, and let $\hat{\Phi}$, $\hat{\Phi}_k$, $k = 1, 2, \ldots$, be the stationary weak solutions of the equation $X_{n+1} = f(X_n, [A_n, S_n])$, $n \in \mathbb{Z}$, of the loss system, cf.* (5.3.1), *corresponding to Φ and Φ_k, $k = 1, 2, \ldots$, respectively.*

Assume $q_k \xrightarrow[k\to\infty]{} q$. *Then*

$$\hat{\Phi}_k \xrightarrow[k\to\infty]{\mathcal{D}} \hat{\Phi}.$$

Remember that if $q = 1$, *then* $\Phi_k \xrightarrow[k\to\infty]{\mathcal{D}} \Phi$ *yields* $q_k = 1$ *for almost all* k, *cf. Remark* 1.7.6.

In the time-stationary situation we have the following result.

8.3.2. Theorem. *Let* $(\overline{\Psi}, \overline{\mathsf{P}})$, $(\overline{\Psi}_k, \overline{\mathsf{P}}_k)$, $k = 1, 2, \ldots$, *be time-stationary, ergodic inputs of a queueing system* G/G/m/0 *satisfying* (8.3.1), (8.3.2), (8.3.3), $m_{Ak} \xrightarrow[k\to\infty]{} m_A$, $\overline{\Psi}_k \xrightarrow[k\to\infty]{\mathcal{D}} \overline{\Psi}$, *and* $q_k \to q$. *Then*

$$[\overline{X}_k(u_1), \ldots, \overline{X}_k(u_r)] \xrightarrow[k\to\infty]{\mathcal{D}} [\overline{X}(u_1), \ldots, \overline{X}(u_r)], \tag{8.3.4}$$

and

$$[\overline{L}_k(u_1), \ldots, \overline{L}_k(u_r)] \xrightarrow[k\to\infty]{\mathcal{D}} [\overline{L}(u_1), \ldots, \overline{L}(u_r)], \tag{8.3.5}$$

$$u_1, \ldots, u_r \in \mathbb{R}, \quad r \geqq 1,$$

where $(\overline{X}(t))$, $(\overline{L}(t))$, $(\overline{X}_k(t))$, $(\overline{L}_k(t))$ *are the time-stationary state processes and the time-stationary processes of the number of customers in the system (corresponding to the (arrival-stationary) weak solutions* $\hat{\Phi}$ *and* $\hat{\Phi}_k$ *respectively).*

We omit the detailed proof of Theorem 8.3.2 here. A complete proof of

$$\overline{X}_k(0) \xrightarrow[k\to\infty]{\mathcal{D}} \overline{X}(0)$$

and

$$\overline{L}_k(0) \xrightarrow[k\to\infty]{\mathcal{D}} \overline{L}(0)$$

is given in Lisek (1981); the proof of the convergence of all finite-dimensional distributions is analogous.

8.4. Approximation of GI/GI/m/0 by GI/GI/∞

What happens when the queueing systems GI/GI/m/0 and GI/GI/∞ work with the same input (Φ, P), $\Phi = ([A_n, S_n])$? In Sections 8.2 and 8.3 we used the same state space for the queueing systems G/G/∞ and G/G/m/0. This allows us to compare directly the states of these systems. Let (Φ, P) be the input of both the GI/GI/m/0 queue and the GI/GI/∞ queue. Assume $m_S < \infty$, $0 < m_A < \infty$, and

$$v^2 = \frac{\sigma_A^2}{m_A^2} = \frac{\mathsf{E}(A_0 - m_A)^2}{m_A^2} < \infty.$$

As in Section 8.2, we denote the arrival-stationary state process of the system GI/GI/∞ by $(x_n(\Phi))$. Let $((\hat{X}_n), \mathsf{Q})$ be the arrival-stationary state process (given by Theorem 5.3.1) of the system GI/GI/m/0. Consider

$p^m(k)$ – the probability that the number of customers in the GI/GI/m/0 queue at the arrival of the 0-th customer is equal to k,

$p^\infty(k)$ – the probability that the number of customers in the GI/GI/∞ queue at the arrival of the 0-th customer is equal to k.

Then

$$\operatorname{Var}\left(p^m(\cdot), p^\infty(\cdot)\right) \leqq 2\mathsf{Q}\left(\hat{X}_0 \neq x_0(\Phi)\right),$$

$$|p^m(k) - p^\infty(k)| \leqq \mathsf{Q}\left(\hat{X}_0 \neq x_0(\Phi)\right), \qquad k = 0, 1, \ldots$$

Hence it is useful to estimate the probability $\mathsf{Q}\left(\hat{X}_0 \neq x_0(\Phi)\right)$.

Let $H(\cdot)$ be the renewal function of the synchronous renewal process (T_n), and $l_j(\Phi) = \#\{k : k < j,\ T_k + S_k > T_j\}$. We obtain

$$\begin{aligned}
\mathsf{Q}\left(\hat{X}_0 \neq x_0(\Phi)\right) &\leqq \mathsf{P}\left(\bigcup_{j<0} \{l_j(\Phi) \geqq m,\ T_j + S_j > 0\}\right) \\
&\leqq \sum_{j=-\infty}^{-1} \mathsf{P}\left(l_j(\Phi) \geqq m,\ T_j + S_j > 0\right) \\
&= \sum_{j=-\infty}^{-1} \int_0^\infty \mathsf{P}\left(l_j(\Phi) \geqq m,\ A_j + \cdots + A_{-1} < s\right) \mathrm{dS}(s) \\
&= \mathsf{P}\left(l_0(\Phi) \geqq m\right) \int_0^\infty H(s)\, \mathrm{dS}(s) \\
&\leqq \frac{1}{m} \mathsf{E} l_0(\Phi) \int_0^\infty H(s)\, \mathrm{dS}(s) \\
&= \frac{\varrho}{m} \int_0^\infty H(s)\, \mathrm{dS}(s) \\
&\leqq \frac{\varrho}{m} (\varrho + v^2),
\end{aligned} \tag{8.4.1}$$

cf. Theorem A 3.2.1.

Now we derive a similar result for the corresponding time-stationary models. Denote the time-stationary input corresponding to (Φ, P) by $(\overline{\Psi}, \overline{\mathsf{P}})$, $\overline{\Psi} = ([\overline{T}_n, \overline{S}_n])$. Let $\overline{p}$ be the (time-stationary) probability that the state of the G/G/m/0 queue at time 0 differs from the state of the G/G/∞ queue at time 0. Then

$$\begin{aligned}
\overline{p} &\leqq \overline{\mathsf{P}}\left(\bigcup_{j \leqq 0} \left\{\#\{k : k < j,\ \overline{T}_k + \overline{S}_k > \overline{T}_j\} \geqq m,\ \overline{T}_j + \overline{S}_j > 0\right\}\right) \\
&= \frac{1}{m_A} \int_0^\infty \mathsf{P}\left(\bigcup_{j \leqq 0} \{A_0 > t,\ l_j(\Phi) \geqq m,\ T_j + S_j > t\}\right) \mathrm{d}t
\end{aligned}$$

$$\leqq \frac{1}{m_A} \int_0^\infty \mathsf{P}\big(A_0 > t, l_0(\Phi) \geqq m\big)\,dt$$

$$+ \frac{1}{m_A} \int_0^\infty \mathsf{P}\left(\bigcup_{j<0} \{A_0 > t, l_j(\Phi) \geqq m, T_j + S_j > t\}\right) dt$$

$$\leqq \mathsf{P}\big(l_0(\Phi) \geqq m\big) + \frac{1}{m_A} \int_0^\infty \mathsf{P}\left(\bigcup_{j<0} \{A_0 > t, l_j(\Phi) \geqq m, T_j + S_j > 0\}\right) dt$$

$$= \mathsf{P}\big(l_0(\Phi) \geqq m\big) + \mathsf{P}\left(\bigcup_{j<0} \{l_j(\Phi) \geqq m, T_j + S_j > 0\}\right)$$

$$\leqq \mathsf{P}\big(l_0(\Phi) \geqq m\big) + \mathsf{P}\big(l_0(\Phi) \geqq m\big) \int_0^\infty H(s)\,d\mathsf{S}(s)$$

$$\leqq \frac{\varrho}{m} (1 + \varrho + v^2), \tag{8.4.2}$$

cf. the verification of (8.4.1).

8.5. The system G/G/1/∞

Consider a queueing system G/G/1/∞ with the stationary input (Φ, P), $\Phi = ([A_n, S_n])$, $\varrho = m_S/m_A < 1$. Let $(V_n) = \big(v_n(\Phi)\big)$, $v_n(\Phi) = v_n^{\min}(\Phi)$, be the unique stationary work load process constructed in Section 5.4. For this process we have the following continuity theorem.

8.5.1. Theorem. (i) *Let* (Φ, P), (Φ_k, P_k), $k = 1, 2, \ldots$, *be arrival-stationary inputs of a queueing system* G/G/1/∞ *satisfying*

$$m_S/m_A < 1, \qquad m_{Sk}/m_{Ak} < 1, \tag{8.5.1}$$

$$m_{Sk} \xrightarrow[k\to\infty]{} m_S, \qquad \frac{1}{n} \sum_{i=1}^n S_i \xrightarrow[n\to\infty]{} m_S \quad \mathsf{P}\text{-a.s.}, \tag{8.5.2}$$

$$\frac{1}{n} \sum_{i=1}^n A_i \xrightarrow[n\to\infty]{} m_A \quad \mathsf{P}\text{-a.s.}, \tag{8.5.3}$$

and

$$\Phi_k \xrightarrow[k\to\infty]{\mathcal{D}} \Phi.$$

Then

$$\big([[A_{kn}, S_{kn}], v_n(\Phi_k)]\big) \xrightarrow[k\to\infty]{\mathcal{D}} \big([[A_n, S_n], v_n(\Phi)]\big). \tag{8.5.4}$$

(ii) *Let* $(\overline{\Psi}, \overline{\mathsf{P}})$, $(\overline{\Psi}_k, \overline{\mathsf{P}}_k)$, $k = 1, 2, \ldots$, *be time-stationary inputs of a queueing system* G/G/1/∞ *satisfying* (8.5.1)–(8.5.3), $m_{A_k} \to m_A$, *and*

$$\overline{\Psi}_k \xrightarrow[k\to\infty]{\mathcal{D}} \overline{\Psi}.$$

Consider the time-stationary models $\bar{\chi}_k$, $\bar{\chi}$,

$$\bar{\chi}_k = \left[(\bar{V}_k(t)), ([\bar{T}_{kn}, [\bar{V}_{kn}, \bar{S}_{kn}]])\right], \qquad k = 1, 2, \ldots,$$

$$\bar{\chi} = \left[(\bar{V}(t)), ([\bar{T}_n, [\bar{V}_n, \bar{S}_n]])\right],$$

corresponding to $([[A_{kn}, S_{kn}], v_n(\Phi_k)])$ *and* $([[A_n, S_n], v_n(\Phi)])$, *respectively. Then*

$$([\bar{T}_{kn}, [\bar{V}_{kn}, \bar{S}_{kn}]]) \xrightarrow[k\to\infty]{\mathcal{D}} ([\bar{T}_n, [\bar{V}_n, \bar{S}_n]]), \tag{8.5.5}$$

$$[\bar{V}_k(u_1), \ldots, \bar{V}_k(u_r)] \xrightarrow[k\to\infty]{\mathcal{D}} [\bar{V}(u_1), \ldots, \bar{V}(u_r)], \quad u_1, \ldots, u_r \in \mathbb{R}, \quad r \geqq 1. \tag{8.5.6}$$

Proof. (i) Lemma A 2.8 yields $\mathsf{E}_{\mathsf{P}_k}(S_{k0} - A_{k0})_+ \xrightarrow[k\to\infty]{} \mathsf{E}_{\mathsf{P}}(S_0 - A_0)_+$, and from Lemma A 2.9 we obtain

$$\limsup_{k\to\infty} \mathsf{P}_k\left(\sup_{j>m} \sum_{l=1}^{j} (S_{k,-l} - A_{k,-l}) \geqq 0\right) \xrightarrow[m\to\infty]{} 0. \tag{8.5.7}$$

For given $\varepsilon > 0$ we can choose m in such a way that

$$\mathsf{P}\left(\sup_{j>m} \sum_{l=1}^{j} (S_{-l} - A_{-l}) \geqq 0\right) < \varepsilon,$$

$$\limsup_{k\to\infty} \mathsf{P}_k\left(\sup_{j>m} \sum_{l=1}^{j} (S_{k,-l} - A_{k,-l}) \geqq 0\right) < \varepsilon.$$

It holds (cf. Theorem 5.4.1) that

$$\begin{aligned} v_0(\Phi_k) &= \left(\sup_{j\geqq 1} \sum_{l=1}^{j} (S_{k,-l} - A_{k,-l})\right)_+, \\ v_0(\Phi) &= \left(\sup_{j\geqq 1} \sum_{l=1}^{j} (S_{-l} - A_{-l})\right)_+. \end{aligned} \tag{8.5.8}$$

Let $x \geqq 0$ be a point of continuity of the distributions of

$$\max_{1\leqq j\leqq s} \sum_{l=1}^{j} (S_{-l} - A_{-l}), \qquad s = 1, 2, \ldots \tag{8.5.9}$$

(Notice that the points of discontinuity of the distribution of (8.5.9) form a countable set.) Then

$$\begin{aligned} \limsup_{k\to\infty} \mathsf{P}_k\big(v_0(\Phi_k) \geqq x\big) &\leqq \varepsilon + \limsup_{k\to\infty} \mathsf{P}_k\left(\max_{1\leqq j\leqq m} \sum_{l=1}^{j} (S_{k,-l} - A_{k,-l}) \geqq x\right) \\ &= \varepsilon + \mathsf{P}\left(\max_{1\leqq j\leqq m} \sum_{l=1}^{j} (S_{-l} - A_{-l}) \geqq x\right) \leqq 2\varepsilon + \mathsf{P}\big(v_0(\Phi) \geqq x\big). \end{aligned}$$

Analogously we get

$$\begin{aligned} \liminf_{k\to\infty} \mathsf{P}_k\big(v_0(\Phi_k) \geqq x\big) &\geqq \liminf_{k\to\infty} \mathsf{P}_k\left(\max_{1\leqq j\leqq m} \sum_{l=1}^{j} (S_{k,-l} - A_{k,-l}) \geqq x\right) \\ &= \mathsf{P}\left(\max_{1\leqq j\leqq m} \sum_{l=1}^{j} (S_{-l} - A_{-l}) \geqq x\right) \geqq \mathsf{P}\big(v_0(\Phi) \geqq x\big) - \varepsilon. \end{aligned}$$

Since these inequalities are true for all $\varepsilon > 0$, we have shown that

$$\mathsf{P}_k\big(v_0(\Phi_k) \geqq x\big) \xrightarrow[k\to\infty]{} \mathsf{P}\big(v_0(\Phi) \geqq x\big)$$

and hence, cf. Lemma A 2.2,

$$v_0(\Phi_k) \xrightarrow[k\to\infty]{\mathfrak{D}} v_0(\Phi).$$

In the same way one can prove the convergence of all finite-dimensional distributions:

$$\big[[A_{k,-n}, S_{k,-n}, v_{-n}(\Phi_k)], \ldots, [A_{kn}, S_{kn}, v_n(\Phi_k)]\big]$$

$$\xrightarrow[k\to\infty]{\mathfrak{D}} \big[[A_{-n}, S_{-n}, v_{-n}(\Phi)], \ldots, [A_n, S_n, v_n(\Phi)]\big], \qquad n \geqq 0.$$

Thus (8.5.4) is true.

(ii) The statements (8.5.5), (8.5.6) follow by means of Theorem 8.1.1. ∎

8.6. The system G/GI/m/∞

Consider a queueing system G/GI/m/∞ with FCFS queueing discipline and with the stationary ergodic input (Φ, P),

$$\Phi = ([A_n, S_n]), \qquad \varrho = m_S/m_A < m.$$

Then, by Theorem 5.5.7, we have a unique stationary work load vector process $(V_n) = \big(v_n(\Phi)\big) = \big([v_{n1}(\Phi), \ldots, v_{nm}(\Phi)]\big)$, $v_{ni}(\Phi) = v_{ni}^{\min}(\Phi)$, $i = 1, \ldots, m$, i.e. a unique stationary sequence (V_n) satisfying

$$V_{n+1} = R(V_n + S_n e_1 - A_n \mathbf{1})_+ = f(V_n, [A_n, S_n]), \qquad n \in \mathbb{Z}. \tag{8.6.1}$$

For this process we can prove the following continuity theorem.

8.6.1. Theorem. (i) *Let* (Φ, P), (Φ_k, P_k), $k = 1, 2, \ldots$, *be arrival-stationary, ergodic inputs of a queueing system* G/GI/m/∞ *satisfying the conditions*

$$m_S/m_A < m, \qquad m_{Sk}/m_{Ak} < m, \tag{8.6.2}$$

$$m_{Sk} \xrightarrow[k\to\infty]{} m_S, \tag{8.6.3}$$

and

$$\Phi_k \xrightarrow[k\to\infty]{\mathfrak{D}} \Phi.$$

Then

$$\big([[A_{kn}, S_{kn}], v_n(\Phi_k)]\big) \xrightarrow[k\to\infty]{\mathfrak{D}} \big([[A_n, S_n], v_n(\Phi)]\big). \tag{8.6.4}$$

(ii) *Let* $(\overline{\Psi}, \overline{\mathsf{P}})$, $(\overline{\Psi}_k, \overline{\mathsf{P}}_k)$, $k = 1, 2, \ldots$, *be time-stationary inputs of a queueing system* G/GI/m/∞ *satisfying* (8.6.2), (8.6.3), *and*

$$\overline{\Psi}_k \xrightarrow[k\to\infty]{\mathfrak{D}} \overline{\Psi}, \qquad m_{Ak} \xrightarrow[k\to\infty]{} m_A.$$

Consider the time-stationary models $\bar{\chi}_k$, $\bar{\chi}$,

$$\bar{\chi}_k = \left[(\bar{V}_k(t)), \left([\bar{T}_{kn}, [\bar{V}_{kn}, \bar{S}_{kn}]]\right)\right], \qquad k = 1, 2, \dots,$$

$$\bar{\chi} = \left[(\bar{V}(t)), \left([\bar{T}_n, [\bar{V}_n, \bar{S}_n]]\right)\right],$$

corresponding to $\left([[A_{kn}, S_{kn}], v_n(\Phi_k)]\right)$ *and* $\left([[A_n, S_n], v_n(\Phi)]\right)$, *respectively. Then*

$$\left([\bar{T}_{kn}, [\bar{V}_{kn}, \bar{S}_{kn}]]\right) \xrightarrow[k\to\infty]{\mathcal{D}} \left([\bar{T}_n, [\bar{V}_n, \bar{S}_n]]\right),$$
$$[\bar{V}_k(u_1), \dots, \bar{V}_k(u_r)] \xrightarrow[k\to\infty]{\mathcal{D}} [\bar{V}(u_1), \dots, \bar{V}(u_r)], \tag{8.6.5}$$

$$u_1, \dots, u_r \in \mathbb{R}, \qquad r \geqq 1. \tag{8.6.6}$$

Proof. (i) Beside the G/GI/m/∞ queue with FCFS queueing discipline we consider an m-server queue with cyclic queueing discipline, cf. Section 5.6, and the same input Φ for both. For this reason we number the servers from 1 up to m. Let J_0 be a random variable uniformly distributed on $\{1, \dots, m\}$ and independent of Φ. The 0-th customer (i.e. the customer with the index 0) will be assigned to the server with number J_0. The customer with the index n will be assigned to the server with the number $J_n = (J_0 + n - 1) \bmod m + 1$. We require that each server serves the assigned customers in the FCFS discipline. For this m-server, cyclic queue there is a unique stationary weak solution $\left([[A_n, S_n], V_n^c]\right)$, $V_n^c = [V_{n1}^c, \dots, V_{nm}^c]$ of the equation

$$V_{n+1}^c = (V_n^c - S_n e(J_n) - A_n 1)_+, \qquad n \in \mathbb{Z}, \tag{8.6.7}$$

where $e(j) = [1_{\{1\}}(j), \dots, 1_{\{m\}}(j)]$, cf. Theorem 5.6.2. Consider the sequences

$$\Phi^{(j)} = ([A_n^{(j)}, S_n^{(j)}]) = \left([A_n, S_n 1_{\{j\}}(J_n)]\right), \qquad j = 1, \dots, m.$$

Obviously, $\Phi^{(j)}$ is a possible description of that part of the input Φ which is served by the j-th server according to the cyclic discipline (all the customers which are served by other servers get the service time zero). The sequences $\Phi^{(j)}$, $j = 1, \dots, m$, are stationary. In the proof of Theorem 5.6.2 we have shown that the following strong law of large numbers is true:

$$\frac{1}{n} \sum_{i=1}^{n} S_i^{(j)} \xrightarrow[n\to\infty]{} \mathsf{E} S_0^{(j)} = m_S/m \quad \text{a.s.} \tag{8.6.8}$$

By $\left(V_n^c(0), n \geqq 0\right)$, $V_n^c(0) = [V_{n1}^c(0), \dots, V_{nm}^c(0)]$, we denote the solution of (8.6.7) with the initial condition $V_0^c(0) = [0, \dots, 0]$. Analogously, let $\left(V_n(0), n \geqq 0\right)$,

$$V_n(0) = v_n(0, [A_0, S_0], \dots, [A_{n-1}, S_{n-1}]) = [V_{n1}(0), \dots, V_{nm}(0)],$$

be the solution of (8.6.1) with the initial condition $V_0(0) = [0, \dots, 0]$ for the m-server queue with FCFS discipline. By Theorem 5.5.7 we have $V_n(0) \xrightarrow[n\to\infty]{\mathcal{D}} v_0(\Phi)$, in particular $V_{nm}(0) \xrightarrow[n\to\infty]{\mathcal{D}} V_{0m}$, and by Theorem 5.6.2

$$V_n^c(0) \xrightarrow[n\to\infty]{\mathcal{D}} V_0^c. \tag{8.6.9}$$

Since the function max $\{x_1, \ldots, x_m\}$ is continuous in $[x_1, \ldots, x_m]$, (8.6.9) leads to

$$\max_{1 \leq j \leq m} V^c_{n,j}(0) \xrightarrow[n\to\infty]{\mathcal{D}} \max_{1 \leq j \leq m} V^c_{0,j}. \tag{8.6.10}$$

Now we use a result by Foss (1980):

$$\mathsf{P}\big(V_{n,m}(0) > x\big) \leqq \mathsf{P}^c\Big(\max_{1 \leq j \leq m} V^c_{n,j}(0) > x\Big). \tag{8.6.11}$$

for all x and all $n \geqq 1$.

Using (8.6.9), (8.6.10), and (8.6.11), we get

$$\mathsf{P}\big(v_{0,m}(\Phi) > x\big) \leqq \mathsf{P}^c\Big(\max_{1 \leq j \leq m} V^c_{0,j} > x\Big)$$

for all common points x of continuity of $v_{0,m}(\Phi)$ and $\max_{1 \leq j \leq m} V^c_{0,j}$. Thus, in view of the stationarity of $(v_n(\Phi))$ and $(V_n{}^c)$:

$$\mathsf{P}\big(v_{n,m}(\Phi) > x\big) \leqq \mathsf{P}^c\Big(\max_{1 \leq j \leq m} V^c_{n,j} > x\Big) \tag{8.6.12}$$

for all x and n. (Here P^c denotes the common distribution of Φ and J_0.)

Consider the stationary strong solutions $(\hat{\Phi}_k, \hat{\mathsf{P}}_k)$, $\hat{\Phi}_k = \big([[A_{kn}, S_{kn}], v_n(\Phi_k)]\big)$, and $(\hat{\Phi}, \hat{\mathsf{P}})$, $\hat{\Phi} = \big([[A_n, S_n], v_n(\Phi)]\big)$, of equation (8.6.1) for the inputs Φ_k, $k = 1, 2, \ldots$, and Φ, respectively. (Notice, $\hat{\mathsf{P}}_k$, $k \geqq 1$, and $\hat{\mathsf{P}}$ are induced by P_k and P, respectively.) We shall show that for every $\varepsilon > 0$ there exist a_n, b_n, c_n, $n \in \mathbb{Z}$, with the property

$$P_k\big(A_{kn} \leqq a_n, S_{kn} \leqq b_n, v_{n,1}(\Phi_k) \leqq c_n, \ldots, v_{n,m}(\Phi_k) \leqq c_n, n \in \mathbb{Z}\big)$$
$$\geqq 1 - \varepsilon \tag{8.6.13}$$

for all k. It is well-known that (8.6.13) means the tightness and thus the relative compactness of the distributions $\hat{\mathsf{P}}_k$ of $\hat{\Phi}_k$, cf. Theorem A 2.5. From $\Phi_k \xrightarrow{\mathcal{D}} \Phi$ it follows that there are $a_n, b_n, n \in \mathbb{Z}$, satisfying

$$\mathsf{P}_k(A_{kn} \leqq a_n, S_{kn} \leqq b_n, n \in \mathbb{Z}) \geqq 1 - \frac{\varepsilon}{2}. \tag{8.6.14}$$

Consider the "partial inputs" $\Phi^{(j)}$ and $\Phi^{(j)}_k$, $j = 1, \ldots, m$, generated by Φ and Φ_k, respectively, by applying the cyclic discipline. It is easy to see that the assumption $\Phi_k \xrightarrow{\mathcal{D}} \Phi$ leads to

$$\Phi_k{}^{(j)} \xrightarrow[k\to\infty]{\mathcal{D}} \Phi^{(j)}, \qquad j = 1, \ldots, m. \tag{8.6.15}$$

We denote the work load vector of an m-server queue with the input Φ_k and cyclic discipline by $V_n{}^{k,c} = [V^{k,c}_{n,1}, \ldots, V^{k,c}_{n,m}]$. In view of (8.6.3), (8.6.8), (8.6.2), (8.6.15), and the ergodicity of (A_n) we can apply Theorem 8.5.1 to the $\Phi^{(j)}, \Phi_k{}^{(j)}$, $j = 1, \ldots, m$, and obtain in particular that there exists a c_0 such that

$$\mathsf{P}_k{}^c(V^{k,c}_{0,j} > c_0) < \frac{\varepsilon}{4m}, \qquad j = 1, \ldots, m, \tag{8.6.16}$$

where $\mathsf{P}_k^{\mathrm{c}}$ denotes the common distribution of Φ_k and J_0. In analogy to (8.6.12) we have

$$\mathsf{P}\big(v_{n,m}(\Phi_k) > x\big) \leqq \mathsf{P}_k^{\mathrm{c}}\left(\max_{1\leqq j\leqq m} V_{n,j}^{k,\mathrm{c}} > x\right)$$

for all x and n. This and (8.6.16) yield

$$\mathsf{P}\big(v_{0,m}(\Phi_k) \leqq c_0\big) \geqq 1 - \frac{\varepsilon}{4}. \tag{8.6.17}$$

In the same way, for every $n \neq 0$ one can find a c_n satisfying

$$\mathsf{P}\big(v_{n,m}(\Phi_k) \leqq c_n\big) \geqq 1 - \frac{\varepsilon}{2^{|n|+3}}. \tag{8.6.18}$$

The desired inequality (8.6.13) follows from (8.6.14), (8.6.17), (8.6.18) by the following calculation:

$$\mathsf{P}\big(A_{kn} \leqq a_n, S_{kn} \leqq b_n, v_{n,1}(\Phi_k) \leqq c_n, \ldots, v_{n,m}(\Phi_k) \leqq c_n, n \in \mathbb{Z}\big)$$

$$= \mathsf{P}\big(A_{kn} \leqq a_n, S_{kn} \leqq b_n, v_{n,m}(\Phi_k) \leqq c_n, n \in \mathbb{Z}\big)$$

$$\geqq -1 + \mathsf{P}(A_{kn} \leqq a_n, S_{kn} \leqq b_n, n \in \mathbb{Z})$$

$$+ \mathsf{P}\big(v_{n,m}(\Phi_k) \leqq c_n, n \in \mathbb{Z}\big)$$

$$\geqq -1 + 1 - \frac{\varepsilon}{2} + 1 - \sum_{n\in\mathbb{Z}} \mathsf{P}\big(v_{n,m}(\Phi_k) > c \;)$$

$$\geqq 1 - \frac{\varepsilon}{2} - 2\sum_{n=1}^{\infty} \frac{\varepsilon}{2^{n+3}} - \frac{\varepsilon}{4} = 1 -$$

Now we show that $\hat{\Phi}_k \xrightarrow{\mathscr{D}} \hat{\Phi}$ holds. Since $(\hat{\mathsf{P}}_k)$ is relatively compact, there exists a convergent subsequence $(\hat{\Phi}_{k_l}, l \geqq 1)$ of $(\hat{\Phi}_k, k \geqq 1)$. Let $\hat{\Phi}^*$ be its limit:

$$\hat{\Phi}_{k_l} \xrightarrow[l\to\infty]{\mathscr{D}} \hat{\Phi}^*.$$

In view of Theorem 2.3 in BILLINGSLEY (1968) it remains to show $\hat{\Phi}^* \overset{\mathscr{D}}{=} \hat{\Phi}$. We first note that without loss of generality in the limiting sequence $\hat{\Phi}^* = \big([[A_n^*, S_n^*], V_n^*]\big)$ we can take $A_n^* = A_n$ and $S_n^* = S_n$. (This follows from $\Phi_{k_l} \xrightarrow[l\to\infty]{\mathscr{D}} \Phi$.) The sequence $\hat{\Phi}^*$ is stationary. Next we show that $\hat{\Phi}^*$ satisfies equation (8.6.1). For this it suffices to prove

$$\hat{\mathsf{P}}^*\big(V_1^* = f(V_0^*, [A_0, S_0])\big) = 1, \tag{8.6.19}$$

where $\hat{\mathsf{P}}^*$ denotes the distribution of $\hat{\Phi}^*$. The function f is continuous. Hence $\{([[x_n, y_n], z_n]) : f(z_0, [x_0, y_0]) = z_1\}$ is a closed subset of $\mathbb{R}^2 \times \mathbb{R}^m$. Using the property

$$\mathsf{P}_{k_l}\Big(v_1(\Phi_{k_l}) = f\big(v_0(\Phi_{k_l}), [A_{k_l,0}, S_{k_l,0}]\big)\Big) = 1, \qquad l = 1, 2, \ldots,$$

and $\hat{\Phi}_{k_l} \xrightarrow[l\to\infty]{\mathfrak{D}} \hat{\Phi}^*$ we obtain

$$1 = \limsup_{l\to\infty} \mathsf{P}_{k_l}\big(v_1(\Phi_{k_l}) = f(v_0(\Phi_{k_l}), [A_{k_l,0}, S_{k_l,0}])\big)$$
$$\leqq \hat{\mathsf{P}}^*\big(V_1^* = f(V_0^*, [A_0, S_0])\big).$$

Thus (8.6.19) is true, and $\hat{\Phi}^*$ satisfies (8.6.1). On the other hand, in view of Theorem 5.5.7, there is only one stationary weak solution of (8.6.19) for Φ. Hence it follows that $\hat{\Phi}^* \overset{\mathfrak{D}}{=} \hat{\Phi}$, and (8.6.4) is true.

(ii) The statement (8.6.5), (8.6.6) follows by means of Theorem 8.1.1. ∎

8.7. Remarks and references

8.2. For the system G/GI/∞ FRANKEN (1969, 1970) proved a continuity theorem similar to Theorem 8.2.6(ii) (time-stationary case). His proof goes along the lines of the general scheme mentioned in Section 1.7. First one proves the relative compactness of the sequence of stationary solutions. Then one can show that the limit distribution of every convergent subsequence of stationary solutions is associated with a stationary solution corresponding to the limit input distribution P. Since the solution for P is unique, the statement follows. In the same way, BOROVKOV (1972a) proved Theorem 8.2.6(i) (arrival-stationary case), cf. also BOROVKOV (1978, 1980). In [FKAS] a proof of the continuity theorem for G/G/∞ is given in the time-stationary case, and then, by means of Theorem 8.1.1, this result is carried over to the arrival-stationary case. For this the condition $m_{Ak} \to m_A$ is needed. Of course, the proofs by Franken and Borovkov are much shorter than our approach via Lemmas 8.2.1–8.2.5. However, we used the (technically complicated) method of metric modification in order to give an illustrative example of this method and to obtain estimates of the rate of convergence of the form (8.2.29), (8.2.30). The idea of using a special metric $\bar{d}_1$ in Section 8.2 is due to LISEK (1981), where the method of metric modifications is explained in the sample space of time-stationary inputs and estimates analogous to (8.2.29), (8.2.30) are obtained in the time-stationary case. ZOLOTAREV (1977a) used several special metrics for proving model continuity of G/G/∞, including estimates of the rate of convergence. However, his assumptions and results are not compatible with the usual weak convergence $\xrightarrow{\mathfrak{D}}$. For other estimates see AKHMAROV (1980).

8.3. Concerning continuity theorems for loss systems, the history is similar to the G/G/∞ case. A continuity theorem for the system G/GI/m/0 in the time-stationary case was proved by FRANKEN (1969, 1970) via relative compactness, cf. the hints above and in Section 1.7, under the assumption that all queueing processes considered have infinitely many empty points, cf. Theorem 5.2.3 and the proof of Theorem 5.3.8. Using the same ideas, BOROVKOV (1972b) proved a continuity theorem in the arrival-stationary case under the assumption that the limit input has some kind of renewing epochs, cf. also BOROVKOV

(1978, 1980). A similar assumption in the time-stationary case led to the continuity theorem of Franken and Kalähne (1978, 1980). In [FKAS] a continuity theorem for G/G/m/0 is proved in the time-stationary case (again under the assumption that "renewing epochs" exist) and is then carried over to the arrival-stationary case (using the condition $m_{Ak} \to m_A$). Estimates of the rate of convergence in the continuity theorem for the time-stationary system G/G/m/0 without renewing epochs were obtained in Lisek (1981). Estimates for the rate of convergence in loss systems under certain assumptions are given in Akhmarov (1979b, 1981) and Foss (1986).

8.4. As expected, the estimate (8.4.2) is not very sharp for the special case of M/GI/m/0 systems. In this case it is known (cf. Gutjahr (1977a)) that

$$\operatorname{Var}\left(p^m(\cdot), p^\infty(\cdot)\right) = 2e^{-\varrho} \sum_{k=m+1}^{\infty} \frac{\varrho^k}{k!}. \tag{8.7.1}$$

Numerical comparisons of (8.4.2) and (8.7.1) in Brandt and Lisek (1980) show that in this special case the approximation (8.4.2) is very bad. On the other hand, Gutjahr (1977a) has shown that the rate of convergence of $\operatorname{Var}\left(p^m(\cdot), p^\infty(\cdot)\right)$ to zero for $m \to \infty$ is only of the order m^{-1} in the general case. For completeness we mention another result of Gutjahr (1977a). Under the additional assumptions $p^m(k) \geqq p^\infty(k)$ for $k = 0, \ldots, m$ we have

$$\operatorname{Var}\left(p^m(\cdot) - p^\infty(\cdot)\right) \leqq \frac{2\varrho}{m+1}.$$

8.5. A continuity theorem for the actual waiting time in the system GI/GI/1/∞ was first proved by Rossberg (1965), see also Kennedy (1972). The proof of Theorem 8.5.1 (i) is due to Borovkov (1980). There one can also find a direct proof of Theorem 8.5.1 (ii). Zolotarev (1976a) used several special metrics for obtaining estimates of the rate of convergence in continuity theorems for G/G/1/∞ queues. Other estimates of the rate of convergence are given in Borovkov (1972a, 1977, 1980), see also Kotzurek and Stoyan (1976), Kotzurek (1978), Kalashnikov (1977).

8.6. A continuity theorem for GI/GI/m/∞ was stated by Borovkov (1972a). However, this proof was not correct, as pointed out by Stoyan (1976). Stoyan tried to correct Borovkov's proof by using an idea suggested by Brumelle (1971a): For every input $\Phi = ([A_n, S_n])$ of a GI/GI/m/∞ queue with $0 < m_A < \infty$, $0 < m_S < \infty$, $\varrho < m$ there exists a stationary sequence (δ_n), where for every n the variable δ_n is independent of Φ, and $\mathsf{P}(\delta_n = 1) = 1 - \mathsf{P}(\delta_n = 0) = 1/m$. Moreover, the arrival-stationary waiting time distribution of the system GI/GI/m/∞ with the input Φ is stochastically not greater than the arrival-stationary waiting time distribution of the G/G/1/∞ queue with the input $([A_n, \delta_n S_n])$. Stoyan assumed erroneously that the random variables δ_n are i.i.d. and that the sequence (δ_n) is independent of Φ, cf. Brandt and Lisek (1983) for a discussion of this error.

The idea of our proof of Theorem 8.6.1 is due to Arndt (1978, 1980), where only the time-stationary case was considered. As a technical tool we used the estimate (8.6.11) for the FCFS system by means of the corresponding cyclic system, which is due to Foss (1980), Gittins (1978), Vasicek (1977), see also Foss (1981) for a discussion of the inequality (8.6.11). However, Arndt assumed that according to the cyclic discipline the partition of an ergodic input leads to m identically distributed ergodic inputs. But this is not true, as a counter-example in Brandt and Lisek (1983) shows, see Example 5.6.1.

Continuity properties of G/G/m/∞ were investigated by Borovkov (1978, 1980) under the assumption that the limit input has some kind of renewing epochs and that these renewing epochs fulfil a certain compactness condition. For completeness we mention Whitt (1974) and Stoyan (1983) concerning the continuity of non-stationary distributions of quantities in waiting systems, and Kalashnikov (1977). Continuity properties for systems G/G/m/r and for waiting systems with impatient customers were investigated by Krupin (1974, 1975), see also Borovkov (1980), Akhmarov (1979a). Concerning estimates of the rate of convergence in ergodic and continuity theorems we refer to Akhmarov (1979b) and Foss (1985, 1986).

For the system GI/GI/m/∞ it is shown in Asmussen and Johansen (1986) that under the assumptions of Theorem 8.6.1 (i) the k-th moment of the arrival-stationary waiting time converges if the $(k+1)$-th moment of the service times converges.

The statements of Theorems 8.2.6 (ii), 8.3.2, 8.5.1 (ii), and 8.6.1 (ii) can be extended, cf. Brandt (1983): If, in addition to the assumptions formulated in the above theorems, the limiting input $\overline{\Psi}$ is a simple MPP, then

$$\left(\overline{X}_k(t)\right) \xrightarrow[k\to\infty]{\mathcal{D}} \left(\overline{X}(t)\right) \quad \text{(in Th. 8.2.6 (ii), 8.3.2)},$$

$$\left(\overline{V}_k(t)\right) \xrightarrow[k\to\infty]{\mathcal{D}} \left(\overline{V}(t)\right) \quad \text{(in Th. 8.5.1, 8.6.1)},$$

respectively. For further continuity results concerning the processes $\left(\overline{L}_k(t)\right)$, $\left(\overline{L}(t)\right)$ of the number of customers in the system we refer to Brandt (1983).

9. Further models

In this chapter we show by means of two examples that the techniques developed in this book can successfully be applied to models outside queueing theory, too. First we investigate the stochastic difference equation $X_{n+1} = A_n X_n + B_n$. Under weak conditions the existence, uniqueness, and continuity of the solution is shown. Finally, we discuss some robust filters for cleaning a stationary sequence from outliers.

9.1. The stochastic equation $X_{n+1} = B_n X_n + C_n$ with stationary coefficients

In this section we deal with the stochastic difference equation

$$X_{n+1} = B_n X_n + C_n, \qquad n \in \mathbb{Z}, \tag{9.1.1}$$

where the sequence (Φ, P), $\Phi = ([B_n, C_n])$, $B_n, C_n \in \mathbb{R}$, is assumed to be stationary and ergodic. Our first aim is to give conditions that ensure the existence of a uniquely determined stationary solution of (9.1.1). Later we shall deal with the continuous dependence of the distribution of the stationary solution on P. In the following we need the n-fold iteration of $f(x) = bx + c$:

$$\begin{aligned} &x_n(x, [B_0, C_0], [B_1, C_1], \ldots, [B_{n-1}, C_{n-1}]) \\ &= \sum_{j=0}^{n-1} \left(\prod_{i=n-j}^{n-1} B_i \right) C_{n-j-1} + \left(\prod_{i=0}^{n-1} B_i \right) x, \qquad n \geqq 1, \end{aligned} \tag{9.1.2}$$

(we set $\prod_{i=n}^{n-1} B_i = 1$).

9.1.1. Theorem. *Suppose the sequence $\Phi = ([B_n, C_n])$ is stationary and ergodic, and one of the conditions*

$$-\infty \leqq \mathsf{E} \log |B_0| < 0 \quad \textit{and} \quad \mathsf{E}(\log |C_0|)_+ < \infty, \tag{9.1.3}$$

$$\mathsf{P}(B_0 = 0) > 0 \tag{9.1.4}$$

is satisfied. Then

$$\begin{aligned} x_n(\Phi) &= \lim_{k \to \infty} x_k(0, [B_{n-k}, C_{n-k}], \ldots, [B_{n-1}, C_{n-1}]) \\ &= \sum_{j=0}^{\infty} \left(\prod_{i=n-j}^{n-1} B_i \right) C_{n-j-1}, \qquad n \in \mathbb{Z}, \end{aligned} \tag{9.1.5}$$

exists almost surely, is finite, and $\big([[B_n, C_n], x_n(\Phi)]\big)$ *is the only stationary weak solution of* (9.1.1). *Furthermore,*

$$\lim_{n\to\infty} |x_n(X, [B_0, C_0], \ldots, [B_{n-1}, C_{n-1}]) - x_n(\Phi)| = 0 \quad a.s. \tag{9.1.6}$$

for an arbitrary random initial condition X, *in particular*

$$x_n(X, [B_0, C_0], \ldots, [B_{n-1}, C_{n-1}]) \xrightarrow[n\to\infty]{\mathcal{D}} x_0(\Phi). \tag{9.1.7}$$

Proof. 1. First we show that the series on the right-hand side of (9.1.5) converges absolutely.

1.1. Assumption (9.1.3) and the strong law of large numbers yield

$$\limsup_{k\to\infty} \frac{1}{k} \left(\sum_{i=-1}^{-k} \log |B_{n+i}| + \log |C_{n-k-1}| \right) < 0 \quad \text{a.s.},$$

i.e.

$$\limsup_{k\to\infty} \log \left(|B_{n-1}| \cdots |B_{n-k}|\, |C_{n-k-1}|\right)^{1/k} < 0 \quad \text{a.s.},$$

and thus

$$\limsup_{k\to\infty} \left(|B_{n-1}| \cdots |B_{n-k}|\, |C_{n-k-1}|\right)^{1/k} < 1 \quad \text{a.s.}$$

From this the convergence of (9.1.5) follows by Cauchy's root criterion.

1.2. If the assumption (9.1.4) is satisfied, we obtain from Theorem A 1.1.2

$$\mathsf{P}(\#\{k < 0: B_k = 0\} = \#\{k \geqq 0: B_k = 0\} = \infty) = 1. \tag{9.1.8}$$

This implies that the sum $x_n(\Phi)$ consists only of finitely many summands, which yields the convergence of (9.1.5).

2. Now we prove (9.1.6). It follows from (9.1.2) and (9.1.5) that

$$\begin{aligned} &|x_n(X, [B_0, C_0], \ldots, [B_{n-1}, C_{n-1}]) - x_n(\Phi)| \\ &= \left| -\sum_{j=n}^{\infty} \left(\prod_{i=n-j}^{n-1} B_i\right) C_{n-j-1} + \left(\prod_{i=0}^{n-1} B_i\right) X \right| \\ &\leqq \left(\prod_{i=0}^{n-1} |B_i|\right) \left(|x_0(\Phi)| + |X|\right). \end{aligned} \tag{9.1.9}$$

By the strong law of large numbers, from (9.1.3) we get

$$\prod_{i=0}^{n-1} |B_i| = \left(\exp\left(\frac{1}{n} \sum_{i=0}^{n-1} \log |B_i|\right)\right)^n \xrightarrow[n\to\infty]{} 0 \quad \text{a.s.}$$

If (9.1.4) is satisfied, then it follows from (9.1.8) that $\prod_{i=0}^{n-1} |B_i| = 0$ a.s. for sufficiently large n. Together with (9.1.9) this implies (9.1.6).

3. The shift applied to Φ on the right-hand side of (9.1.5) produces the shift of $\big(x_n(\Phi)\big)$ on the left-hand side, so $\big(x_n(\Phi)\big)$ is stationary. From (9.1.6) we obtain

$$|x_n(X, [B_0, C_0], \ldots, [B_{n-1}, C_{n-1}]) - x_n(\Phi)| \xrightarrow{\mathsf{P}} 0$$

(convergence in probability). Together with $x_n(\Phi) \stackrel{\mathcal{D}}{=} x_0(\Phi)$ we thus obtain the assertion (9.1.7) (cf. BILLINGSLEY (1968), Theorem 4.1).

Moreover,

$$B_n x_n(\Phi) + C_n = B_n \left(\sum_{j=0}^{\infty} \left(\prod_{i=n-j}^{n-1} B_i \right) C_{n-j-1} \right) + C_n$$

$$= \sum_{j=1}^{\infty} \left(\sum_{i=(n+1)-j}^{(n+1)-1} B_i \right) C_{(n+1)-j-1} + C_n = x_{n+1}(\Phi) \text{ a.s.},$$

i.e. the difference equation (9.1.1) is almost surely satisfied. Thus $\big([B_n, C_n], x_n(\Phi)]\big)$ is a stationary strong solution of (9.1.1). We complete the proof by showing that it is the only stationary weak solution. Assume that $\big([[B_n', C_n'], X_n']\big)$ is another stationary weak solution of (9.1.1). In the following we omit the primes for convenience. Then

$$|X_n - x_n(\Phi)| = |B_{n-1}| \, |X_{n-1} - x_{n-1}(\Phi)|$$

$$= \left| \prod_{i=1}^{k} B_{n-i} \right| |X_{n-k} - x_{n-k}(\Phi)|$$

$$\leq \left| \prod_{i=1}^{k} B_{n-i} \right| |X_{n-k}| + \left| \prod_{i=1}^{k} B_{n-i} \right| |x_{n-k}(\Phi)|. \tag{9.1.10}$$

If (9.1.3) is satisfied, then, according to the strong law of large numbers, we obtain

$$\left| \prod_{i=1}^{k} B_{n-i} \right| = \left(\exp \left(\frac{1}{k} \sum_{i=1}^{k} \log |B_{n-i}| \right) \right)^k \xrightarrow[k \to \infty]{} 0 \quad \text{a.s.},$$

and if (9.1.4) is satisfied, then by (9.1.8)

$$\lim_{k \to \infty} \prod_{i=1}^{k} |B_{n-i}| = 0 \quad \text{a.s.}$$

Due to this and because (X_n) and $\big(x_n(\Phi)\big)$ are stationary sequences, we have

$$\left| \prod_{i=1}^{k} B_{n-i} \right| |X_{n-k}| \xrightarrow[k \to \infty]{\mathsf{P}} 0 \quad \text{and} \quad \left| \prod_{i=1}^{k} B_{n-i} \right| |x_{n-k}(\Phi)| \xrightarrow[k \to \infty]{\mathsf{P}} 0.$$

From this and (9.1.10) we obtain $X_n - x_n(\Phi) = 0$, $n \in \mathbb{Z}$, a.s. and the uniqueness statement follows. ∎

The form of the stationary solution (9.1.5) immediately yields the following corollary, cf. also Lemma 1.1 and Theorem 1.6 in VERVAAT (1979).

9.1.2. Corollary. *Let $\Phi = ([B_n, C_n])$ satisfy the assumptions of Theorem* 9.1.1 *and assume that the pairs $[B_n, C_n]$, $n \in \mathbb{Z}$, are i.i.d. Then $x_n(\Phi)$ and $[B_n, C_n]$ are independent, and*

$$x_0(\Phi) \stackrel{\mathcal{D}}{=} B_0 x_0(\Phi) + C_0 .$$

9.1.3. Remark. The proof of Theorem 9.1.1 shows that under appropriate conditions the results (9.1.5), (9.1.6), and (9.1.7) can also be extended to non-ergodic Φ.

Now we deal with the continuous dependence of the distribution of the stationary solution $\left([[B_n, C_n], x_n(\Phi)]\right)$ on the distribution P of $\Phi = ([B_n, C_n])$.

9.1.4. Theorem. *Consider stationary and ergodic sequences* (Φ, P), $\Phi = ([B_n, C_n])$, (Φ_k, P_k), $\Phi_k = ([B_{kn}, C_{kn}])$, $k \geqq 1$, *satisfying the following conditions:*

$$\Phi_k \xrightarrow[k\to\infty]{\mathcal{D}} \Phi, \tag{9.1.11}$$

$$b_k \xrightarrow[k\to\infty]{} b, \qquad \mathsf{E}(\log |B_{k0}|)_+ \xrightarrow[k\to\infty]{} \mathsf{E}(\log |B_0|)_+, \tag{9.1.12}$$

$$c_k \xrightarrow[k\to\infty]{} c, \qquad \mathsf{E}(\log |C_{k0}|)_+ \xrightarrow[k\to\infty]{} \mathsf{E}(\log |C_0|)_+, \tag{9.1.13}$$

where the moments $b = \mathsf{E} \log |B_0|, b_k = \mathsf{E} \log |B_{k0}|$, $c = \mathsf{E} \log |C_0|$, *and* $c_k = \mathsf{E} \log |C_{k0}|$ *are finite and* $-\infty < b < 0$, $-\infty < b_k < 0$, $k \geqq 1$. *Let* $\left([[B_n, C_n], x_n(\Phi)]\right)$, $\left([[B_{kn}, C_{kn}], x_n(\Phi_k)]\right)$, $k \geqq 1$, *be the stationary solutions of* (9.1.1) *for* Φ, Φ_k, $k \geqq 1$. *Then*

$$\left([[B_{kn}, C_{kn}], x_n(\Phi_k)], n \in \mathbb{Z}\right) \xrightarrow[k\to\infty]{\mathcal{D}} \left([[B_n, C_n], x_n(\Phi)], n \in \mathbb{Z}\right). \tag{9.1.14}$$

Proof. It is sufficient to show that

$$[\Phi_k, x_0(\Phi_k)] \xrightarrow[k\to\infty]{\mathcal{D}} [\Phi, x_0(\Phi)]. \tag{9.1.15}$$

(The convergence of all finite-dimensional distributions follows from (9.1.15) by means of the continuous mapping theorem.) By the continuous mapping theorem,

$$\left[\Phi_k, \sum_{j=0}^{r} C_{k,-j-1} \prod_{i=-j}^{-1} B_{ki}\right] \xrightarrow[k\to\infty]{\mathcal{D}} \left[\Phi, \sum_{j=0}^{r} C_{-j-1} \prod_{i=-j}^{-1} B_i\right], \qquad r \geqq 0.$$

Since

$$x_0(\Phi_k) - \sum_{j=0}^{r} C_{k,-j-1} \prod_{i=-j}^{-1} B_{ki} = \sum_{j=r+1}^{\infty} C_{k,-j-1} \prod_{i=-j}^{-1} B_{ki} \xrightarrow[r\to\infty]{} 0 \quad \mathsf{P}_k\text{-a.s.}$$

and

$$x_0(\Phi) - \sum_{j=0}^{r} C_{-j-1} \prod_{i=-j}^{-1} B_i = \sum_{j=r+1}^{\infty} C_{-j-1} \prod_{i=-j}^{-1} B_i \xrightarrow[r\to\infty]{} 0 \quad \mathsf{P}\text{-a.s.},$$

by Theorem 4.2 of Billingsley (1968) it is sufficient for (9.1.15) to prove that for all $\varepsilon > 0$

$$\lim_{r\to\infty} \limsup_{k\to\infty} \mathsf{P}_k \left(\left|\sum_{j=r+1}^{\infty} C_{k,-j-i} \prod_{i=-j}^{-1} B_{k,i}\right| > \varepsilon\right) = 0. \tag{9.1.16}$$

From $b_k \to b$ and Theorem A 2.11, for all $\delta > 0$ we obtain

$$\lim_{r\to\infty} \limsup_{k\to\infty} \mathsf{P}_k \left(\left|\sum_{j=r+1}^{\infty} C_{k,-j-1} \prod_{i=-j}^{-1} B_{k,i}\right| > \sum_{j=r+1}^{\infty} \exp\left(j(b + 3\delta)\right)\right)$$

$$\leqq \lim_{r\to\infty} \limsup_{k\to\infty} \mathsf{P}_k \left(\sum_{j=r+1}^{\infty} |C_{k,-j-1}| \prod_{i=-j}^{-1} |B_{k,i}| \right.$$

$$\left. > \sum_{j=r+1}^{\infty} \exp\left(j(b_k + \delta)\right) \exp\left(j\delta\right)\right)$$

$$\leq \limsup_{r\to\infty} \sup_k \mathsf{P}_k \left(\sup_{j>r} \left| \frac{1}{j} \sum_{i=-j}^{-1} \log |B_{k,i}| - b_k \right| > \delta \right)$$

$$+ \limsup_{r\to\infty} \sup_k \mathsf{P}_k \left(\sup_{j>r} \frac{1}{j} \log |B_{k,-j-1}| > \delta \right) = 0. \qquad (9.1.17)$$

For $b + 3\delta < 0$ and r sufficiently large,

$$\sum_{j=r+1}^{\infty} \exp \big(j(b + 3\delta) \big) < \varepsilon$$

and thus the left-hand side of (9.1.17) majorizes the left-hand side of (9.1.16). Hence (9.1.16) is true. ∎

If the conditions (9.1.12) or (9.1.13) in Theorem 9.1.4 are not satisfied, then the assertion (9.1.14) does not hold in general, as the following examples show (cf. also Remark A 2.10).

9.1.5. Example. Consider (Φ, P), $\Phi = ([B_n, C_n])$, defined by $\mathsf{P}(B_n = 1/2) = \mathsf{P}(C_n = 1) = 1$ and the sequences (Φ_k, P_k), $\Phi_k = ([B_{kn}, C_{kn}])$, of i.i.d. random pairs with $\mathsf{P}_k(B_{kn} = \mathrm{e}^{k/2}) = 1 - \mathsf{P}_k(B_{kn} = 1/2) = 1/k$, and $\mathsf{P}_k(C_{kn} = 1) = 1$. Then the assumptions of Theorem 9.1.4 are satisfied except for (9.1.12). By means of easy calculations we obtain

$$\mathsf{P}_k \left(x_0(\Phi_k) \geqq 1 + 2 \left(\frac{\mathrm{e}}{2} \right)^{k/2} \right) \geqq 1 - \left(\frac{k-1}{k} \right)^{k/2+1} \xrightarrow[k\to\infty]{} 1 - 1/\sqrt{\mathrm{e}},$$

and $\mathsf{P}\big(x_0(\Phi) = 2\big) = 1$, i.e. (9.1.14) does not hold.

9.1.6. Example. Let (Φ, P), $\Phi = ([B_n, C_n])$, be as in Example 9.1.5, and let the sequences (Φ_k, P_k), $\Phi_k = ([B_{kn}, C_{kn}])$, be i.i.d. random pairs with $\mathsf{P}_k(B_{kn} = 1/2) = 1$ and $\mathsf{P}_k(C_{kn} = \mathrm{e}^k) = 1 - \mathsf{P}_k(C_{kn} = 1) = 1/k$. Then the assumptions of Theorem 9.1.4 are satisfied except for (9.1.13). As

$$\mathsf{P}_k \left(x_0(\Phi_k) \geqq 1 + \left(\frac{\mathrm{e}}{2} \right)^{k} \right) \geqq 1 - \left(\frac{k-1}{k} \right)^{k+1} \xrightarrow[k\to\infty]{} 1 - 1/\mathrm{e},$$

(9.1.14) does not hold.

9.2. Stability of robust filter cleaners

9.2.1. Introduction

In this section we study the dependence on initial conditions of two recursive filters for cleaning a contaminated time series from additive outliers. Although the functions in the recursive equations are not contractive in general, we show that under weak assumptions there exists a stationary strong solution, and two iterates with different initial conditions coincide after some random time N_0. However N_0 may be quite large.

A simple outlier model, which is often used, is the following additive outlier model (AO-model):

$$Y_n = Z_n + V_n, \qquad n \in \mathbb{Z}.$$

Here (Z_n) is considered as the clean process, and V_n are the additive outliers which are non-zero a small fraction of the time, i.e. $P(V_n = 0) = 1 - \gamma$, $0.01 \leqq \gamma \leqq 0.25$. We shall assume that (Z_n) and (V_n) are independent and that the V_n are i.i.d. r.v.'s. One has to estimate the distribution of the clean process (Z_n) or some parameters of this distribution from the data Y_n. Unfortunately, conventional spectrum estimators are very sensitive concerning outliers. For this reason robust-resistant methods of spectrum estimation, which involve "data cleaning" operations, have been developed. In order to explain the idea of some robust filters for cleaning the contaminated time series (Y_n) from outliers, we model the clean process (Z_n) as a stationary autoregressive process of order p (AR(p)-process) with coefficients $a_1, \ldots, a_p$:

$$Z_n = \sum_{j=1}^{p} a_j Z_{n-j} + \varepsilon_n, \qquad n \in \mathbb{Z}. \tag{9.2.1}$$

Assume that innovations ε_n are i.i.d. r.v.'s with mean 0 and variance 1. If the ε_n are normally distributed, then the clean process (Z_n) is a Gaussian AR(p)-process.

The idea for obtaining a cleaned sequence $(\hat{Z}_n)$ is to put $\hat{Z}_n = Y_n$ if Y_n is close to a prediction in the AR(p)-model and to replace Y_n otherwise by some value closer to this prediction.

Kleiner et al. (1979) proposed the following robust filter:

$$\hat{Z}_n = \sum_{j=1}^{p} a_j \hat{Z}_{n-j} + h\left(Y_n - \sum_{j=1}^{p} a_j \hat{Z}_{n-j}\right) \tag{9.2.2}$$

with $h(t) = \max\big(-c, \min(c, t)\big)$. Here $\hat{Z}_n = Y_n$ if the difference between Y_n and the prediction

$$\sum_{j=1}^{p} a_j \hat{Z}_{n-j}$$

is not greater than c, and

$$\hat{Z}_n = \sum_{j=1}^{p} a_j \hat{Z}_{n-j} \pm c$$

otherwise. (We use Huber's h for simplicity, but all our results remain correct if h is odd, $0 \leqq h(t) \leqq t$ for $t \geqq 0$, $h(t) = t$ for $|t| \leqq c$, and $\lim_{t\to\infty} h(t) < \infty$.) (9.2.2) can be written in the usual vector form

$$\hat{\boldsymbol{Z}}_n = A\hat{\boldsymbol{Z}}_{n-1} + \boldsymbol{h}\big(Y_n - (A\hat{\boldsymbol{Z}}_{n-1})_1\big), \tag{9.2.3}$$

where $\hat{\boldsymbol{Z}}_n = [\hat{Z}_n, \hat{Z}_{n-1}, \ldots, \hat{Z}_{n-p+1}]^{\mathsf{T}}$, $\boldsymbol{h}(t) = [h(t), 0, \ldots, 0]^{\mathsf{T}}$,

$$A = \begin{pmatrix} a_1 & a_2 & \ldots & & a_p \\ 1 & 0 & \ldots & & 0 \\ \vdots & & & & \vdots \\ 0 & & \ldots & 1 & 0 \end{pmatrix}$$

and $(b)_1$ is the first component of the vector b. The recursion (9.2.3) is started with p initial values $\hat{\boldsymbol{Z}}_0 = [\hat{Z}_0, \ldots, \hat{Z}_{-p+1}]$.

Martin and Thomson (1982) have proposed an improvement of (9.2.3) by introducing a variable scale as follows:

$$\hat{\boldsymbol{Z}}_n = A\hat{\boldsymbol{Z}}_{n-1} + h\left(\left(Y_n - (A\hat{\boldsymbol{Z}}_{n-1})_1\right)/S_n\right) \boldsymbol{m}_n/S_n, \tag{9.2.4a}$$

$$M_{n+1} = A\left(M_n - w\left(\left(Y_n - (A\hat{\boldsymbol{Z}}_{n-1})_1\right)/S_n\right) \boldsymbol{m}_n\boldsymbol{m}_n^{\mathsf{T}}/S_n^2\right) A^{\mathsf{T}} + Q, \tag{9.2.4b}$$

where $\boldsymbol{m}_n$ is the first column of the $p \times p$ matrix M_n, $S_n^2 = M_{n,11}$, $w(t) = h(t)/t$, and the $p \times p$ matrix Q is given by $Q_{11} = 1$, $Q_{ij} = 0$ otherwise. Notice that in (9.2.1), (9.2.3), and (9.2.4a, b) we used the same parameters $a_1, \ldots, a_p$, i.e. we suppose that $a_1, \ldots, a_p$ are known. (In practice these parameters have to be estimated. This leads to further problems.)

Both the filters (9.2.3) and (9.2.4a, b) are recursive. We shall study here whether there exists a stationary solution of (9.2.3) or (9.2.4a, b), and whether, and how fast, iterations of (9.2.3) or (9.2.4a, b) with arbitrary initial conditions converge to it. These questions are important for the following reasons: Changing the Y_n-value at some instant n means starting at n with a new initial condition. If the filter is robust, only a few $\hat{\boldsymbol{Z}}_j$'s should be affected by such a change. Iterations of (9.2.3) or (9.2.4a, b) with the same Y_n but different initial conditions should thus come close together quickly. Moreover, in order to investigate asymptotic properties of estimators based on $\hat{\boldsymbol{Z}}_n$ one must known how quickly $\hat{\boldsymbol{Z}}_n$ converges to the stationary solution and how strongly this stationary solution depends on the past of (Y_n). Throughout Section 9.2 we assume that the observable sequence (Y_n) is stationary. Thus the clean sequence (Z_n) has to be stationary too. Hence the AR(p)-model (9.2.1) should have a stationary solution in order to be a suitable approximation of the clean process. For this reason we assume that the spectral radius of the matrix A is less than 1, i.e. the eigenvalues of A are less than 1 in absolute value.

As we shall show in Section 9.2.2, in general one cannot find a norm for which (9.2.3) or (9.2.4a, b) become contractive. Moreover a small percentage of suitably placed outliers can prevent the coincidence of two iterations with different initial conditions as n tends to infinity, see Fig. 4. The problem studied here is thus rather delicate. In Sections 9.2.4, 9.2.5 we shall show that a stationary strong solution of (9.2.3) and (9.2.4a, b) exists and iterates of (9.2.3) and (9.2.4a, b) coincide a.s. with this solution from some random time N_0 on, provided

(A 9.2.1) $\mathsf{P}([Y_1, \ldots, Y_n] \in G) > 0$ for all n and all measurable $G \subseteq \mathbb{R}^n$ with non-empty interior.

This is proved by checking the conditions for a general theorem on recursive stochastic equations given in Section 9.2.3. (A 9.2.1) holds for most models of contaminated time series, and we leave it to the reader to give explicit conditions which imply (A 9.2.1).

9.2.2. Non-contractivity of the filters

Consider a general recursive equation

$$X_n = f(X_{n-1}, Y_n), \qquad n \in \mathbb{Z}, \tag{9.2.5}$$

where X_n and Y_n take values in the Polish spaces $\mathbb{X}$ and $\mathbb{Y}$, respectively, and $f: \mathbb{X} \times \mathbb{Y} \to \mathbb{X}$ is measurable. We say that (9.2.5) is contractive if there is a norm $\|\cdot\|$ on $\mathbb{X}$ and $\alpha < 1$ such that for all $y \in \mathbb{Y}$, $x, x' \in \mathbb{X}$

$$\|f(x, y) - f(x', y)\| \leqq \alpha \|x - x'\|. \tag{9.2.6}$$

Clearly, (9.2.6) implies that the iterates X_n, $n \geqq 0$, and X_n', $n \geqq 0$, of (9.2.5) forget their initial conditions X_0 and X_0', respectively, at exponential speed, no matter what the Y_n look like. Moreover, $(\|X_n - X_n'\|)$ converges to zero uniformly in (Y_n).

9.2.1. Lemma. (i) *Consider the filter* (9.2.3). *If*

$$\sum_{j=1}^{p} |a_j| < 1,$$

then (9.2.3) *is contractive.*

(ii) *In general, neither* (9.2.3) *nor* (9.2.4a, b) *are contractive.*

Proof. (i) On $\mathbb{R}^p$ we consider the norm

$$\|z\| = \sum_{j=1}^{p} \gamma_j |z_j|$$

with $\gamma_1 = 1$, $\gamma_j = 1 - \sum_{k=1}^{j-1} (|a_k| + \varepsilon)$, $j > 1$, where ε is so small that $\gamma_p > 0$ and $1 - \sum_{k=1}^{j} |a_k| \geqq p\varepsilon$. Since the mapping $u \to h(u) - u$ is Lipschitz-continuous with Lipschitz constant 1, we have

$$|\hat{Z}_{n+1} - \hat{Z}'_{n+1}| \leqq \left| \sum_{j=1}^{p} a_j(\hat{Z}_{n+1-j} - \hat{Z}'_{n+1-j}) \right|.$$

Hence

$$\begin{aligned} \|\hat{\mathbf{Z}}_{n+1} - \hat{\mathbf{Z}}'_{n+1}\| &\leqq |a_p| \, |\hat{Z}_{n+1-p} - \hat{Z}'_{n+1-p}| \\ &\quad + \sum_{j=1}^{p-1} (\gamma_{j+1} + |a_j|) \, |\hat{Z}_{n+1-j} - \hat{Z}'_{n+1-j}| \\ &\leqq \sum_{j=1}^{p} (\gamma_j - \varepsilon) \, |\hat{Z}_{n+1-j} - \hat{Z}'_{n+1-j}|, \end{aligned}$$

and contractivity follows.

(ii) First consider (9.2.3) in the case $p = 2$. If $Y_n = \hat{Z}_n = \hat{Z}_n'$ and $Y_k > \max(a_1\hat{Z}_{k-1} + a_2\hat{Z}_{k-2}, a_1\hat{Z}'_{k-1} + a_2\hat{Z}'_{k-2}) + c$, $k = n + 1, n + 2$, then it can easily be checked that $\hat{Z}_{n+2} - \hat{Z}'_{n+2} = a_1 a_2(\hat{Z}_{n-1} - \hat{Z}'_{n-1})$. Thus for $|a_1 a_2| > 1$, which is possible for a stationary AR(2), $\hat{Z}_n - \hat{Z}_n'$ cannot converge to zero for arbitrary values Y_n.

Finally consider (9.2.4a, b) for $p = 1$ and $a_1 = a > 0$. We assume $S_n \leqq S_n'$ and $\hat{Z}_{n-1} < \hat{Z}'_{n-1}$ other cases being similar. If $Y_n = a\hat{Z}'_{n-1} + cS_n'$, then $\hat{Z}_n' = Y_n$, $S'_{n+1} = 1$, $\hat{Z}_n = a\hat{Z}_{n-1} + cS_n < \hat{Z}_n'$, $S_{n+1} > 1$. Hence for $Y_{n+1} \leqq a\hat{Z}_n - cS_{n+1}$ we obtain

$$\hat{Z}'_{n+1} - \hat{Z}_{n+1} = a\hat{Z}_n' - c - a\hat{Z}_n + cS_{n+1}$$
$$= a^2(\hat{Z}'_{n-1} - \hat{Z}_{n-1}) + c(S_{n+1} - 1).$$

Now using $S_n' \geqq S_n \geqq 1$,

$$S^2_{n+1} = 1 + a^2S_n^2(1 - cS_n/|Y_n - a\hat{Z}_{n-1}|)$$
$$\geqq 1 + a^2S_n^2\left(1 - cS_n/\left(a(\hat{Z}'_{n-1} - \hat{Z}_{n-1}) + cS_n\right)\right)$$
$$\geqq 1 + a^3(\hat{Z}'_{n-1} - \hat{Z}_{n-1})/\left(a(\hat{Z}'_{n-1} - \hat{Z}_{n-1}) + c\right).$$

Hence

$$S_{n+1} - 1 \geqq a^3(\hat{Z}'_{n-1} - \hat{Z}_{n-1})/2c + O\left((\hat{Z}'_{n-1} - \hat{Z}_{n-1})^2\right),$$

and we obtain (for $|\hat{Z}'_{n-1} - \hat{Z}_{n-1}|$ small)

$$|\hat{Z}'_{n+1} - \hat{Z}_{n+1}| \geqq \left(a^2 + \left(\frac{1}{2} - \varepsilon\right) a^3\right) |\hat{Z}'_{n-1} - \hat{Z}_{n-1}|$$

for sufficiently small ε. This contradicts contractivity if a is close enough to 1. ∎

An example where the differences between filtered values with different initial conditions do not converge to zero is given in Fig. 4. It has been constructed by adding outliers to an AR(2)-process at selected instants in order to keep $\hat{Z}_n - \hat{Z}_n'$ away from zero. However, in doing so we violate (A 9.2.1), and this condition becomes crucial for the arguments in Sections 9.2.4, 9.2.5.

In the next section we provide a general result on recursive stochastic equations of the type (9.2.5), which will help us to obtain stationary strong solutions of (9.2.3) and (9.2.4a, b). This method is, in some sense, similar to the method of a solution generating system described in Section 1.4.

9.2.3. A further method for constructing solutions of recursive stochastic equations

Consider the equation (9.2.5) where the sequence $\Phi = (Y_n, n \in \mathbb{Z})$ is assumed to be stationary and ergodic. Here we denote the n-th iterate of f by

$$x_n(x, y_1, \ldots, y_n) = f\left(x_{n-1}(x, y_1, \ldots, y_{n-1}), y_n\right);$$
$$x_1(x, y) = f(x, y).$$

Further let $X_n = x_n(x_0, Y_1, \ldots, Y_n)$ and $X_n' = x_n(x_0', Y_1, \ldots, Y_n)$ be two iterates with different initial conditions.

9.2.2. Theorem. *Assume:*

(A 9.2.2) *There exists a non-empty, measurable set D such that $f(D, Y_n) \subseteqq D$ a.s.*

(A 9.2.3) *$P(N_0 < \infty) > 0$, where $N_0 = \inf\{n: n \geqq 1,\ X_n = X_n'$ for all $x_0, x_0' \in D\}$.*

Then

(i) *there is a measurable function $g\colon \mathbb{Y}^{\mathbb{Z}_+} \to D$ such that $x_n(\Phi) = g(Y_n, Y_{n-1}, \ldots)$ defines a stationary strong solution of* (9.2.5).

(ii) *$N_0 < \infty$ almost surely.*

Note that by (i) *and the definition of N_0 we have $X_n = x_n(\Phi)$ for $n \geqq N_0$ and arbitrary $x_0 \in D$.*

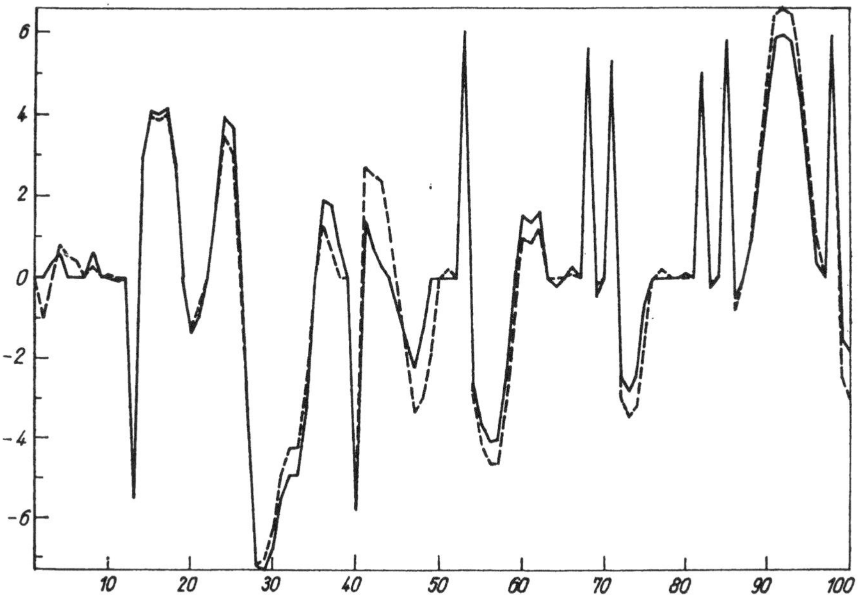

Fig. 4. Plot of $Y_n - \hat{Z}_n$ and $Y_n - \hat{Z}_n'$ for the filter (9.2.3) with two different initial conditions. (Y_n) is a simulated AR (2) process with $a_1 = 1.85$, $a_2 = -0.9$ and innovation variance one to which outliers were added at $n = 13, 40, 53, 68, 71, 82, 85, 98$. In (9.2.3) the above values for a_1, a_2 and $c = 1$ were used.

Proof. The variable N_0 defined in (A 9.2.3) is measurable if the underlying probability space is complete, which we will assume here, for convenience, cf. Dellacherie and Meyer (1978). Let

$$N_n = \inf\{k: k \geqq 1,\ x_k(x, Y_{n+1}, \ldots, Y_{n+k}) = x_k(x', Y_{n+1}, \ldots, Y_{n+k}) \text{ for all } x, x' \in D\}.$$

By the construction of N_n, the stationarity and ergodicity of Φ imply that $([Y_n, N_n],\ n \in \mathbb{Z})$ is stationary and ergodic too. From this and since (A 9.2.3) implies $\mathsf{P}(N_0 < k^*) > 0$ for k^* sufficiently large, we get a decreasing sequence of random indices $K_i \to -\infty$ such that $K_{i+1} + k^* < K_i$, $N_{K_i} < k^*$, $i \in \mathbb{Z}_+$. Applying (A 9.2.2) we obtain that the intersections

$$B_n(\Phi) = \bigcap_{k<n} x_{n-k+1}(D, Y_k, \ldots, Y_n)$$

are non-empty and that they consist of a single element, which will be denoted by $x_n(\Phi)$. (Notice the analogy to the proof of Theorem 1.4.3.) It follows by the construction that $(x_n(\Phi),\ n \in \mathbb{Z})$ is stationary and ergodic and that $x_n(\Phi)$ satisfies (9.2.5). Moreover, for every fixed element $x \in D$ it holds that

$$\lim_{k \to -\infty} x_{n-k+1}(x, Y_k, \ldots, Y_n) \in B_n(\Phi) = \{x_n(\Phi)\}.$$

Since the functions arising in this limit are measurable,

$$g(Y_n, Y_{n-1}, \ldots) = \lim_{k \to -\infty} x_{n-k+1}(x, Y_k, \ldots, Y_n)$$

has the properties stated in (i). By the same arguments as above, we obtain an increasing sequence of random indices $J_i \to \infty$ such that $J_i + k^* < J_{i+1}$ and $N_{J_i} < k^*$, $i \in \mathbb{Z}_+$. This yields $N_0 < \infty$ in view of (A 9.2.2). ∎

9.2.4. The filter with fixed scale

Equation (9.2.3) is a special case of the general recursive equation (9.2.5) if we put $X_n = \hat{Z}_n$, $\mathbb{X} = \mathbb{R}^p$, $\mathbb{Y} = \mathbb{R}$, and

$$f(z, y) = Az + h(y - (Az)_1).$$

Remember that the eigenvalues of A are assumed to be less than one in absolute value. As is well-known, see e.g. Householder (1964, p. 46), there exists a $p \times p$ matrix F such that

$$\|A\|_* < 1, \tag{9.2.7}$$

where

$$\|A\|_* = \sup \{\|Az\|_* : \|z\|_* = 1\}$$

and

$$\|z\|_* = \|F^{-1}z\|_1 = \sum_{j=1}^{p} |(F^{-1}z)_j|.$$

We first state some elementary properties of f leaving the proof to the reader.

9.2.3. Lemma. *Let $d = c\,\|[1, 0, \ldots, 0]^\top\|_*/(1 - \|A\|_*)$. If $\|z\|_* \leq d$, then $\|f(z, y)\|_* \leq d$ for all $y \in \mathbb{R}$.*

9.2.4. Lemma. *If $\hat{Z}_0, \hat{Z}_0'$ are such that $|Y_n - (A\hat{Z}_{n-1})_1| \leq c$ and $|Y_n - (A\hat{Z}'_{n-1})_1| \leq c$ for $n = 1, \ldots, p$, then $\hat{Z}_n = \hat{Z}_n'$ for all $n \geq p$.*

The main result of this section is

9.2.5. Theorem. *If (A 9.2.1) holds, then there is a measurable function g such that $g(Y_n, Y_{n-1}, \ldots)$ defines a stationary strong solution of (9.2.3), and to any initial condition $\hat{Z}_0$ with $\|\hat{Z}_0\|_* \leq d$ there is a finite random time N_0 such that $\hat{Z}_n = g(Y_n, Y_{n-1}, \ldots)$ for $n \geq N_0$.*

Proof. We are going to check conditions (A 9.2.2), (A 9.2.3) of Theorem 9.2.2 given in the previous section. (A 9.2.2) follows from Lemma 9.2.3 with $D = \{z : \|z\|_* \leq d\}$. Because of (A 9.2.1), (A 9.2.3) follows if there are some k

and $G \subseteqq \mathbb{R}^k$ with non-empty interior such that

$$[Y_1, \ldots, Y_k] \in G, \qquad \hat{Z}_0, \hat{Z}_0' \in D \text{ implies } \hat{Z}_k = \hat{Z}_k'. \tag{9.2.8}$$

First we show that there is an n^* and $G^* \subseteqq \mathbb{R}^{n^*}$ with non-empty interior such that

$$\left.\begin{array}{l}[Y_1, \ldots, Y_{n^*}] \in G^*, \\ \hat{Z}_0, \hat{Z}_0' \in D \text{ implies } |(A(\hat{Z}_n - \hat{Z}_n'))_1| < c/2 \\ \text{for } n = n^*, \ldots, n^* + p + 1.\end{array}\right\} \tag{9.2.9}$$

If $Y_n \geqq \sup\{|(Az)_1| : z \in D\} + c$ for $n = 1, \ldots, k$, then for any $\hat{Z}_0 \in D$ we have

$$\hat{Z}_k = A^k\hat{Z}_0 + \left(\sum_{n=0}^{k-1} A^n\right)[c, 0, \ldots, 0]^{\mathsf{T}}.$$

Hence $\|\hat{Z}_k - \hat{Z}_k'\|_* \leqq \|A^k(\hat{Z}_0 - \hat{Z}_0')\|_* \leqq 2d\,\|A\|_*^{\,k} \to 0$, and (9.2.9) is seen to hold for n^* large enough and $G^* = [\sup\{|(Az)_1| : z \in D\} + c, \infty)^{n^*}$. Denote by $\hat{Z}_n^0$ the n-th iterate of (9.2.3) with $\hat{Z}_0^0 = 0$. Then the set

$$G = \{[Y_1, \ldots, Y_{n^*+p}] : [Y_1, \ldots, Y_{n^*}] \in G^*,$$
$$|(A\hat{Z}^0_{n^*+i-1})_1 - Y_{n^*+i}| \leqq c/2,\ i = 1, \ldots, p\}$$

has a non-empty interior and satisfies (9.2.8), in view of Lemma 9.2.4. ∎

9.2.6. Remark. If $p = 1$, then (9.2.3) is contractive in view of Lemma 9.2.1 (i). This implies immediately

(i) the existence of a stationary strong solution and
(ii) that two iterates with different initial conditions will meet with exponential speed.

Hence for $p = 1$ condition (A 9.2.1) is not needed.

9.2.5. The filter with variable scale

The filter (9.2.4a, b) is of the form (9.2.5) if we put $X_n = [\hat{Z}_n, M_{n+1}]$ and $f(\hat{Z}_{n-1}, M_n, Y_n) = [f_1(\hat{Z}_{n-1}, M_n, Y_n), f_2(\hat{Z}_{n-1}, M_n, Y_n)]$ with f_1 and f_2 defined by (9.2.4a) and (9.2.4b), respectively. We introduce the matrix

$$R = \sum_{j=0}^{\infty} A^j Q (A^{\mathsf{T}})^j \tag{9.2.10}$$

which is none other than the covariance matrix of $[Z_1, \ldots, Z_p]$ in the AR-model (9.2.1). For two symmetric matrices A, B let us define $A \leqq B$ if $B - A$ is positive semidefinite.

9.2.7. Lemma. *If M is symmetric and positive semidefinite, then so is $f_2(z, M, y)$, and if $M \leqq R$, then $Q \leqq f_2(z, M, y) \leqq R$ for all $z \in \mathbb{R}^p$, $y \in \mathbb{R}$.*

Proof. Let $e = [1, 0, \ldots, 0]^{\mathsf{T}}$. Then $m = Me$ (the first column of M) and thus for any a

$$a^{\mathsf{T}} m m^{\mathsf{T}} a / M_{11} = (a^{\mathsf{T}} M e)^2 / (e^{\mathsf{T}} M e) \leqq a^{\mathsf{T}} M a$$

by Cauchy-Schwartz. Thus, since $w(\cdot) \leqq 1$, $f_2(z, M, y) \geqq Q \geqq 0$, where 0 denotes the matrix of order p in which all elements are zero. If $M \leqq R$, then $f_2(z, M, y) \leqq AMA^\top + Q \leqq ARA^\top + Q = R$ because $w(\cdot) \geqq 0$. ∎

Hence f_2 leaves the set

$$\mathbf{M} = \{M: M \text{ is symmetric}, Q \leqq M \leqq R\} \tag{9.2.11}$$

invariant. On $\mathbf{M}$ we shall use the norm

$$\|M\|_2 = \sup \{z^\top Mz/z^\top z: z \in \mathbb{R}^p, z \neq 0\}. \tag{9.2.12}$$

Since $\mathbf{M}$ is compact and $M_{11} \geqq 1$,

$$\beta_1 = c \sup \{\|m\|_*/M_{11}: M \in \mathbf{M}\} < \infty,$$

where m denotes the first column of M.

9.2.8. Lemma. *Let $\beta = \beta_1/(1 - \|A\|_*)$. If $\|z\|_* \leqq \beta$, then $\|f_1(z, M, y)\|_* \leqq \beta$ for all $M \in \mathbf{M}$, $y \in \mathbb{R}$.*

The proof is straightforward. Finally we have the analogue of Lemma 9.2.4.

9.2.9. Lemma. *If $\hat{Z}_0$, $\hat{Z}_0'$, M_1, M_1' are such that $|Y_n - (A\hat{Z}_{n-1})_1| \leqq cS_n$ and $|Y_n - (A\hat{Z}'_{n-1})_1| \leqq cS_n'$ for $n = 1, \ldots, p$, then $\hat{Z}_n = \hat{Z}_n'$, $M_{n+1} = M'_{n+1}$ for all $n \geqq p$.*

Proof. If $w(\cdot) = 1$, (9.2.4b) is the Kalman filter recursion with zero observation noise. From this it is clear that $M_{n,ij} = M'_{n,ij} = 0$ for $[i, j] \in \{1, \ldots, n\} \times \{1, \ldots, n\} \setminus \{[1, 1]\}$ and $n = 2, \ldots, p$. It can also be checked by direct calculation. This implies $\hat{Z}_p = \hat{Z}_p'$ and $M_{p+1} = M'_{p+1} = Q$. ∎

Now we can show the following

9.2.10. Theorem. *If* (A 9.2.1) *holds, then there is a measurable function g: $\mathbb{R}^{\mathbb{Z}_+} \to \mathbb{R}^p \times \mathbf{M}$ such that $g(Y_n, Y_{n-1}, \ldots)$ defines a stationary strong solution of* (9.2.4a, b), *and to any initial condition $[\hat{Z}_0, M_1]$ with $\|\hat{Z}_0\|_* \leqq \beta$ and $M_1 \in \mathbf{M}$ there is a finite random time N_0 such that $[\hat{Z}_n, M_{n+1}] = g(Y_n, Y_{n-1}, \ldots)$ for $n \geqq N_0$*

Proof. As in Theorem 9.2.5 we are going to apply Theorem 9.2.2 from Section 9.2.3. (A 9.2.2) follows from Lemma 9.2.7 and 9.2.8 by putting $D = \{z: \|z\|_* \leqq \beta\} \times \mathbf{M}$. Next we have to check (A 9.2.3). First we construct an n^* and $G^* \subseteqq \mathbb{R}^{n^*}$ with non-empty interior such that

$$\begin{gathered}[Y_1, \ldots, Y_{n^*}] \in G^*, \quad [\hat{Z}_0, M_1] \in D, \quad [\hat{Z}_0', M_1'] \in D \text{ implies} \\ |(A(\hat{Z}_n - \hat{Z}_n')_1)| < c/2 \quad \text{for } n = n^*, \ldots, n^* + p - 1.\end{gathered} \tag{9.2.13}$$

By the continuity of (9.2.4a, b) there are $\varepsilon_1, \varepsilon_2$ such that (9.2.13) follows from

$$\begin{gathered}[Y_1, \ldots, Y_{n^*}] \in G^*, \quad [\hat{Z}_0, M_1] \in D, \quad [\hat{Z}_0', M_1'] \in D \text{ implies} \\ \|\hat{Z}'_{n^*} - \hat{Z}_{n^*}\|_* \leqq \varepsilon_1, \quad \|M_{n^*+1} - R\|_2 \leqq \varepsilon_2, \quad \|M'_{n^*+1} - R\|_2 \leqq \varepsilon_2.\end{gathered} \tag{9.2.14}$$

But for any $K > c$ and $Y_n > K\sqrt{R_{11}} + \sup\{|(Az)_1| : \|z\|_* \leqq \beta\}$ for $n = 1, \ldots, k$, we have

$$\|M_{k+1} - R\|_* \leqq \|A^k(M_1 - R)(A^\top)^k\|_2 + c \sum_{j=1}^{k} \|A^j m_{k+1-j} m_{k+1-j}^\top (A^\top)^j\|_2 / K \leqq \Delta_k + \text{const}/K, \quad \text{where } \Delta_k \to 0,$$

because $M_n \in \mathbf{M}$ and $\mathbf{M}$ is compact. Similarly

$$\|M'_{k+1} - R\|_2 < \Delta_k + \text{const}/K.$$

Using this, after some algebra we get

$$\|\hat{Z}_k - \hat{Z}_k'\|_* \leqq \|A^k(\hat{Z}_0 - \hat{Z}_0')\|_* + c \sum_{j=0}^{k-1} \|A^j(m_{k-j}/S_{k-j} - m'_{k-j}/S'_{k-j})\|_* \leqq \tilde{\Delta}_k + \text{const}/K \text{ with } \tilde{\Delta}_k \to 0.$$

Thus (9.2.14) follows if we choose both n^* and K large and

$$G^* = \left(K\sqrt{R_{11}} + \sup\{|(Az)_1| : \|z\|_* \leqq \beta\}, \infty\right)^{n^*}.$$

Denote by $[\hat{Z}_n{}^0, M_{n+1}^0]$ the n-th iterate of (9.2.4a, b) with $\hat{Z}_0{}^0 = 0$ and $M_1{}^0 = Q$. If $[Y_1, \ldots, Y_{n^*}] \in G^*$ and $|(A\hat{Z}^0_{n^*+i-1})_1 - Y_{n^*+i}| < c/2$, $i = 1, \ldots, p$, then for arbitrary $[\hat{Z}_0, M_1] \in D$, $[\hat{Z}_0', M_1'] \in D$ we have $|Y_n - (A\hat{Z}_{n-1})_1| \leqq cS_n$ and $|Y_n - (A\hat{Z}'_{n-1})_1| \leqq cS_n'$, $n = n^* + 1, \ldots, n^* + p$. Thus (A 9.2.3) follows from Lemma 9.2.9 and (A 9.2.1). ∎

Looking more closely at the proofs of Theorems 9.2.5 and 9.2.10, one sees that the time N_0 after which $\hat{Z}_n$ and $\hat{Z}_n'$ coincide may be quite large. This is partly due to technical reasons, but also because we considered arbitrary initial conditions from a very large set D. If $\hat{Z}_0$ and $\hat{Z}_0'$ do not differ by very much, we do not need many outliers before Lemmas 9.2.4 and 9.2.9 are brought into action. Moreover, once $\hat{Z}_n$ and $\hat{Z}_n'$ are close they remain so for some time, which makes the probability of coincidence quite high. Nevertheless, if p and/or $\sum |a_j|$ is large, changing one Y_n may affect quite a few of the following $\hat{Z}_n$. In such a case these filters have to be used with care.

9.3. Remarks and references

9.1. The material is from Brandt (1986). An extensive discussion of equation (9.1.1) is given in Vervaat (1979). In this paper and also in numerous references given therein the pairs $[B_n, C_n]$ are assumed to be i.i.d. $\mathbb{R}^2$-valued r.v.'s. The assumptions of Theorem 9.1.1 are the same as those of Theorem 1.6 in Vervaat (1979). The proof of the finiteness of $x_n(\Phi)$ given in the proof of Theorem 9.1.1 is similar to that of Lemma 1.7 in Vervaat (1979). The difference is that the $x_n(\Phi)$ are given by means of a backward construction, whereas in VERVAAT (1979) a corresponding forward construction is considered. The multi-dimensional case of equation (9.1.1) (i.e. if B_n are matrices, and C_n and X_n are vectors) can be treated analogously to the approach given in 9.1. Concerning stationary solutions of the multi-dimensional continuous time analogue of equation (9.1.1), see e.g. HASMINSKIJ (1969) and ARNOLD and WIHSTUTZ (1982).

9.2. The results are from BRANDT and KÜNSCH (1988). For the high sensitivity of conventional spectrum estimators concerning outliers, see KLEINER et al. (1979).

Appendix

A 1. Stationary sequences

In this section we summarize the results concerning stationary sequences which are needed in the book. Some of them are well-known and thus will not be proved here.

If not stated otherwise, let $(\mathbb{Y}, \mathcal{Y})$ be an arbitrary measurable space. Denote by $\mathcal{Y}_{\mathbb{Z}}$ the σ-field generated by the measurable cylinders of the space $\mathbb{Y}^{\mathbb{Z}}$ of all double infinite sequences $(y_i, i \in \mathbb{Z})$. Define the (two-sided) shift operator θ on $\mathbb{Y}^{\mathbb{Z}}$ by

$$\theta(y_n, n \in \mathbb{Z}) = (y_{n+1}, n \in \mathbb{Z})$$

and its iterates θ^k by $\theta^k = \theta^{k-1}\theta$, $k \geqq 1$, $\theta^0(y_n) = (y_n)$. A sequence (Φ, P), $\Phi = (Y_n, n \in \mathbb{Z})$ of $\mathbb{Y}$-valued r.v.'s is called *stationary* if

$$(Y_n) \overset{\mathcal{D}}{=} (Y_{n+1}),$$

i.e.

$$\mathsf{P}\big((Y_n) \in A\big) = \mathsf{P}\big(\theta(Y_n) \in A\big) \quad \text{for all } A \in \mathcal{Y}_{\mathbb{Z}}.$$

Analogously, a one-sided sequence $(Y_n, n \geqq 0)$ with distribution P^+ of $\mathbb{Y}$-valued r.v.'s is stationary if

$$(Y_n, n \geqq 0) \overset{\mathcal{D}}{=} (Y_{n+1}, n \geqq 0),$$

i.e.

$$\mathsf{P}^+\big((Y_n, n \geqq 0) \in A\big) = \mathsf{P}^+\big(\theta(Y_n, n \geqq 0) \in A\big) \quad \text{for all } A \in \mathcal{Y}_{\mathbb{Z}_+}$$

where $\mathcal{Y}_{\mathbb{Z}_+}$ is the σ-field generated by the measurable cylinders of the set $\mathbb{Y}^{\mathbb{Z}_+}$ of all one-sided sequences $(y_i, i \geqq 0)$, and θ is the one-sided shift $\theta(y_n, n \geqq 0) = (y_{n+1}, n \geqq 0)$. (Notice that the two-sided shift is a one-to-one mapping whereas the one-sided shift is not a one-to-one mapping.)

A 1.1. Basic results

From Kolmogorov's extension theorem it follows immediately that every one-sided stationary sequence can be extended to a two-sided stationary sequence:

A 1.1.1. Lemma. *Let $(Y_n, n \geqq 0)$ be a one-sided stationary sequence with distribution P^+. Then there is a stationary sequence $(Y_n', n \in \mathbb{Z})$ with the distribution*

P *such that*

$$(Y_n, n \geqq 0) \stackrel{\mathcal{D}}{=} (Y_n', n \geqq 0),$$

and the distribution P *is uniquely determined.*

A 1.1.2. Theorem. *Every stationary sequence* (Y_n) *takes either infinitely often or never values in a fixed set* $L \in \mathcal{Y}$:

$$\mathsf{P}(\#\{n \leqq 0 : Y_n \in L\} = \#\{n > 0 : Y_n \in L\} = \infty) \\ + \mathsf{P}(\#\{n \leqq 0 : Y_n \in L\} = \#\{n > 0 : Y_n \in L\} = 0) = 1$$

for every $L \in \mathcal{Y}$.

Proof. Define the sets H_k by

$$H_k = \{(y_n) : \#\{n : k \leqq n, y_n \in L\} = 0\}, \qquad k \in \mathbb{Z}.$$

Since $\theta^s H_k = H_{k-s}$, by the stationarity we get that the probabilities $\mathsf{P}(H_k)$ are independent of k, though the sets H_k are increasing with respect to k. Thus

$$\mathsf{P}\left(\bigcup_{k=1}^{\infty} H_k\right) = \mathsf{P}(H_0) = \mathsf{P}\left(\bigcap_{k=1}^{\infty} H_{-k}\right).$$

On the other hand, we have

$$H = \bigcup_{k=1}^{\infty} H_k = \{(y_n) : \#\{n > 0 : y_n \in L\} < \infty\},$$

$$G = \bigcap_{k=1}^{\infty} H_{-k} = \{(y_n) : \#\{n : y_n \in L\} = 0\},$$

and $G \subseteqq H$. Thus $\mathsf{P}(H \setminus G) = 0$. Analogously we obtain $\mathsf{P}(H' \setminus G) = 0$, where

$$H' = \{(y_n) : \#\{n \leqq 0 : y_n \in L\} < \infty\}.$$

Summarizing these conditions we get

$$\mathsf{P}((H \cup H') \setminus G) = 0,$$

finishing the proof. ∎

A 1.1.3. Corollary. *Let* (Y_n) *be a stationary sequence of real-valued r.v.'s. Then*

$$\mathsf{P}\left(\bigcup_{k=1}^{\infty} \{\#\{n \leqq 0 : Y_n \leqq k\} = \#\{n > 0 : Y_n \leqq k\} = \infty\}\right) = 1.$$

Proof. We have for $k \in \mathbb{Z}_+$

$$\mathsf{P}(\#\{n \leqq 0 : Y_n \leqq k\} = 0) = \mathsf{P}\left(\bigcap_{n \leqq 0} \{Y_n > k\}\right) \leqq \mathsf{P}(Y_0 > k),$$

and the right-hand side tends to zero as $k \to \infty$. Now the assertion follows immediately from Theorem A 1.1.2. ∎

A 1.1.4. Lemma. *A stationary sequence* (Y_n) *of real-valued r.v.'s with non-increasing (non-decreasing) sample paths is almost surely constant, i.e.*

$$Y_n \geqq Y_{n+1} \ (Y_n \leqq Y_{n+1}), \qquad n \in \mathbb{Z}, \quad a.s.$$

implies

$$Y_n = Y_0, \qquad n \in \mathbb{Z}, \quad a.s.$$

Proof. Assume (Y_n) is non-increasing. (The case of a non-decreasing sequence can be treated analogously.) For $c > 0$ define

$$Z_n = \begin{cases} c & \text{if } Y_n > c, \\ Y_n & \text{if } Y_n \in [-c, c], \\ -c & \text{if } Y_n < -c. \end{cases}$$

Then (Z_n) is a stationary, non-increasing sequence too, and $\mathsf{E}\,|Z_n| < c$. Thus we get

$$\mathsf{E}\,|Z_n - Z_{n+1}| = \mathsf{E}(Z_n - Z_{n+1}) = 0,$$

and hence $Z_n = Z_{n+1}$ a.s. Letting $c \to \infty$ we obtain $Y_n = Y_{n+1}$ a.s. which immediately yields the assertion. ∎

A 1.1.5. Lemma. *Let* (Y_n) *be a stationary sequence of random, non-negative integers, and*

$$L_n = \sum_{k \leqq n} 1\{k + Y_k > n\}.$$

Then $\mathsf{E}L_n = \mathsf{E}Y_0$.

Proof. Since $L_{n+1} = \sum_{k \leqq n} 1\{k + Y_{k+1} > n\}$, (L_n) is stationary too.

$$\mathsf{E}L_n = \mathsf{E}L_0 = \sum_{k \leqq 0} \mathsf{P}(k + Y_k > 0) = \sum_{k=0}^{\infty} \mathsf{P}(Y_0 > k) = \mathsf{E}Y_0. \quad ∎$$

A 1.2. Ergodicity and mixing

A set $A \in \mathcal{Y}_{\mathbb{Z}}$ is said to be *invariant* (with respect to θ) if $A = \theta A$. It is easy to see that the family $\mathcal{J} \subseteqq \mathcal{Y}_{\mathbb{Z}}$ of all invariant sets, i.e.

$$\mathcal{J} = \{A : A \in \mathcal{Y}_{\mathbb{Z}}, A = \theta A\},$$

is a σ-field. Analogously, a set $A \in \mathcal{Y}_{\mathbb{Z}_+}$ of one-sided sequences is said to be *invariant* if $A = \theta A$. The family $\mathcal{J}^+ \subseteqq \mathcal{Y}_{\mathbb{Z}_+}$ of all invariant sets, i.e.

$$\mathcal{J}^+ = \{A : A \in \mathcal{Y}_{\mathbb{Z}_+}, A = \theta A\},$$

is a σ-field.

A 1.2.1. Theorem. (Individual ergodic theorem, Birkhoff-Khintchin theorem, cf. Ash and Gardner (1975, 3.3.6, 3.3.9)). *Let* $\Phi = (Y_n)$ *be a stationary sequence and* $f \in F(\mathbf{Y}^{\mathbb{Z}})$ *with* $\mathsf{E}\,|f(\Phi)| < \infty$. *Then*

$$\lim_{n \to \infty} n^{-1} \sum_{i=0}^{n-1} f(\theta^i \Phi) = \mathsf{E}\big(f(\Phi) \mid \mathcal{J}\big) \quad a.s.$$

A stationary, two-sided sequence (Y_n) (one-sided sequence $(Y_n, n \geqq 0)$) with distribution P (P^+) is said to be *ergodic* if for every invariant set $A \in \mathcal{J}$ ($A \in \mathcal{J}^+$) either $\mathsf{P}(A) = 0$ $(\mathsf{P}^+(A) = 0)$ or $\mathsf{P}(A) = 1$ $(\mathsf{P}^+(A) = 1)$. There are various ways of expressing ergodicity, cf. e.g. Ash and Gardner (1975). Let $\mathcal{Y}^c_{\mathbb{Z}}$ ($\mathcal{Y}^c_{\mathbb{Z}_+}$) be the set of measurable cylinders of $\mathcal{Y}_{\mathbb{Z}}$ ($\mathcal{Y}_{\mathbb{Z}_+}$).

A 1.2.2. Theorem. *Let $\Phi = (Y_n)$ be a stationary sequence with distribution* P. *Then the following conditions are equivalent:*

(i) *(Y_n) is ergodic.*

(ii) $\lim_{n\to\infty} n^{-1} \sum_{i=0}^{n-1} f(\theta^i\Phi) = \mathsf{E}f(\Phi)$ *a.s. for all $f \in F(\mathbb{Y}^{\mathbb{Z}})$ with* $\mathsf{E}\,|f(\Phi)| < \infty$.

(iii) $\lim_{n\to\infty} n^{-1} \sum_{i=0}^{n-1} 1_A(\theta^i\Phi) = \mathsf{P}(A)$ *a.s. for each $A \in \mathcal{Y}^c_{\mathbb{Z}}$.*

(iv) $\lim_{n\to\infty} n^{-1} \sum_{i=0}^{n-1} \mathsf{P}(A \cap \theta^{-i}B) = \mathsf{P}(A)\,\mathsf{P}(B)$ *for all $A, B \in \mathcal{Y}^c_{\mathbb{Z}}$.*

(v) *The one-sided sequence $(Y_n, n \geqq 0)$ is ergodic.*

(vi) $\lim_{n\to\infty} n^{-1} \sum_{i=0}^{n-1} 1_A\big(\theta^i(Y_k, k \geqq 0)\big) = \mathsf{P}(A)$ *a.s. for each $A \in \mathcal{Y}^c_{\mathbb{Z}_+}$.*

(vii) $\lim_{n\to-\infty} (-n)^{-1} \sum_{i=n}^{-1} f(\theta^i\Phi) = \mathsf{E}f(\Phi)$ *a.s. for all $f \in F(\mathbb{Y}^{\mathbb{Z}})$ with* $\mathsf{E}\,|f(\Phi)| < \infty$.

Sometimes the following criterion for ergodicity is useful.

A 1.2.3. Theorem. (Ash and Gardner (1975, p. 135)). *A stationary sequence (Y_n) with distribution P is ergodic iff every representation of P as a mixture*

$$\mathsf{P} = \alpha \mathsf{P}_1 + (1 - \alpha)\,\mathsf{P}_2, \qquad 0 \leqq \alpha \leqq 1,$$

of probability distributions $\mathsf{P}_1, \mathsf{P}_2$ of stationary sequences $(Y_n^1), (Y_n^2)$ is trivial, i.e. $\alpha = 0$ or $\alpha = 1$ or $\mathsf{P}_1 = \mathsf{P}_2$.

A stationary sequence (Y_n) is said to be *mixing* if

$$\mathsf{P}\big((Y_n) \in A, \theta^k(Y_n) \in B\big) \xrightarrow[k\to\infty]{} \mathsf{P}\big((Y_n) \in A\big)\,\mathsf{P}\big((Y_n) \in B\big) \qquad \text{(Ap 1.1)}$$

for all $A, B \in \mathcal{Y}_{\mathbb{Z}}$.

A 1.2.4. Theorem. (Ash and Gardner (1975, p. 120, 3.2.6)). *A mixing sequence is ergodic.*

Theorem A 1.2.4 follows from (Ap 1.1) and Theorem A 1.2.2, (iv). Note that the reverse of Theorem A 1.2.4 is not true, i.e. mixing is a stronger property than ergodicity.

A 1.2.5. Theorem. (Ash and Gardner (1975, p. 120, 3.2.7)): *Let (Y_n) be a stationary sequence and let $\mathcal{S}$ be a field of subsets of $\mathbb{Y}^{\mathbb{Z}}$ such that the σ-field generated by $\mathcal{S}$ is $\mathcal{Y}_{\mathbb{Z}}$. If the mixing condition* (Ap 1.1) *holds for all $A, B \in \mathcal{S}$ then (Y_n) is mixing.*

Consider a stationary sequence $(Y_n) = ([Y_n^1, Y_n^2])$ with independent marginal sequences (Y_n^1) and (Y_n^2). By means of simple examples it is easy to see that the ergodicity of the two marginal sequences does not imply the ergodicity of (Y_n). However, if one of them is mixing, ergodicity follows, cf. e.g. BROWN (1976).

A 1.2.6. Theorem. *Let (Y_n^1) and (Y_n^2) be two independent stationary sequences. If (Y_n^1) is ergodic and (Y_n^2) is mixing, then $(Y_n) = ([Y_n^1, Y_n^2])$ is stationary and ergodic.*

Concerning further conditions ensuring the ergodicity of product measures we refer e.g. to BROWN (1976).

We often use the following simple criterion.

A 1.2.7. Lemma. *Let $\Phi = (X_n)$ be a stationary and ergodic sequence of $\mathbb{X}$-valued r.v.'s and $g_n : \mathbb{X}^{\mathbb{Z}} \to \mathbb{Y}$, $n \in \mathbb{Z}$, a sequence of measurable functions with the property $g_{n+1}(\varphi) = g_n(\theta\varphi)$, $n \in \mathbb{Z}$. Then $\Psi = \big(g_n(\Phi), n \in \mathbb{Z}\big)$ is stationary and ergodic, too.*

Proof. The stationarity of Ψ is obvious. Let $A \in \mathcal{Y}_{\mathbb{Z}}$ be an invariant set. Then

$$\begin{aligned}\theta\{\varphi: \big(g_n(\varphi)\big) \in A\} &= \{\varphi: \big(g_n(\theta^{-1}\varphi)\big) \in A\} = \{\varphi: \big(g_{n-1}(\varphi)\big) \in A\} \\ &= \{\varphi: \big(g_n(\varphi)\big) \in \theta A\} = \{\varphi: \big(g_n(\varphi)\big) \in A\},\end{aligned}$$

and thus $\{\varphi: \big(g_n(\varphi)\big) \in A\}$ is invariant, too. Now it follows that $\mathsf{P}(\Psi \in A) = \mathsf{P}\big(\varphi: \big(g_n(\varphi)\big) \in A\big) = 0$ or 1, in view of the ergodicity of Φ.

A 1.3. Conditional distributions

Following WIRTH (1982), in this section we provide some useful relationships between the distribution of a stationary sequence and its particular conditional distributions, cf. also ROLSKI (1981, Chapter 2) and [FKAS, 1984, Section 1.7.3]. Consider a stationary sequence (Φ, P), $\Phi = (Y_n)$, of $\mathbb{Y}$-valued r.v.'s. Let L be a measurable subset of $\mathbb{Y}$ with the property

$$\mathsf{P}(\# \{n: Y_n \in L\} \geqq 1) = 1. \tag{Ap 1.2}$$

Then Theorem A 1.1.2 implies $\mathsf{P}(Y_0 \in L) > 0$. Thus the conditional distribution

$$\tilde{\mathsf{P}}(\cdot) = \mathsf{P}(\cdot \mid Y_0 \in L) \tag{Ap 1.3}$$

is well-defined. Let $\tilde{\Phi} = (\tilde{Y}_n)$ be a sequence of $\mathbb{Y}$-valued r.v.'s distributed according to $\tilde{\mathsf{P}}$.

A 1.3.1. Theorem. *Let (Φ, P), $\Phi = (Y_n)$, be a stationary sequence and $L \in \mathcal{Y}$ satisfying* (Ap 1.2). *Then the conditional distribution $\tilde{\mathsf{P}}$ defined by* (Ap 1.3) *has the following properties:*

(i) *The first passage time $\tilde{\nu} = \min \{j \geqq 1: \tilde{Y}_j \in L\}$ is a proper r.v. with finite mean*

$$\mathsf{E}\tilde{\nu} < \infty. \tag{Ap 1.4}$$

(ii) $$\tilde{\mathsf{P}}\big(\tilde{\Phi} \in (\cdot)\big) = \tilde{\mathsf{P}}\big(\theta^{\tilde{\nu}}\tilde{\Phi} \in (\cdot)\big). \qquad \text{(Ap 1.5)}$$

(iii) *The distribution* P *can be expressed by* $\tilde{\mathsf{P}}$ *via*

$$\mathsf{P}\big((Y_n) \in (\cdot)\big)$$

$$= (\mathsf{E}\tilde{\nu})^{-1} \sum_{j=0}^{\infty} \tilde{\mathsf{P}}\big(\tilde{\nu} > j, (\tilde{Y}_{n+j}, n \in \mathbb{Z}) \in (\cdot)\big) \qquad \text{(Ap 1.6a)}$$

$$= (\mathsf{E}\tilde{\nu})^{-1} \sum_{j=1}^{\infty} \tilde{\mathsf{P}}\big(\tilde{\nu} \geqq j, (\tilde{Y}_{n+j}, n \in \mathbb{Z}) \in (\cdot)\big). \qquad \text{(Ap 1.6b)}$$

In particular,

$$\mathsf{P}(Y_0 \in L) = (\mathsf{E}\tilde{\nu})^{-1}. \qquad \text{(Ap 1.7)}$$

Proof. By Theorem A 1.1.2, condition (Ap 1.2) is equivalent to

$$\mathsf{P}(\text{there is an } n \geqq 1 \text{ with } Y_n \in L) = 1 \qquad \text{(Ap 1.8)}$$

and also to

$$\mathsf{P}(\# \{n : n \leqq -1, Y_n \in L\} = \infty) = 1. \qquad \text{(Ap 1.9)}$$

From (Ap 1.3) and the stationarity of (Y_n), we obtain

$$\mathsf{P}\big(\Phi \in (\cdot)\big)$$

$$= \mathsf{P}(Y_0 \in L)\, \mathsf{P}\big(\Phi \in (\cdot) \mid Y_0 \in L\big) + \mathsf{P}\big(\Phi \in (\cdot), Y_0 \notin L\big)$$

$$= \mathsf{P}(Y_0 \in L)\, \tilde{\mathsf{P}}\big(\tilde{\Phi} \in (\cdot)\big) + \mathsf{P}\big(\theta\Phi \in (\cdot), Y_1 \notin L\big)$$

$$= \mathsf{P}(Y_0 \in L) \Big(\tilde{\mathsf{P}}\big(\tilde{\Phi} \in (\cdot)\big) + \tilde{\mathsf{P}}\big(\theta\tilde{\Phi} \in (\cdot), \tilde{Y}_1 \notin L\big)\Big)$$

$$+ \mathsf{P}\big(\theta^2\Phi \in (\cdot), Y_1 \notin L, Y_2 \notin L\big)$$

$$= \mathsf{P}(Y_0 \in L) \left(\sum_{j=0}^{m-1} \tilde{\mathsf{P}}\big(\theta^j\tilde{\Phi} \in (\cdot), \tilde{Y}_1 \notin L, \ldots, \tilde{Y}_j \notin L\big)\right)$$

$$+ \mathsf{P}\big(\theta^m\Phi \in (\cdot), Y_1 \notin L, \ldots, Y_m \notin L\big). \qquad \text{(Ap 1.10)}$$

Now

$$\mathsf{P}\big(\theta^m\Phi \in (\cdot), Y_1 \notin L, \ldots, Y_m \notin L\big) \leqq \mathsf{P}(Y_1 \notin L, \ldots, Y_m \notin L).$$

Moreover, by (Ap 1.8) we get

$$\lim_{m\to\infty} \mathsf{P}(Y_1 \notin L, \ldots, Y_m \notin L) = \mathsf{P}(Y_n \notin L, n \geqq 1) = 0.$$

Thus, $\mathsf{P}\big(\theta^m\Phi \in (\cdot), Y_1 \notin L, \ldots, Y_m \notin L\big)$ converges uniformly to zero, and (Ap 1.10) yields

$$\mathsf{P}\big(\Phi \in (\cdot)\big) = \mathsf{P}(Y_0 \in L) \sum_{j=0}^{\infty} \tilde{\mathsf{P}}\big(\tilde{\nu} > j, \theta^j\tilde{\Phi} \in (\cdot)\big). \qquad \text{(Ap 1.11)}$$

Applying (Ap 1.11) to the event $\mathbb{Y}^{\mathbb{Z}}$, we get

$$1 = \mathsf{P}(Y_0 \in L)\, \mathsf{E}\tilde{\nu},$$

and (Ap 1.4), (Ap 1.7) follow in view of $\mathsf{P}(Y_0 \in L) > 0$. Now (Ap 1.6a) follows from (Ap 1.11) and (Ap 1.7). Applying (Ap 1.6a) to $\theta^{-1}(\cdot)$ we get (Ap 1.6b). It remains to show (Ap 1.5). We have

$$\tilde{\mathsf{P}}(\theta^{\tilde{\nu}}\tilde{\Phi} \in (\cdot)) = \sum_{j=1}^{\infty} \tilde{\mathsf{P}}(\tilde{\nu} = j, \theta^j\tilde{\Phi} \in (\cdot))$$

$$= \sum_{j=1}^{\infty} \tilde{\mathsf{P}}(\tilde{Y}_1 \notin L, \ldots, \tilde{Y}_{j-1} \notin L, \tilde{Y}_j \in L, \theta^j\tilde{\Phi} \in (\cdot)).$$

Taking into account (Ap 1.3) and the stationarity of Φ, we can continue as follows:

$$\tilde{\mathsf{P}}(\theta^{\tilde{\nu}}\tilde{\Phi} \in (\cdot))$$

$$= \sum_{j=1}^{\infty} \mathsf{P}(Y_1 \notin L, \ldots, Y_{j-1} \notin L, Y_j \in L, \theta^j\Phi \in (\cdot) \mid Y_0 \in L)$$

$$= \big(\mathsf{P}(Y_0 \in L)\big)^{-1} \sum_{j=1}^{\infty} \mathsf{P}(Y_{-j} \in L, Y_{-j+1} \notin L, \ldots, Y_{-1} \notin L, Y_0 \in L, \Phi \in (\cdot))$$

$$= \big(\mathsf{P}(Y_0 \in L)\big)^{-1} \mathsf{P}(Y_0 \in L, \Phi \in (\cdot)) = \tilde{\mathsf{P}}(\tilde{\Phi} \in (\cdot)),$$

in view of (Ap 1.9). ∎

The distribution $\tilde{\mathsf{P}}$ of a sequence $\tilde{\Phi}$ with the properties (Ap 1.4) and (Ap 1.5) is the conditional distribution (Ap 1.3) of a stationary sequence with the property (Ap 1.2):

A 1.3.2. Theorem. *For every fixed $L \in \mathcal{Y}$ formulae* (Ap 1.3), (Ap 1.6) *define a one-to-one mapping between the families of probability distributions*

$\mathbf{P}_L = \{\mathsf{P}$: P *is the probability distribution of a stationary sequence $\Phi = (Y_n)$ with the property* (Ap 1.2)$\}$

and

$\tilde{\mathbf{P}}_L = \{\tilde{\mathsf{P}}$: $\tilde{\mathsf{P}}$ *is the probability distribution of a sequence $\tilde{\Phi} = (\tilde{Y}_n)$ with the properties* (Ap 1.4) *and* (Ap 1.5)$\}$.

Proof. In view of Theorem A 1.3.1 we have to prove that every $(\tilde{\Phi}, \tilde{\mathsf{P}})$ satisfying (Ap 1.4) and (Ap 1.5) corresponds to a stationary (Φ, P) satisfying (Ap 1.2), and that $\tilde{\mathsf{P}}$ is just the conditional distribution (Ap 1.3).

Consider a sequence $(\tilde{\Phi}, \tilde{\mathsf{P}})$, $\tilde{\Phi} = (\tilde{Y}_n)$, satisfying (Ap 1.4) and (Ap 1.5). Let P be the distribution defined according to (Ap 1.6):

$$\mathsf{P}(\cdot) = (\mathsf{E}\tilde{\nu})^{-1} \sum_{j=0}^{\infty} \tilde{\mathsf{P}}(\tilde{\nu} > j, \theta^j\tilde{\Phi} \in (\cdot)), \tag{Ap 1.12}$$

and let $\Phi = (Y_n)$ be a sequence with distribution P. The series (Ap 1.12) converges uniformly in view of (Ap 1.4). The stationarity of $\Phi = (Y_n)$ follows from

$$\mathsf{P}(\Phi \in (\cdot)) - \mathsf{P}(\theta\Phi \in (\cdot))$$

$$= (\mathsf{E}\tilde{\nu})^{-1} \sum_{j=0}^{\infty} \big(\tilde{\mathsf{P}}(\tilde{\nu} > j, \theta^j\tilde{\Phi} \in (\cdot)) - \tilde{\mathsf{P}}(\tilde{\nu} > j, \theta^{j+1}\tilde{\Phi} \in (\cdot))\big)$$

$$= (\mathsf{E}\tilde{\nu})^{-1} \Big(\tilde{\mathsf{P}}(\tilde{\nu} > 0, \tilde{\Phi} \in (\cdot))$$

$$+ \sum_{j=0}^{\infty} \Big(\tilde{\mathsf{P}}(\tilde{\nu} > j+1, \theta^{j+1}\tilde{\Phi} \in (\cdot)) - \tilde{\mathsf{P}}(\tilde{\nu} > j, \theta^{j+1}\tilde{\Phi} \in (\cdot))\Big)\Big)$$

$$= (\mathsf{E}\tilde{\nu})^{-1} \Big(\tilde{\mathsf{P}}(\tilde{\Phi} \in (\cdot)) - \sum_{j=0}^{\infty} \tilde{\mathsf{P}}(\tilde{\nu} = j+1, \theta^{j+1}\tilde{\Phi} \in (\cdot))\Big)$$

$$= (\mathsf{E}\tilde{\nu})^{-1} \Big(\tilde{\mathsf{P}}(\tilde{\Phi} \in (\cdot)) - \tilde{\mathsf{P}}(\theta^{\tilde{\nu}}\tilde{\Phi} \in (\cdot))\Big) = 0.$$

Property (Ap 1.2) follows immediately from

$$\mathsf{P}\left(\bigcap_{n=0}^{\infty} \{Y_n \notin L\}\right) = (\mathsf{E}\tilde{\nu})^{-1} \sum_{j=0}^{\infty} \tilde{\mathsf{P}}\left(\tilde{\nu} > j, \bigcap_{n=j}^{\infty} \{\tilde{Y}_n \notin L\}\right)$$

$$\leqq (\mathsf{E}\tilde{\nu})^{-1} \sum_{j=0}^{\infty} \tilde{\mathsf{P}}(\tilde{\nu} > j, \tilde{Y}_{\tilde{\nu}} \notin \tilde{L}) = 0.$$

It remains to show that $\tilde{\mathsf{P}}$ is the conditional distribution (Ap 1.3):

$$\mathsf{P}(Y_0 \in L) = (\mathsf{E}\tilde{\nu})^{-1} \sum_{j=0}^{\infty} \tilde{\mathsf{P}}(\tilde{\nu} > j, \tilde{Y}_j \in L)$$

$$= (\mathsf{E}\tilde{\nu})^{-1} \tilde{\mathsf{P}}(\tilde{\nu} > 0, \tilde{Y}_0 \in L) = (\mathsf{E}\tilde{\nu})^{-1} > 0.$$

$$\mathsf{P}\big(\Phi \in (\cdot) \mid Y_0 \in L\big) = \big(\mathsf{P}(Y_0 \in L)\big)^{-1} \mathsf{P}\big(\Phi \in (\cdot), Y_0 \in L\big)$$

$$= \sum_{j=0}^{\infty} \tilde{\mathsf{P}}\big(\tilde{\nu} > j, \theta^j\tilde{\Phi} \in (\cdot), \tilde{Y}_j \in L\big)$$

$$= \tilde{\mathsf{P}}\big(\tilde{\nu} > 0, \tilde{\Phi} \in (\cdot), \tilde{Y}_0 \in L\big) = \tilde{\mathsf{P}}\big(\tilde{\Phi} \in (\cdot)\big).$$

The theorem has completely been proved. ∎

A 1.3.3. Remark. Every stationary sequence $\Phi = (Y_n)$ can be considered as a special stationary marked point process $([n, Y_n])$ on the discrete axis $\mathbb{Z}$ with the mark space $\mathbb{Y}$ (each point n of the discrete axis $\mathbb{Z}$ is a point of this marked point process). Then $\tilde{\mathsf{P}}$ appears as the so-called *Palm distribution* $\mathsf{P}_L{}^0$ of P with respect to the subset L of $\mathbb{Y}$, cf. [KMN], [FKAS] in case of continuous time.

Applying standard integration techniques to (Ap 1.6), we get

$$\mathsf{E}h(\Phi) = (\mathsf{E}\tilde{\nu})^{-1} \mathsf{E} \sum_{j=0}^{\tilde{\nu}-1} h(\theta^j\tilde{\Phi})$$

for every $h \in F_+(\mathbb{Y}^{\mathbb{Z}})$. In particular,

$$\mathsf{E}h(Y_0) = (\mathsf{E}\tilde{\nu})^{-1} \mathsf{E} \sum_{j=0}^{\tilde{\nu}-1} h(\tilde{Y}_j) \qquad \text{(Ap 1.13)}$$

for every $h \in F_+(\mathbb{Y})$. It is worth mentioning that (Ap 1.13) appears as a generalization of the well-known Wald's identity, cf. Section 1.6 and Example 2.2.6.

Let us denote by $\cdots < \nu_{-1} \leqq 0 < \nu_0 < \nu_1 < \cdots$ the (random) indices with the property $Y_{\nu_j} \in L$ (condition (Ap 1.2) implies that all the ν_j, $j \in \mathbb{Z}$, are proper). The following ergodic statement is true.

A 1.3.4. Theorem. *Consider a stationary ergodic sequence* (Φ, P), $\Phi = (Y_n)$, *and a set* $L \subseteqq \mathbb{Y}$ *such that* (Ap 1.2) *holds. Let* $(\tilde{L}_n)$ *be a sequence with the distribution* $\tilde{\mathsf{P}}(\cdot) = \mathsf{P}(\cdot \mid Y_0 \in L)$. *Then for every function* $h \in F_+(\mathbb{Y})$

$$n^{-1} \sum_{j=0}^{n-1} h(\tilde{Y}_j) \xrightarrow[n \to \infty]{} \mathsf{E}h(Y_0) \qquad \tilde{\mathsf{P}}\text{-}a.s.$$

and

$$n^{-1} \sum_{j=0}^{n-1} h(Y_{m+\nu_j}) \xrightarrow[n \to \infty]{} \mathsf{E}h(\tilde{Y}_m) \qquad \mathsf{P}\text{-}a.s.,$$

where m is an arbitrary integer.

Proof. The first assertion follows from Theorem A 1.2.2, (ii) and the fact that $\tilde{\mathsf{P}}(\tilde{\Phi} \in D) = \mathsf{P}(\Phi \in D \mid Y_0 \in L) = 1$ if $\mathsf{P}(\Phi \in D) = 1$.

Let $\cdots < \tilde{\nu}_{-1} = 0 < \tilde{\nu}_0 < \tilde{\nu}_1 < \cdots$ be the (random) indices with the property $\tilde{Y}_{\tilde{\nu}_j} \in L$. From Theorems A 1.3.1 and A 1.3.2 we have that the sequence $([\underline{\tilde{Y}}_n, \tilde{\nu}_n])$, $\underline{\tilde{Y}}_n = [\tilde{Y}_{\tilde{\nu}_n}, \ldots, \tilde{Y}_{\tilde{\nu}_{n+1}-1}]$, is stationary and ergodic too. Then, for $h \in F_+(\mathbb{Y})$, the ergodic Theorem A 1.2.2 (ii) yields

$$n^{-1} \sum_{j=0}^{n-1} h(\tilde{Y}_{m+\tilde{\nu}_j}) = \mathsf{E}h(\tilde{Y}_m) \qquad \tilde{\mathsf{P}}\text{-a.s.} \tag{Ap 1.14}$$

for every m ($\tilde{\nu}_{-1} = 0$). Now the second assertion follows from (Ap 1.14) and (Ap 1.6a). ∎

A 1.4. Stationary sequences of grouped individuals

This section is based on WIRTH (1986a).

In many applications, in particular in queueing and reliability theory, sequences (Y_n) of r.v.'s occur, where some members of the sequence (successive or not) may belong to a common group or class, e.g. customers of a batch, customers finding the system in the same state, etc. There can be interest in characteristics related to single members of the sequence (Y_n) or to whole groups of members. In this section we confine ourselves to groups of finite size and generalize the results given in Section A 1.3.

Consider a sequence $(Y_n, n \in \mathbb{Z})$ of $\mathbb{Y}$-valued r.v.'s. The members of the sequence (Y_n) can belong to groups of finite size. We write

$$Y_n \sim Y_m \text{ iff } Y_n \text{ and } Y_m \text{ belong to the same group.} \tag{Ap 1.15}$$

Define the group-predecessor Pre (Y_n) and the group-successor Suc (Y_n) of Y_n by

$$\text{Pre}(Y_n) = Y_p \quad \text{with } p = \begin{cases} \max\{m\colon m < n,\ Y_m \sim Y_n\} \\ \qquad \text{if } \{m\colon m < n,\ Y_m \sim Y_n\} \neq \emptyset, \\ n \qquad \text{otherwise} \end{cases} \tag{Ap 1.16}$$

and

$$\operatorname{Suc}(Y_n) = Y_s \quad \text{with } s = \begin{cases} \min\{m : m > n,\ Y_m \sim Y_n\} \\ \qquad \text{if } \{m : m > n,\ Y_m \sim Y_n\} \neq \emptyset, \\ n \quad \text{otherwise}. \end{cases} \tag{Ap 1.17}$$

We can define the "distances" B_n and C_n between Y_n and its group-predecessor and its group-successor, respectively:

$$B_n = n - p \quad \text{iff} \quad Y_p = \operatorname{Pre}(Y_n), \tag{Ap 1.18}$$

$$C_n = s - n \quad \text{iff} \quad Y_s = \operatorname{Suc}(Y_n). \tag{Ap 1.19}$$

The sequence (K_n), $K_n = [Y_n, B_n, C_n]$, contains the original sequence (Y_n) and, in addition, includes the group-memberships of its members. The group $G(Y_n)$ containing the member Y_n is given by

$$G(Y_n) = \{Y_m : Y_n \sim Y_m\}. \tag{Ap 1.20}$$

By assumption the sets $G(Y_n)$ are finite and contain at least the element Y_n.

In the following we suppose that the sequence $(K_n) = ([Y_n, B_n, C_n])$ is stationary and denote its distribution by P. Analogously to (Ap 1.16), (Ap 1.17), and (Ap 1.20) we define

$$\left.\begin{array}{l} \operatorname{Pre}(K_n) = K_p \quad \text{iff} \quad Y_p = \operatorname{Pre}(Y_n), \\ \operatorname{Suc}(K_n) = K_s \quad \text{iff} \quad Y_s = \operatorname{Suc}(Y_n), \\ G(K_n) = \{K_m : Y_m \in G(Y_n)\}. \end{array}\right\} \tag{Ap 1.21}$$

Further we define

$$G_n = \#\{m : K_m \in G(K_n)\} = \#\{m : Y_m \sim Y_n\}.$$

Since we have supposed that the groups are finite, it holds that

$$\mathsf{P}(G_n < \infty, n \in \mathbb{Z}) = 1. \tag{Ap 1.22}$$

The stationary sequence (K_n) can be interpreted as the description of a sequence of grouped individuals in steady state seen from the point of view of an arbitrarily chosen individual.

To characterize a group we use particular members, namely the first and the last member of the group. (If we knew that every group consists of at least three members, we could also use the second or third member of a group to describe it. What follows would run analogously. Here we use the first and the last member.) The member K_n is a first (last) member of its group iff $B_n = 0$ ($C_n = 0$). Since all groups are finite, there are infinitely many groups in the sequence (K_n) and, consequently, infinitely many first and infinitely many last members of groups. Thus

$$\mathsf{P}(B_0 = 0) > 0, \qquad \mathsf{P}(C_0 = 0) > 0. \tag{Ap 1.23}$$

Define the conditional probabilities

$$\tilde{\mathsf{P}}(\cdot) = \mathsf{P}\big((K_n) \in (\cdot) \mid B_0 = 0\big), \ \check{\mathsf{P}}(\cdot) = \mathsf{P}\big((K_n) \in (\cdot) \mid C_0 = 0\big). \tag{Ap 1.24}$$

By $(\tilde{K}_n)$ and $(\check{K}_n)$ we denote sequences of $\mathbb{Y}\times\mathbb{Z}_+\times\mathbb{Z}_+$-valued r.v.'s distributed according to $\tilde{\mathsf{P}}$ and $\check{\mathsf{P}}$, respectively. All the r.v.'s derived from $(\tilde{K}_n)$ and $(\check{K}_n)$ are endowed with the sign $\sim$ or $\vee$, respectively. From Theorem A 1.3.1 and Theorem A 1.3.2 we obtain immediately the following results:

(i) The index $\tilde{\nu}_b = \min\{j \geqq 1 : \tilde{B}_j = 0\}$ $(\check{\nu}_c = \min\{j \geqq 1 : \check{C}_j = 0\})$ is a proper r.v. with $\mathsf{E}\tilde{\nu}_b < \infty$ $(\mathsf{E}\check{\nu}_c < \infty)$.

(ii) $((\tilde{K}_n), \tilde{\mathsf{P}})$ and $((\check{K}_n), \check{\mathsf{P}})$ have the following invariance properties:

$$\tilde{\mathsf{P}}\big((\tilde{K}_n) \in (\cdot)\big) = \tilde{\mathsf{P}}\big((\tilde{K}_{n+\tilde{\nu}_b}) \in (\cdot)\big),$$
$$\check{\mathsf{P}}\big((\check{K}_n) \in (\cdot)\big) = \check{\mathsf{P}}\big((\check{K}_{n+\check{\nu}_c}) \in (\cdot)\big). \qquad \text{(Ap 1.25)}$$

(iii) There exists a one-to-one mapping between the families of probability distributions

$\mathbf{P} = \{\mathsf{P} : \mathsf{P}$ is the probability distribution of a stationary sequence (K_n), $K_n = [Y_n, B_n, C_n]$, with the property $\mathsf{P}(\# \{n : B_n = 0\} \geqq 1) = 1$ (resp. $\mathsf{P}(\# \{n : C_n = 0\} \geqq 1) = 1)\}$

and

$\tilde{\mathbf{P}} = \{\tilde{\mathsf{P}} : \tilde{\mathsf{P}}$ is the probability distribution of a sequence $(\tilde{K}_n)$, $\tilde{K}_n = [\tilde{Y}_n, \tilde{B}_n, \tilde{C}_n]$, with the properties (i) and (ii)$\}$

(resp. $\check{\mathbf{P}} = \{\check{\mathsf{P}} : \text{(i), (ii)}\}$).

This mapping is given by equation (Ap 1.24) as well as by

$$\mathsf{P}\big((K_n) \in (\cdot)\big) = (\mathsf{E}\tilde{\nu}_b)^{-1} \sum_{j=0}^{\infty} \tilde{\mathsf{P}}\big(\tilde{\nu}_b > j, (\tilde{K}_{n+j}) \in (\cdot)\big)$$
$$= (\mathsf{E}\check{\nu}_c)^{-1} \sum_{j=0}^{\infty} \check{\mathsf{P}}\big(\check{\nu}_c > j, (\check{K}_{n+j}) \in (\cdot)\big). \qquad \text{(Ap 1.26)}$$

In view of the invariance properties (Ap 1.25) we call both sequences $((\tilde{K}_n), \tilde{\mathsf{P}})$ and $((\check{K}_n), \check{\mathsf{P}})$ *group-stationary*. $((K_n), \mathsf{P})$ is called *individual-stationary*.

Formula (Ap 1.26) is useful if the members of a group have successive numbers without gaps. In that case, the formula shows that the individual-stationary distribution is a certain average of the shifted group-stationary distributions. In the following we shall show that such a result holds also for arbitrarily spread groups.

Besides the usual shift operator

$$\theta(K_n, n \in \mathbb{Z}) = (K_{n+1}, n \in \mathbb{Z}) \qquad \text{(Ap 1.27)}$$

we define the shift operators $\tilde{\theta}$ and $\check{\theta}$ as follows:

$$\tilde{\theta}(K_n, n \in \mathbb{Z}) = \theta^{C_0}(K_n, n \in \mathbb{Z}), \qquad \text{(Ap 1.28)}$$

$$\check{\theta}(K_n, n \in \mathbb{Z}) = \theta^{-B_0}(K_n, n \in \mathbb{Z}). \qquad \text{(Ap 1.29)}$$

Notice that for $B_0 > 0$ it holds that $\tilde{\theta}\check{\theta}(K_n) = (K_n)$, and for $C_0 > 0$ we have $\check{\theta}\tilde{\theta}(K_n) = (K_n)$.

A 1.4.1. Theorem. *For every stationary sequence* $(K_n, n \in \mathbb{Z})$, $K_n = [Y_n, B_n, C_n]$, *satisfying* (Ap 1.22) *it holds that*

$$\mathsf{E}\tilde{G}_0 = \mathsf{E}\check{G}_0 < \infty, \tag{Ap 1.30}$$

$$\mathsf{P}\big((K_n) \in (\cdot)\big) = (\mathsf{E}\tilde{G}_0)^{-1} \sum_{j=0}^{\infty} \tilde{\mathsf{P}}\big(\tilde{G}_0 > j, \tilde{\theta}^j(\tilde{K}_n) \in (\cdot)\big)$$

$$= (\mathsf{E}\check{G}_0)^{-1} \sum_{j=0}^{\infty} \check{\mathsf{P}}\big(\check{G}_0 > j, \check{\theta}^j(\check{K}_n) \in (\cdot)\big). \tag{Ap 1.31}$$

In particular,

$$\mathsf{P}(B_0 = 0) = \mathsf{P}(C_0 = 0) = (\mathsf{E}\tilde{G}_0)^{-1} = (\mathsf{E}\check{G}_0)^{-1}. \tag{Ap 1.32}$$

Proof. It holds that

$$\mathsf{P}\big((K_n) \in (\cdot)\big) = \mathsf{P}\big((K_n) \in (\cdot), B_0 = 0\big) + R_0,$$

where

$$R_0 = \sum_{l=1}^{\infty} \mathsf{P}\big((K_n) \in (\cdot), B_0 = l\big).$$

Since $\tilde{\theta}\check{\theta}(K_n) = (K_n)$ for $B_0 > 0$, and using the stationarity of (K_n), we get

$$R_0 = \sum_{l=1}^{\infty} \mathsf{P}\big((K_n) \in (\cdot), C_{-l} = l\big)$$

$$= \sum_{l=1}^{\infty} \mathsf{P}\big(\theta^l(K_n) \in (\cdot), C_0 = l\big) = \mathsf{P}\big(\tilde{\theta}(K_n) \in (\cdot), C_0 > 0\big).$$

Using the notation $c_0\big((K_n)\big) = C_0$ we get $c_0\big(\theta^i(K_n)\big) = C_i$. Applying the above arguments successively, we obtain

$$\mathsf{P}\big((K_n) \in (\cdot)\big) = \sum_{j=0}^{m} \mathsf{P}\Big(\tilde{\theta}^j(K_n) \in (\cdot), c_0\big(\tilde{\theta}^{j-1}(K_n)\big) > 0, \ldots, c_0\big((K_n)\big) > 0, B_0 = 0\Big) + R_m,$$

where

$$R_m = \mathsf{P}\Big(\tilde{\theta}^m(K_n) \in (\cdot), c_0\big(\tilde{\theta}^{m-1}(K_n)\big) > 0, \ldots, c_0\big((K_n)\big) > 0, B_0 > 0\Big).$$

Since

$$R_m \leq \mathsf{P}\Big(c_0\big(\tilde{\theta}^{m-1}(K_n)\big) > 0, \ldots, c_0\big((K_n)\big) > 0, B_0 > 0\Big)$$

$$= \mathsf{P}(G_0 \geq m + 2),$$

assumption (Ap 1.22) implies

$$\lim_{m\to\infty} R_m = 0 \quad \text{a.s.}$$

Hence

$$\mathsf{P}\big((K_n) \in (\cdot)\big) = \sum_{j=0}^{\infty} \mathsf{P}\Big(\tilde{\theta}^j(K_n) \in (\cdot), c_0\big(\tilde{\theta}^{j-1}(K_n)\big) > 0, \ldots, c_0\big((K_n)\big) > 0, B_0 = 0\Big)$$

$$= \sum_{j=0}^{\infty} \mathsf{P}\big(\tilde{\theta}^j(K_n) \in (\cdot), G_0 > j, B_0 = 0\big)$$

$$= \mathsf{P}(B_0 = 0) \sum_{j=0}^{\infty} \tilde{\mathsf{P}}\big(\tilde{G}_0 > j, \tilde{\theta}^j(\tilde{K}_n) \in (\cdot)\big). \tag{Ap 1.33}$$

Analogously, we obtain

$$\mathsf{P}\big((K_n) \in (\cdot)\big) = \mathsf{P}(C_0 = 0) \sum_{j=0}^{\infty} \check{\mathsf{P}}\big(\check{G}_0 > j, \check{\theta}^j(\check{K}_n) \in (\cdot)\big). \tag{Ap 1.34}$$

Inserting the sure event in (Ap 1.33), (Ap 1.34), we get

$$1 = \mathsf{P}(B_0 = 0)\, \mathsf{E}\tilde{G}_0 = \mathsf{P}(C_0 = 0)\, \mathsf{E}\check{G}_0, \tag{Ap 1.35}$$

and thus (Ap 1.31) follows from (Ap 1.33) and (Ap 1.23). Formula (Ap 1.33) yields

$$\mathsf{P}(C_0 = 0) = \mathsf{P}(B_0 = 0)\, \mathsf{E} \sum_{j=0}^{\tilde{G}_0 - 1} 1\big\{c_0\big(\tilde{\theta}^j(\tilde{K}_n)\big) = 0\big\}$$

$$= \mathsf{P}(B_0 = 0),$$

since $c_0\big(\tilde{\theta}^j(\tilde{K}_n)\big) = 0$ holds if and only if $j = \tilde{G}_0 - 1$ (the last member of the group). This, (Ap 1.35), and (Ap 1.23) imply (Ap 1.30) and (Ap 1.32). ∎

A2. Convergence in distribution of stationary sequences. Uniform law of large numbers

Let $Y, Y_1, Y_2, \ldots$ be r.v.'s of an arbitrary metric space $\mathbf{Y}$. The sequence $(Y_n, n \geqq 1)$ converges in distribution to Y (abbr. $Y_n \xrightarrow{\mathcal{D}} Y$) iff

$$\mathsf{E}h(Y_n) \to \mathsf{E}h(Y)$$

for all real valued, bounded continuous functions on $\mathbf{Y}$.

Let (Y_1, P_1), (Y_2, P_2) be two random elements of a Polish space $\mathbf{Y}$ endowed with the metric d. The Prokhorov distance $\pi(\mathsf{P}_1, \mathsf{P}_2)$ of the distributions P_1 and P_2 is defined by

$$\pi(\mathsf{P}_1, \mathsf{P}_2) = \max\{\pi_1(\mathsf{P}_1, \mathsf{P}_2), \pi_1(\mathsf{P}_2, \mathsf{P}_1)\},$$

where

$$\pi_1(\mathsf{P}_1, \mathsf{P}_2) = \inf\{\varepsilon : \varepsilon > 0, \quad \mathsf{P}_1(B) < \mathsf{P}_2\Big(\bigcup_{y \in B} \{z : d(z, y) < \varepsilon\}\Big) + \varepsilon$$

for all closed subsets $B \subseteqq \mathbf{Y}\}$.

The Prokhorov distance π makes the space of all probability distributions on the Borel σ-field $\mathcal{Y}$ of $\mathbf{Y}$ to a Polish space. The convergence with respect to π coincides with the convergence in distribution of the corresponding r.v.'s.

Let us first state some well-known results.

A 2.1. Lemma. *Let $X, X_1, X_2, \ldots$ be non-negative real-valued r.v.'s with $X_n \xrightarrow{\mathcal{D}} X$ and $\mathsf{E}X_n \to \mathsf{E}X$. Then the X_n, $n \geqq 1$, are uniformly integrable, i.e.*

$$\sup_n \mathsf{E}(X_n 1\{X_n \geqq s\}) \xrightarrow[s \to \infty]{} 0.$$

In particular,

$$\sup_n \int_s^\infty \mathsf{P}(X_n \geqq t)\, \mathrm{d}t \xrightarrow[s \to \infty]{} 0.$$

For a proof of the first assertion we refer to Billingsley (1968, Theorem 5.4). The second follows from the first one because

$$\begin{aligned}\mathsf{E}(X_n 1\{s \leqq X_n\}) &= \int_0^\infty \mathsf{E}(1\{t \leqq X_n\}\, 1\{s \leqq X_n\})\, \mathrm{d}t \\ &= \int_0^s \mathsf{E}1\{s \leqq X_n\}\, \mathrm{d}t + \int_s^\infty \mathsf{E}1\{t \leqq X_n\}\, \mathrm{d}t \\ &\geqq \int_s^\infty \mathsf{P}(X_n \geqq t)\, \mathrm{d}t. \quad \blacksquare\end{aligned}$$

From BILLINGSLEY (1968, Section 3) we have

A 2.2. Lemma. *Let X, X_n, $n \geqq 1$, be real-valued r.v.'s with the d.f.'s $\mathsf{F}(x)$ and $\mathsf{F}_n(x)$, respectively. Then*

$$X_n \xrightarrow{\mathcal{D}} X \quad \textit{iff there is a dense set } D \subseteqq \mathbb{R} \textit{ such that}$$

$$\mathsf{F}_n(x) \to \mathsf{F}(x) \textit{ for all } x \in D.$$

A 2.3. Lemma. *Let (Φ_k, P_k), $\Phi_k = (Y_{kn}, n \in \mathbb{Z})$, $k = 1, 2, \ldots$, (Φ, P), $\Phi = (Y_n, n \in \mathbb{Z})$, be arbitrary sequences of $\mathbb{Y}$-valued r.v.'s, and consider the metric*

$$\tilde{d}\big((y_n), (z_n)\big) = \sum_n 2^{-|n|} d(y_n, z_n)$$

(or any other metric which is equivalent to this) on $\mathbb{Y}^{\mathbb{Z}}$. Then

$$\Phi_k \xrightarrow{\mathcal{D}} \Phi \quad \textit{iff } [Y_{kn_1}, \ldots, Y_{kn_m}] \xrightarrow{\mathcal{D}} [Y_{n_1}, \ldots, Y_{n_m}]$$

$$\textit{for all } m \geqq 1 \textit{ and all } n_1, \ldots, n_m \in \mathbb{Z}.$$

The following result is well-known under the name "continuous mapping theorem", cf. e.g. BILLINGSLEY (1968, Section 5).

A 2.4. Theorem. *Let $\mathbb{Y}, \mathbb{W}$ be two metric spaces, $h\colon \mathbb{Y} \to \mathbb{W}$ a measurable mapping and $D_h \subseteqq \mathbb{Y}$ the set of all discontinuity points of h. Consider some $\mathbb{Y}$-valued r.v.'s (Y, P), (Y_n, P_n), $n \geqq 1$, with $Y_n \xrightarrow{\mathcal{D}} Y$ and $\mathsf{P}(Y \in D_h) = 0$. Then $h(Y_n) \xrightarrow{\mathcal{D}} h(Y)$.*

The following criterion by Prokhorov is used in Section 8.6, cf. BILLINGSLEY (1968).

A 2.5. Theorem. *A family E of probability measures on a Polish space is relatively compact with respect to the Prokhorov metric π if and only if E is tight, i.e. for each $\varepsilon > 0$ there is a compact set K_ε such that $\mathsf{P}(K_\varepsilon) \geqq 1 - \varepsilon$ for all $\mathsf{P} \in E$.*

A 2.6. Representation Theorem by Skorokhod (1956). *Let $\mathbb{Y}$ be a Polish space with the Borel σ-field $\mathcal{Y}$, and $(Y_n, n \geqq 1)$ a sequence of $\mathbb{Y}$-valued random elements converging in distribution to the random element Y,*

$$Y_n \xrightarrow{\mathcal{D}} Y.$$

Then there are random elements Y', Y_n', $n \geqq 1$, such that

$$Y \stackrel{\mathcal{D}}{=} Y', \qquad Y_n \stackrel{\mathcal{D}}{=} Y_n', \qquad n \geqq 1,$$

and

$$Y_n' \xrightarrow[n\to\infty]{} Y' \quad a.s.$$

The following two auxiliary results are due to BOROVKOV (1972c, 1980).

A 2.7. Lemma. *Let (Y_i) be an arbitrary stationary sequence. Then for any* $c > 0$

$$\mathsf{P}\left(\sup_{r\geq 0} \sum_{i=-r}^{-1} Y_i > 0\right) \leqq c^{-1}\mathsf{E}(Y_0 1\{Y_0 \geqq 0\}) + \mathsf{P}(Y_0 > -c). \quad \text{(Ap 2.1)}$$

Proof. We can restrict ourselves to the case of

$$\mathsf{E}(Y_0 1\{Y_0 > 0\}) < \infty,$$

because the assertion is trivially true otherwise. Define the r.v.'s

$$Y_j^* = \begin{cases} Y_j & \text{if } Y_j \in (-A, -c] \cup [0, \infty), \\ 0 & \text{if } Y_j \in (-c, 0), \\ -A & \text{if } Y_j \in (-\infty, -A], \end{cases}$$

where $A > c$ is a fixed number. Then (Y_j^*) is stationary, and $\mathsf{E}\,|Y_j^*| < \infty$. Putting

$$Z_r^* = \sum_{j=1}^{r} Y_{-j}^*, \qquad Z^* = \sup_{r\geq 0} Z_r^*$$

we get the inequality (Doob (1953, p. 418))

$$\mathsf{E}(Y_{-1}^* 1\{Z^* > 0\}) \geqq 0. \quad \text{(Ap 2.2)}$$

Since $\mathsf{P}\big(Y_{-1}^* \in (-c, 0)\big) = 0$, from (Ap 2.2) we get

$$\begin{aligned} \mathsf{E}(Y_{-1}^* 1\{Y_{-1}^* \geqq 0, Z^* > 0\}) &\geqq -\mathsf{E}(Y_{-1}^*\, 1\{Y_{-1}^* \leqq -c, Z^* > 0\}) \\ &\geqq c\mathsf{P}(Y_{-1}^* \leqq -c, Z^* > 0). \end{aligned}$$

Consequently,

$$\begin{aligned} \mathsf{P}(Z^* > 0) &= \mathsf{P}(Y_{-1}^* \geqq 0, Z^* > 0) + \mathsf{P}(Y_{-1}^* \leqq -c, Z^* > 0) \\ &\leqq \mathsf{P}(Y_{-1}^* \geqq 0) + c^{-1}\mathsf{E}(Y_{-1}^* 1\{Y_{-1}^* \geqq 0\}), \end{aligned} \quad \text{(Ap 2.3)}$$

which yields assertion (Ap 2.1), because

$$\mathsf{P}\left(\sup_{r\geq 0} \sum_{i=-r}^{-1} Y_i > 0\right) \leqq \mathsf{P}(Z^* > 0),$$

$$\mathsf{P}(Y_{-1}^* \geqq 0) = \mathsf{P}(Y_{-1} > -c),$$

and

$$\mathsf{E}(Y_{-1}^* 1\{Y_{-1}^* \geqq 0\}) = \mathsf{E}(Y_{-1} 1\{Y_{-1} \geqq 0\}) = \mathsf{E}(Y_0 1\{Y_0 \geqq 0\}). \quad \blacksquare$$

A 2.8. Lemma. *If the vectors* $\mathbf{X} = [X_1, \ldots, X_r]$, $\mathbf{X}_1 = [X_{11}, \ldots, X_{1r}]$, $\mathbf{X}_2 = [X_{21}, \ldots, X_{2r}]$, ... *of* $\mathbb{R}$*-valued r.v.'s satisfy the conditions*

$$\mathbf{X}_k \xrightarrow[k\to\infty]{\mathcal{D}} \mathbf{X},$$

$$\mathsf{E}(X_{ki})_+ \xrightarrow[k\to\infty]{} \mathsf{E}(X_i)_+, \qquad i = 1, \ldots, r, \quad \text{(Ap 2.4)}$$

then

$$\mathsf{E}\left(\sum_{i=1}^{r} X_{ki}\right)_+ \xrightarrow[k\to\infty]{} \mathsf{E}\left(\sum_{i=1}^{r} X_i\right)_+ .$$

Proof. It suffices to show the assertion for $r = 2$. Let $Z = X_1 + X_2$, $Z_k = X_{k1} + X_{k2}$. For fixed $c > 0$ it holds that

$$\begin{aligned}
&|\mathsf{E}(Z_k)_+ - \mathsf{E}Z_+| \\
&= |\mathsf{E}(Z_k 1\{Z_k \geqq 0\}) - \mathsf{E}(Z 1\{Z \geqq 0\})| \\
&\leqq |\mathsf{E}(X_{k1} 1\{Z_k \geqq 0, |X_{k1}| < c\} + X_{k2} 1\{Z_k \geqq 0, |X_{k2}| < c\}) \\
&\quad - \mathsf{E}(X_1 1\{Z \geqq 0, |X_1| < c\} + X_2 1\{Z \geqq 0, |X_2| < c\})| \\
&\quad + \mathsf{E}(|X_{k1}|\, 1\{Z_k \geqq 0, |X_{k1}| \geqq c\}) + \mathsf{E}(|X_{k2}|\, 1\{Z_k \geqq 0, |X_{k2}| \geqq c\}) \\
&\quad + \mathsf{E}(|X_1|\, 1\{Z \geqq 0, |X_1| \geqq c\}) + \mathsf{E}(|X_2|\, 1\{Z \geqq 0, |X_2| \geqq c\}).
\end{aligned} \tag{Ap 2.5}$$

By means of the inequality

$$|x_1|\, 1\{x_1 + x_2 \geqq 0, |x_1| \geqq c\} \leqq x_1 1\{x_1 \geqq c\} + x_2 1\{x_2 \geqq c\} \tag{Ap 2.6}$$

and since $(X_{ki})_+ \xrightarrow{\mathcal{D}} (X_i)_+$, from Theorem A 2.1 we obtain

$$\sup_k \mathsf{E}\big(|X_{ki}|\, 1\{Z \geqq 0, |X_{ki}| \geqq c(\varepsilon)\}\big) < \varepsilon/4, \qquad i = 1, 2, \tag{Ap 2.7}$$

and

$$\mathsf{E}\big(|X_i|\, 1\{Z \geqq 0, |X_i| \geqq c(\varepsilon)\}\big) < \varepsilon/4, \qquad i = 1, 2, \tag{Ap 2.8}$$

for every $\varepsilon > 0$ and $c(\varepsilon)$ sufficiently large. Now consider an arbitrary $\varepsilon > 0$. We choose $c(\varepsilon)$ large enough such that (Ap 2.7) and (Ap 2.8) are satisfied and such that $c(\varepsilon)$ is a point of continuity of the (marginal) distributions $\mathsf{P}(X_1 \leqq t)$ and $\mathsf{P}(X_2 \leqq t)$. It is easy to check that the function

$$\begin{aligned}
h(x_1, x_2) &= x_1 1\{x_1 + x_2 \geqq 0, |x_1| < c(\varepsilon)\} \\
&\quad + x_2 1\{x_1 + x_2 \geqq 0, |x_2| < c(\varepsilon)\}
\end{aligned}$$

is bounded and almost surely continuous. Thus, the first difference on the right-hand side of (Ap 2.5) converges to zero as $k \to \infty$ (cf. e.g. Theorem 5.2 in BILLINGSLEY (1968)), and we get

$$\lim_{k\to\infty} |\mathsf{E}(Z_k)_+ - \mathsf{E}(Z)_+| \leqq \varepsilon$$

in view of (Ap 2.7) and (Ap 2.8). Since ε was arbitrary,

$$\lim_{k\to\infty} \mathsf{E}(Z_k)_+ = \mathsf{E}(Z)_+. \quad \blacksquare$$

The following profound result by BOROVKOV (1972c, 1980) plays a key rule in proving continuity theorems for functionals of stationary sequences.

A 2.9. Theorem. *Let (Φ, P), $\Phi = (Y_n)$, (Φ_k, P_k), $\Phi_k = (Y_{kn})$, $k \geqq 1$, be stationary sequences of $\mathbb{R}$-valued r.v.'s. Assume*

(AA 1) *For (Y_n) the strong law of large numbers is true, i.e.*

$$\lim_{n\to\infty} n^{-1} \sum_{i=1}^{n} Y_i = \mathsf{E} Y_0 \quad a.s.,$$

and

$$\mathsf{E} Y_0 < 0.$$

(AA 2) $\Phi_k \xrightarrow[k\to\infty]{\mathcal{D}} \Phi$.

(AA 3) $\mathsf{E}(Y_{kn})_+ \xrightarrow[k\to\infty]{} \mathsf{E}(Y_n)_+ < \infty$.

Then

$$\sup_{k\geq 1} \mathsf{P}_k \left(\sup_{r\geq n} \sum_{i=m-r}^{m-1} Y_{ki} \geqq 0 \right) \xrightarrow[n\to\infty]{} 0, \qquad m \in \mathbb{Z}. \tag{Ap 2.9}$$

Proof. By the stationarity of Φ_k, $k \geqq 1$, the probabilities

$$\mathsf{P}_k \left(\sup_{r\geq n} \sum_{i=m-r}^{m-1} Y_{ki} \geqq 0 \right)$$

do not depend on m. Therefore we can assume $m = 0$. Define

$$Z_k^r = \sum_{i=-r}^{-1} Y_{ki}, \qquad Z^r = \sum_{i=-r}^{-1} Y_i, \qquad k, r \geqq 1,$$

$$Z_k^0 = Z^0 = 0.$$

For arbitrary $s > 0$ and $j > 0$ it holds that

$$\begin{aligned}
&\mathsf{P}_k \left(\sup_{r\geq n} Z_k^r \geqq 0 \right) \\
&\leqq \mathsf{P}_k \left(Z_k^n < -sn, \sup_{r\geq n} Z_k^r \geqq 0 \right) + \mathsf{P}_k (Z_k^n \geqq -sn) \\
&\leqq \mathsf{P}_k \left(Z_k^n < -sn, \sup_{r\geq n} Z_k^r \geqq 0, \bigcap_{i=0}^{\infty} \{Z_k^{n+ij} - Z_k^n \leqq -sij\} \right) \\
&\quad + \mathsf{P}_k \left(\bigcup_{i=0}^{\infty} \{Z_k^{n+ij} - Z_k^n > -sij\} \right) + \mathsf{P}_k (Z_k^n \geqq -sn).
\end{aligned} \tag{Ap 2.10}$$

We denote the terms on the right-hand side of (Ap 2.10) by Σ_1, Σ_2, and Σ_3, respectively. Using Lemma A 2.7 for $c = sj$ and taking into account the stationarity of Φ_k, we get

$$\begin{aligned}
\Sigma_2 = \mathsf{P}_k \left(\sup_{i\geq 0} (Z_k^{ij} + sij) > 0 \right) &\leqq \mathsf{P}_k (j^{-1} Z_k^j > -2s) \\
&\quad + s^{-1} \mathsf{E}\left(j^{-1} (Z_k^j + sj)_+\right).
\end{aligned}$$

For the first term Σ_1 we have

$$\Sigma_1 \leqq \mathsf{P}_k\left(\bigcup_{q=0}^{\infty} \bigcup_{i=qj+1}^{(q+1)j} \{Y_{k,-i} > j^{-1}(ns + qjs)\}\right)$$

$$\leqq j \sum_{q=0}^{\infty} \mathsf{P}_k(Y_{k1} > qs + ns/j) \leqq (j/s) \int_{ns/j-s}^{\infty} \mathsf{P}_k(Y_{k1} > t)\,dt.$$

From Theorem A 2.1 we have that

$$\int_u^{\infty} \mathsf{P}_k(Y_{k1} > t)\,dt$$

converges uniformly with respect to k as $u \to \infty$. (Notice that $(Y_{k1})_+ \xrightarrow[k\to\infty]{\mathcal{D}} (Y_1)_+$, $\mathsf{E}(Y_{k1})_+ \xrightarrow[k\to\infty]{} \mathsf{E}(Y_1)_+$.) Thus, there is a function $g(u)$ with $g(u) \to 0$ as $u \to \infty$ such that

$$\Sigma_1 \leqq (j/s)\, g(ns/j).$$

Since for every j and for $n \to \infty$ we have $j\, g(n/j) \to 0$, there exists an increasing function $h(j)$ such that $j\ g\big(h(j)/j\big) \to 0$, $h(j) \to \infty$, as $j \to \infty$. Henceforth we shall assume that j and n are connected by $n = h(j)/s$. For every n (or j) and s we get that

$$\mathsf{P}_k\left(\sup_{r\geqq n} Z_k^r \geqq 0\right) \leqq \mathsf{P}_k(Z_k^n/n \geqq -s) + \mathsf{P}_k(Z_k^j/j > -2s)$$
$$+ s^{-1}\mathsf{E}\big((Z_k^j + sj)_+/j\big) + (j/s)\, g(ns/j).$$

For those s (an everywhere dense set) which are points of continuity for the distributions of $-Z^n/n$ and $-Z^j/(2j)$, in view of Lemma A 2.8 we have

$$\limsup_{k\to\infty} \mathsf{P}_k\left(\sup_{r\geqq n} Z_k^r \geqq 0\right) \leqq H(n, s) + (j/s)\, g(ns/j), \qquad \text{(Ap 2.11)}$$

where

$$H(n, s) = \mathsf{P}(Z^n/n \geqq -s) + \mathsf{P}(Z^j/j > -2s) + s^{-1}\mathsf{E}\big((Z^j + sj)_+/j\big).$$

Now we show that the right-hand side of (Ap 2.11) becomes arbitrarily small with suitably chosen n and s. Let $\varepsilon > 0$. For $s = -a/3$, $a = \mathsf{E}Y_0 < 0$, the last term of $H(n, s)$ satisfies

$$s^{-1}\mathsf{E}\big((Z^j + sj)_+/j\big) \leqq s^{-1}\mathsf{E}(Z^j/j - a)_+ \leqq s^{-1}\mathsf{E}\,|Z^j/j - a|$$

and thus

$$H(n, s) \leqq \mathsf{P}(Z^n/n \geqq a/3) + \mathsf{P}(Z^j/j > 2a/3) + s^{-1}\mathsf{E}\,|Z^j/j - a|.$$

Via the law of large numbers we get that $\mathsf{E}\,|Z^j/j - a| \xrightarrow[j\to\infty]{} 0$. Then we may choose n (and consequently j) so large that

$$H(n, s) \leqq \varepsilon/2, \qquad (j/s)\, g(ns/j) \leqq \varepsilon/2.$$

Therefore the left-hand side of (Ap 2.11) becomes arbitrarily small, and (Ap 2.9) has been proved. ∎

A 2.10. Remark. Assumption (AA 3) cannot be omitted in Theorem A 2.9. Even the condition $\mathsf{E}Y_{kn} \xrightarrow[k\to\infty]{} \mathsf{E}Y_n$ is not sufficient for (Ap 2.9), as can be seen by elementary examples, cf. Borovkov (1980, p. 207).
Conditions (AA 1)–(AA 3) imply convergence (as $k \to \infty$) of the distributions of a whole class of functionals on

$$\left(\sum_{i=-r}^{-1} Y_{ki}, r \geqq 0\right)$$

(the so-called V-continuous functionals, see Borovkov (1972c) and Sakhanenko (1974)). Condition (AA 3) is a compactness condition on the sequence of distributions of these functionals.

By means of Theorem A 2.9 one can easily show that under natural convergence assumptions the law of large numbers holds uniformly.

A 2.11. Theorem. *Let* (Φ, P), $\Phi = (Y_n)$, (Φ_k, P_k), $\Phi_k = (Y_{kn})$, $k \geqq 1$, *be stationary and ergodic sequences of real-valued r.v.'s and assume that the expectations* $a = \mathsf{E}Y_0$, $a_k = \mathsf{E}Y_{k0}$, $k \geqq 1$, *exist and are finite. If*

$$\Phi_k \xrightarrow{\mathfrak{D}} \Phi, \tag{Ap 2.12}$$

$$a_k \to a, \qquad \mathsf{E}(Y_{k0})_+ \to \mathsf{E}(Y_0)_+, \tag{Ap 2.13}$$

then

$$\sup_k \mathsf{P}_k\left(\sup_{j>n}\left|j^{-1}\sum_{i=m-j}^{m-1} Y_{ki} - a_k\right| \geqq \delta\right) \xrightarrow[n\to\infty]{} 0 \tag{Ap 2.14}$$

and

$$\sup_k \mathsf{P}_k\left(\sup_{j>n} j^{-1}\,|Y_{k,m-j}| \geqq \delta\right) \xrightarrow[n\to\infty]{} 0, \tag{Ap 2.15}$$

where δ *is an arbitrary positive real number and* $m \in \mathbb{Z}$.

Proof. First we shall prove (Ap 2.14). From (Ap 2.12) and (Ap 2.13) we obtain

$$\left[(Y_{k0})_+ + (a_k)_-, -\left((Y_{k0})_- + (a_k)_+ + \delta\right)\right]$$
$$\xrightarrow[k\to\infty]{\mathfrak{D}} \left[(Y_0)_+ + a_-, -\left((Y_0)_- + a_+ + \delta\right)\right]$$

and, in view of Lemma A 2.8,

$$\mathsf{E}\left((Y_{k0})_+ + (a_k)_-\right)_+ \xrightarrow[k\to\infty]{} \mathsf{E}\left((Y_0)_+ + a_-\right)_+,$$
$$\mathsf{E}\left(-\left((Y_{k0})_- + (a_k)_+ + \delta\right)\right)_+ \xrightarrow[k\to\infty]{} \mathsf{E}\left(-\left((Y_0)_- + a_+ + \delta\right)\right)_+,$$

where $x_- = -\min(0, x)$, $x_+ = \max(0, x)$ for $x \in \mathbb{R}$.
Applying Lemma A 2.8 we get

$$\mathsf{E}(Y_{k0} - a_k - \delta)_+ \xrightarrow[k\to\infty]{} \mathsf{E}(Y_0 - a - \delta)_+.$$

Now, applying Theorem A 2.9 we obtain

$$\sup_k \mathsf{P}_k\left(\sup_{j>n} \sum_{i=m-j}^{m-1} (Y_{ki} - a_k - \delta) \geqq 0\right) \xrightarrow[n\to\infty]{} 0$$

and thus

$$\sup_k \mathsf{P}_k \left(\sup_{j>n} j^{-1} \sum_{i=m-j}^{m-1} (Y_{ki} - a_k - \delta) \geqq 0 \right)$$

$$= \sup_k \mathsf{P}_k \left(\sup_{j>n} j^{-1} \sum_{i=m-j}^{m-1} (Y_{ki} - a_k) \geqq \delta \right) \xrightarrow[n\to\infty]{} 0. \qquad \text{(Ap 2.16)}$$

Analogously,

$$\sup_k \mathsf{P}_k \left(\sup_{j>n} j^{-1} \sum_{i=m-j}^{m-1} (-Y_{ki}) - (-a_k) \geqq \delta \right) \xrightarrow[n\to\infty]{} 0. \qquad \text{(Ap 2.17)}$$

Combining (Ap 2.16) and (Ap 2.17) we obtain the first assertion (Ap 2.14).

Now we prove (Ap 2.15). Set

$$Z^j_{km} = j^{-1} \sum_{i=m-j}^{m-1} Y_{ki} - a_k.$$

Then

$$j^{-1} Y_{k,m-j} = Z^j_{km} - (1 - j^{-1})\, Z^{j-1}_{km} + a_k/j,$$

and (Ap 2.15) follows easily from (Ap 2.14) and $a_k \to a$. ∎

A much stronger type of convergence than convergence with respect to the Prokhorov distance is the convergence in variation:

$$\mathsf{P}_k \xrightarrow{\mathrm{Var}} \mathsf{P} \text{ iff } \mathrm{Var}(\mathsf{P}_k, \mathsf{P}) = \sup_{\substack{\bigcup_i Y_i = \mathbf{Y} \\ Y_i \cap Y_j = \emptyset \text{ for } i \neq j}} \sum_i |\mathsf{P}_k(Y_i) - \mathsf{P}(Y_i)| \xrightarrow[k\to\infty]{} 0.$$

A 3. Miscellaneous topics

A 3.1. Binomial moments

The r-th binomial moment B_r of a probability distribution $(p_n, n \geqq 0)$, concentrated on the non-negative integers $\{0, 1, 2, \ldots\}$, is defined by the series

$$B_r = \sum_{i=r}^{\infty} \binom{i}{r} p_i, \qquad r \geqq 1, \qquad B_0 = 1.$$

Obviously,

$$B_r = \lim_{z \uparrow 1} \frac{1}{r!} \frac{\mathrm{d}^r}{\mathrm{d}z^r} \Pi(z), \tag{Ap 3.1}$$

where

$$\Pi(z) = \sum_{n=0}^{\infty} p_n z^n, \qquad -1 \leqq z \leqq 1,$$

is the probability generating function (PGF) of $(p_n, n \geqq 0)$.

A 3.1.1. Theorem. (Ch. Jordan). *Let* $(p_n, n \geqq 0)$ *be a probability distribution on* $\{0, 1, 2, \ldots\}$. *If*

$$\sum_{r=i}^{\infty} \binom{r}{i} B_r < \infty, \tag{Ap 3.2}$$

then

$$p_i = \sum_{r=i}^{\infty} (-1)^{r-i} \binom{r}{i} B_r. \tag{Ap 3.3}$$

Proof. Taking into account (Ap 3.2), the assertion (Ap 3.3) follows from

$$\begin{aligned}
\sum_{r=i}^{\infty} (-1)^{r-i} \binom{r}{i} B_r &= \sum_{r=i}^{\infty} (-1)^{r-i} \binom{r}{i} \sum_{j=r}^{\infty} \binom{j}{r} p_j \\
&= \sum_{j=i}^{\infty} \sum_{r=i}^{j} (-1)^{r-i} \binom{r}{i} \binom{j}{r} p_j \\
&= p_i + \sum_{j=i+1}^{\infty} \left(\sum_{r=i}^{j} (-1)^{r-i} \binom{r}{i} \binom{j}{r} \right) p_j \\
&= p_i + \sum_{j=i+1}^{\infty} \binom{j}{i} \left(\sum_{l=0}^{j-i} (-1)^l \binom{j-i}{l} \right) p_j = p_i. \quad \blacksquare
\end{aligned}$$

Let $\boldsymbol{v} = (v_i, i \geqq 0)$, $\boldsymbol{w} = (w_i, i \geqq 0)$ be two sequences of real numbers. Then the sequence $\boldsymbol{t} = (t_i, i \geqq 0)$,

$$t_i = \sum_{j=0}^{i} v_{i-j} w_j , \qquad i \geqq 0,$$

is called the *convolution* of $\boldsymbol{v}$ and $\boldsymbol{w}$. We write $\boldsymbol{t} = \boldsymbol{v} * \boldsymbol{w}$.

A 3.1.2. Theorem. *Let $(a_n, n \geqq 0)$ and $(b_n, n \geqq 0)$ be two probability distributions on $\{0, 1, 2, \ldots\}$, and $(c_n, n \geqq 0)$ their convolution, i.e.*

$$c_n = \sum_{i=0}^{n} a_i b_{n-i} . \tag{Ap 3.4}$$

Then the binomial moments A_r, B_r, and C_r of $(a_n, n \geqq 0)$, $(b_n, n \geqq 0)$, and $(c_n, n \geqq 0)$, respectively, satisfy

$$C_r = \sum_{j=0}^{r} A_j B_{r-j} , \tag{Ap 3.5}$$

i.e. $(C_r, r \geqq 0)$ appears as the convolution of $(A_r, r \geqq 0)$ and $(B_r, r \geqq 0)$.

Proof. Consider the PGF's

$$A(z) = \sum_{i=0}^{\infty} a_i z^i , \qquad B(z) = \sum_{i=0}^{\infty} b_i z^i ,$$

and

$$C(z) = \sum_{i=0}^{\infty} c_i z^i .$$

From (Ap 3.4) we get $C(z) = A(z)\, B(z)$. Thus,

$$\frac{1}{r!} \frac{\mathrm{d}^r}{\mathrm{d}z^r} C(z) = \sum_{i=0}^{r} \frac{1}{r!} \binom{r}{i} \left(\frac{\mathrm{d}^i}{\mathrm{d}z^i} A(z)\right) \left(\frac{\mathrm{d}^{r-i}}{\mathrm{d}z^{r-i}} B(z)\right)$$
$$= \sum_{i=0}^{r} \frac{1}{i!} \left(\frac{\mathrm{d}^i}{\mathrm{d}z^i} A(z)\right) \frac{1}{(r-i)!} \left(\frac{\mathrm{d}^{r-i}}{\mathrm{d}z^{r-i}} B(z)\right),$$

which yields the assertion in view of (Ap 3.1). ∎

A 3.1.3. Lemma. *Let $Y_1, Y_2, \ldots$ be independent non-negative integer-valued r.v.'s, and*

$$S = \sum_{i=1}^{\infty} Y_i .$$

Then for the r-th binomial moment $\mathsf{E}\binom{S}{r}$ of S the following identity holds:

$$\mathsf{E}\binom{S}{r} = \sum_{l=1}^{r} \sum_{\substack{k_1+\cdots+k_l=r\\ k_i \geqq 1}} \sum_{i_1=1,\ldots,i_l=1}^{\infty,\ldots,\infty} \mathsf{E}\binom{Y_{i_1}}{k_1} \mathsf{E}\binom{Y_{i_1+i_2}}{k_2} \cdots \mathsf{E}\binom{Y_{i_1+\cdots+i_l}}{k_l} .$$

Proof. By means of Theorem A 3.1.2 we get

$$\mathsf{E}\binom{Y_1+\cdots+Y_n}{r} = \sum_{\substack{r_1+\cdots+r_n=r\\ r_i\geqq 0}} \mathsf{E}\binom{Y_1}{r_1}\mathsf{E}\binom{Y_2}{r_2}\cdots\mathsf{E}\binom{Y_n}{r_n}. \qquad \text{(Ap 3.6)}$$

In view of $\mathsf{E}\binom{Y_i}{0} = 1$, the sum of the right-hand side of (Ap 3.6) can be expressed in the following manner:

$$\mathsf{E}\binom{Y_1+\cdots+Y_n}{r}$$
$$= \sum_{l=1}^{r} \sum_{\substack{k_1+\cdots+k_l=r\\ k_i\geqq 1}} \sum_{\substack{i_1\geqq 1,\ldots,i_l\geqq 1\\ i_1+\cdots+i_l\leqq n}} \mathsf{E}\binom{Y_{i_1}}{k_1}\mathsf{E}\binom{Y_{i_1+i_2}}{k_2}\cdots\mathsf{E}\binom{Y_{i_1+\cdots+i_l}}{k_l}. \qquad \text{(Ap 3.7)}$$

From (Ap 3.7) the assertion follows by means of monotone convergence as $n\to\infty$. ∎

In Chapter 7 we need the convergence of particular series:

A 3.1.4. Lemma. *Let $(a_r, r\geqq 1)$ be a non-increasing sequence of non-negative real numbers with limit 0 and $a_r < 1$. For a fixed positive number k consider the sequence $(b_r, r\geqq 0)$ defined by*

$$b_0 = 1, \qquad b_r = \frac{a_r}{1-a_r}\sum_{j=(r-k)_+}^{r-1}\binom{k}{r-j} b_j, \qquad r\geqq 1.$$

Then

$$\sum_{r=i}^{\infty}\binom{r}{i} b_r < \infty, \qquad i\geqq 0.$$

Proof. 1. Consider the column vectors $z_r = [b_r, \ldots, b_{r-k+1}]^{\mathsf{T}}$, $r\geqq 0$, where $b_{-1} = \cdots = b_{-k+1} = 0$, i.e. $z_0 = [1, 0, \ldots, 0]^{\mathsf{T}}$, $z_1 = [b_1, 1, 0, \ldots, 0]^{\mathsf{T}}, \ldots$ The sequence $(z_r, r\geqq 0)$ satisfies the matrix equations

$$z_{r+1} = H_{r+1} z_r, \qquad r\geqq 0, \qquad \text{(Ap 3.8)}$$

where H_r is the following $k\times k$-matrix:

$$H_r = \begin{pmatrix} \frac{a_r}{1-a_r}\binom{k}{1} & \frac{a_r}{1-a_r}\binom{k}{2} & & \frac{a_r}{1-a_r}\binom{k}{k} \\ 1 & 0 & & 0 \\ 0 & 1 & & 0 \\ \vdots & & \ddots & \vdots \\ 0 & & 1 & 0 \end{pmatrix}.$$

We shall show that

$$\sum_{r=i}^{\infty}\binom{r}{i} z_r < \infty,$$

which yields the assertion of the lemma immediately.

2. By an elementary calculation we obtain the characteristic polynomial of H_r:

$$0 = x^k - \frac{a_r}{1 - a_r}\left((1 + x)^k - x^k\right).$$

Thus, the eigenvalues of H_r are

$$x_j = \frac{\sqrt[k]{a_r}\,\varepsilon_j^{(k)}}{1 - \sqrt[k]{a_r}\,\varepsilon_j^{(k)}}, \qquad j = 0, \ldots, k - 1,$$

where $\varepsilon_0^{(k)}, \ldots, \varepsilon_{k-1}^{(k)}$ are the k roots of 1 (i.e.

$$\varepsilon_j^{(k)} = \exp\left(i^* \frac{2\pi j}{k}\right), \qquad j = 0, 1, \ldots, k - 1; \; i^* = \sqrt{-1}).$$

Let $\varrho(H_r)$ be the spectral radius of H_r, i.e.

$$\varrho(H_r) = \max_{j=0,\ldots,k-1} |x_j|.$$

Since $a_r \xrightarrow[r\to\infty]{} 0$, we get $\varrho(H_r) \xrightarrow[r\to\infty]{} 0$. Thus there is a number $r^* > \max(k, i)$ and an $\varepsilon > 0$ such that

$$q = \varrho(H_{r^*}) + \varepsilon < 1. \tag{Ap 3.9}$$

3. Let $\mathbb{C}$ be the set of complex numbers and $M(k, k)$ the set of all matrices of order k with complex members. In view of (Ap 3.9), there is a norm $\|\cdot\|_*$ on $\mathbb{C}^k$ such that

$$\|H_{r^*}\|_* \leqq q < 1, \tag{Ap 3.10}$$

where

$$\|C\|_* = \sup\{\|Cz\|_* : \|z\|_* = 1, z \in \mathbb{C}^k\}, \qquad C \in M(k, k),$$

is the norm in the set $M(k, k)$ induced by $\|\cdot\|_*$. Since a_r is non-increasing,

$$0 \leqq H_r \leqq H_{r^*}, \qquad r \geqq r^*, \tag{Ap 3.11}$$

where 0 denotes the matrix of order k in which all elements are zero. (Here $\leqq$ is the usual semi-ordering relation for matrices with real members, i.e. for $A = (a_{ij})$, $B = (b_{ij})$ with $a_{ij}, b_{ij} \in \mathbb{R}$: $A \leqq B$ iff $a_{ij} \leqq b_{ij}$ for all i, j.) From (Ap 3.8) and (Ap 3.11) for $r \geqq r^*$ we obtain

$$z_r = H_r H_{r-1} \cdots H_{r^*} H_{r^*-1} \cdots H_1 z_0 \leqq (H_{r^*})^{r-r^*+1} H_{r^*-1} \cdots H_1 z_0. \tag{Ap 3.12}$$

By means of (A 3.12), the series

$$\sum_{r=i}^{\infty} \binom{r}{i} z_r$$

can be majorized as follows:

$$\sum_{r=i}^{\infty} \binom{r}{i} z_r \leq \frac{1}{i!} \sum_{r=i}^{r^*-1} r^i H_r \cdots H_1 z_0$$

$$+ \frac{1}{i!} \sum_{r=r^*}^{\infty} r^i (H_{r^*})^{r-r^*+1} H_{r^*-1} \cdots H_1 z_0 . \qquad \text{(Ap 3.13)}$$

Now we show the finiteness of the right-hand side of (Ap 3.13). Obviously, the first summand is finite. Because of (Ap 3.10), for the second summand we get

$$\left\| \frac{1}{i!} \sum_{r=r^*}^{\infty} r^i (H_{r^*})^{r-r^*+1} H_{r^*-1} \cdots H_1 z_0 \right\|_*$$

$$\leq \frac{1}{i!} \|H_{r^*-1} \cdots H_1 z_0\|_* \sum_{r=r^*}^{\infty} r^i q^{r-r^*+1} . \qquad \text{(Ap 3.14)}$$

Since $q < 1$, we conclude by D'Alembert's criterion that the series of the right-hand side of (Ap 3.14) is finite. Since all norms in $\mathbb{R}^k$ are equivalent, the series

$$\sum_{r=i}^{\infty} \binom{r}{i} z_r$$

is finite, which finishes the proof. ∎

A 3.2. Lorden's renewal inequality

Lorden (1970) gave the following estimate for the renewal function.

A 3.2.1. Theorem. *Suppose* $(T_n, n \geqq 1)$, $T_0 = 0$, $T_{n+1} - T_n = A_n$, $n \geqq 0$, *is a renewal process generated by the sequence* $(A_n, n \geqq 0)$ *of non-negative i.i.d. r.v.'s with mean* $m_A = \mathsf{E}A_0 < \infty$ *and finite second moment* $\mathsf{E}A_0^2 < \infty$. *Then we have the following upper bound for the renewal function* $H(t) = \mathsf{E}N(t)$, $N(t) = \max \{n: T_n \leqq t\}$:

$$H(t) \leqq \frac{t}{m_A} + \frac{\mathsf{E}A_0^2}{m_A^2} - 1 .$$

Proof. We follow Carlsson and Nerman (1986). Let $\tilde{N}(t) = \big(N(t) + 1\big) \times 1\{t \geqq 0\} = \# \{n \geqq 0: T_n \leqq t\}, t \in \mathbb{R}$, and $\tilde{H}(t) = \mathsf{E}\tilde{N}(t) = \big(H(t) + 1\big) 1\{t \geqq 0\}$. It is easy to see that

$$\tilde{H}(t + s) \leqq \tilde{H}(t) + \tilde{H}(s) \qquad \text{(Ap 3.15)}$$

(i.e. $\tilde{H}(t)$ is subadditive). Namely, the conditional expectation of $\tilde{N}(t + u) - \tilde{N}(t)$ given $\tilde{N}(t) = n$, $A_0 = a_0, \ldots, A_{n-1} = a_{n-1}$ equals $\mathsf{E}\tilde{N}(u - r)$, where $r = a_0 + \cdots + a_{n-1} - t > 0$ and $\tilde{N}(u - r)$ is zero if $r > u$, so that $\mathsf{E}\tilde{N}(u - r) \leqq \mathsf{E}\tilde{N}(u)$ in any case, cf. Lorden (1970).

For simplicity assume that $m_A = 1$. Let $D^{(1)}, D^{(2)}$ be independent r.v.'s with the stationary delay distribution

$$F_0(x) = m_A^{-1} \int_0^x \big(1 - F(u)\big) \, du .$$

By Theorem 3.3.1 (b) we have

$$\mathsf{E}\tilde{H}(t - D^{(i)}) = t, \qquad t \geqq 0, \tag{Ap 3.16}$$

where we use the convention $\tilde{H}(u) = 0$ if $u < 0$. By the subadditive property (Ap 3.15)

$$\begin{aligned}\tilde{H}(t) &= \mathsf{E}\big(\tilde{H}(t + D^{(1)} - D^{(2)} + D^{(2)} - D^{(1)})\big)\\ &\leqq \mathsf{E}\big(\tilde{H}(t + D^{(1)} - D^{(2)})\big) + \mathsf{E}\big(\tilde{H}(D^{(2)} - D^{(1)})\big).\end{aligned}$$

Together with (Ap 3.16) this implies

$$\begin{aligned}\tilde{H}(t) &\leqq \mathsf{E}\left(\mathsf{E}\big(\tilde{H}(t + D^{(1)} - D^{(2)}) \mid D^{(1)}\big)\right)\\ &\qquad + \mathsf{E}\left(\mathsf{E}\big(\tilde{H}(D^{(2)} - D^{(1)}) \mid D^{(2)}\big)\right)\\ &= \mathsf{E}(t + D^{(1)}) + \mathsf{E}D^{(2)} = t + 2\mathsf{E}D^{(1)}\\ &= t + \mathsf{E}A_0^{\,2}\end{aligned}$$

as desired. ∎

A 3.2.2. Remark. For extensions to random walks with positive drift (i.e. the A_i are allowed to be negative) and for estimates of the forward recurrence time cf. LORDEN (1970).

A 3.3. The Markov renewal equation

A 3.3.1. Theorem. *Suppose* $([T_n, K_n], n \geqq 1)$ *is a Markov renewal process with a countable state space* $\mathbb{K}$ *and parameters* $q_i, p_{ij}, F_{ij}^0(x), F_{ij}(x), i, j \in \mathbb{K}$. *Then the renewal functions*

$$H_j(x) = \mathsf{E}\,\#\{n \geqq 1: T_n \leqq x, K_n = j\}, \qquad j \in \mathbb{K},$$

satisfy the following Markov renewal equations:

$$H_j(x) = \sum_{i\in\mathbb{K}} q_i p_{ij} F_{ij}^0(x) + \sum_{l\in\mathbb{K}} p_{lj} \int_0^x F_{lj}(x - u)\,\mathrm{d}H_l(u), \qquad j \in \mathbb{K}.$$

Proof. We set $G_{mn}(u) = p_{mn}F_{mn}(u)$,

$$G_m{}^0(u) = \sum_{i\in\mathbb{K}} q_i p_{im} F_{im}^0(u)$$

and

$$J(r, k, l) = \left\{\mathbf{l} = [l_1, \ldots, l_r] \in \mathbb{K}^r: l_r = l, \sum_{s=1}^{r} \mathbf{1}\{l_s = l\} = k\right\}$$

for $r, k \geqq 1$ and $l \in \mathbb{K}$.

Then we have

$$\sum_{i\in\mathbb{K}} q_i p_{ij} F_{ij}^0(x) + \sum_{l\in\mathbb{K}} p_{lj} \int_0^x F_{lj}(x-u)\,\mathrm{dH}_l(u)$$

$$= G_j^0(x) + \sum_{l\in\mathbb{K}} \int_0^x H_l(x-u)\,\mathrm{d}G_{lj}(u)$$

$$= G_j^0(x) + \sum_{l\in\mathbb{K}} \int_0^x \sum_{k=1}^{\infty} \mathsf{P}\big(N_l(x-u) \geqq k\big)\,\mathrm{d}G_{lj}(u)$$

$$= G_j^0(x) + \sum_{k=1}^{\infty} \sum_{l\in\mathbb{K}} \int_0^x \sum_{r=1}^{\infty} \sum_{l\in J(r,k,l)} G_{l_1}^0 * G_{l_1 l_2} * \cdots * G_{l_{r-1} l_r}(x-u)\,\mathrm{d}G_{lj}(u)$$

$$= G_j^0(x) + \sum_{k=1}^{\infty} \sum_{l\in\mathbb{K}} \sum_{r=1}^{\infty} \sum_{l\in J(r,k,l)} G_{l_1}^0 * G_{l_1 l_2} * \cdots * G_{l_{r-1} l_r} * G_{l_r j}(x)$$

$$= G_j^0(x) + \sum_{r=1}^{\infty} \sum_{l_1,\ldots,l_r\in\mathbb{K}} G_{l_1}^0 * G_{l_1 l_2} * \cdots * G_{l_{r-1} l_r} * G_{l_r j}(x)$$

$$= G_j^0(x) + \sum_{r=1}^{\infty} \sum_{l_1,\ldots,l_{r+1}\in\mathbb{K},\, l_{r+1}=j} G_{l_1}^0 * G_{l_1 l_2} * \cdots * G_{l_r l_{r+1}}(x)$$

$$= G_j^0(x) + \sum_{r=2}^{\infty} \sum_{l_1,\ldots,l_r\in\mathbb{K},\, l_r=j} G_{l_1}^0 * G_{l_1 l_2} * \cdots * G_{l_{r-1} l_r}(x)$$

$$= \sum_{r=1}^{\infty} \sum_{l_1,\ldots,l_r\in\mathbb{K},\, l_r=j} G_{l_1}^0 * G_{l_1 l_2} * \cdots * G_{l_{r-1} l_r}(x)$$

$$= \sum_{k=1}^{\infty} \sum_{r=1}^{\infty} \sum_{l\in J(r,k,j)} G_{l_1}^0 * G_{l_1 l_2} * \cdots * G_{l_{r-1} l_r}(x)$$

$$= \sum_{k=1}^{\infty} \mathsf{P}\big(N_j(x) \geqq k\big) = H_j(x). \quad \blacksquare$$

A 3.4. Stopped sequences of i.i.d. variables

In order to prove that a shifted renewal process is again a renewal process with the same underlying distribution function $F(x)$, cf. Section 3.3, one needs the following theorem, cf. Borovkov (1976).

A 3.4.1. Theorem. *Let $(X_n, n \geqq 1)$ be a sequence of independent r.v.'s, identically distributed for $n \geqq 2$ and ν a stopping time, i.e. the event $\{\nu \leqq n\}$ belongs to the σ-field generated by $X_1, \ldots, X_n$, $n = 1, 2, \ldots$ Then $[\nu, X_1, \ldots, X_\nu]$ and $(X_{\nu+n}, n \geqq 1)$ are independent, and the sequences $(X_{\nu+n}, n \geqq 1)$ and $(X_n, n \geqq 2)$ have the same distribution.*

Proof. For all $k \geqq 1$ and all measurable sets $C_1, \ldots, C_k, B_1, B_2, \ldots$ we have

$$\begin{aligned}
&\mathsf{P}\big(\nu \in (\cdot), X_1 \in B_1, \ldots, X_\nu \in B_\nu, X_{\nu+1} \in C_1, \ldots, X_{\nu+k} \in C_k\big)\\
&= \sum_{j \in (\cdot)} \mathsf{P}(\nu = j, X_1 \in B_1, \ldots, X_j \in B_j, X_{j+1} \in C_1, \ldots, X_{j+k} \in C_k)\\
&= \sum_{j \in (\cdot)} \mathsf{P}(\nu = j, X_1 \in B_1, \ldots, X_j \in B_j)\, \mathsf{P}(X_{j+1} \in C_1, \ldots, X_{j+k} \in C_k),
\end{aligned}$$

since (by the definition of a stopping time) the event $\{\nu = j, X_1 \in B_1, \ldots, X_j \in B_j\}$ is independent of $\{X_{j+1} \in C_1, X_{j+2} \in C_2, \ldots\}$. Now we can continue as follows:

$$\begin{aligned}
&= \sum_{j \in (\cdot)} \mathsf{P}(\nu = j, X_1 \in B_1, \ldots, X_\nu \in B_\nu)\, \mathsf{P}(X_2 \in C_1, \ldots, X_{k+1} \in C_k)\\
&= \mathsf{P}\big(\nu \in (\cdot), X_1 \in B_1, \ldots, X_\nu \in B_\nu\big)\, \mathsf{P}(X_2 \in C_1, \ldots, X_{k+1} \in C_k).
\end{aligned}$$

From this we obtain $[X_{\nu+1}, \ldots, X_{\nu+k}] \stackrel{\mathfrak{D}}{=} [X_2, \ldots, X_{k+1}]$ and thus $(X_{\nu+n}, n \geqq 1) \stackrel{\mathfrak{D}}{=} (X_n, n \geqq 2)$ and

$$\begin{aligned}
&\mathsf{P}\big(\nu \in (\cdot), X_1 \in B_1, \ldots, X_\nu \in B_\nu, X_{\nu+1} \in C_1, \ldots, X_{\nu+k} \in C_k\big)\\
&= \mathsf{P}\big(\nu \in (\cdot), X_1 \in B_1, \ldots, X_\nu \in B_\nu\big)\, \mathsf{P}(X_{\nu+1} \in C_1, \ldots, X_{\nu+k} \in C_k). \quad \blacksquare
\end{aligned}$$

References

AFANAS'EVA, L. G. (1965). On the existence of a limit distribution in a queueing system with bounded sojourn time (Russian). Teor. Veroyatnost. i Primenen. **10**, 570 to 578.

AFANAS'EVA, L., G., and A. V. MARTYNOV (1969). The ergodic properties of a queueing system with bounded sojourn time (Russian). Teor. Veroyantnost. i Primenen. **14**, 102–112.

AKHMAROV, I. (1979a). Ergodicity and stability of multichannel queueing systems with a limited waiting time (Russian). Sibirsk. Mat. Zh. **20**, 911–916.

— (1979b). The rate of convergence in the ergodicity and continuity theorems for multichannel queueing systems (Russian). Teor. Veroyatnost. i Primenen. **24**, 418–424.

— (1980). The convergence rate in continuity theorems for systems with an infinite number of service channels (Russian). Sibirsk. Mat. Zh. **21**, 16–21.

— (1981). The rate of convergence in ergodicity and continuity theorems for systems with refusals (Russian). Teor. Veroyatnost. i Primenen. **26**, 182–189.

AKHMAROV, I., and N. P. LEONTYEVA (1976). Conditions for convergence to limit processes and the strong law of large numbers for queueing systems (Russian). Teor. Veroyatnost. i Primenen. **21**, 559–570.

ARNDT, K., and P. FRANKEN (1977). Random point processes applied to availability analysis of redundant systems with repair. IEEE Trans. Reliability **R-26**, 266–269.

— (1979). Construction of a class of stationary processes with applications in reliability. Zastos. mat. **16**, 379–393.

ARNDT, U. (1978). A continuity theorem for stationary distributions in general m server queues. Elektron. Informationsverarb. Kybernetik **14**, 7/8, 385–391.

— (1980). A remark on the paper "A continuity theorem for stationary distributions in general m server queues". Elektron. Informationsverarb. Kybernetik **16**, 5/6, 301–302.

ARNOLD, L., and V. WIHSTUTZ (1982). Stationary solutions of linear systems with additive and multiplicative noise. Stochastics **7**, 133–155.

ASH, R. B., and M. F. GARDNER (1975). Topics in Stochastic Processes. Academic Press, New York.

ASMUSSEN, S., and H. JOHANSEN (1986). Über eine Stetigkeitsfrage betreffend das Bedienungssystem GI/GI/s. Elektron. Informationsverarb. Kybernetik **22**, 565–570.

ATHREYA, K. B., D. MCDONALD, and P. NEY (1978). Limit theorems for semi-Markov processes and renewal theory for Markov chains. Ann. Probab. **6**, 788–797.

BACCELLI, F. (1989). Ergodic theory of stochastic Petri networks. Technical report No. 1037, INRIA — Sophia Antipolis.

BACCELLI, F., P. BOYER, and G. HEBUTERNE (1984). Single server queues with impatient customers. Adv. in Appl. Probab. **16**, 887–904.

BACCELLI, F., and P. BRÉMAUD (1987). Palm Probabilities and Stationary Queues. Lecture Notes in Statistics, vol. 41, Springer-Verlag, Berlin—Heidelberg—New York.

BARLOW, R. E., and F. PROSCHAN (1975). Statistical Theory of Reliability and Life Testing. Probability Models. Holt, Rinehart and Winston Inc. (German Edition, Akademie-Verlag, Berlin, 1978).

BAUER, H, (1974). Wahrscheinlichkeitstheorie und Grundzüge der Maßtheorie. 2nd ed. Walter de Gruyter, Berlin, New York.

BEICHELT, F., and P. FRANKEN (1983). Zuverlässigkeit und Instandhaltung. VEB Verlag Technik, Berlin. (Carl Hanser Verlag, München—Wien, 1984; Russian Edition, revised, Radio i Svyaz, Moscow, 1988).

BEREZNER, S. A., and V. A. MALYSHEV (1989). The stability of infinite-server networks with random routing. J. Appl. Probab. **26**, 363—371.

BILLINGSLEY, P. (1968). Convergence of Probability Measures. Wiley, New York.

BOROVKOV, A. A. (1972a). Stochastic Processes in Queueing Theory (Russian). Nauka, Moscow (Engl. Edition, revised, Springer-Verlag, New York—Berlin, 1976).

— (1972b). Continuity theorems for multichannel systems with losses (Russian). Teor. Veroyatnost. i Primenen. **17**, 458—468.

— (1972c). Convergence of the distributions of functionals of sequence and processes given on the entire axis (Russian). Trudy Mat. Inst. Steklov **128**, 41—65.

— (1976). Wahrscheinlichkeitstheorie. Akademie-Verlag, Berlin.

— (1977). Some estimates of the convergence rate in stability theorems (Russian). Teor. Veroyatnost. i Primenen. **22**, 689—699.

— (1978). Ergodic and stability theorems for a class of stochastic equations and their applications (Russian). Teor. Veroyatnost. i Primenen. **23**, 241—262.

— (1980). Asymptotic Methods in Queueing Theory (Russian). Nauka, Moscow. (Engl. Transl., Wiley, Chichester, 1984.)

— (1986). Limit theorems for queueing networks I (Russian). Teor. Veroyatnost. i Primenen. **31**, 474—490.

— (1987). Limit theorems for queueing networks. II (Russian). Teor. Veroyatnost. i. Primenen. **32**, 282—298.

— (1988). On ergodicity and stability properties of the sequence $w_{n+1} = f(w_n, \xi_n)$. Applications to communication networks (Russian). Theor. Veroyatnost. i Primenen. **33**, 641—658.

— (1989). A phenomenon of asymptotic stabilization for the decentralized ALOHA-algorithm. Diffusion approximation (Russian). Problemy Peredači Informacii. **25**, 55—64.

BRANDT, A. (1983). Qualitative Untersuchungen der schwachen Lösungen und Stetigkeitssätze für das m-linige Wartesystem. Diss. A, Humboldt-Universität zu Berlin.

— (1985a). On stationary waiting times and limiting behaviour of queues with many servers I: The general G/G/m/∞ case. Elektron. Informationsverarb. Kybernetik **21**, 47—64.

— (1985b). On stationary waiting times and limiting behaviour of queues with many serves II: The G/GI/m/∞ case. Elektron. Informationsverarb. Kybernetik **21**, 151—162.

— (1986). The stochastic equation $Y_{n+1} = A_n Y_n + B_n$ with stationary coefficients. Adv. in Appl. Probab. **18**, 211—220.

— (1987a). On the GI/M/∞ queue with batch arrivals. Preprint Nr. 134/1987, Sektion Mathematik, Humboldt-Universität zu Berlin.

— (1987b). On stationary queue length distributions for G/M/s/r queues. Queueing Systems **2**, 321—332.

Brandt A. (1989). On the GI/M/∞ queue with batch arrivals and several kinds of service distributions. Queueing Systems 4, 351–365.

Brandt, A., P. Franken, and B. Lisek (1984). Ergodicity and steady state existence. Continuity of stationary distributions of queueing characteristics, Lecture Notes in Control and Information Sciences, vol. 60, Springer-Verlag, Berlin—Heidelberg—New York, 275–296.

— (1986). Stationary queueing systems. Existence and stability. Invited Paper held on 1. World-Congress of the Bernoulli-Society, Taschkent, September 1986, VNU Science Press 1988.

Brandt, A., and H. R. Künsch (1988). On the stability of robust filter-cleaners. Stoch. Process. Appl. 30, 253–262.

Brandt, A., and B. Lisek (1980). On the approximation of GI/GI/m/0 by means of GI/GI/∞. Elektron. Informationsverarb. Kybernetik 16, 597–600.

Brandt, A., and B. Lisek (1983). On the continuity of G/GI/m/∞ queues. Math. Operationsforsch. Statist., Ser. Statist. 14, 577–587.

Brandt, A., B. Lisek, and O. Nerman (1986). On stationary Markov chains and independent random variables. Preprint 17/86, Dept. of Mathematics, Chalmers University of Technology and The University of Göteborg.

Brandt, A., and H. Sulanke (1986). On the calculation and approximation of the mean time between failure of a system with two and three component subsystems. Elektron. Informationsverarb. Kybernetik 22, 635–645.

— (1987). On the GI/M/∞ queue with batch arrivals of constant size. Queueing Systems 2, 187–200.

— (1988). Reliability analysis of a standby system with repair and intermediate stocks for failed and repaired units. (Submitted to Applied Stochastic Models and Data Analysis.)

Brémaud, P. (1981). Point Processes and Queues: martingale dynamics. Springer-Verlag, Berlin—Heidelberg—New York.

— (1989). Characteristics of queueing systems observed at events and at the connection between stochastic intensity and Palm probability. Queueing Systems, to appear.

— (1990). Necessary and sufficient condition for the equality of event averages and time averages. (Submitted for publication.)

Brown, J. R. (1976). Ergodic Theory and Topological Dynamics. Academic Press, New York.

Brown, M., and S. M. Ross (1972). Asymptotic properties of cumulative processes. SIAM J. Appl. Math. 22, 93–105.

Brumelle, S. L. (1971a). Some inequalities for parallel-server queues. Oper. Res. 19, 402–413.

— (1971b). On the relation between customer and time averages in queues. J. Appl. Probab. 8, 508–520.

— (1972). A generalization of $L = \lambda W$ to moments of queue length and waiting times. Oper. Res. 20, 1127–1136.

Burke, P. J. (1975). Delays in single-server queues with batch input. Oper. Res. 23, 830–833.

— (1976). Proof of a conjecture on the interarrival time distribution in an M/M/1 queue with feedback. IEEE Trans. Comm. C-24, 575–576.

Carlson, H., and O. Nerman (1986). An alternative proof of Lorden's renewal inequality. Adv. in Appl. Probab. 18, 1015–1016.

Chaudhry, M. L. (1979). The queueing system M^X/G/1 and its ramifications. Naval Res. Logist. Quart. 26, 667–674.

CHAUDHRY, M. L., and J. G. C. TEMPLETON (1983). A First Course in Bulk Queues. Wiley, New York.

ÇINLAR, E. (1969). On semi-Markov processes on arbitrary spaces. Proc. Cambridge Philos. Soc. **66**, 381–392.

— (1975). Introduction to Stochastic Processes. Prentice-Hall, Englewood Cliffs, New Jersey.

COHEN, J. W. (1976a). On a single server queue with group arrivals. J. Appl. Probab. **13**, 619–622.

— (1976b). On Regenerative Processes in Queueing Theory. Lecture Notes in Economics Math. Systems, vol. 121, Springer-Verlag, Berlin–Heidelberg–New York.

— (1977). On up- and downcrossings. J. Appl. Probab. **14**, 405–410.

CRABILL, T. B. (1968). Sufficient conditions for positive recurrence and recurrence of specially structured Markov chains. Oper. Res. **16**, 858–867.

DELLACHERIE, C., and P.-A. MEYER (1978). Probabilities and Potential. North-Holland, Amsterdam.

DEUL, N. (1982). The influence of the preserverance function in queueing systems with repeated calls. Elektron. Informationsverarb. Kybernetik **18**, 587–594.

DEWESS, M. (1975). Erhaltungssätze im Wartemodell G/G/1 mit „Erwärmung". Math. Operationsforsch. Statist. **6**, 427–436.

DISNEY, R. L., and D. KÖNIG (1985). Queueing networks: a survey of their random processes. SIAM Rev. **27**, 335–403.

DISNEY, R. L., D. C. MCNICKLE, and B. SIMON (1980). The M/G/1 queue with instantaneous Bernoulli feedback. Naval Res. Logist. Quart. **27**, 635–644.

DO LE MINH (1980). Analysis of the exceptional queueing system by the use of regenerative processes and analytical methods. Math. Oper. Res. **5**, 147–159.

DOOB, J. L. (1953). Stochastic Processes. Wiley, New York.

DOORN, E. VAN (1981). Renewal traffic with batch arrivals and exponential holding times. Dr. Neher Lab. TR INF/122.

DOORN, E. VAN, and G. J. K. REGTERSCHOT (1988). Conditional PASTA. Oper. Res. Letters **7**, 229–232.

DOSHI, B. T. (1985). A note on stochastic decomposition in a GI/G/1 queue with vacations or set up times. J. Appl. Probab. **22**, 419–428.

— (1986). Queueing systems with vacations — a survey. Queueing Systems **1**, 29–66.

FALIN, G. I. (1987). On the convergence of queueing processes in composite systems to stationary ones. Teor. Veroyatnost. i Primenen. **32**, 577–580.

FASSLER, R. (1982). Offene Systeme und ihre Anwendungen in der Bedienungstheorie. Diss. A, Friedrich-Schiller-Universität, Jena.

FELLER, W. (1971). An Introduction to Probability Theory and its Applications. Vol. I, II. 3rd ed., Wiley, New York.

FICHTNER, K.-H. (1979a). Stationäre Folgen von Zufallsvariablen als Lösung unendlicher Gleichungssysteme. Math. Nachr. **88**, 255–278.

— (1979b). Abhängige Verschiebungen von Punktprozessen I. Math. Nachr. **91**, 197 bis 244.

— (1979c). Abhängige Verschiebungen von Punktprozessen II. Math. Nachr. **92**, 31–89.

FICHTNER, K.-H., and W. FREUDENBERG (1980). Asymptotic behaviour of time evolutions of infinite particle systems. Z. Wahrsch. verw. Gebiete **54**, 141–159.

FINCH, P. D. (1959a). A probability limit theorem with applications to a generalization of queueing theory. Acta Math. Acad. Sci. Hung. **10**, 317–325.

— (1959b). On the distribution of queue size in queueing problems. Acta Math. Acad. Sci. Hung. **10**, 327–336.

[FKAS], cf. FRANKEN, P., D. KÖNIG, U. ARNDT, and V. SCHMIDT (1981).

FOLEY, R. D., and R. L. DISNEY (1983). Queues with delayed feedback. Adv. in Appl. Probab. **15**, 162–182.

FOSS, S. G. (1980). On the approximation of multichannel service systems (Russian). Sibirsk. Mat. Zh. **21**, 132–140.

— (1981). Comparison of queueing disciplines in multichannel systems with waiting (Russian). Sibirsk. Mat. Zh. **22**, 190–197.

— (1983). Ergodicity conditions in multiserver queueing systems with waiting time (Russian). Sibirsk. Mat. Zh. **24**, 168–175.

— (1985). On a method of estimating the convergence rate for the ergodicity and continuity for many server queueing systems. In: Limit Theorems of Probability Theory; Proceedings of the Institute of Mathematics. Acad. of Sciences of USSR, Sibirian Branch. Novosibirsk. 126–137.

— (1986). The method of renovating events and its applications in queueing theory. In: Proceedings of an International Symposium on Semi-Markov Processes and their Applications. Ed. J. Janssen, Plenum Press, New York and London, 337 to 350.

FOX, B. L., and P. W. GLYNN (1987). Estimating time averages via randomly spaced observations. SIAM J. Appl. Math. **47**, 186–200.

FRANKEN, P. (1969). Beiträge zur Bedienungstheorie. Habilitationsschrift, Friedrich-Schiller-Universität, Jena.

— (1970). Ein Stetigkeitssatz für Verlustsysteme. In: Operationsforschung und Mathematische Statistik, vol. 2, ed. O. BUNKE, Akademie-Verlag, Berlin, pp. 9–23.

— (1975). Stationary probabilities of states of queueing systems at different times. Engrg. Cybernetics **13**, 1, 84–89.

— (1976). Einige Anwendungen der Theorie zufälliger Punktprozesse in der Bedienungstheorie I., Math. Nachr. **70**, 303–319.

— (1982). The Point Process Approach to Queueing Theory and Related Topics. Seminarbericht Nr. 43, Sektion Mathematik, Humboldt-Universität zu Berlin.

FRANKEN, P., and U. KALÄHNE (1978). Existence, uniqueness and continuity of stationary distributions for queueing systems without delay. Math. Nachr. **86**, 97–115.

— (1980): A remark on the paper "Existence, uniqueness and continuity of stationary distributions for queueing systems without delay". Math. Nachr. **97**, 344.

FRANKEN, P., and J. KERSTAN (1968). Bedienungssysteme mit unendlich vielen Bedienungsapparaten. In: Operationsforschung und Mathematische Statistik, vol. 1, ed. O. BUNKE, Akademie-Verlag, Berlin, pp. 67–76.

FRANKEN, P., B.-M. KIRSTEIN, and A. STRELLER (1984). Reliability analysis of complex systems with repair. Elektron. Informationsverarb. Kybernetik **20**, 407 to 422.

FRANKEN, P., D. KÖNIG, U. ARNDT, and V. SCHMIDT (1981). Queues and Point Processes. Akademie-Verlag, Berlin; Wiley, Chichester, 1982. (Russian Edition, revised and enlarged, Naukova Dumka, Kiev, 1984).

FRANKEN, P., and B. LISEK (1982). On Wald's identity for dependent variables. Z. Wahrsch. verw. Gebiete **60**, 143–150.

FRANKEN, P., and A. STRELLER (1979). Stationary generalized regenerative processes (Russian). Teor. Veroyatnost. i Primenen. **24**, 78–90.

— (1980). Reliability analysis of complex repairable systems by means of marked point processes. J. Appl. Probab. **17**, 154–167.

FUHRMANN, S. W. (1984). A note on the M/G/1 queue with server vacations. Oper. Res. **32**, 1368–1373.

FUHRMANN, S. W., and R. B. COOPER (1985). Stochastic decompositions in the M/G/1 queue with generalized vacations. Oper. Res. **33**, 1117–1129.

GALLIHER, H. P., P. M. MORSE, and M. SIMOND (1959). Dynamics of two classes of continuous review inventory systems. Oper. Res. **7**, 362–384.

GITTINS, J. C. (1978). A comparison of service disciplines for GI/G/m queues. Math. Operationsforsch. Statist., Ser. Optim. **9**, 255–260.

GNEDENKO, B. W., and D. KÖNIG (ed.) (1984). Handbuch der Bedienungstheorie, vol. II. Akademie-Verlag, Berlin.

GRANDELL, J. (1976). Doubly Stochastic Poisson Processes. Lecture Notes in Mathematics, vol. 529, Springer-Verlag, Berlin–Heidelberg–New York.

— (1977). Point processes and random measures. Adv. in Appl. Probab. **9**, 502–506.

GUTJAHR, R. (1977a). Beiträge zur Theorie von Bedienungssystemen mit allgemeinem einlaufenden Verkehr. Diss. A, Friedrich-Schiller-Universität, Jena.

— (1977b). Eine Bemerkung zur Existenz unendlich vieler „Leerpunkte" in Bedienungssystemen mit unendlich vielen Bedienungsgeräten. Math. Operationsforsch. Statist., Ser. Optim. 8, 245–251.

HALFIN, S. (1983). Batch delays versus customer delays. Bell System Tech. J. **62**, 2011–2015.

HANSON, D. L. (1963). On the representation problem for stationary stochastic processes with trivial tail field. J. Math. Mech. **12**, 293–301.

HARDY, G. H., I. E. LITTLEWOOD, and G. POLYA (1952). Inequalities. University Press, Cambridge.

HARRIS, R. (1974). The expected number of idle servers in a queueing system. Oper. Res. **22**, 1258–1259.

HASMINSKIJ, R. Z. (1969). Stochastic Stability of Differential Equations (Russian). Moscow. (Engl. Transl., Sijthoff and Noordhoff, 1980).

HEYMAN, D., and S. STIDHAM jr. (1980). The relation between customer and time averages in queues. Oper. Res. **28**, 983–993.

HOUSEHOLDER, A. S. (1964). The Theory of Matrices in Numerical Analysis. Blaisdell, New York.

HOLMAN, D. F., M. L. CHAUDHRY, and B. R. K. KASHYAP (1982). On the number in the system $GI^X/M/\infty$. Sankhya **44**, Ser. A Pt. 1, 294–297.

ISAACSON, D., and R. L. TWEEDIE (1978). Criteria for strong ergodicity of Markov chains. J. Appl. Probab. **15**, 87–95.

IVANOFF, B. G. (1980). The functions space $D([0, \infty)^q, E)$. Canad. J. Statist. 8, 179–191.

IV'NICKIJ, V. A. (1971). On a single-server queue (Russian). Izv. Akad. Nauk SSSR, Tekhn. Kibernet. **6**, 110–120.

JACOBSEN, M. (1982). Statistical Analysis of Counting Processes. Lecture Notes in Statistics, vol. 12, Springer-Verlag, Berlin–Heidelberg–New York.

JAGERS, P. (1973). On Palm probabilities. Z. Wahrsch. verw. Gebiete **26**, 17–32.

JEWELL, W. (1967). A simple proof of $L = \lambda W$. Oper. Res. **15**, 1109–1116.

JOFFE, A., and P. E. NEY (1963). Convergence theorem for multiple channel loss probabilities. Ann. Math. Statist. **34**, 260–273.

KALÄHNE, U. (1976). Existence, uniqueness and some invariance properties of stationary distributions for general single server queues. Math. Operationsforsch. Statist. **7**, 557–575.

KALASHNIKOV, V. V. (1977). Analysis of stability in queuing problems by a method of trial functions (Russian). Teor. Veroyatnost. i Primenen. **22**, 89–105.

— (1983). A complete metric in the function space $D(0, \infty)$ and its applications. Lecture Notes in Mathematics, vol. 982, 60–76. Springer-Verlag, Berlin–Heidelberg–New York–Tokyo.

KALLENBERG, O. (1975). Random Measures. Akademie-Verlag, Berlin; Academic Press, London, 1976; Third Edition, revised and enlarged, Akademie-Verlag, Berlin, and Academic Press, London 1983.

KAPLAN, N. (1975). Limit theorems for a GI/G/∞ queue. Ann. Probab. **3**, 780–789.

KARR, A. F. (1986). Point Processes and their Statistical Inference. Marcel Dekker Inc., New York–Basel.

KEILSON, J. (1974). Monotonicity and Convexity in System Survival Functions and Metabolic Disappearance Curves. In: Reliability and Biometry, SIAM, Philadelphia.

KELBERT, M. YA., and YU. M. SUKHOV (1983). Existence and uniqueness conditions for a random field describing the state of a switching network. Problems Inf. Transmission **19**, 289–304.

— (1985). Weak dependence of random field describing state of a switching network with small transit flows. Problems Inf. Transmission **21**, 237–245.

KENNEDY, D. P. (1972). The continuity of the single server queue. J. Appl. Probab. **9**, 370–381.

KERSTAN, J., K. MATTHES, and J. MECKE (1974). Unbegrenzt teilbare Punktprozesse. Akademie-Verlag, Berlin. (Engl. Edition, revised and enlarged; MATTHES, K., J. KERSTAN, and J. MECKE: Infinitely Divisible Point Processes. Wiley-Intersciences, New York–London–Sydney–Toronto, 1978; Russian Edition, revised and enlarged, Nauka, Moscow, 1982.)

KHINCHIN, A. YA. (1955). Mathematical methods in queueing theory (Russian). Trudy Mat. Inst. Steklov **49**, 1–122. (Engl. transl.: Mathematical Methods in the Theory of Queueing. Griffin, London, 1960.)

KIEFER, J., and J. WOLFOWITZ (1955). On the theory of queues with many servers. Trans. Amer. Math. Soc. **78**, 1–18.

KLEINER, B., R. D. MARTIN, and D. J. THOMSON (1979). Robust estimation of power spectra (with discussion). J. Royal Statist. Soc., Ser. B **41**, 313–351.

[KMM], cf. KERSTAN, J., K. MATTHES, and J. MECKE (1974).

[KMN], cf. KÖNIG, D., K. MATTHES, and K. NAWROTZKI (1971).

KÖNIG, D., K. MATTHES, and K. NAWROTZKI (1967). Verallgemeinerungen der Erlangschen und Engsetschen Formeln (Eine Methode in der Bedienungstheorie). Akademie-Verlag, Berlin.

KÖNIG, D., K. MATTHES, and K. NAWROTZKI (1971, 1974). Unempfindlichkeitseigenschaften von Bedienungsprozessen. Appendix to: GNEDENKO, B. W., and I. N. KOWALENKO. Einführung in die Bedienungstheorie. Akademie-Verlag, Berlin.

KÖNIG, D., T. ROLSKI, V. SCHMIDT, and D. STOYAN (1978). Stochastic processes with imbedded marked point processes (PMP) and their applications in queueing. Math. Operationsforsch. Statist., Ser. Optim. **9**, 125–141.

KÖNIG, D., and V. SCHMIDT (1978). Relationships between time- and customer-stationary characteristics of service systems. Proc. Conf. on Point Processes and Queueing Theory. Keszthely. North-Holland Publ. Comp., Amsterdam, 181–225.

— (1980a). Imbedded and non-imbedded stationary characteristics of queueing systems with varying service rate and point processes. J. Appl. Probab. **17**, 753–767.

— (1980b). Stochastic inequalities between customer-stationary and time-stationary characteristics of queueing systems with point processes. J. Appl. Probab. **17**, 768–777.

KÖNIG, D., V. SCHMIDT and E. A. VAN DOORN (1989). On the "PASTA" property and a further relationship between customer and time averages in stationary queueing systems. Stochastic Models, to appear.

König, D., V. Schmidt, and D. Stoyan (1976). On some relations between stationary distributions of queue lengths and imbedded queue lengths in G/G/s queueing systems. Math. Operationsforsch. u. Statist. 7, 577–586.

Kotzurek, M. (1978). Beiträge zum Problem der Stabilität in der Bedienungstheorie. Diss. A, Bergakademie Freiberg.

Kotzurek, M., and D. Stoyan (1976). A quantitative continuity theorem for the mean stationary waiting time in GI/G/1. Math. Operationsforsch. Statist. 7, 595–599.

Krakowski, M. (1973). Conservation methods in queueing theory. Rev. franc. automat. inform. rech. oper. 7, 63–83.

Krupin, B. G. (1974). On the continuity of characteristics of multi-server queues with bounded delay (Russian). Izv. Akad. Nauk SSSR, Tekhn. Kibernet. 6, 132–138.

— (1975). A continuity theorem for many server queues with limited queue length (Russian). Queueing Theory. Proc. Seminar Probabil. Methods in Engineering. Moscow State University Press, Moscow.

Küchler, I., and A. Semenov (1979). Die Waldsche Fundamentalidentität und ein sequentieller Quotiententest für eine zufällige Irrfahrt über einer homogenen irreduziblen Markovschen Kette mit endlichem Zustandsraum. Math. Operationsforsch. Statist., Ser. Statist. 10, 319–331.

Lamperti, J. (1963). Criteria for stochastic processes II: passage-time moments. J. Math. Anal. Appl. 7, 127–145.

Lang, R. (1979). On the asymptotic behaviour of infinite gradient systems. Comm. Math. Phys. 65, 129–149.

Laslett, G. M., D. B. Pollard, and R. L. Tweedie (1978). Techniques for establishing ergodic and recurrence properties of continuous valued Markov chains. Naval. Res. Logist. Quart. 25, 455–472.

Last, G. (1990). Some remarks on conditional distributions for point processes. Stoch. Process. Appl., to appear.

Lemoine, A. J. (1974). On two stationary distributions for the stable GI/G/1 queue. J. Appl. Probab. 11, 849–852.

Levy, H., and L. Kleinrock (1986). A queue with starter and a queue with vacations: delay analysis by decomposition. Oper. Res. 34, 426–436.

Lewis, P. A. W. (ed.) (1972). Stochastic Point Processes: Statistical Analysis, Theory and Applications. Wiley-Intersciences, New York–London–Sydney–Toronto.

Lindley, D. V. (1952). The theory of queues with a single server. Proc. Cambridge Philos. Soc. 48, 277–289.

Lindvall, T. (1973). Convergence of probability measures and random functions in the function space $D[0, \infty)$. J. Appl. Probab. 10, 109–121.

Liptser, R. S., and A. N. Shiryayev (1978). Statistics of Random Processes. Vol. I, II. Springer-Verlag, Berlin.

Lisek, B. (1979a). Existenz, Eindeutigkeit und Stabilität von Bedienungsprozessen. Diss. A, Humboldt-Universität zu Berlin.

— (1979b). Construction of stationary state distributions for loss systems. Math. Operationsforsch. Statist., Ser. Statist. 10, 561–581.

— (1981). Stability theorems for queueing systems without delay. Elektron. Informationsverarb. Kybernet. 17, 259–278.

— (1982). A method for solving a class of recursive stochastic equations. Z. Wahrsch. verw. Gebiete 60, 151–161.

— (1985a). Zur Theorie und Anwendung rekursiver stochastischer Gleichungen. Diss. B, Humboldt-Universität zu Berlin.

— (1985b). Stochastic equations $X_{n+1} = f(X_n, U_n)$. Properties of solutions and applications. Preprint Nr. 103/1985, Sektion Mathematik, Humboldt-Universität zu Berlin.

LITTLE, J. (1961). A proof of the queueing formula: $L = \lambda W$. Oper. Res. **9**, 383–387.

LORDEN, G. (1970). On excess over the boundary. Ann. Math. Statist. **41**, 520–527.

LOYNES, R. (1962). The stability of a queue with non-independent inter-arrival and service times. Proc. Cambridge Philos. Soc. **58**, 497–520.

MANTHEY, R. (1982). Generalized solutions of stochastic differential equations for infinite particle systems. Math. Nachr. **108**, 129–152.

MARSHALL, K. T. (1968). Bounds for some generalizations of the GI/G/1 queue. Oper. Res. **16**, 841–848.

MARTIN, R. D., and D. J. THOMSON (1982). Robust-resistant spectrum estimation. Proc. IEEE **70**, 1097–1115.

MATTHES, K. (1963). Stationäre zufällige Punktfolgen I. Jahresbericht der DMV **66**, 66–79.

MECKE, J. (1967). Stationäre zufällige Maße auf lokalkompakten Abelschen Gruppen. Z. Wahrsch. verw. Gebiete **9**, 36–58.

MELAMED, B., and W. WHITT (1989). On arrivals that see time averages: A martingale approach. J. Appl. Probab., to appear.

— (1990). On arrivals that see time averages. Oper. Res., to appear.

MILLER, L. (1964). Alternating priorities in multi-class queues. Ph. D. Th., Cornell University, Ithaca, New York.

MILLS, R. C. (1980). Models of stochastic service systems with batched arrivals. Ph. D. Th., Columbia University, Columbia.

MIYAZAWA, M. (1976). Stochastic order relations among GI/G/1 queues with a common traffic intensity. J. Oper. Res. Japan **19**, 193–208.

— (1977). Time and customer processes in queues with stationary inputs. J. Appl. Prob. **14**, 349–357.

— (1979). A formal approach to queueing processes in the steady state and their applications. J. Appl. Probab. **16**, 332–346.

— (1985). The intensity conservation law for queues with randomly changed service rate. J. Appl. Probab. **22**, 408–418.

MOGULSKI, A. A., and K. V. TROFIMOV (1977). Wald's Identity and coding costs for Markov chains (Russian). Proc. 7th All-Union Symposium on Redundancy Problem in Information Systems, Leningrad.

MOUSTAFA, M. D. (1957). Input-output Markov processes. Proc. Koninkijke Nederlande Akad. Wetenschappen **60**, 112–118.

MURARI, K. (1969). A queueing problem with arrivals in batches of variable size and service rate depending on queue length. Z. Angew. Math. Mech. **49**, 157–162.

NAKATSUKA, T. (1986). The substability and ergodicity of complicated queueing systems. J. Appl. Probab. **23**, 193–200.

NAWROTZKI, K. (1975). Markovian random marked sequences and their applications to queueing theory (Russian). Math. Operationsforsch. Statist. **6**, 445–477.

— (1978). Einige Bemerkungen zur Verwendung der Palmschen Verteilung in der Bedienungstheorie. Math. Operationsforsch. Statist., Ser. Optim. **9**, 241–253.

— (1981a). Discrete open systems or Markov chains in a random environment I. Elektron. Informationsverarb. Kybernetik **17**, 569–599.

— (1981b). An ergodic theorem for stochastic processes with an embedded homogeneous random sequence of points (Russian). Teor. Veroyatnost. i Primenen. **26**, 395–399.

NEVEU, J. (1976). Sur les measures de Palm de deuse processus ponctuels stationaires. Z. Wahrsch. verw. Gebiete **34**, 199–203.

NEVEU, J. (1977). Processus ponctuels. In: Lecture Notes in Mathematics, vol. 598, Springer-Verlag, Berlin—Heidelberg—New York, pp. 249—447.

— (1984). Construction de files d'attente stationaires. Lecture Notes in Control and Information Sciences vol. 60, Springer-Verlag, Berlin—Heidelberg—New York, pp. 31—41.

NUMMELIN, E. (1984). General Irreducible Markov Chains and Non-Negative Operators. Cambridge University Press, Cambridge.

PALM, C. (1943). Intensitätsschwankungen im Fernsprechverkehr. Ericsson Technics **44**, 1—189.

PAPANGELOU, F. (1974). On the Palm probabilities of processes of points and processes of lines. In: Stochastic Geometry and Analysis (ed. D. G. KENDALL and E. F. HARDING). Wiley, London—New York—Sydney—Toronto, pp. 114—147.

PORT, S. C., and C. J. STONE (1973). Infinite particle systems. Trans. Amer. Math. Soc. **178**, 307—340.

PYKE, R., and R. SCHAUFELE (1966). The existence and uniqueness of stationary measures for Markov renewal processes. Ann. Math. Statist. **37**, 1439—1462.

RACHEV, S. T. (1981). Minimal metric in the random variables space (Russian). Soviet Mat. Dokl. **257**, 1067—1070.

— (1982). Minimal metrics in the random variables space. Pub. Inst. Stat. Univ. of Sofia **27**, 27—47.

RACHEV, S. T., and V. V. KALASHNIKOV (1988). Mathematical methods for constructing queueing models (Russian). Nauka, Moscow.

RAMALHOTO, M. F., J. A. AMARAL, and M. T. COCHITO (1983). A survey of J. Little's formula. Internat. Statist. Rev. **51**, 255—278.

REYNOLDS, J. F. (1968). Some results for the bulk-arrival infinite server Poisson queue. Oper. Res. **16**, 186—189.

RICE, S. O. (1962). The single-server system — 1. relations between some averages, 2. busy periods. Bell System Techn. J. **41**, 269—309.

ROLSKI, T. (1981a). Queue with non-stationary input stream: Ross's conjecture. Adv. in Appl. Probab. **13**, 603—618.

— (1981b). Stationary Random Processes Associated with Point Processes. Lecture Notes in Statistics, vol. 5, Springer-Verlag, Berlin—Heidelberg—New York.

ROSENBLATT, M. (1960). Stationary Markov chains and independent random variables. J. Math. Mech. **9**, 945—949.

— (1962). Abbendum to: "Stationary Markov chains and independent random variables". J. Math. Mech. **11**, 317.

— (1971). Markov Processes. Structure and Asymptotic Behaviour. Springer-Verlag, Berlin—Heidelberg—New York.

ROSS, S. M. (1970). Applied Probability Models with Optimization Applications. Holden-Day. San Francisco.

ROSS, S. M. (1975). On the calculation of asymptotic system reliability characteristics. In: Reliability and Fault Tree Analysis, SIAM, Philadelphia.

ROSSBERG, H.-J. (1965). Über die Verteilung von Wartezeiten. Math. Nachr. **30**, 1—16.

ROSSBERG, H.-J., and G. SIEGEL (1974). The GI/G/1 model with warming-up time. Zastos Mat. **14**, 17—26.

RYLL-NARDZEWSKI, C. (1961). Remarks on processes of calls. Proc. 4th Berkeley Sympos. Math. Statist. Probab., vol. 2, pp. 455—465, Berkeley—Los Angeles.

SAKHANENKO, A. I. (1974). The convergence of the distributions of functionals of processes given on the whole axis (Russian). Sibirsk. Mat. Zh. **15**, 102—119.

SCHÄL, M. (1970). Markov renewal processes with auxiliary paths. Ann. Math. Statist. **41**, 1604—1623.

SCHÄL, M. (1971). The analysis of queues with state-dependent parameters by Markov renewal processes. Adv. in Appl. Probab. 3, 155—175.

SCHASSBERGER, R. (1973). Warteschlangen. Springer-Verlag, Wien—New York.

SCHMIDT, V. (1978). On some relations between stationary time and customer state probabilities for queueing systems G/GI/s/r. Math. Operationsforsch. Statist., Ser. Optim. 9, 261—272.

SERFOZO, R. E. (1972). Semistationary processes. Z. Wahrsch. verw. Gebiete 23, 125 bis 132.

— (1989a). Poisson functionals of Markov processes and queueing networks. Adv. in Appl. Probab., to appear.

— (1989b). Markov network processes: Congestion-dependent routing and processing. Queueing Systems, to appear.

SERFOZO, R. E., and S. STIDHAM (1978). Semi-stationary clearing processes. Stoch. Proc. Appl. 6, 165—178.

SHANBHAG, D. N., and D. G. TAMBOURATZIS (1973). Erlang's formula and some results on the departure process for a loss system. J. Appl. Probab. 10, 233—240.

SIEGEL, G. (1974). Verallgemeinerung einer Irrfahrt im $\mathbb{R}^1$ und deren Bedeutung für Bedienungssysteme mit verzögertem Bedienungsbeginn. Math. Operationsforsch. Statist. 5, 465—486.

SKOROKHOD, A. V. (1956). Limit theorems for stochastic processes (Russian). Teor. Veroyatnost. i Primenen. 1, 289—319.

SLIVNYAK, I. M. (1962). Some properties of stationary streams of homogeneous random events (Russian). Teor. Veroyatnost. i Primenen. 7, 347—352.

STIDHAM jr., S. (1972). $L = \lambda W$: A discounted analogue and a new proof. Oper. Res. 20, 1115—1126.

— (1974). A last word on $L = \lambda W$. Oper. Res. 22, 417—421.

STIDHAM, S., and M. EL TAHA (1989). Sample-path analysis of processes with imbedded point processes. Queueing systems, to appear.

STÖRMER, H. (1970). Semi-Markoff Prozesse mit endlich vielen Zuständen. Lecture Notes in Operations Research and Mathematical Systems, vol. 34, Springer-Verlag, Berlin.

STOYAN, D. (1976). A critical remark on a system approximation in queueing theory. Math. Operationsforsch. Statist. 7, 953—956.

— (1977). Further stochastic order relations among GI/GI/1 queues with a common traffic intensity. Math. Operationsforsch. Statist., Ser. Optim. 8, 541—548.

— (1983). Comparison Methods for Queues and Other Stochastic Models. (ed. D. J. DALEY). Wiley, Chichester. (German Edition: Qualitative Eigenschaften und Abschätzungen stochastischer Modelle. Akademie-Verlag, Berlin, 1977).

STOYAN, D., W. S. KENDALL, and J. MECKE (1987): Stochastic Geometry. Akademie-Verlag, Berlin, and Wiley, Chichester.

STRELLER, A. (1980). A generalization of cumulative processes. Elektron. Informationsverarb. Kybernetik 16, 449—460.

— (1982). On stochastic processes with an embedded marked point process. Math. Operationsforsch. Statist., Ser. Statist. 13, 561—576.

SZCZOTKA, W. (1986a). Stationary representation of queues I. Adv. in Appl. Probab. 18, 815—848.

— (1986b). Stationary representation of queues II. Adv. in Appl. Probab. 18, 849—859.

TAKÁCS, L. (1956). On the generalization of Erlang's formula. Acta Math. Acad. Sci. Hung. 7, 419—433.

— (1962). Introduction to the theory of queues. Oxford University Press. New York.

— (1963a). A single server queue with feedback. Bell System Tech. J. 42, 509—515.

TAKÁCS, L. (1963b). The limiting distribution of the virtual waiting time and the queue size for a single server queue with recurrent input and general service times. Sankhya A25, 91–100.

TWEEDIE, R. L. (1975). Relations between ergodicity and mean drift for Markov chains. Austral. J. Statist. 17, 96–102.

VASICEK, O. A. (1977). Inequality for the variance of waiting time under general queueing discipline. Oper. Res. 25, 879–884.

VERVAAT, W. (1979). On a stochastic difference equation and a representation of non-negative infinitely divisible random variables. Adv. in Appl. Probab. 11, 750–783.

WALD, A. (1947). Sequential Analysis. Wiley, New York.

WELCH, P. D. (1964). On a generalized M/G/1 queueing process in which the first customer of each busy period receives exceptional service. Oper. Res. 12, 736–752.

WHITT, W. (1972). Embedded renewal processes in the GI/G/s queue. J. Appl. Probab. 9, 650–658.

— (1974). Continuity of queues. Adv. in Appl. Probab. 6, 175–183.

— (1980). Some useful functions for functional limit theorems. Math. Oper. Res. 5, 67–81.

— (1983). Comparing batch delays and customer delays. Bell System Tech. J. 62, 2001–2009.

WIRTH, K.-D. (1982). On stationary queues with batch arrivals. Elektron. Informationsverarb. Kybernetik 18, 603–619.

— (1983). Einlinige Bedienungssysteme mit Besonderheiten. Diss. A, Humboldt-Universität zu Berlin.

— (1984a). Some remarks on feedback queues. Elektron. Informationsverarb. Kybernetik 20, 55–64.

— (1984b). A new approach to queues with warming-up times. Elektron. Informationsverarb. Kybernetik 20, 426–434.

— (1985). A remark on relationships between batch delays and customer delays. Elektron. Informationsverarb. Kybernetik 21, 65–67.

— (1986a). Stationary sequences with groups. Unpublished manuscript.

— (1986b). Stability conditions for single server queues with varying service rate. Elektron. Informationsverarb. Kybernetik 22, 105–123.

WOLFF, R. W. (1982). Poisson arrivals see time averages. Oper. Res. 30, 223–231.

YEO, G. F. (1962). Single server queues with modified service mechanisms. J. Austral. Math. Soc. 3, 499–507.

ZÄHLE, M. (1980). Ergodic properties of general Palm measures. Math. Nachr. 95, 93–106.

ZOLOTAREV, V. M. (1976a). On stochastic continuity of queueing systems of type G/G/1 (Russian). Teor. Veroyatnost. i Primenen. 21, 260–279.

— (1976b). Metric distances in spaces of random variables and their distributions. Math. USSR Sbornik 30, 3, 373–401.

— (1977a). Quantitative estimates for the continuity property of queueing systems of type G/G/∞ (Russian). Teor. Veroyatnost. i Primenen. 22, 700–711.

— (1977b). General problems of the stability of mathematical models. Proc. 41st Session Intern. Statist. Inst., New Delhi.

— (1979). Ideal metrics in the problems of probability theory and mathematical statistics. Austral. J. Statist. 21, 193–208.

Symbol Index

A_n ($\hat{A}_n$)	interarrival times between customers (batches)
$\mathsf{A}(t)$	d.f. of interarrival times
$a_n(\varphi) = a_n$ for $\varphi = ([a_n, k_n]) \in S_K{}^*$	
$a_{\mathsf{P}} = \mathsf{E}_{\mathsf{P}} A_0$	
$A(n, \varphi) = \left\{x(1, n, \varphi), \ldots, x\left(\# A(n, \varphi), n, \varphi\right)\right\}$	set of potential states
$\mathcal{A}(\mathcal{S}) = \left\{([u_n, x_n]) : x_n \in A\left(n, \left((u_k)\right), n \in \mathbb{Z}\right\}\right.$	
$B(n, \varphi)$	reduced set of $A(n, \varphi)$, cf. Section 1.4
$D\left((-\infty, \infty), \mathbb{X}\right)$ $\left(D([0, \infty) \times \mathbb{X})\right)$	Skorokhod space, 122
E	expectation
E_{P} ($\mathsf{E}_{\bar{\mathsf{P}}}$, $\mathsf{E}_{\tilde{\mathsf{P}}}$, $\mathsf{E}_{\hat{\mathsf{P}}}$)	expectation with respect to P ($\bar{\mathsf{P}}$, $\tilde{\mathsf{P}}$, $\hat{\mathsf{P}}$)
$e(j) = [0, \ldots, \underset{j\text{-th}}{1}, \ldots, 0]$	
f ($\hat{f}$, $\bar{f}$)	function describing the system dynamics step by step (batch by batch, in continuous time)
$F(\mathbb{Y})$, $\left(F_+(\mathbb{Y})\right)$	set of all (non-negative) measurable functions on $\mathbb{Y}$
$\mathsf{F}_1 * \mathsf{F}_2$	convolution of the d.f.'s F_1 and F_2
F^{*k}	k-th convolution power of the d.f. F
$\hat{G}_n$	size of the n-th batch (in the batch-stationary model)
$\hat{g}_i = \mathsf{P}(\hat{G}_0 = i)$	batch size distribution
G_0	size, of the batch that contains the customer with number 0 (in the arrival-stationary model)
$g_n = \mathsf{P}(G_0 = n)$	
$I(n, \varphi)$	set of indices, cf. e.g. Chapter 1
$i(\varphi) = \# B(0, \varphi)$	cf. Sect. 1.4
$\mathcal{J}$ ($\mathcal{J}^+$)	family of invariant subsets of $\mathbb{Y}^{\mathbb{Z}}$ ($\mathbb{Y}^{\mathbb{Z}_+}$)
$[\mathbb{K}, \mathcal{K}]$	mark space
$k_n(\varphi) = k_n$	n-th mark of the sequence $\varphi = ([a_n, k_n]) \in S_K{}^*$
$k_n(\psi) = k_n$	n-th mark of $\psi = ([t_n, k_n]) \in M_K{}^*$
L_n ($L_n{}^*$, $\hat{L}_n$)	arrival (departure, batch-arrival)-stationary number of customers in the system
$L(t)$ $\left(\bar{L}(t)\right)$	number of customers at time t in the arrival (time)-stationary model
$M_K{}^*$	space of all sequences $\psi = ([t_n, k_n])$ with $t_n \leqq t_{n+1}$, $n \in \mathbb{Z}$
$M_K = M_K{}^* \cap \left\{\psi : \lim\limits_{n \to \pm\infty} t_n = \pm\infty\right\}$	
m_X (m_S, m_A, $m_{\hat{G}}$)	expectation of the r.v. X_0 (S_0, A_0, $\hat{G}_0$)
N_0	position of the customer with number 0 in his own batch (in the arrival-stationary model)

$N_L(t) = \# \{n: 0 < T_n \leqq t, K_n \in L\}$,
$\bar{N}_L(t) = \# \{n: 0 < \bar{T}_n \leqq t, \bar{K}_n \in L\}$
$N(t) = N_{\mathbb{K}}(t), \bar{N}(t) = \bar{N}_{\mathbb{K}}(t)$
$\tilde{\mathsf{P}}(\cdot) = \mathsf{P}(\cdot \mid Y_0 \in L)$, Chapter 7: $\tilde{\mathsf{P}}(\cdot) = \mathsf{P}(\cdot \mid A_{-1} > 0)$

p_i ($\bar{p}_i, p_i^*, \hat{p}_i$)	customer-arrival (time, departure, batch-arrival)-stationary probabilities of the number of customers
Q	distribution of a particular weak solution, cf. Sect. 1.4
$q = i(\Phi)$	for ergodic Φ, cf. Sect. 1.4.
Q_n $(\bar{Q}(t))$	arrival (time)-stationary number of customers in the queue
$Q(t)$	number of customers in the queue at time t in the arrival-stationary model
$\mathbb{R}$ ($\mathbb{R}_+$)	real line (non-negative real line)
$\bar{R}(x)$ $(R(x))$	operator of reordering the components of a vector in descending (increasing) order
$r_n = \mathsf{P}(N_0 = n)$	
$R(s)$	PGF of $(r_n, n \geqq 1)$
S_n ($\hat{S}_n$)	service time of the n-th customer (batch $\hat{S}_n^1 + \cdots + \hat{S}_n^{\hat{G}_n}$)
$\underline{\hat{S}}_n = [\hat{S}_{n,1}, \ldots, \hat{S}_{n,\hat{G}_n}]$	vector of service times of the $\hat{G}_n$ customers in the n-th batch
$\mathsf{S}(t) = \mathsf{P}(S_0 \leqq t)$, $\hat{\mathsf{S}}(t) = \mathsf{P}(\hat{S}_0 \leqq t)$	
$\mathsf{S}_R(t)$ $(\hat{\mathsf{S}}_R(t))$	d.f. of the residual service time of a customer (batch)
$\bar{S}(0)$	time-stationary residual service time
S_K^*	set of all sequences $\varphi = ([a_n, k_n])$ with $a_n \geqq 0$
$S_K^{\cdot} = S_K^* \cap \left\{\varphi: \sum_{n=-\infty}^{-1} a_n = \sum_{n=0}^{\infty} a_n = \infty\right\}$	
$\mathcal{S} = \{A(n, \varphi): n \in \mathbb{Z}, \varphi \in M\}$	solution generating system (SGS)
$t_n(\varphi)$	85
T_n	85
U_n	the n-th input variable
$[\mathbb{U}, \mathcal{U}]$	space of the input variables U_n
$\mathcal{U}_{\mathbb{Z}}$	σ-field of measurable cylinders of $\mathbb{U}^{\mathbb{Z}}$
V_n ($V_n^*, \hat{V}_n$)	arrival (departure, batch-arrival)-stationary work load at the n-th customer (at the n-th departure, at the n-th batch-arrival)
$V(t)$ $(\bar{V}(t))$	work load at time t in the arrival (time)-stationary model
$\bar{\mathsf{V}}(x)$ $(\mathsf{V}(x), \hat{\mathsf{V}}(x))$	time (customer-arrival, batch-arrival)-stationary d.f. of work load
W_n $(\bar{W}(t))$	waiting time of the n-th customer (at time t) in the arrival (time)-stationary model
$\bar{\mathsf{W}}(t)$	d.f. of $\bar{W}(t)$
$[\mathbb{X}, \mathcal{X}]$	state space
X_n $(X(t), \hat{X}_n)$	system state at the n-th step (at t, at the n-th batch-arrival)
$X_n^{\min} = x_n^{\min}(\Phi)$	minimal stationary state process
$X_n^{\max} = x_n^{\max}(\Phi)$	maximal stationary state process
$\hat{x}_n(\Phi)$	state process constructed by means of renewing epochs
$\hat{x}_n(x, \Phi)$	state process constructed by means of (x, l)-renewing epochs
$x(j, n, \varphi)$	elements of $A(n, \varphi)$ (potential states)
$x(j, \varphi)$	elements of $B(0, \varphi)$
$[\mathbb{Y}, \mathcal{Y}]$	measurable space
$\mathcal{Y}_{\mathbb{Z}}$	σ-field of measurable cylinders of $\mathbb{Y}^{\mathbb{Z}}$
$\mathbb{Z} = \{\ldots, -1, 0, 1, \ldots\}$	

$\mathbb{Z}_+ = \{0, 1, \ldots\}$	
$\hat{\Gamma}(s)$	PGF of the batch size distribution $(\hat{g}_i, i \geqq 1)$
θ (θ_t)	discrete (continuous) time shift operator
λ, $\lambda_{\overline{\mathsf{P}}}$	intensities
$\hat{\lambda}$, (λ^*)	intensity of arriving batches (of departing customers)
$\lambda_{\overline{\mathsf{P}}}(L) = \nu_{\overline{\mathsf{P}}}\big((0, 1] \times L\big)$, $L \in \mathcal{K}$	measure on the mark space $[\mathbb{K}, \mathcal{K}]$
μ	service rate
ν_{R}	intensity measure, cf. Definition 3.2.3
$\xi = \breve{\xi}(\chi)$	sequence of cycles of a PEMP χ
$\Pi(s)$ $\big(\overline{\Pi}(s), \Pi^*(s), \hat{\Pi}(s)\big)$	PGF of the distribution p_i $(\overline{p}_i, p_i^*, \hat{p}_i)$, $i \geqq 0$
$\pi(\mathsf{P}_1, \mathsf{P}_2)$	Prokhorov distance of the distributions P_1 and P_2
$\Phi = (U_n)$	input of a stochastic system
$\Phi = ([A_n, Y_n])$,	in particular $\Phi = ([A_n, S_n])$ input of a queueing system
$\overline{\Phi}$ $(\tilde{\Phi}, \hat{\Phi})$	time (batch-arrival)-stationary input
$\hat{\Phi} = \hat{\varphi}(\tilde{\Phi})$, $\tilde{\Phi} = \tilde{\varphi}(\hat{\Phi})$	Sect. 7.1
$\Phi' = ([U_n', X_n'])$	weak solution of (1.1.1)
$\varphi(\psi) = ([t_{n+1} - t_n, k_n])$, $\psi = ([t_n, k_n]) \in M_K$	
$\chi = \big[\big(X(t)\big), \Psi\big]$, $\Psi = \big(\big[T_n, [X_n, Y_n]\big]\big)$	process with an embedded marked point process (PEMP)
$\overline{\chi}$	stationary PEMP
$\chi = \Big[\big(X(t)\big), \big(\big[T_n, [X_n, Y_n]\big]\big)\Big]$	arrival-stationary model
$\overline{\chi} = \Big[\big(\overline{X}(t)\big), \big(\big[\overline{T}_n, [\overline{X}_n, \overline{Y}_n]\big]\big)\Big]$	time-stationary model
$\chi(\xi)$	synchronous PEMP generated by the sequence of cycles ξ
$\Psi = ([T_n, K_n])$	marked point process (MPP)
$\overline{\Psi} = ([\overline{T}_n, \overline{K}_n])$	stationary marked point process
$\psi(\chi)$	representation of a PEMP χ as an MPP
$\psi(\varphi) = \big([t_n(\varphi), k_n(\varphi)]\big)$, $\varphi \in S_K$	
$n \bmod m = r$	iff $n = mc + r$, $0 \leqq r \leqq m - 1$.
$[n/m] = \max\{x \in \mathbb{Z}: n/m \leqq x\}$	
$x_+ = \max(0, x)$, $x \in \mathbb{R}$,	
$\mathbf{1}\{x \in A\}$, $\mathbf{1}_A(x)$	indicator function of the set A
$\mathbf{1} = [1, \ldots, 1]$	
$\# C$	number of elements of C
A^c	complement of the set A
$\mathrm{cl}\, A$	closure of the set A
∂A	boundary of the set A
$\lvert\cdot\rvert$	Lebesgue measure
P^+	restriction of a distribution P on $\mathbb{Y}^{\mathbb{Z}}$ to $\mathbb{Y}^{\mathbb{Z}_+}$
$\overset{\mathcal{D}}{=}$	equality in distribution
$\overset{\mathcal{D}}{\leqq}$	stochastic ordering
$\overset{\mathcal{D}}{\longrightarrow}$	convergence in distribution
$\overset{\mathsf{P}}{\longrightarrow}$	convergence in probability
$\Rightarrow$	weak convergence
$\mathcal{U} \otimes \mathcal{V}$	product σ-field of the σ-fields $\mathcal{U}$ and $\mathcal{V}$
$\mu \otimes \nu$	product measure of the measures μ and ν
∎	end of proof

Subject Index

Bei Fragen zur Produktsicherheit wenden Sie sich bitte an:
If you have any questions regarding product safety,
please contact:

Walter de Gruyter GmbH
Genthiner Straße 13
10785 Berlin
productsafety@degruyterbrill.com